AAPG Treatise of Petroleum Geology Reprint Series

The American Association of Petroleum Geologists
gratefully acknowledges and appreciates the leadership and support
of the AAPG Foundation in the development of the
Treatise of Petroleum Geology.

GEOPHYSICS III

GEOLOGIC INTERPRETATION OF SEISMIC DATA

COMPILED BY
EDWARD A. BEAUMONT
AND
NORMAN H. FOSTER

TREATISE OF PETROLEUM GEOLOGY
REPRINT SERIES, NO. 14

PUBLISHED BY
THE AMERICAN ASSOCIATION OF PETROLEUM GEOLOGISTS
TULSA, OKLAHOMA 74101-0979, U.S.A.

Copyright © 1989
The American Association of Petroleum Geologists
All Rights Reserved

Library of Congress Cataloging-in-Publication Data

Geophysics III.

(Treatise of petroleum geology reprint series; no. 14
Includes bibliographical references.
1. Petroleum—Prospecting. 2. Seismic interpretation.
I. Beaumont, E.A. (Edward A.) II. Foster, Norman H.
III. Series.
TN271.P4G47   1989   622'.1828   89-18158

ISBN 0-89181-413-2
ISSN 1046-0144

AMERICAN ASSOCIATION OF PETROLEUM GEOLOGISTS FOUNDATION
TREATISE OF PETROLEUM GEOLOGY FUND*

Major Corporate Contributors
($25,000 or more)

Chevron Corporation
Mobil Oil Corporation
Oryx Energy Company
Pennzoil Exploration and Production Company
Shell Oil Company
Union Pacific Foundation

Other Corporate Contributors
($5,000 to $25,000)

Cabot Energy Corporation
Canadian Hunter Exploration Ltd.
Conoco Inc.
Marathon Oil Company
The McGee Foundation, Inc.
Phillips Petroleum Company
Texaco Philanthropic Foundation
Transco Energy Company

Major Individual Contributors
($1,000 or more)

C. Hayden Atchison
Richard R. Bloomer
A.S. Bonner, Jr.
David G. Campbell
Herbert G. Davis
Paul H. Dudley, Jr.
Lewis G. Fearing
James A. Gibbs
George R. Gibson
William E. Gipson
Robert D. Gunn
Cecil V. Hagen
Frank W. Harrison
William A. Heck
Roy M. Huffington
Harrison C. Jamison
Thomas N. Jordan, Jr.
Hugh M. Looney
John W. Mason
George B. McBride
Dean A. McGee
John R. McMillan
Rudolf B. Siegert
Robert M. Sneider
Jack C. Threet
Charles Weiner
Harry Westmoreland
James E. Wilson, Jr.

The Foundation also gratefully acknowledges the many who have supported this endeavor with additional contributions, which now total more than $12,000.

*Contributions received as of November 1, 1989.

Treatise of Petroleum Geology
Advisory Board

W.O. Abbott
Robert S. Agatston
John J. Amoruso
J.D. Armstrong
George B. Asquith
Colin Barker
Ted L. Bear
Edward A. Beaumont
Robert R. Berg
Richard R. Bloomer
Louis C. Bortz
Donald R. Boyd
Robert L. Brenner
Raymond Buchanan
Daniel A. Busch
David G. Campbell
J. Ben Carsey
Duncan M. Chisholm
H. Victor Church
Don Clutterbuck
Robert J. Cordell
Robert D. Cowdery
William H. Curry, III
Doris M. Curtis
Graham R. Curtis
Clint A. Darnall
Patrick Daugherty
Herbert G. Davis
James R. Davis
Gerard J. Demaison
Parke A. Dickey
F.A. Dix, Jr.
Charles F. Dodge
Edward D. Dolly
Ben Donegan
Robert H. Dott, Sr.*
John H. Doveton
Marlan W. Downey
John G. Drake
Richard J. Ebens
William L. Fisher
Norman H. Foster
Lawrence W. Funkhouser
William E. Galloway
Lee C. Gerhard
James A. Gibbs
Arthur R. Green
Robert D. Gunn
Merrill W. Haas
Robert N. Hacker
J. Bill Hailey
Michel T. Halbouty
Bernold M. Hanson
Tod P. Harding
Donald G. Harris
Frank W. Harrison, Jr.
Ronald L. Hart
Dan J. Hartmann
John D. Haun
Hollis D. Hedberg*
James A. Helwig
Thomas B. Henderson, Jr.
Francis E. Heritier
Paul D. Hess
Mason L. Hill
David K. Hobday
David S. Holland
Myron K. Horn
Michael E. Hriskevich
J.J.C. Ingels
Michael S. Johnson
Bradley B. Jones
R.W. Jones
John E. Kilkenny
H. Douglas Klemme
Allan J. Koch
Raden P. Koesoemadinate
Hans H. Krause
Naresh Kumar
Rolf Magne Larsen
Jay E. Leonard
Ray Leonard
Howard H. Lester
Detlev Leythaeuser
John P. Lockridge
Tony Lomando
John M. Long
Susan A. Longacre
James D. Lowell
Peter T. Lucas
Harold C. MacDonald
Andrew S. Mackenzie
Jack P. Martin
Michael E. Mathy
Vincent Matthews, III
James A. McCaleb
Dean A. McGee
Philip J. McKenna
Robert E. Megill
Fred F. Meissner
Robert K. Merrill
David L. Mikesh
Marcus E. Milling
George Mirkin
Richard J. Moiola
D. Keith Murray
Norman S. Neidell
Ronald A. Nelson
Charles R. Noll
Clifton J. Nolte
Susan E. Palmer
Arthur J. Pansze
John M. Parker
Alain Perrodon
James A. Peterson
R. Michael Peterson
David E. Powley
A. Pulunggono
Donald L. Rasmussen
R. Randolf Ray
Dudley D. Rice
Edward C. Roy, Jr.
Eric A. Rudd
Floyd F. Sabins, Jr.
Nahum Schneidermann
Peter A. Scholle
George L. Scott, Jr.
Robert T. Sellars, Jr.
John W. Shelton
Robert M. Sneider
Stephen A. Sonnenberg
William E. Speer
Ernest J. Spradlin
Bill St. John
Philip H. Stark
Richard Steinmetz
Per R. Stokke
Donald S. Stone
Doug K. Strickland
James V. Taranik
Harry TerBest, Jr.
Bruce K. Thatcher, Jr.
M. Raymond Thomasson
Bernard Tissot
Donald Todd
M.O. Turner
Peter R. Vail
Arthur M. Van Tyne
Harry K. Veal
Richard R. Vincelette
Fred J. Wagner, Jr.
Anthony Walton
Douglas W. Waples
Harry W. Wassall, III
W. Lynn Watney
N.L. Watts
Koenradd J. Weber
Robert J. Weimer
Dietrich H. Welte
Alun H. Whittaker
James E. Wilson, Jr.
Martha O. Withjack
P.W.J. Wood
Homer O. Woodbury
Mehmet A. Yukler
Zhai Guangming

*Deceased

IN APPRECIATION...

The American Association of Petroleum Geologists and the AAPG Foundation gratefully acknowledge the contributions of the Society of Exploration Geophysicists to the Treatise of Petroleum Geology Reprint Series volumes on geophysics. The crucial role played by SEG in advancing exploration geophysics is universally recognized in the petroleum industry and is plainly documented by the many papers from *Geophysics* and *Geophysics: The Leading Edge of Exploration* reproduced in these volumes. The spirit of advancement and expansion in the science of geophysical exploration for hydrocarbons as well as the continuing synergism that melds the professions of geology and geophysics is exemplified by the permission granted by SEG to reproduce its papers in this series.

Although SEG and AAPG both have a long history of independent activities and autonomous operation, there have been, and there continue to be, cooperative efforts to improve the professionalism of both geologists and geophysicists. Previous activities, such as joint meetings, research conferences, and publications, exemplify the ongoing cooperative efforts which create results far more valuable than independent efforts by either group would have achieved. SEG's contributions to the earth sciences and to the Treatise of Petroleum Geology Reprint Series volumes on geophysics are gratefully acknowledged by AAPG, the AAPG Foundation, and the Advisory Board of the Treatise of Petroleum Geology.

INTRODUCTION

This reprint volume belongs to a series of that is part of the *Treatise of Petroleum Geology*. The *Treatise of Petroleum Geology* was conceived during a discussion we had at the 1984 AAPG Annual Meeting in San Antonio. When our discussion ended, we had decided to write a state-of-the-art textbook in petroleum geology, directed not at the student, but at the practicing petroleum geologist. The project to put together one textbook gradually evolved into a series of three different publications: the Reprint Series, the Atlas of Oil and Gas Fields, and the Handbook of Petroleum Geology; collectively these publications are known as the *Treatise of Petroleum Geology*. With the help of the Treatise of Petroleum Geology Advisory Board, we designed this set of publications to represent the cutting edge in petroleum exploration knowledge and application. The Reprint Series provides previously published landmark literature; the Atlas collects detailed field studies to illustrate the various ways oil and gas are trapped; and the Handbook is a professional explorationist's guide to the latest knowledge in the various areas of petroleum geology and related fields.

The papers in the various volumes of the Reprint Series complement the different chapters of the Handbook. Papers were selected on the basis of their usefulness today in petroleum exploration and development. Many "classic papers" that led to our present state of knowledge have not been included because of space limitations. In some cases, it was difficult to decide in which Reprint volume a particular paper should be published because that paper covers several topics. We suggest, therefore, that interested readers become familiar with all the Reprint volumes if they are looking for a particular paper.

Geophysics is an indispensable tool for geologists looking for and developing oil and gas fields. Because it lets us "see" into the subsurface, geophysics allows petroleum geologists to build better images of the subsurface than is possible using only surface geology and information from well bores. In the past, geophysics was the domain of the geophysicist, and the geophysicist alone acquired, processed, and interpreted geophysical data. During the past two decades, however, the technology of geophysics has exploded; at the same time, the petroleum industry has been forced to look for more and more subtle traps in more and more difficult terrain. This placed a tremendous burden on geophysicists, and they naturally looked to their colleagues, the geologists, for relief. At first, geologists only helped with interpretation. Today, however, geologists are also involved in helping geophysicists make decisions regarding acquisition and processing of data.

The choice of papers in these geophysics reprint volumes reflects this evolution. The papers were chosen to help geologists, not geophysicists, enhance their knowledge of geophysics. Math-intensive papers were excluded because those papers are relatively esoteric and have limited applicability for most geologists. Many of the papers included do contain mathematical equations, but they were selected because they are germane, and the math is presented at a level that, we trust, the majority of geologists are now comfortable with.

The number and distribution of the papers reprinted in these volumes reflect the current importance and uses of the different geophysical methods described in the papers. We have divided the topic of geophysics into four volumes. The first volume contains papers on Seismic Methods. Papers in this volume are concerned with seismic theory and are grouped into six sections: Seismic Methods, Seismic Rock Properties, Seismic Acquisition, Seismic Processing and Display, Seismic Velocities, and Migration. Volume II is subtitled Tools for Seismic Interpretation. Section titles in this volume are Synthetic Seismograms and Velocity Inversion; Seismic Modeling; Seismic Attributes: Amplitude, Frequency, Phase, Velocity; Shear Waves; Amplitude Variation with Offset; and Vertical Seismic Profiling. Volume III, Geologic Interpretation of Seismic Data, contains sections on Structural Interpretation and Stratigraphic Interpretation. The last volume is on Gravity, Magnetic, and Magnetotelluric Methods. It contains two sections: Gravity and Magnetic Methods, and Magnetotelluric Methods.

We would like to thank the various societies and publishers who gave us permission to reprint these papers, especially the Society of Exploration Geophysicists. We also wish to thank the members of Advisory Board of the Treatise of Petroleum Geology who suggested papers for these volumes, especially R. Randy Ray. Randy Ray is a geophysicist who was trained initially as a geologist. From a large list of proposed papers, he helped us select papers that would be both understandable and useful to a geologist exploring for and developing oil and gas fields.

Edward A. Beaumont Norman H. Foster
Tulsa, Oklahoma Denver, Colorado

TABLE OF CONTENTS

GEOPHYSICS III
GEOLOGIC INTERPRETATION OF SEISMIC DATA

STRUCTURAL INTERPRETATION

A process of seismic reflection interpretation   J. G. Hagedoorn 3

Interactive seismic mapping of net producible gas sand in the Gulf of Mexico   Alistair R. Brown, Roger M. Wright, Keith D. Burkhart, and William L. Abriel 54

Interactive interpretation of seismic data   Anthony C. Gerhardstein and Alistair R. Brown 83

STRATIGRAPHIC INTERPRETATION

Inferring stratigraphy from seismic data   R. E. Sheriff 96

Seismic stratigraphy, a fundamental exploration tool   G. R. Ramsayer 111

Seismic facies analysis concepts   M. M. Roksandic 121

Seismic signatures of sedimentation models   J. C. Harms and P. Tackenberg 137

Integration of biostratigraphy and seismic stratigraphy: Pliocene-Pleistocene, Gulf of Mexico
John M. Armentrout 152

The role of horizontal seismic sections in stratigraphic interpretation   Alastair R. Brown 161

Seismic stratigraphy and global changes of sea level, part 10: seismic recognition of carbonate buildups   J. N. Bubb and W. G. Hatlelid 173

Seismic interpretation of carbonate depositional environments   J. M. Fontaine, R. Cussey, J. Lacaze, R. Lanaud, and L. Yapaudjian 193

Seismic expression of carbonate build-ups, northwest Java basin   J. E. Burbury 211

Field development with three-dimensional seismic methods in the Gulf of Thailand—a case history   C. G. Dahm and R. J. Graebner 241

Aspects of seismic reflection prospecting for oil and gas   P. N. S. O'Brien 269

Predictive isopach mapping of gas sands from seismic impedance: modeled and empirical cases from Ship Shoal Block 134 field   Robert D. Woock and Alan R. Kin 301

New seismic technology can guide field development   J. P. Lindsey, M. W. Schramm, Jr., and L. K. Nemeth 311

How hydrocarbon reserves are estimated from seismic data   J. P. Lindsey and C. I. Craft 317

Progress in stratigraphic seismic exploration and the definition of reservoirs   Norman S. Neidell and John H. Beard 321

Interpretation of depositional facies from seismic data   J. B. Sangree and J. M. Widmier 339

Seismic stratigraphic model of depositional platform margin, eastern Anadarko basin, Oklahoma
William E. Galloway, Marshall S. Yancey, and Arthur P. Whipple 369

Exploration for oil accumulations in Entrada Sandstone, San Juan basin, New Mexico
Richard R. Vincelette and William E. Chittum 380

TABLE OF CONTENTS

Geophysics I
Seismic Methods

Seismic Waveforms and Resolution

Aspects of seismic resolution.
R. E. Sheriff.

How thin is a thin bed?
M. B. Widess.

Complex seismic trace analysis of thin beds.
James D. Robertson and Henry H. Nogami.

The limits of resolution of zero-phase wavelets.
R. S. Kallweit and L. C. Wood.

Resolution comparison of minimum-phase and zero-phase signals.
M. Schoenberger.

Seismic Rock Properties

Formation velocity and density—the diagnostic basics for stratigraphic traps.
G. H. F. Gardner, L. W. Gardner, and A. R. Gregory.

Effect of water saturation on seismic reflectivity of sand reservoirs encased in shale.
S. N. Domenico.

Seismic Acquisition

Whatever happened to ground roll?
Nigel A. Anstey.

Field techniques for high resolution.
Nigel A. Anstey.

Vibroseis' gentle massage obtains structural data safely, economically.
N. A. Anstey.

The Vibroseis system of seismic mapping.
Robert L. Geyer.

Vibroseis parameter optimization.
Robert L. Geyer.

Seismic data enhancement—a case history.
R. J. Graebner.

Seismic Processing and Display

Common reflection point horizontal data stacking techniques.
W. Harry Mayne.

Correlation techniques—a review.
N. A. Anstey.

The digital processing of seismic data.
Daniel Silverman.

Seismic data display and reflection perceptibility.
Frank J. Feagin.

Semblance and other coherency measures for multichannel data.
N. S. Neidell and M. Turhan Taner.

Predictive deconvolution: theory and practice.

K. L. Peacock and Sven Treitel.

Estimation and correction of near-surface time anomalies.
M. Turhan Taner, F. Koehler, and K. A. Alhilali.

Seismic signal processing.
Lawrence C. Wood and Sven Treitel.

SEISMIC VELOCITIES

Seismic velocities from surface measurements.
C. Hewitt Dix.

An analysis of stacking, rms, average, and interval velocities over a horizontally layered ground.
M. Al-Chalabi.

Time-depth and velocity-depth relations in western Canada.
C. H. Acheson.

Apparent velocity from dipping interface reflections.
F. K. Levin.

A velocity function including lithologic variation.
L. Y. Faust.

Seismic data indicate depth, magnitude of abnormal pressures.
E. S. Pennebaker, Jr.

Synthetic sonic logs—a process for stratigraphic interpretation.
R. O. Lindseth.

Velocity spectra—digital computer derivation and applications of velocity functions.
M. Turhan Taner and Fulton Koehler.

The effects of cracks on the compressibility of rock.
J. B. Walsh.

MIGRATION

Migration.
P. Hood.

Two-dimensional and three-dimensional migration of model-experiment reflection profiles.
William S. French.

A simple theory for seismic diffractions.
A. W. Trorey.

Migration of seismic data from inhomogeneous media.
Les Hatton, Ken Larner, and Bruce S. Gibson.

Time migration—some ray theoretical aspects.
P. Hubral.

The wave equation applied to migration.
D. Loewenthal, L. Lu, R. Roberson, and J. Sherwood.

Wave-front charts and three dimensional migrations.
Albert W. Musgrave.

Migration by Fourier transform.
R. H. Stolt.

Geophysics II
Tools for Seismic Interpretation

Synthetic Seismograms and Velocity Inversion

The synthesis of seismograms from well log data.
R. A. Peterson, W. R. Fillippone, and F. B. Coker.

Inversion of seismograms and pseudo velocity logs.
M. Lavergne and C. Willm.

Well log editing in support of detailed seismic studies.
Brian E. Ausburn.

Seismic Modeling

Stratigraphic modeling: a step beyond bright spot.
E. V. Dedman, J. P. Lindsey, and M. W. Schramm, Jr.

Three-dimensional seismic modeling.
Fred J. Hilterman.

Interpretive lessons from three-dimensional modeling.
Fred J. Hilterman.

Stratigraphic modeling and interpretation—geophysical principals and techniques.
Norman S. Neidell and Elio Poggiagliolmi.

Synthetic seismic sections of typical petroleum traps.
Bruce T. May and Franta Hron.

Seismic Attributes: Amplitude, Frequency, Phase, Velocity

Application of amplitude, frequency, and other attributes to stratigraphic and hydrocarbon determination.
M. T. Taner and R. E. Sheriff.

Reflections on amplitudes.
R. F. O'Doherty and N. A. Anstey.

Velocity spectra and their use in stratigraphic and lithologic differentiation.
Ernest E. Cook and M. Turhan Taner.

Outlining of shale masses by geophysical methods.
A. W. Musgrave and W. G. Hicks.

Three-dimensional seismic monitoring of an enhanced oil recovery process.
Robert J. Greaves and Terrance J. Fulp.

Shear Waves

Basis for interpretation of Vp/Vs ratios in complex lithologies.
Raymond L. Eastwood and John P. Castagna.

Evaluation of direct hydrocarbon indicators through comparison of compressional- and shear-wave seismic data: a case study of the Myrnam gas field, Alberta.
Ross Alan Ensley.

Relationships between compressional-wave and shear-wave velocities in clastic silicate rocks.
J. P. Castagna, M. L. Batzle, and R. L. Eastwood.

Direct hydrocarbon detection using comparative P-wave and S-wave seismic sections.
James D. Robertson and William C. Pritchett.

Vp/Vs and lithology.
R. S. Tatham.

Table of Contents

Geophysics IV
Gravity, Magnetic, and Magnetotelluric Methods

Gravity and Magnetic Methods

Gravity and magnetics for geologists and seismologists.
L. L. Nettleton.

Exploring for stratigraphic traps with gravity gradients.
Sigmund Hammer and Rodolfo Anzoleaga.

Measurement of gravity at sea and in the air.
Lucien J. B. LaCoste.

An approximate solution of the problem of maximum depth in gravity interpretation.
D. C. Skeels.

Use of gravity, magnetic, and electrical methods in stratigraphic-trap exploration.
L. L. Nettleton.

The direct approach to magnetic interpretation and its practical application.
Leo J. Peters.

Magnetotelluric Methods

Basic theory of the magneto-telluric method of geophysical prospecting.
Louis Cagniard.

Processing and interpretation of magnetotelluric soundings.
G. Kunetz.

The magnetotelluric method in the exploration of sedimentary basins.
Keeva Vozoff.

Magnetotelluric responses of three-dimensional bodies in layered earths.
Philip E. Wannamaker, Gerald W. Hohmann, and Stanley H. Ward.

STRUCTURAL INTERPRETATION

A PROCESS OF SEISMIC REFLECTION INTERPRETATION [*]

BY

J. G. HAGEDOORN [**]

Abstract

A process is described whereby the interpretation of seismic reflection data is carried out by a preliminary two-dimensional plotting procedure followed by a three-dimensional migration. The concept of a surface of maximum convexity is introduced as an integral part of the process of migration. The procedures for deriving the necessary charts of curves are considered and a number of serviceable charts presented.

Contents:

Introduction 86
I. *Surfaces of equal reflection time and surfaces of maximum convexity*
   1. Vertical plotting determines the surfaces of equal reflection time 89
   2. Surfaces of equal reflection time as surfaces of maximum concavity . . 91
   3. The conception of surfaces and curves of maximum convexity 92
   4. Observation of curves of maximum convexity in practice 94
   5. Relationship between a reflector and a vertically plotted surface 95
II. *Two-dimensional migration*
   1. Two-dimensional migration utilizing curves of maximum convexity . . . 97
   2. Comparison of two-dimensional migration methods 98
   3. Two-dimensional migration applied to three-dimensional cases 99
III. *Three-dimensional migration with the aid of charts of curves*
   1. Migration in space visualized with the aid of families of surfaces 101
   2. Charts to be used as templates for three-dimensional migration 104
   3. Migration of contour lines with the aid of a pair of charts 106
   4. Practical application of the proposed method of migration 107
IV. *Wavefront charts for vertical plotting, based on time-depth relationships*
   1. Centres of curvature of the wavefronts at the central axis 110
   2. Elucidation of time-distance graphs by using oblique coordinates 112
   3. Derivation of wavefront charts from observed time-depth values 114
V. *Approximation of time-depth values by a smooth velocity function*
   1. Line of approach when approximating by a smooth velocity function . . 116
   2. Derivation of a serviceable smooth velocity function 118
   3. A template for direct determination of a smooth velocity function . . . 120
VI. *Derivation of the charts to be used for the purpose of migration*
   1. Use of wavefronts instead of curves of equal reflection time 121
   2. Charts based on velocities linear with depth or with vertical time . . . 123
Conclusion 125
References 127
Enclosures:
   A1, Wavefront chart for the velocity distribution linear with depth
   A2, Chart of curves of maximum convexity derived from chart A1
   A3, Contour maps migrated with the aid of the charts A1 and A2
   B1, Wavefront chart for the velocity distribution linear with time
   B2, Chart of curves of maximum convexity derived from chart B1
   C , Template for direct determination of a smooth velocity function

[*] Thesis, University of Utrecht, Holland, May 24, 1954.
[**] N.V. de Bataafsche Petroleum Maatschappij, The Hague.

Introduction

Interpretations of physical phenomena are almost always carried out by processes of abstraction by which pertinent data are reduced to those essentials which admit of mathematical treatment. The extent to which such reductions are permissible is conditioned by the accuracy required of the results to be obtained, which accuracy is limited by the reliability of the data upon which one has to work.

Both the more theoretical and the more practical seismologists aim to simplify their interpretational methods, the former in order to be able to use the mathematical tools at his disposal while the latter does so to achieve fast routine methods. The three main criteria by which the relative values of different interpretational methods must be judged are reliability, speed and clarity. A critical analysis of the existing interpretational methods, as laid down in textbooks (Nettleton, Dix[3], etc.) and publications (Geophysics, etc.) and further adapted to routine work by the practicists, obviously does not fall within the scope of this thesis. It might even seem superfluous to wish to add yet a new line of thought to these established methods which have already led to so many satisfactory results in practice. The method introduced in this thesis, however, brings forward some new concepts which may lead to an improvement in reliability and speed in interpretation. Furthermore, an almost purely geometrical line of thought is followed which may appeal, particularly, to the less analytically minded geophysicists.

As already done, for example, by Thornburgh and Riznichenko, the concept wavefront is taken as the point of departure and not the concept trajectory as is done in most methods. Choice of either concept depends, naturally, upon the way in which the abstractions wavefront and trajectory are formulated and are to be used. A general picture of the manner in which a seismic wave is propagated can be obtained by briefly considering the physical conditions, and at the same time an idea is gained of the degree of vagueness of the concepts wavefront and trajectory.

In geometrical optics the concept of a trajectory or ray as a narrow beam along which energy is transported has a visual and physical significance because long series of waves, with wavelengths in the order of fractions of a thousandth of a millimeter, are involved. Only when the dimensions involved in optical research are in this same small order of magnitude, as in microscopy, do complications like diffraction arise.

In the usual seismic work, however, the energy is transported in the form of a short compressional wavelet which does not change its form very rapidly as it travels along. In its simplest form (Dix[1, 2], Ricker) it will roughly resemble the one illustrated in fig. 1. A compression is followed by a rarefaction and not much else. In reflection work, where such a pulse is registered against a background of disturbance of the same order of magnitude, it may be the first compression from A to B with its peak at P which is observed in practice when only a small amount of filtering is applied.

Even for such a short pulse it is possible to speak of a wavelength and a frequency in a very broad sense. Analogous to the principle of Huygens-Fresnel, an energy quantum from the source S in fig. 2 can contribute to the first compression received in R if its trajectory does not exceed the minimum path by more than a half wavelength, corresponding roughly to the distance from A to B in fig. 1. From these considerations the order of magnitude of the width of the beam between source and receiver can be evaluated in the manner shown in fig. 2. Within the usual range of distances, velocities and frequencies

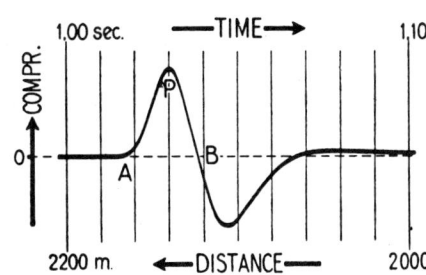

Fig. 1. A simple type of waveform from a seismic shot.

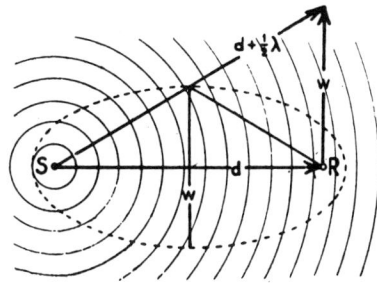

$$d^2 + w^2 = (d + \tfrac{1}{2}\lambda)^2 = d^2 + d\lambda + \tfrac{1}{4}\lambda^2$$
$$w \approx \sqrt{d\lambda}$$

distance: d from 500 to 10000 m.
velocity: V „ 1000 „ 6000 m/sec.
frequency: n „ 10 „ 100 cycles/sec.
wavelength: $\lambda = V/n$ „ 10 „ 600 m.
width: w „ 70 „ 2500 m.

Fig. 2. The width of a beam between shotpoint and receiver for the practical magnitudes of distance and frequency.

encountered in seismic work, it is clear that it is quite impossible to speak of trajectories in a physical sense of narrow beams.

On the other hand a wavefront, the front surface of the wave travelling away from the explosive source at a certain moment, can be physically visualized and is determined within relatively narrow limits. In reflection work where, in contrast to refraction work, the first compression is observed, at P in figure 1, the wavefront must be defined as the surface to which this first peak has travelled and this surface is also well defined physically. This can be illustrated by dropping a pebble into a pond. The wave with its front, first peak and following troughs and peaks can be seen travelling outwards but trajectories are not seen. This illustration, even though referring to a two-dimensional transverse wave, more closely resembles a seismic occurrence than any analogy from geometrical optics where it is fundamentally impossible for the eye to follow wavefronts moving with the speed of light.

In seismic interpretation a trajectory or ray can only be visualized as a mathematical abstraction, either as the line between source and receiver along which the traveltime is a minimum, or, except in case of anisotropy, as the line perpendicular to the wavefront at each intermediate time.

It must be borne in mind that the frequency range, as referred to in fig. 2, with its associated broad beam, is the only part of the sonic spectrum which, in practice, could be expected to give rise to observable reflections over the distances involved. This is due to the high degree of inhomogeneity of the transmitting earth material as compared, for example, with the atmosphere and water. The smaller the samples within which velocities are measured, the larger will be the variation in velocities found. Even when determining velocities in samples measuring a few inches (Baule), which is large compared to the size of individual grains, astonishingly large variations are found. Because of the levelling effect, due to the energy between source and receiver having travelled in a broad beam, the overall impression of the medium is one of homogeneity for which the conception of a wavelet travelling along without changing its form or being scattered too rapidly becomes possible. The longer the wavelengths the greater is the levelling effect, and the smaller is the scattering creating the background of disturbance against which a reflection must be observed. On the other hand the wavelengths must be short enough to enable determination, within reasonable limits, of the location of a reflector.

Figure 3 is an idealized picture of the extents of the regions involved when a reflection time is measured. It is clear that also a certain volume of the reflector, to depths of penetration of roughly one quarter of the wavelengths, is involved in the time measurement. This volume introduces a fundamental uncertainty in the location of the reflector even when data are obtained at a number of closely spaced receivers.

From these considerations it is understandable, as was also pointed out by Clewell and Simon, why the frequencies used in seismic work must lie in the frequency range of roughly 10 to 100 cycles/sec, if seismic reflections are to be obtained at all. It must be considered a stroke of fortune that, under favourable circumstances, the possibility of observing reflections actually lies within the borderline of perceptibility.

The simplifications and approximations assumed in this thesis are the usual ones which are applied as a rule in most of the seismic interpretational methods used in routine work. The media in which the seismic pulses travel are taken to be sufficiently homogeneous so that the velocity of the wavefront does not change appreciably with lapse of time or on account of change of frequency content. The velocity is assumed to be a function of depth only, thereby excluding anisotropy and appreciable lateral changes over short distances.

It must be clearly realized that these simplifications are not applied only for the sake of brevity or to arrive at a simple method adapted to fast routine work. It is not so very difficult to treat cases involving changes in velocity, other than as a function of depth, theoretically. But it is very seldom that these complications can actually be evaluated or even recognized from the data obtained in practice (Hagedoorn), to make it possible to take their influence into

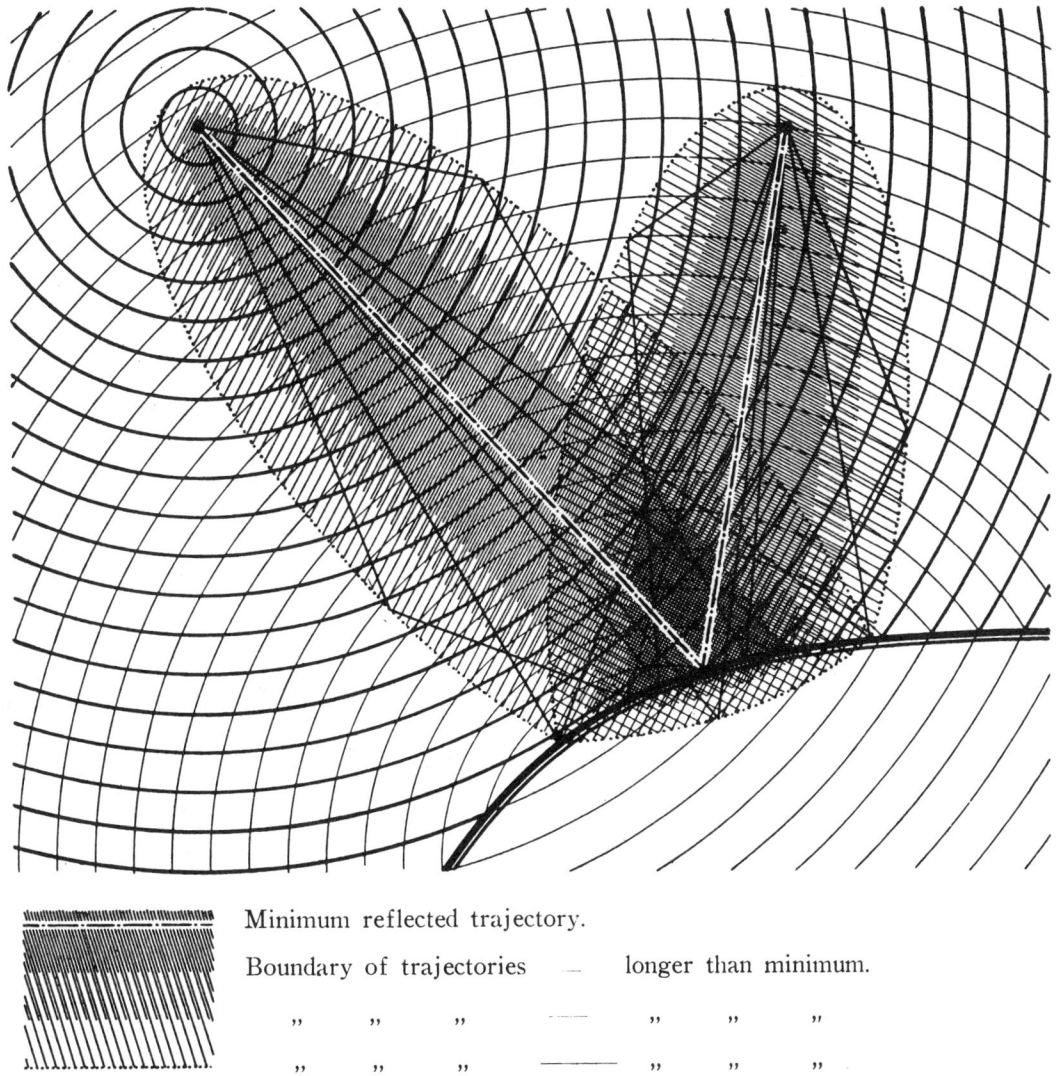

Fig. 3. Volume involved when a reflection time is measured.

account. The possibility that complications leading to grave errors may exist must always be borne in mind, but there is no sense in assuming their presence, without tangible evidence, when carrying out a routine exploration program, where a balance between speed and reliability must be maintained.

I. Surfaces of equal reflection time and surfaces of maximum convexity

1. Vertical plotting determines the surfaces of equal reflection time.

In fig. 4 a vertical section has been drawn through a shotpoint S and one receiver of reflected energy R, together with the intersections with a set of wavefront surfaces centred at S and a set centred at R, both spaced at equal time intervals. A reflection time of 2T seconds, observed at R, will refer to a reflection from a surface that can be tangential, according to the principle of Fermat, at any point to a surface of equal reflection times consisting of the lines of intersection between wavefront surfaces at times T + t from S with wavefront surfaces at times T-t from R, for all values of t.

The intersection of such a surface of equal reflection times with the vertical plane through S and R is shown in figure 4 as the line through P and Q. The "point" of reflection, the point where the actual reflecting surface is tangential to the surface of equal reflection times, will, however, most probably

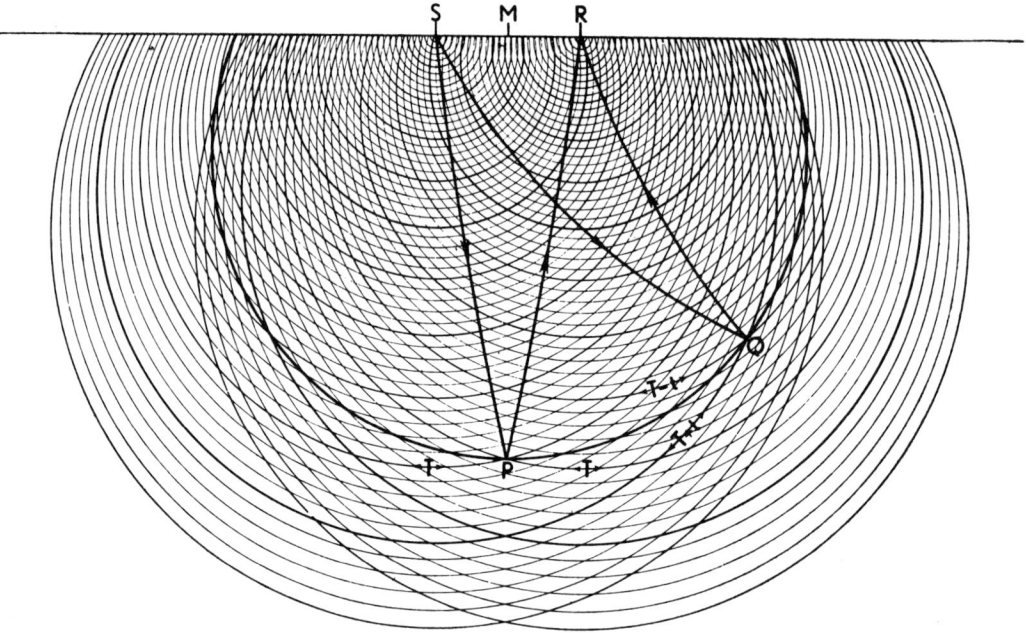

Fig. 4. One single reflection time value determines a surface of equal reflection times.

not lie in this particular plane of drawing. The orientation of one small area of the reflector in space can only be determined, in principle, by at least three surfaces of equal reflection times, found by obtaining the reflection times at at least three receivers located close together at the surface of the ground.

A surface of equal reflection times is, of course, primarily determined by the location of the actual reflector and the location of the shotpoint S and the receiver R. As the position of the reflector is, however, not yet known from one single observation, the surface of equal reflection times must be determined from the one observed reflection time 2T, but it is also fully determined by any arbitrary point on it. The obvious point to choose is the point of symmetry P, which is also the deepest point on this surface.

It is this particular point which is plotted when the procedure called "vertical plotting" is carried out on a two-dimensional section. In this procedure a wavefront chart with wavefront curves drawn at small equal time intervals, for example every 5 milliseconds, is centred at the shotpoint S and the point is plotted at the intersection of the interpolated wavefront curve at half the total reflection time with the vertical through M, midway between S and R. This manner of vertically plotting all time values obtained by shooting from a shotpoint to a series or "spread" of receivers is the most straightforward procedure and is very fast in practice when a wavefront chart is made available with the double times marked at the curves and with vertical lines drawn in at half the customary distances between shotpoint and receiver.

The great advantage of vertical plotting is that it can be carried out on a two-dimensional vertical section, as a first step in determining the position of the actual reflector. This is a logical consequence of the fact that a certain available number of receivers is, in practice, commonly set out in line and recorded simultaneously in order to facilitate the fieldwork and to be able to distinguish even fairly weak reflections by their correlation on a number of adjacent traces. These considerations lead to the practice of shooting seismic "lines" in preference to "patterns". However, this also leads to two-dimensional lines of interpretation, which remain quite correct only as long as vertical plotting is carried out.

It must always be clearly borne in mind that a vertically plotted point has no other significance than that of being one point determining a surface of equal reflection times, which surface is known to be tangential to the actual reflector at some point in space.

2. *Surfaces of equal reflection time are surfaces of maximum concavity.*

"Migration" is the procedure of determining the true reflecting surface from a surface determined by a number of vertically plotted points. This true surface can be found, in principle, as the envelope to all surfaces of equal reflection times determined by the vertically plotted points. This is essentially a three-dimensional procedure, setting out from the usually two-dimensionally, vertically plotted points. Only in the exceptional case when the vertical section through shotpoint and receivers lies in the direction of maximum dip of the reflector, can migration be carried out in the same vertical plane as the vertical plotting. Because of its greater simplicity this circumstance is aimed at quite often in practice, but it obviously excludes, for example, the possibility of shooting on a system of intersecting lines.

In fig. 5 such a two-dimensional picture is shown, representing a vertical section perpendicular to the axis of a cylindrical trough. The concavity of this trough has been chosen so great that it is possible to locate a surface of equal reflection times tangential to this reflecting surface at two points, P and Q. The reflecting curve APMQB would result in the dot-dash line when the reflection times observed along the line XY on the surface of the earth are vertically plotted. In most cases in actual practice only the times resulting in the parts A'P' and B'Q' would be observed as reflection times. The reflected impulses received from between P and Q would most probably be masked by the tail ends of the reflections received from A to P and from Q to B, as the distance between P' and M will be relatively small in most cases.

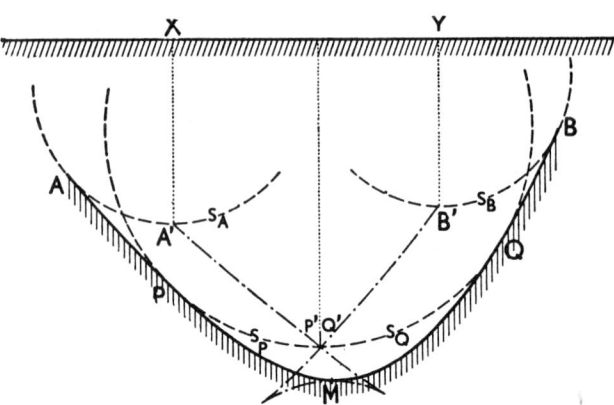

Fig. 5. A surface of equal reflection times is a surface of maximum observable concavity.

From these considerations it follows that, as a rule, a reflecting surface cannot be more concave than a surface of equal reflection times if each point is to result in a vertically plotted point or even in a relevant time observation. Consequently the surfaces of equal reflection times can be considered as the surfaces of maximum observable concavity for reflecting surfaces. A region of higher concavity can hardly ever be determined but its existence will be marked by a sharp bend upwards, at P' in figure 5, of the apparent surface found by plotting vertically.

This fundamental limitation to the interpretation is not so very important when seen from the viewpoint that in exploring for oil a culmination or "high" is the desired feature. It can, however, lead to interpretational errors in regions with poor reflections because a wrong correlation from trace to trace can quite easily be carried out across such a feature.

3. *The conception of surfaces and curves of maximum convexity.*

Migration is the transformation of a horizon, found by vertical plotting, to the true reflecting horizon. This can be carried out by determining the surfaces of equal reflection times, or surfaces of maximum concavity, belonging to each vertically plotted point.

Inverse migration would be the transformation of an actual reflecting surface to the apparent horizon that would have been obtained by vertically plotting the reflection times. Again, as a first step, a single reflecting point can be considered and the corresponding vertically plotted representation of this point can be determined.

In figure 6 a vertical section through a single reflecting point P has been drawn. The reflection times have been obtained from a number of pairs of shotpoint and receiver. By plotting these times vertically an apparent horizon has been obtained whose intersection with the section drawn is the curve c. This surface is determined by the fact that any point Q on this surface must

be the deepest point on a surface of equal reflection times u through P.
If, by vertically plotting, an apparent horizon were found that coincided with c in figure 6, then all points on this horizon would be migrated to the

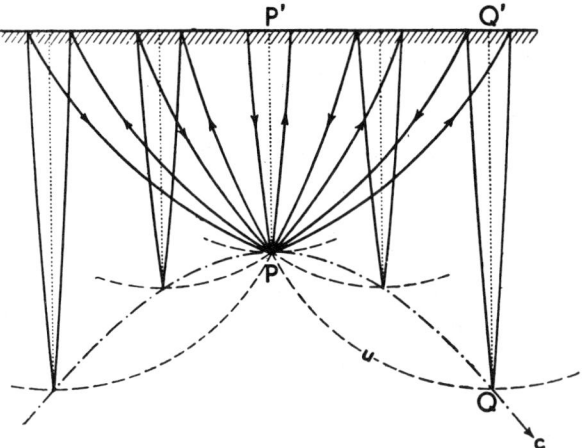

Fig. 6. A vertically plotted horizon from one reflecting point is a surface of maximum obtainable convexity.

one point P. This means that this apparent horizon from one reflecting point is a surface of maximum convexity for vertically plotted horizons.

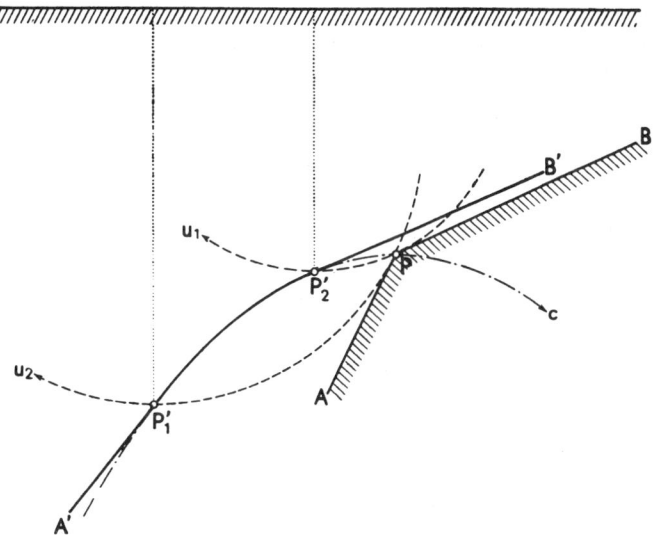

Fig. 7. An abrupt downward bend is vertically plotted as part of a curve of maximum convexity.

In fig. 7 this is illustrated on a two-dimensional section. The true horizon APB has a sharp bend at P. The part PB would be vertically plotted as $P_2'B'$ and the part AP as $A'P_1'$. Between P_1' and P_2' the vertically plotted horizon would consist of part of the curve of maximum convexity c through P, which curve consists of the vertically plotted reflections from P.

This means that a sharp bend downwards of the reflecting horizon will show up on a vertically plotted section as a curved part, corresponding to part of a curve of maximum convexity. It would be possible to detect such sharp bends with the aid of a chart with a number of closely spaced curves of maximum convexity, by observing whether any part of a vertically plotted horizon will fit part of one of these curves.

From these considerations it is clear that there is no fundamental limitation to the interpretation when a reflector has a sharp bend downwards, in contrast to the case discussed in the preceding chapter, when a surface is more concave than a surface of equal reflection times.

4. *Observation of curves of maximum convexity in practice.*

Seen from the viewpoint that reflections must rigidly follow the laws of geometrical optics, curves of maximum convexity could not be expected in actual practice because the one point P in figure 7 would not be able to reflect any appreciable amount of energy. Reflection from one point is, however, a mathematical abstraction but the abstraction has as much significance as when a reflecting point on a smooth surface is considered. The apparent difficulties are cleared up by going back to the fundamental principles of Huygens-Fresnel in their application to seismic work, as was done in the Introduction.

The existence of these diffractions occurring at faults was also recognized by Krey. By using a method of plotting only migrated straight line segments, the essentially curved nature of these events, as can be seen on the seismograms and also when vertically plotted, was not taken into account, so that the clear picture of a curve migrating to a point was not realized. Krey worked out the mathematical problem of the energy decay as a function of distance and frequency, showing how the actual observation of these phenomena can also be explained theoretically.

Fig. 8 shows a record section obtained in Holland in 1948. The original recorded traces have been individually traced and shifted in the direction of the time axis to correct for elevation, weathering and also for the horizontal distance between shotpoint and receivers of from 300 to 600 m. This last correction to the vertical, or coincidence of shotpoint and receiver, is of course a function that varies with depth, but it can be taken to be constant in the small depth region considered in this picture.

The curved event on this vertically plotted record section can be identified, to a high degree of probability, to be a curve of maximum convexity by virtue of the fact that it can so very closely be fitted by a curve of maximum convexity. In this case a good fit is obtained by the dashed line, which is a curve of maximum convexity with its culmination at P, the point where a sharp break or bend of the reflector probably occurs.

The apparent variation of the energy along the curved event is certainly no measure of its real energy content. The automatic volume control of the recording instrument reduces the amplitude to a certain level within a certain time lag. The event will appear strong only when it is strong relative to the noise background. This means that the observed fluctuating variation in energy refers to an inverse fluctuation of the recorded noise and not, necessarily, to a variation of the energy of the event.

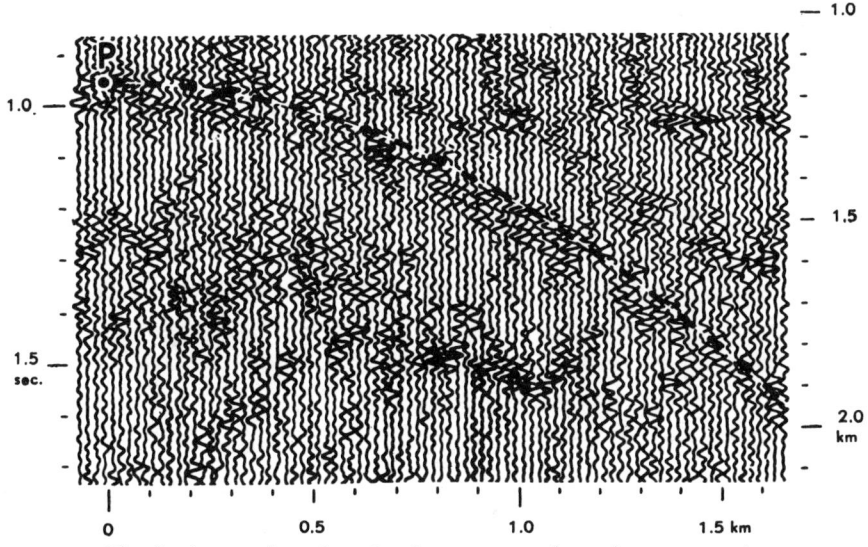

Fig. 8. A record section showing a curve of maximum convexity.

5. *Relationship between a reflector and a vertically plotted surface.*

In fig. 9 a stereometric picture is presented as an aid in visualizing the significance of surfaces of equal reflection times and surfaces of maximum convexity in the relationship between a vertically plotted surface and the corresponding actual reflector.

The actual reflecting point P would be vertically plotted at the axial point Q on the surface of equal reflection times, the apparently spherical surface in figure 9, which is tangential to the actual reflecting surface at P.

On the other hand the point P could be the result of any vertically plotted point on a surface of maximum convexity, the apparently hyperboloidal surface in fig. 9, with its apex at P. This surface must be tangential, at Q, to the surface that would be found by vertically plotting the results obtained from the actual reflector in the vicinity of P.

The vertically plotted point Q is the deepest point or the point on the central axis of that particular surface of equal reflection times which is tangential to the actual reflector at the particular reflecting point P. Conversely the actual reflecting point P is the highest point or the point on the axis of that particular surface of maximum convexity which is tangential to the vertically plotted surface at the particular vertically plotted point Q.

The actual reflecting surface is the envelope to all surfaces of equal reflection times with their axial points on the surface found by vertically plotting. Conversely the vertically plotted surface is the envelope to all surfaces of maximum convexity with their axial points on the actual reflecting surface.

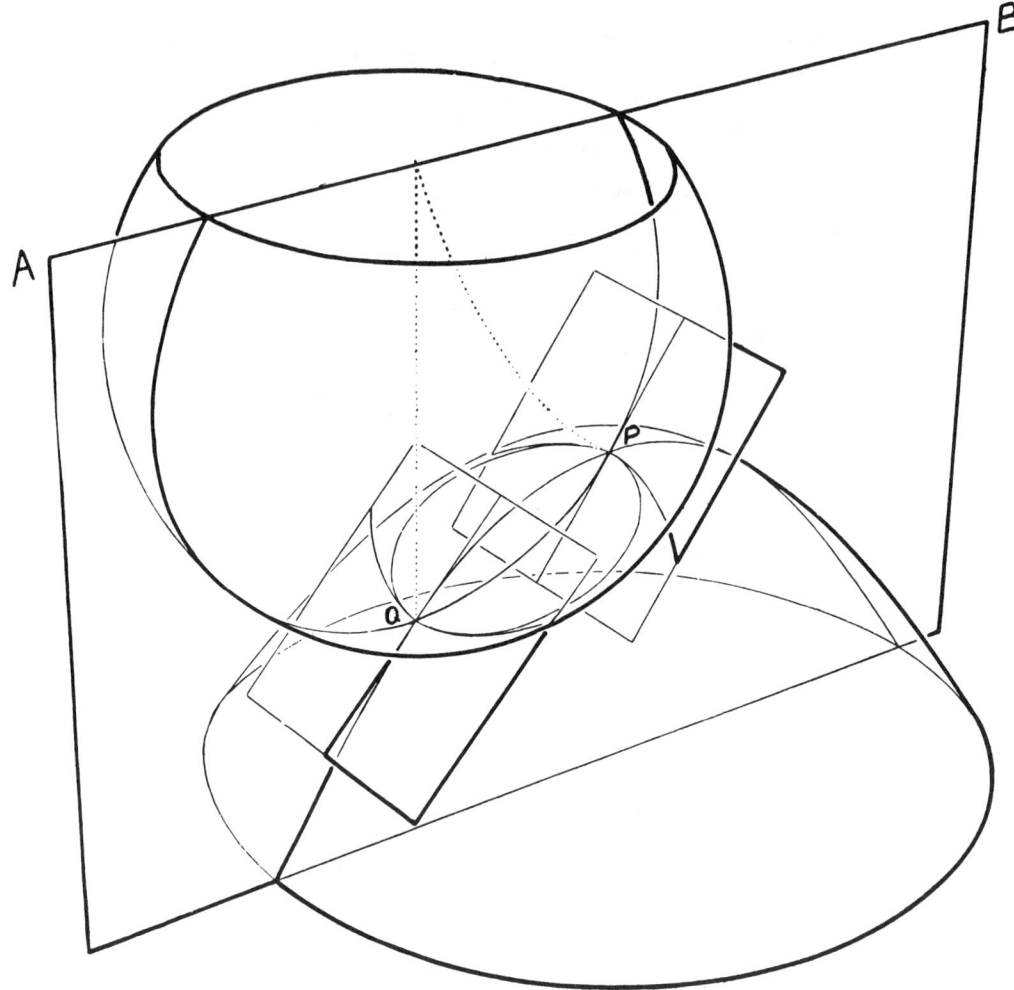

Fig. 9. The relationship between a surface of equal reflection times and a surface of maximum convexity.

Thus, the surfaces of equal reflection times and the surfaces of maximum convexity play quite analogous parts in the relationship between a vertically plotted surface and the corresponding actual reflecting surface. It is clear that a vertically plotted surface has as much fundamental significance, mathematically speaking, as an actual reflecting surface. Seen from this viewpoint it is wrong to consider it as an approximation of the actual surface, which approximation becomes better the less the surface deviates from the horizontal. The vertically plotted surface must be seen as a transformation of the actual surface with a continuous point to point relationship.

The advantage of using a vertically plotted surface, as an intermediate step in determining the actual reflector, lies in the fact that it can be determined directly point by point from the individual time measurements, either by plotting or as a record section as in fig. 8. The location of one point on this surface or its intersection with a two-dimensional vertical section through a line of shotpoints and receivers can be carried out without taking into account the orientation in space of the actual reflector. By combining a number of lines the orientation in space of the vertically plotted surface can be determined and only then can the actual surface be found by a process of transformation known as migration.

II. Two-dimensional Migration

1. *Two-dimensional migration utilizing curves of maximum convexity.*

A vertical plane AB is shown in fig. 9, through the axis of the surface of equal reflection times and the axis of the surface of maximum convexity. Both the vertically plotted surface at Q and the actual reflecting surface at P are perpendicular to this vertical plane which lies in the direction of maximum dip of the reflector at P. By working in this plane a two-dimensional procedure of migration can be followed to determine the position of P from the position of Q.

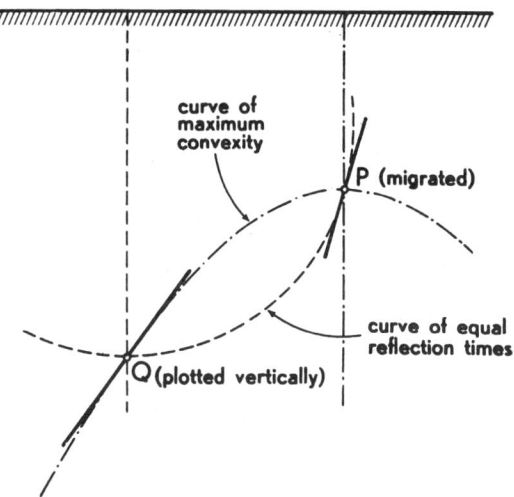

Fig. 10. Two-dimensional migration with the aid of two curves.

Figure 10 is such a two-dimensional representation. It is clear that for the purpose of migration two curves are available; in the first place the curve of equal reflection times through Q with the vertical through Q as axis of symmetry and in the second place the particular curve of maximum convexity through Q which is tangential to the vertically plotted horizon through Q.

This leads to a practical procedure of two-dimensional migration illustrated in figure 11. A chart of curves of equal reflection times is centred on the vertical through Q and a chart of curves of maximum convexity is moved to a position where the best tangential fit to the vertically plotted horizon at Q is obtained.

Of course Q does not necessarily lie either on one of the curves of equal reflection times or on one of the curves of maximum convexity of the charts used. In practice the position of Q in regard to the adjacent curves of maximum convexity will determine an imaginary interpolated curve that can be followed to the intersection with the axis of the chart of curves of maximum convexity; this intersection P then is the migrated point. Obviously, from these considerations, the position of the migrated point P can be determined uniquely by use of the chart of curves of maximum convexity alone, because the intersection P of the curve of maximum convexity with the curve of equal reflection times through Q must lie on the central axis of the chart of curves of maximum convexity. The curve of equal reflection times through Q and P, however,

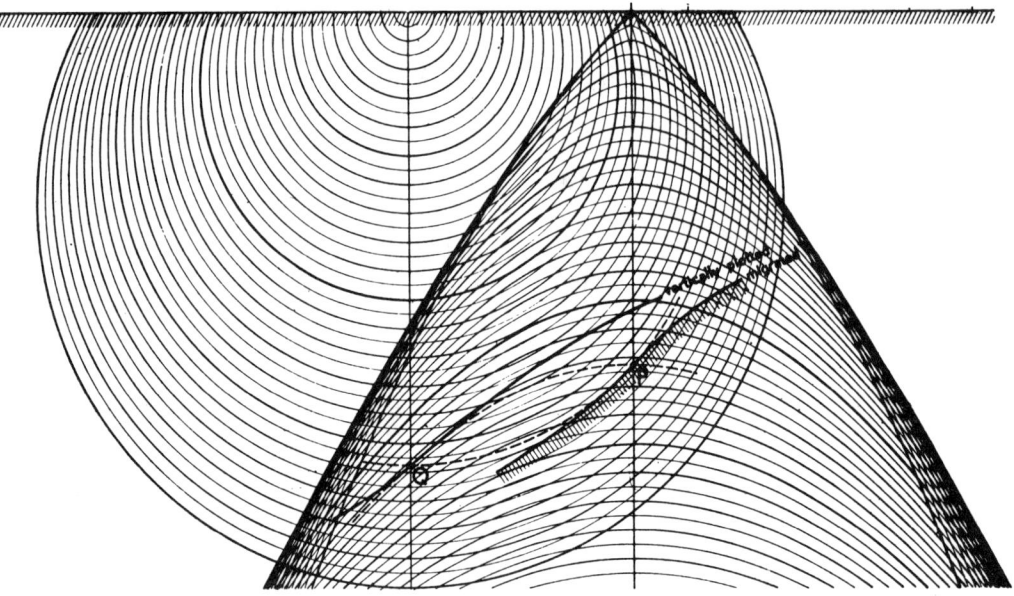

Fig. 11. Practical procedure of two-dimensional migration with the aid of two charts of curves.

determines the dip at P, for which purpose the chart of curves of equal reflection times is indispensable.

For the practical application of this method both charts must be mounted in a device which makes it possible to move them independently in a horizontal direction underneath the transparent section in such a way that the lines of zero depth on all three drawings always coincide.

2. *Comparison of two-dimensional migration methods.*

In practice a wide diversity of methods are in use for determining a reflecting horizon from reflection time data. Most methods are two-dimensional because the three-dimensional methods evolved were cumbersome and time consuming. Very often plotting devices (Daly) or charts (Musgrave) are used to plot the data as a depth section in one single manipulation. These methods are based on the assumption that the location and orientation of a short length of a reflecting horizon is a function of two variables: a mean total travel time t and the change of travel time with distance, dt/dx. It can be seen that this means the determination of a curve of equal reflection times and the direction of the trajectory or the orientation of the wavefront arriving at the receivers.

The mechanical devices are, in practice, limited in their application to "straddle-shooting" and to velocity distributions linear with depth, because only in that case are wavefronts and trajectories circular (van Melle).

The use of charts is not restricted, in principle, to any particular velocity distributions. For any velocity distribution a chart could be prepared with curves of equal reflection times and curves of equal change in reflection time. These are relatively simple to make for a velocity distribution linear with depth but already fairly complicated even when other simple smooth velocity distributions can be used (Musgrave).

These last considerations would seem to apply also to the use of pairs of charts of both curves of equal reflection times and curves of maximum convexity. However, these are used after first vertically plotting the reflection points. In this procedure, shown in figure 11, the shape of the curves is the only essential consideration, whereas for plotting the horizon directly, without first plotting vertically, the time values of the curves of equal reflection times must also be correct. This imposes a stronger limitation to the possibility of using a certain smooth velocity distribution upon which to base a chart. This point will be further elaborated in chapter VI, 1.

A fundamental disadvantage of directly plotting short migrated straight line segments, is that even small uncertainties in determining changes of time with distance will result in a correlating event on a series of records no longer correlating on the section. The horizon is broken up into disconnected parts. In the extreme case where a curve of maximum convexity occurs, this will result in a concentration of short line segments at different angles. The advantage of migrating in the manner shown in fig. 11 is that a curve of maximum convexity is recognized immediately.

3. *Two-dimensional migration applied to three-dimensional cases.*

Fig. 12 is a stereometric view of two sections I and II, along which reflections are obtained from a flat reflector RAB. For the sake of simplicity it has been assumed that the velocity between the surface of the ground and the reflector is constant and that shotpoint and receiver were always coincident (normal incidence).

SP is perpendicular to the reflector and is the trajectory of the reflected impulse received at the intersection S of the two lines. The point Q, vertically plotted below S, is the axial point on the spherical wavefront surface with its centre at S and which is tangential to the reflector RAB at P. By vertically plotting on sections I and II the apparent horizons QA and QB would be found. These must be migrated to PA and PB, which means, actually, that the depths vertically below the lines TA and TB have been determined.

It is clear that by two-dimensional migration of the apparent horizons QA and QB on the vertical sections I and II respectively, the true horizons RA and RB would certainly not be found. As the spherical wavefront through P

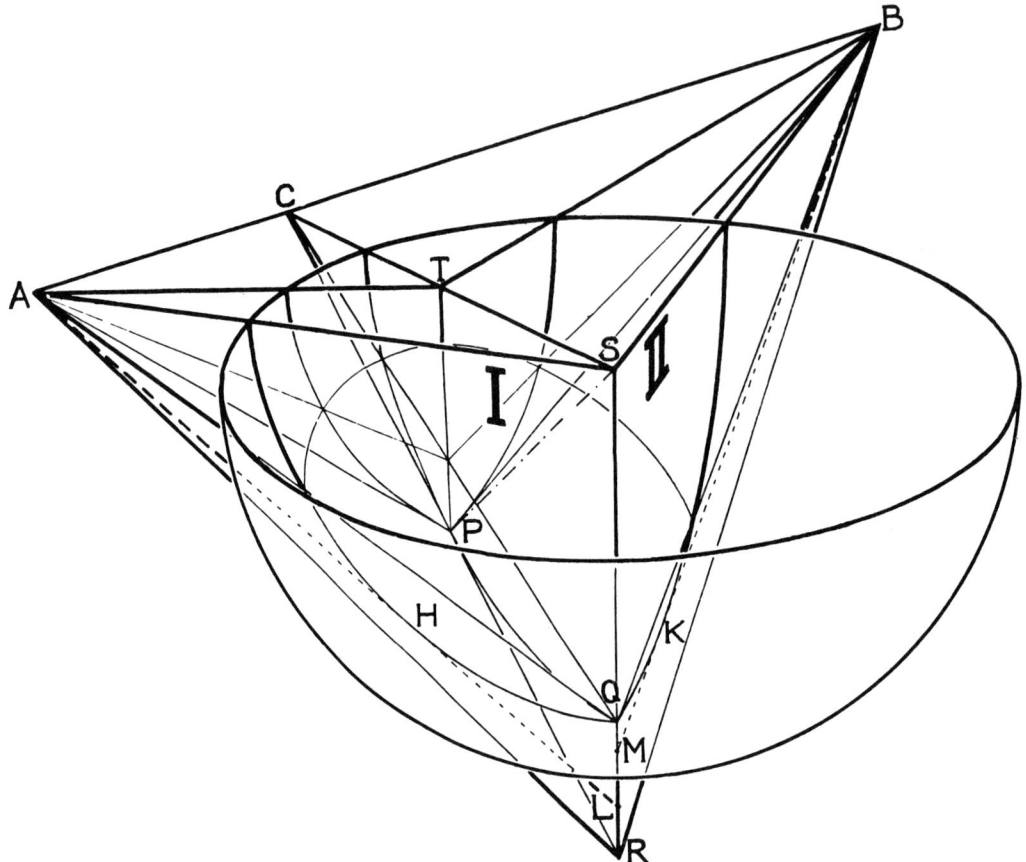

Fig. 12. Steriometric view of two intersecting seismic sections.

and with S as centre is only tangential to the true surface RAB at the one point P, the two horizons found, LA and MB, are always shallower than RA and RB, because they would be found as tangents to the circles of intersection of the sphere with the sections I and II. Furthermore, these two horizons do not cut the vertical SR at the same point, except in the symmetrical case where the two sections I and II are situated at equal angles to the direction of maximum dip CS.

Similar considerations obviously also hold when the velocity is not constant and when the reflector is not flat. It is clear that by any process of two-dimensional migration which does not take into account the direction of maximum dip in regard to the section, horizons are found that, in general, are too shallow and which cannot be correlated at intersections of sections. In fact, the result obtained at point S in fig. 12 would be plotted at two widely different points H and K on sections I and II respectively, instead of actually at P.

Quite often, in general practice, reflection time data are plotted in their migrated position, as a matter of general routine, as though the section actually did run approximately in the direction of maximum dip, without first ascertaining whether this assumption is correct. This procedure is then defended by the assertion that the picture obtained by two-dimensional migration will be closer to reality than the picture obtained by vertical plotting alone, irrespective of the true direction of maximum dip in regard to the section. Even this belief is, however, certainly not always justified, as the procedure of automatically migrating two-dimensional sections can lead to grave errors.

If, for example, the section shown in figure 8 had been shot at an oblique angle to an edge at P, the curved horizon would have been less strongly curved, being an oblique intersection with a cylindrical surface of maximum convexity originating at the edge P. This horizon would then not migrate entirely to P by two-dimensional migration and a curved horizon extending beyond P could have been found. This "true" horizon would have no real significance but would only result in an inexplicable discrepancy when checked by actual drilling.

Strictly speaking, a horizon on a vertically plotted two-dimensional section is worthless without further evidence of its orientation in space. This further evidence can either be acquired from geological data or from cross-sections. A line shot with sufficient "cross-spreads" can give definite information, if the quality of the reflections obtained is sufficient to give reliable dip determinations over the short distances covered by the cross-spreads.

Any curved horizon, either on a vertically plotted or a migrated section should always be regarded with suspicion so long as a three-dimensional picture has not been obtained.

III. THREE-DIMENSIONAL MIGRATION WITH THE AID OF CHARTS OF CURVES

1. *Migration in space visualized with the aid of families of surfaces.*

Three-dimensional migration of a vertically plotted point, Q in figure 9, can be reduced to a two-dimensional migration in the vertical plane in the direction of maximum dip, the plane through A and B in figure 9. This is the procedure which has usually been followed in practice. It means, however, that a great number of auxiliary sections in the direction of maximum dip must be drawn.

Purely three-dimensional migration could be carried out, in principle, by centering a family of surfaces of equal reflection times at the vertical through a vertically plotted point Q and moving a family of surfaces of maximum convexity to the best tangential fit to the vertically plotted horizon at that point Q. By following the surface of maximum convexity through Q to the axis of the family of surfaces of maximum convexity the migrated point P would be found. The surface of equal reflection times through Q and P would then determine the inclination of the true reflecting surface at P.

In order to be practicable, a migration method must, however, be based on a two-dimensional representation of the facts. A reflecting surface can be represented by a system of contour lines, being the projection on a horizontal datum surface of the intersection lines between the reflecting surface and a set of horizontal planes at regular depth intervals. A vertically plotted surface, represented by its contour lines would have to be migrated, in principle, with the aid of surfaces of maximum convexity and surfaces of equal reflection times also represented by their contour lines.

This is shown in fig. 13 which is the same case as shown stereometrically in fig. 9. The contour lines of the actual reflecting surface are seen to fit the contour lines of the surface of equal reflection times tangentially at P and the contour lines of the vertically plotted surface are seen to fit the contour lines of the surface of maximum convexity tangentially at Q.

This tangential coincidence of contour lines takes place on the line PQ, the straight line which is perpendicular to the contour lines at both P and Q. Only the points on this straight line are necessary for the purpose of migration and the rest of the contour lines of both the surface of maximum convexity and of the surface of equal reflection times may be dispensed with.

In the case shown in figure 9, shotpoint and receiver have been chosen to coincide. A surface of equal reflection times then is the same as a wavefront originating at this coincident point of shotpoint and receiver. If the velocity distribution is a function of depth only, such a wavefront is axially symmetrical and the same applies to a surface of maximum convexity.

Such a surface consisting of contour lines which are concentric circles, a two-dimensional representation, is fully determined by the points of intersection with a line through the centre, a one-dimensional representation. Each surface, either a wavefront surface or a surface of maximum convexity, can be represented by such a row of points on a straight line. All these rows of points can then be combined into a pair of two-dimensional templates, one representing all wavefront surfaces and the other all surfaces of maximum convexity.

In this way the use of a prohibitively large number of contour maps of surfaces to fit all possible depths is avoided, so that the method of fitting surfaces at all depths and inclinations becomes feasible. Only two templates must be prepared and these turn out to be the same pair of charts as used for two-dimensional migration in the manner shown in fig. 11.

It must be clearly borne in mind that the representation of an infinite number of surfaces by a template in the manner described can only be carried out for axially symmetrical surfaces. This means that the three-dimensional migra-

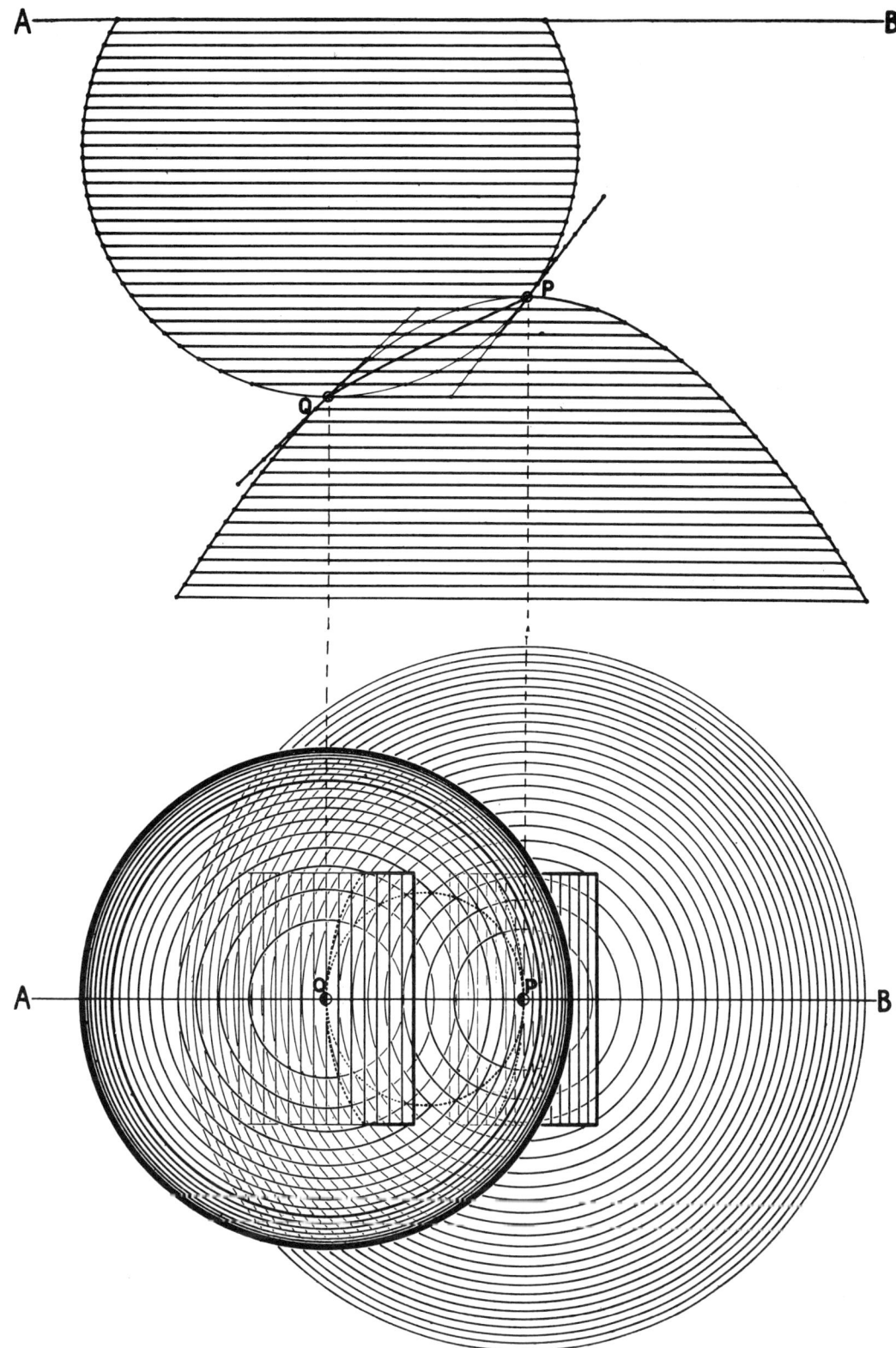

Fig. 13. Representation of the case of fig. 9 by contour lines.

tion method here described can, strictly, only be used in the case of normal incidence, or coincidence of shotpoint and receiver. This point will be further elaborated in paragraph VI, 1.

2. *Charts to be used as templates for three-dimensional migration.*

By the following line of reasoning it can be seen how these templates can be arrived at.

The simplest derivation of a curve of maximum convexity from the wavefront chart, without using auxiliary construction lines, is shown in fig. 14A, comparable to figure 6, but with shotpoint and receiver coinciding. The

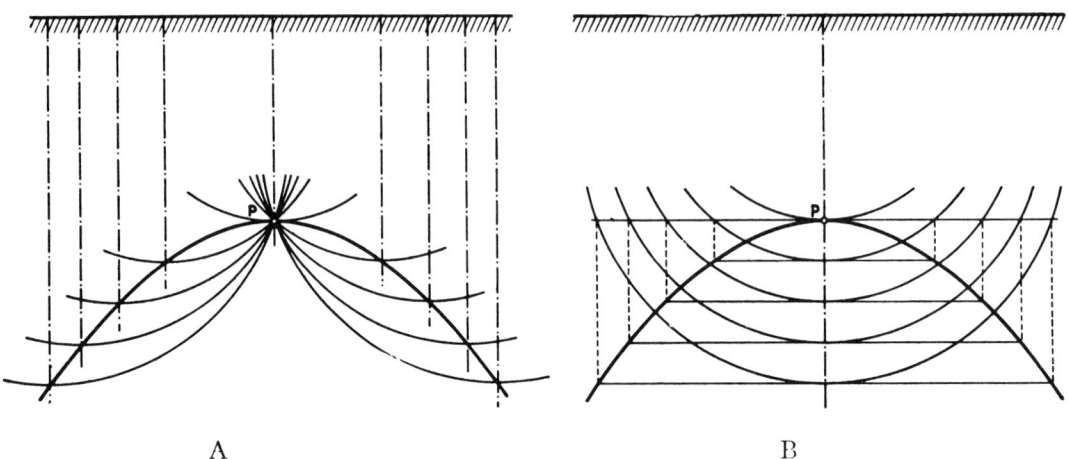

Fig. 14. Two procedures for deriving a curve of maximum convexity from the wavefront chart.

wavefront chart is shifted in a horizontal direction underneath a transparent overlay until a wavefront runs through P. The intersection point of this wavefront with the axis of the chart is then plotted, and that of other wavefronts in similar manner. It can also be carried out as shown in figure 14B, leaving the wavefront chart in the central position. The intersection of a horizontal line through P with a wavefront is projected downwards to the maximum depth of that wavefront. This construction is obviously the same, in principle, as the one shown in fig. 14A.

It is clear that a curve of maximum convexity, obtained in whatever manner, depends only on the depth of its intersection with the axis and, obviously, not on the spacing of the wavefronts on the chart from which it has been derived.

In fig. 15 a wavefront chart has been drawn with the wavefronts spaced at regular maximum depth intervals and the curves of maximum convexity, which cut the axis at the same equally spaced depth values, have been derived from these wavefronts.

At the right hand side of the axis in fig. 15 the same method of constructing the curve of maximum convexity c is shown as in fig. 14B. Fig. 15 can be regarded as a vertical plane through the axis of a set of wavefront surfaces and surfaces of maximum convexity spaced at equal depth intervals at the axis. The circles of intersection between the horizontal plane h and the wavefront surfaces are then seen to be the contour lines representing the surface of maximum convexity c. In the plane of drawing the intersection points of the line h with the wavefronts then are a linear representation of the contour lines

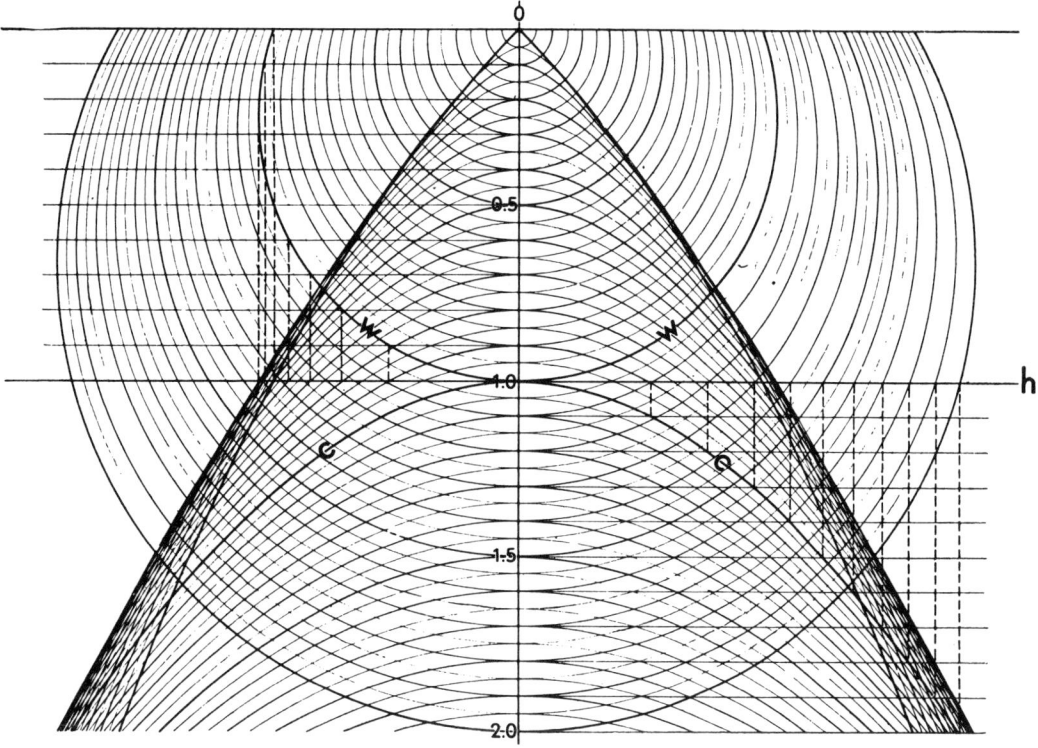

Fig. 15. Mutual projective relationship between wavefronts and curves of maximum convexity.

of the surface of maximum convexity c with its top at the depth of h. This means that the wavefront chart can be used as a template representing all possible surfaces of maximum convexity.

At the left hand side of the axis in figure 15 the intersections of the horizontal line h with the curves of maximum convexity are seen to be the vertical projections of points of the wavefront w at consecutively shallower depths. If figure 15 is again regarded as a plane through the axis of a set of wavefront surfaces and surfaces of maximum convexity, a circle formed by the intersection of the horizontal plane h with a surface of maximum convexity is a contour line on the wavefront surface w. Consequently, the chart of curves of maximum convexity can be used as a template representing all wavefront surfaces.

3. *Migration of contour lines with the aid of a pair of charts.*

Fig. 16 illustrates a procedure for migrating a vertically plotted surface represented by contour lines spaced at the same depth intervals as the wavefronts and the curves of maximum convexity on the charts used as templates.

A certain point Q, midway between the contour lines 28 and 29, is to be migrated. The chart of curves of maximum convexity is orientated underneath the transparent contour map, keeping its axis parallel to the mean direction

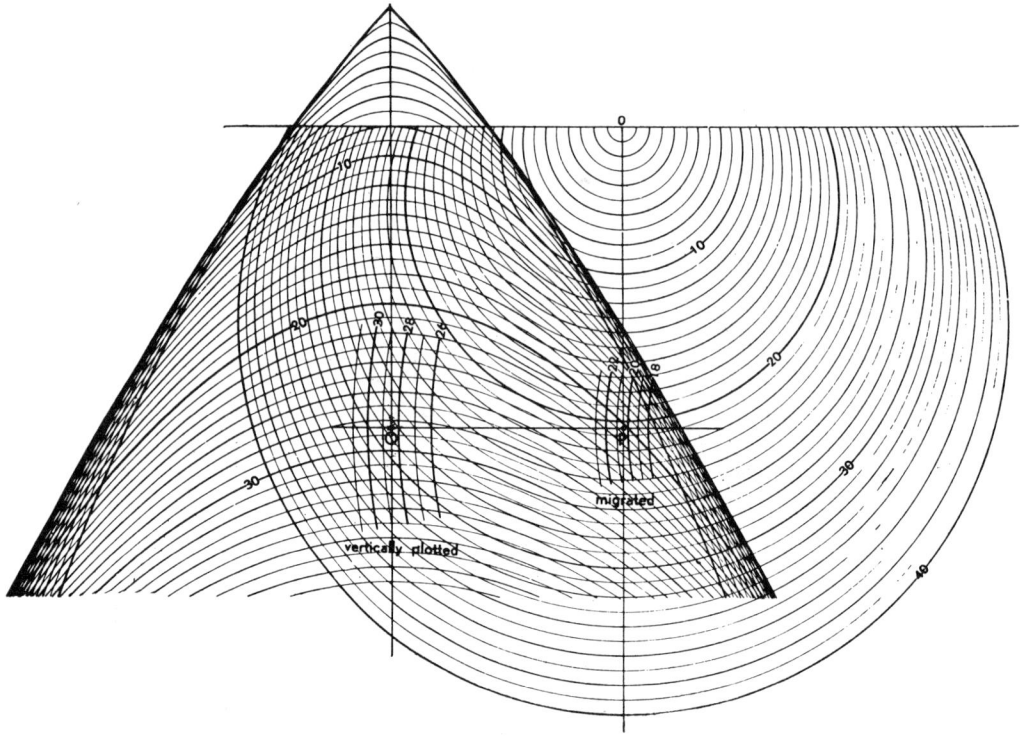

Fig. 16. Migration of contour lines with the aid of a wavefront chart and a chart of curves of maximum convexity employed as templates.

of the contour lines adjacent to Q, to a position where Q lies on the axis and midway between the curves of maximum convexity 28 and 29. The wavefront chart is then orientated, also underneath the contour map, and also keeping its axis parallel to the mean direction of the contour lines adjacent to Q. It is moved to a position where the intersection between the wavefront 28 and the contour line 28 and also the intersection between the wavefront 29 and the contour line 29 both lie on the line through Q which is perpendicular to the mean direction of the contour lines adjacent to Q. The point P where this perpendicular line then intersects the axis of the wavefront chart will be the migrated position of Q. The curves of maximum convexity adjacent to P will intersect the line QP at the same points as the migrated contour lines adjacent to P, which are, of course, also perpendicular to QP.

Thus, the wavefront chart determines the position of P on the map and its depth, and the chart of curves of maximum convexity determines the dip at P by giving the position of the contour lines adjacent to P.

Because of the finite spacing of the contour lines the assumption has been made, during this construction, that the curvature of the vertically plotted surface at Q in the vertical plane through QP is the same as the curvature of the curve of maximum convexity through Q, which curve is essentially convex. At the same time the assumption has been made that the curvature of the actual reflecting surface at P is the same as that of the wavefront through P, which curve is essentially concave. These assumptions are obviously contradictional and introduce a certain error which, however, can be kept reasonably small by choosing a sufficiently close spacing of the contour lines and, correspondingly, also of the curves on the charts employed.

On enclosure A3 two examples are shown where this method of migration has been carried out with the aid of the enclosed wavefront chart A1 and the corresponding chart of curves of maximum convexity A2. These examples correspond to a velocity Vz linear with depth z: $Vz/Vo = 1 + z/L$. All distances have been scaled with L as unit and the contour lines on enclosure A3 have been spaced at intervals of 0.01 L to correspond with the axial intervals on both charts. The reader can familiarize himself with the method of migration by enlarging enclosures A1, A2 and A3 on separate sheets of transparent material to a conveniently large scale where L is exactly the same, for example 10 cm, on all three sheets.

The examples chosen on enclosure A3 are purely hypothetical cases. The shooting system included the shooting of cross-spreads, each comprising two full spreads across the line. If, in practice, this system yields sufficiently clear reflections, a reliable three-dimensional picture of a vertically plotted horizon is obtained in a band below the line itself, and the migrated horizon may be derived as the corresponding band. In this way a clear picture is obtained of those parts of the reflecting horizon that have been actually determined.

4. Practical application of the proposed method of migration.

By referring back to figure 16 it is clear that for migrating contour lines, obtained by vertically plotting reflection points, three separate things are required in a particular disposition in relation to the small portion of the contour map to the migrated. These three separate things are:

(a) the straight line PQ, along which migration is to take place,

(b) a chart of curves of maximum convexity, representing wavefront surfaces and

(c) a wavefront chart, representing surfaces of maximum convexity.

Their particular disposition must always satisfy the condition that the axes of symmetry of (b) and (c) are perpendicular to (a).

One way of achieving any desired disposition of (a), (b) and (c) is to draw them on separate transparent sheets and determine their relative movements by a device where, one chart (b) taken as position of reference, the sheet with the straight line (a) can move in one direction, a slide, and the other chart (c) can move so that the axes of the two charts remain parallel.

This arrangement is shown diagrammatically in fig. 17. The sheet with the line of migration (dashed lines) can slide up and down with respect to the sheet containing the chart of curves of maximum convexity (dot-dash lines) and the sheet with the wavefront chart (full lines) is connected to the other chart with a simple linkage to keep their axes parallel.

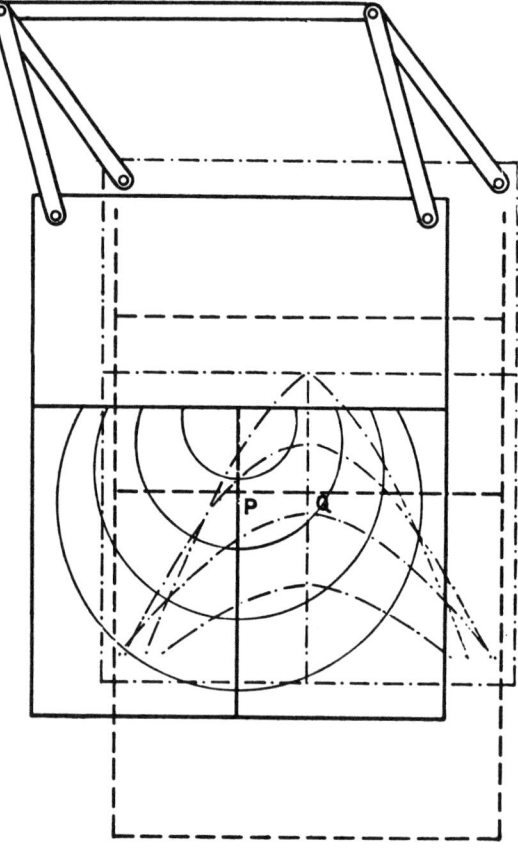

Fig. 17. Relative disposition of the three sheets employed in migrating contour lines.

The manipulation of this system underneath a contour map can be carried out in practice with the aid of a commercial drawing machine, but this means that the contour map cannot remain in a fixed position in regard to the drawing board.

A suitable arrangement is illustrated in figure 18. Two strips of stiff white cardboard are affixed to the drawing board so that a long third strip, the width of the chart of curves of maximum convexity, can slide between them. The line of migration is drawn across this sliding strip. The transparent chart of curves of maximum convexity is attached to the fixed strips with tape after

Fig. 18. Illustration of a practical application of the three-dimensional migration method.

ensuring perpendicularity of its axis to the line on the sliding strip. The wavefront chart must be printed at the lower end of a fairly long strip of transparent film so that when attached, correctly aligned with the chart of curves of maximum convexity, to the drawing machine, its points of attachment will not interfere with the overlying contour map when the chart is moved about beneath it.

With this apparatus the procedure of migration is as follows:

1. The contour map is orientated so that the point to be migrated lies at the correct depth on the axis of the chart of curves of maximum convexity and so that the unmigrated contours at the point are parallel to the axis.

2. The slide is moved to bring the line of migration to pass through the point to be migrated.

3. The wavefront chart is moved to the position in which the corresponding wavefronts and contours, near the point to be migrated, cut one another on the line of migration.

4. The migrated point is located on the axis of the wavefront chart on the line of migration, and short portions of the migrated contours should be drawn close to this point, perpendicular to the line of migration and their depth values marked to satisfy the intersections of adjacent curves of maximum convexity with the line of migration.

IV. WAVEFRONT CHARTS FOR VERTICAL PLOTTING, BASED ON TIME-DEPTH RELATIONSHIPS

1. *Centres of curvature of the wavefronts at the vertical axis.*

When a wavefront chart is used for the purpose of vertical plotting of

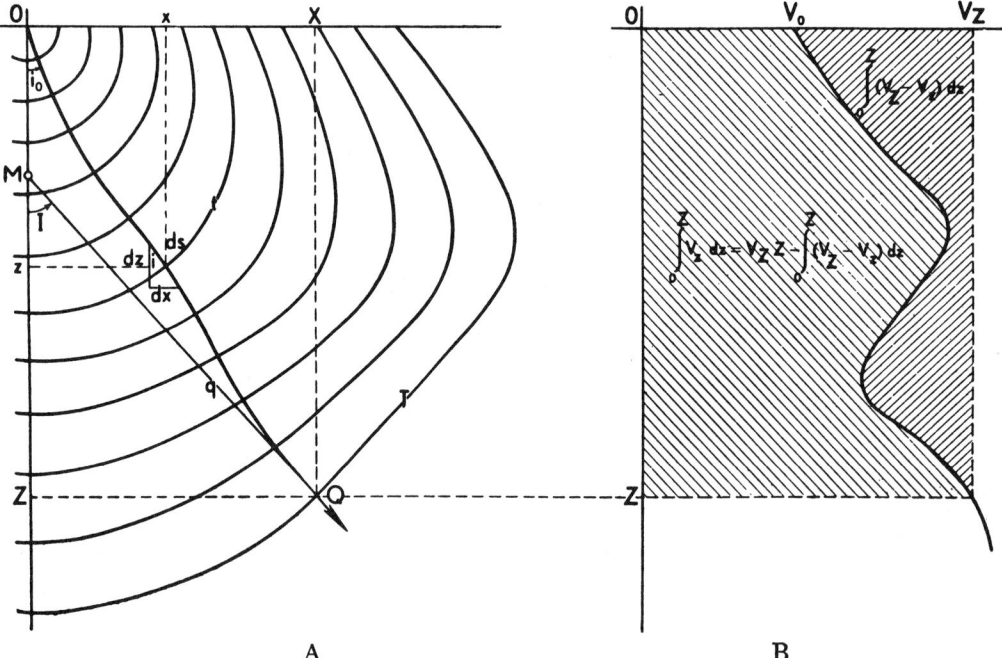

Fig. 19. The curvature of a wavefront is a function of depth and velocity distribution to that depth.

reflection times, in the manner described in paragraph I,1, this chart is only used in a narrow band near the axis, up to a distance from this axis of half the distance between shotpoint and furthest receiver. As a first step in approximating these central parts of the wavefront curves, they can be represented by segments of circles even though the velocity does not increase linearly with depth. The depth to the centre of one of these approximating circles must be some function of the velocity distribution down to the maximum depth of the particular wavefront.

In fig. 19A a number of wavefronts at equally spaced time intervals and also one trajectory originating at O have been drawn, corresponding to the velocity distribution shown in figure 19B. The form of such a trajectory, a line orthogonal to the wavefronts, is fixed by the condition, Snell's law, that $V_z/\sin i$ remains constant. The distance $MQ = q$ can be found in the following way:

$$\frac{dx}{ds} = \sin i; \quad X = \int_{"O"}^{"Q"} \sin i \, ds$$

$$q = \frac{X}{\sin I} = \int_{"O"}^{"Q"} \frac{\sin i}{\sin I} \, ds = \int_{"O"}^{"Q"} \frac{V_z}{V_Z} \, ds = \frac{1}{V_Z} \int_{"O"}^{"Q"} V_z \, ds$$

The distance q is not necessarily the radius of curvature of the wavefront at Q. The centre of curvature at Q would be the intersection point of two lines perpendicular to the wavefront at points at infinitesimally small distances on both sides of Q, and this intersection point will not necessarily lie on the axis. Because of the symmetry at the point where a wavefront cuts the axis, the centre of curvature of the arc at this point must lie on the axis and then the radius of curvature will be the value of q for $i_o = 0$ or for $ds = dz$:

$$R_Z = q \, (i_o = 0) = \frac{1}{V_Z} \int_0^Z V_z \, dz$$

The depth to the centre of curvature is:

$$Z_c = Z - R_Z = \frac{1}{V_Z} \int_0^Z (V_Z - V_z) \, dz$$

The integrals in these equations for the radius of curvature and for the depth to the centre are illustrated in figure 19B as the shaded areas at both sides of the curve representing V_z as a function of z. These areas must be divided by V_Z, the velocity attained, in order to arrive at the values of R_Z and Z_c. One important fact is directly evident from this representation: the depth to the centre of curvature remains constant for any depth interval where the velocity remains constant.

2. *Elucidation of time-distance graphs by using oblique coordinates.*

A wavefront surface is determined by the travel time and by the velocity distribution. This velocity distribution is usually derived from a vertical time-depth relation obtained directly by carrying out measurement in wells or indirectly from seismic reflection or refraction data.

On a time-distance or t, x graph, whether it is a time-depth graph from a well survey or a refraction time-distance graph, the slope of the curve at a certain point, dx/dt, represents a velocity at a certain depth. A curve built up of straight line segments implies the existence of layers with constant velocities, and a gradually changing slope implies a gradual change of velocity with depth. A clear analysis of these graphs, however, is often difficult because the changes in velocity are often small and the observational errors relatively large.

Distance values can always be measured to an accuracy which makes their uncertainty negligible as compared to that in the measurement of travel times, for which the uncertainty of both the moment of the explosion and the moment of arrival is at least one millisecond in practice. To express this in a graph means that the time scale must be chosen sufficiently large, for example one millisecond equal to one half millimeter, and this leads to very large graphs in practice. These considerations, however, do not apply, in principle, to the choice of the distance scale.

This leads to the two lines of approach illustrated in fig. 20. The original small graph does not show the uncertainties of the time measurements because the scale is too small. As a first line of approach, only the time scale can be enlarged but then a very steep curve is obtained where the uncertainties in the time measurements are not clearly seen because they occur in a vertical direction and not in a direction perpendicular to a mean curve through the points. The only possible way to improve this picture, using rectangular coordinates, is, as shown in fig. 20, by also enlarging the distance scale, even though this is not necessary for the purpose of observing irregularities in the distance values.

A more profitable line of approach is by not only enlarging the time scale but by also inclining the distance axis to where the mean curve, defined by the points, will be approximately vertical to the time axis. The deviations of the points from this curve then refer to uncertainties in the time measurements only. Further advantages are that the graphs can be confined to small areas and that, by using a smaller distance scale, a change in velocity results in a more marked change in slope than on the graphs with rectangular coordinates.

The increased facility in analysing time-distance relations, by using this procedure of plotting on oblique coordinates, must outweigh the difficulty of converting to an unusual operation. Plotting on oblique coordinates can also be conceived, algebraically, by subtracting from each time value an amount proportional to the corresponding distance value and then plotting on rectangular coordinates. This means that the variations of the time values from a straight line, the line $t = x/2500$ in fig. 20, are plotted; this procedure

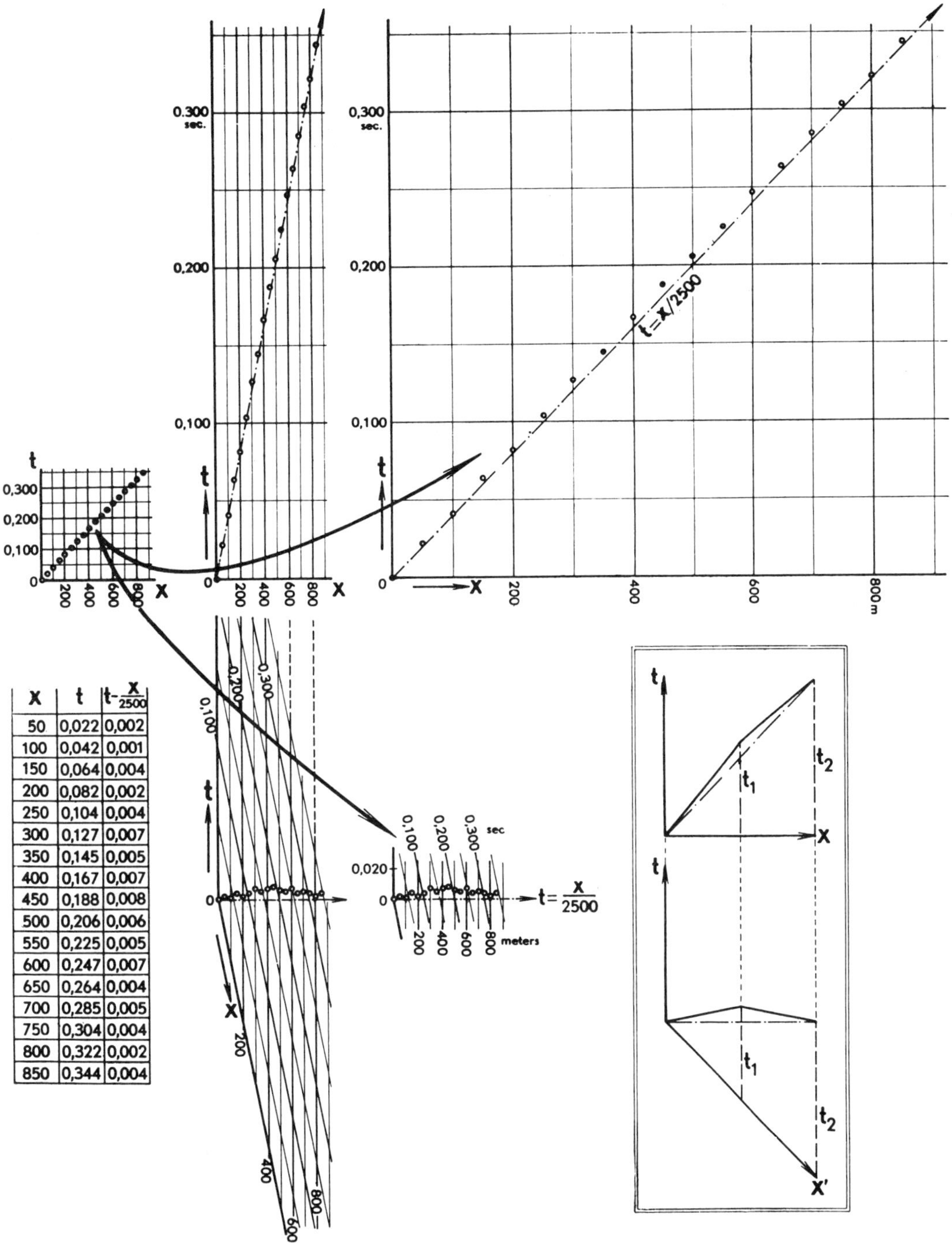

x	t	$t-\frac{x}{2500}$
50	0,022	0,002
100	0,042	0,001
150	0,064	0,004
200	0,082	0,002
250	0,104	0,004
300	0,127	0,007
350	0,145	0,005
400	0,167	0,007
450	0,188	0,008
500	0,206	0,006
550	0,225	0,005
600	0,247	0,007
650	0,264	0,004
700	0,285	0,005
750	0,304	0,004
800	0,322	0,002
850	0,344	0,004

Fig. 20. The analysis of a time-distance curve is facilitated by plotting on a graph with inclined distance axis.

is exactly the same, in principle, as plotting on oblique coordinates but may be found easier to carry out in practice. It is helpful actually to draw a few inclined lines of equal time at unit intervals in order to be able to evaluate at once the velocity corresponding to a certain slope.

3. Derivation of wavefront charts from observed time-depth values.

In fig. 21 a practical example has been worked out. The time-depth values from a well survey have been plotted in fig. 21A on oblique coordinates, where the line $z/t = 2.5$ Km/sec is perpendicular to the time axis.

The interval velocities between successive points were computed and plotted in fig. 21B as the full line. From these the depths Z_c to the centres of cur-

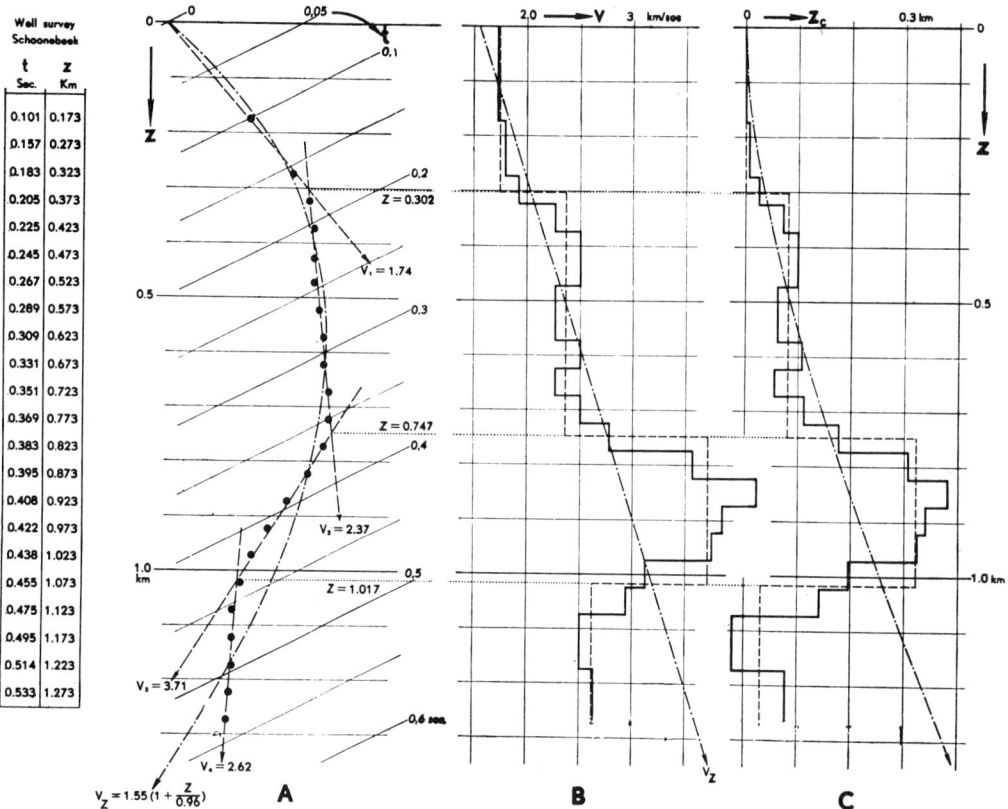

Fig. 21. Different approximations to a velocity distribution and corresponding depths to centres of curvature.

vature were computed for each velocity interval by summation of the values $(V_Z - V_z) \Delta z$, where Δz is the thickness of the layer with velocity V_z:

$$Z_c = \frac{1}{V_Z} \sum (V_Z - V_z) \Delta z, \text{ for all values of } V_z \text{ and } \Delta z \text{ down to } Z.$$

In this manner a step-graph, the full line in fig. 21C, is obtained which is very similar to the graph of interval velocities, the full line in fig. 21B.

Up to this point a mechanical procedure has been followed in which only the actually measured values have been used. This particular case of fig. 21, however, most probably concerns four layers, each with a constant velocity because the time-depth values in fig. 21A can be approximated to within two milliseconds by four straight line segments, the dashed lines in fig. 21A. The corresponding velocity distribution and depths to the centres of curvature are shown as dashed lines in fig. 21B and 21C respectively.

The time-depth values in fig. 21A can also be approximated by a linear velocity distribution, $V_z = 1.55 (1 + z/0.96)$, the dot-dash line, which approximates the actual time-depth values to within 10 m down to a depth of 0.850 Km and to within 40 m below that depth. This velocity distribution and the depths to the centres of curvature of the corresponding wavefronts are shown as dot-dash lines in fig. 21B and 21C respectively.

In fig. 22 a wavefront chart corresponding to this practical example has been drawn. The full curves are the wavefronts corresponding to the four layers

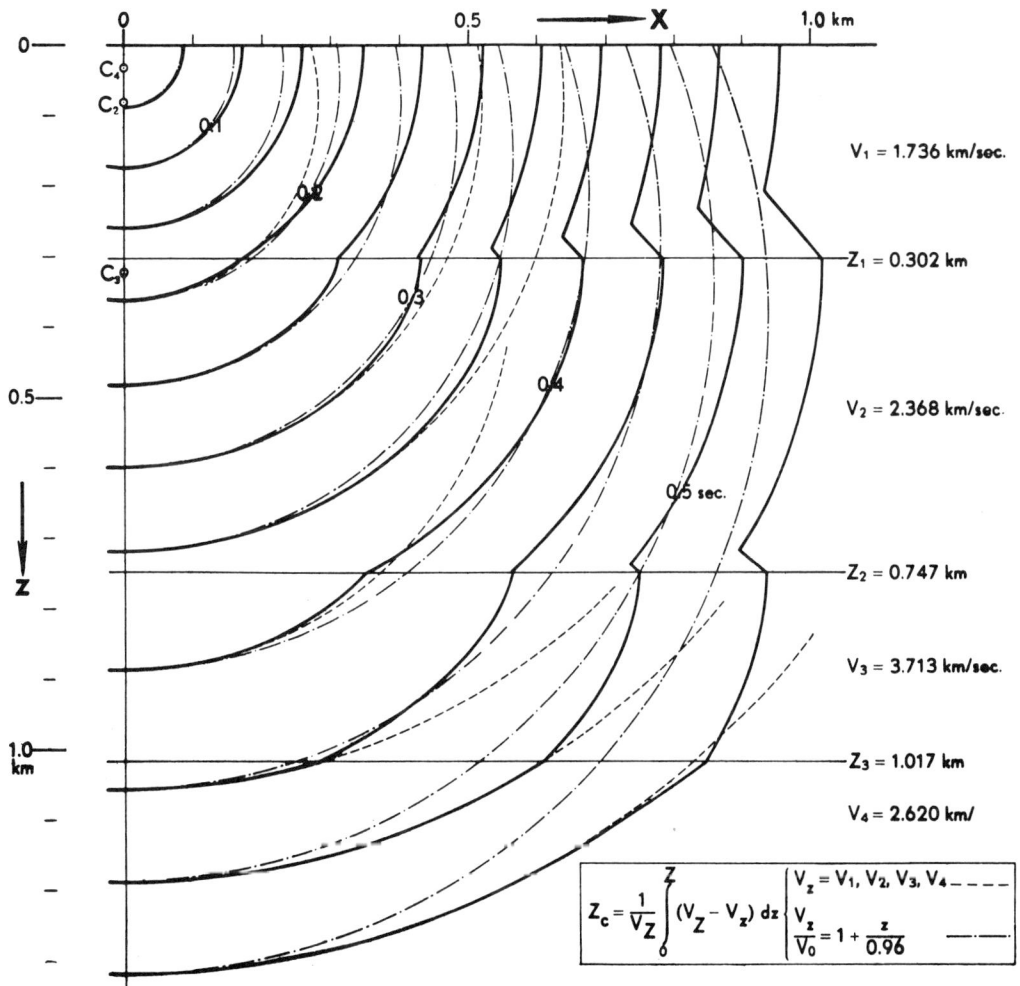

Fig. 22. Wavefronts corresponding to the velocity approximations in fig. 21.

with constant velocity as they most probably actually exist. The centres of curvature of these wavefronts at the axis are O, C_2, C_3, C_4 and the circles with these points as centres have been drawn as dashed lines.

It is clear from fig. 22 that in this case these approximate wavefronts can be used for vertical plotting up to shooting distances of at least 400 m without incurring observable errors.

A different line of approach would be by approximating the relation between the depth to the centre of curvature Z_c and the depth Z by a smooth curve as shown in fig. 21C, where the dot-dash line corresponds to a linear relation between depth and velocity. By using these centres of curvature the dot-dash lines in fig. 22 have been obtained. These do not approximate the actual wavefronts near the axis as well as the dashed lines, but they do give a better overall approximation of the full wavefronts.

These considerations can lead to a procedure of splitting the derivation of a wavefront chart for vertical plotting into two parts. First the actual vertical time-depth relation found is used to determine the intersections of the wavefronts with the axis, so that the correct depths will be plotted. A certain amount of smoothing out can be applied to the measured time-depth relationship, because its accuracy can be trusted at the place of measurement whereas, elsewhere in the area in which the wavefront chart is to be used the depths to the velocity layers may be different. The extent to which smoothing out can be applied is conditioned by the variations in depths, the velocity contrasts and also by the size of the area for which one wavefront chart is to be used. And second, the actual wavefronts can then be approximated, for the sake of convenience in draughting, by segments of circles. The centres of curvature of these circles are derived from the velocity distribution approximated by a more or less smooth curve, depending on the shooting distance employed. The smoothing out can usually be carried out to a much greater degree than the smoothing out of the time-depth relation.

Only in exceptional cases, where it is impossible to approximate the wavefronts by segments of circles will a more complicated procedure have to be followed, but the determination of the curvature at the axis can always be taken as point of departure for constructing non-circular wavefronts.

V. APPROXIMATION OF TIME-DEPTH VALUES BY A SMOOTH VELOCITY FUNCTION

1. *Line of approach when approximating by a smooth velocity function.*

In order to facilitate the construction of a wavefront chart it is advantageous to try to approximate the velocity distribution by a smooth function that lends itself to simple draughting and easy mathematical treatment. In this respect the linear velocity function $Vz/Vo = 1 + z/L$ is outstanding because it results in circular wavefronts (van Melle).

The approximation must, of course, be within reasonable bounds, depending

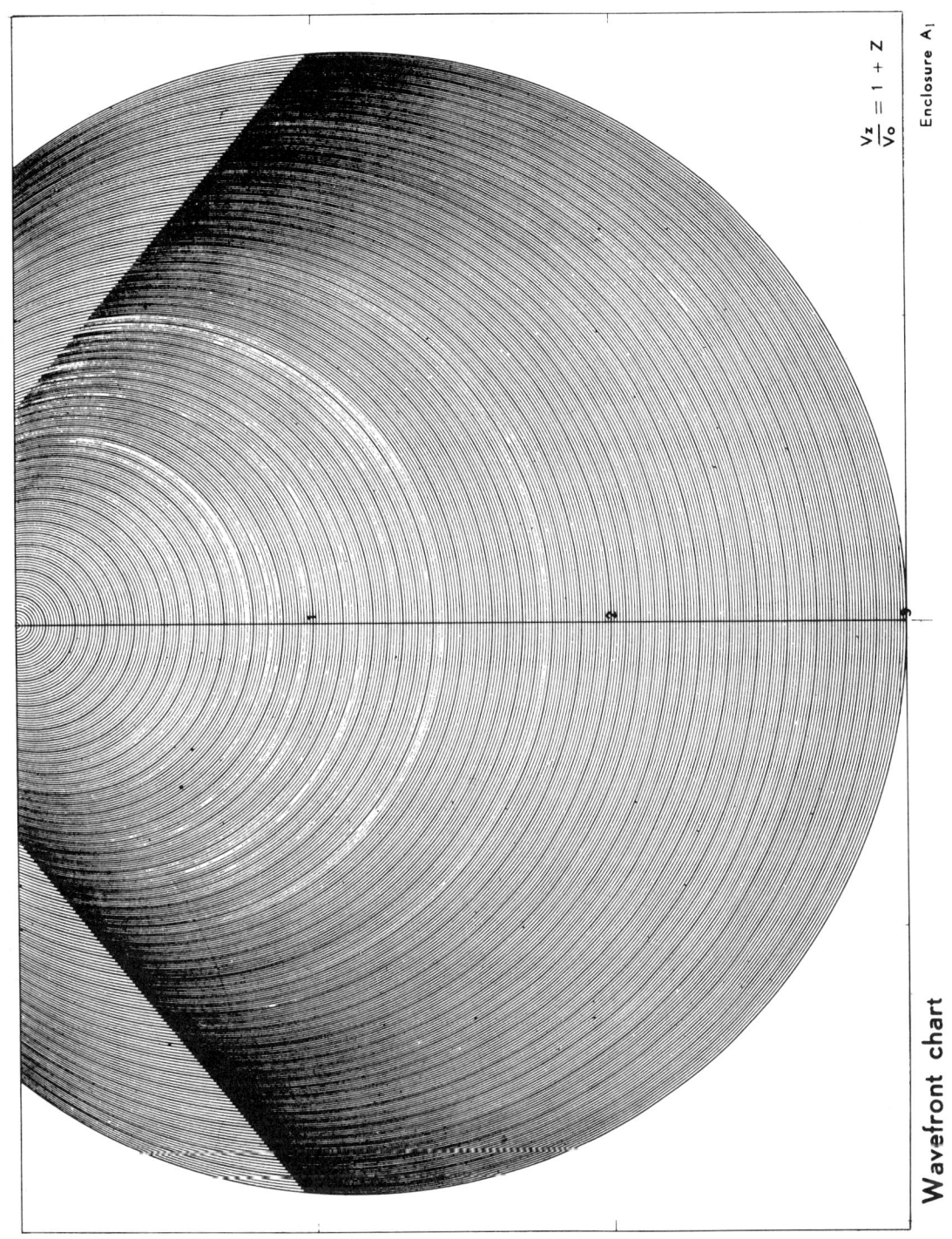

Wavefront chart

$\dfrac{V_z}{V_0} = 1 + Z$

Enclosure A₁

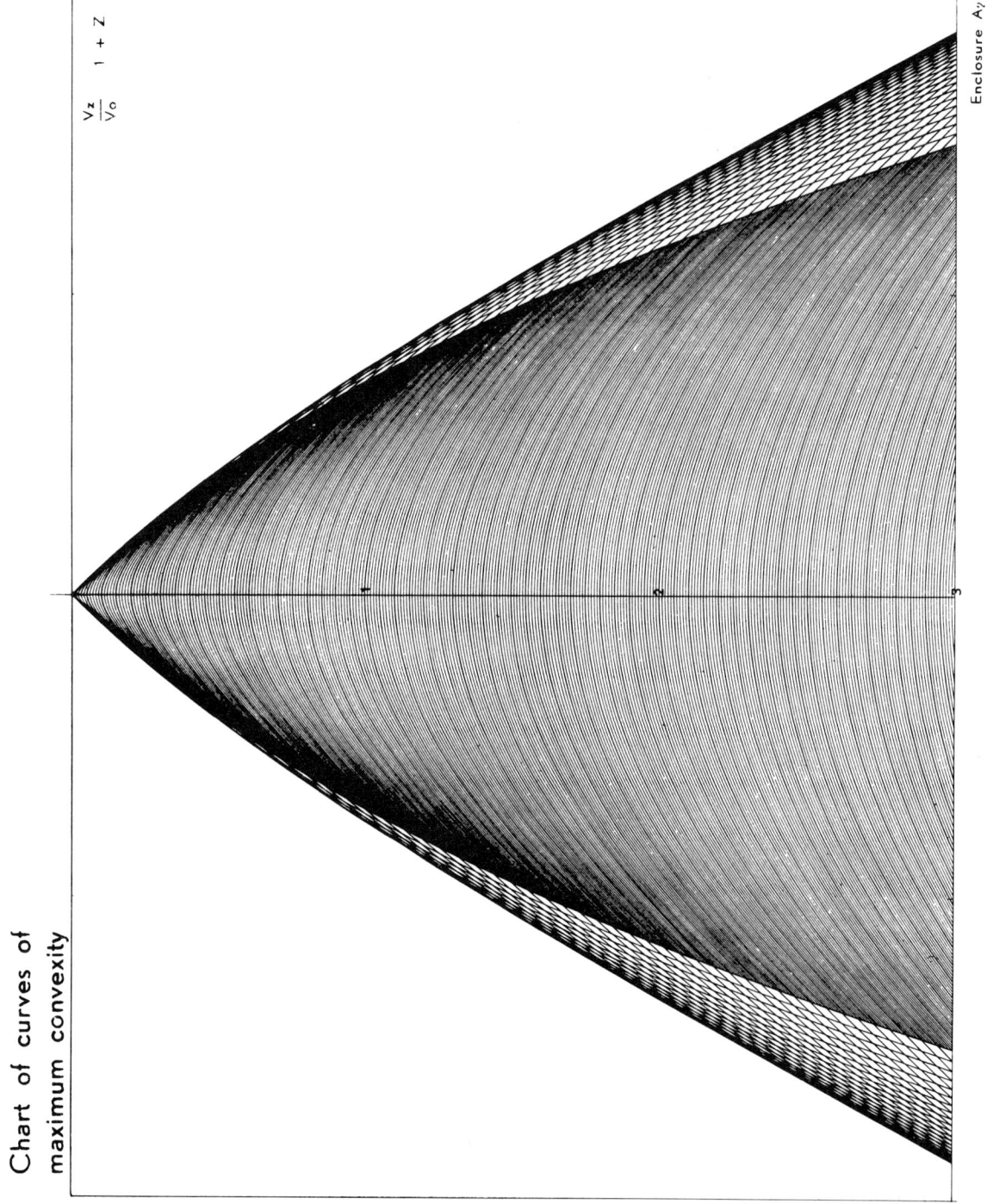

Chart of curves of maximum convexity

$\dfrac{V_z}{V_o} \quad 1 + Z$

Enclosure A7

Contour maps migrated with the aid of the charts A₁ and A₂

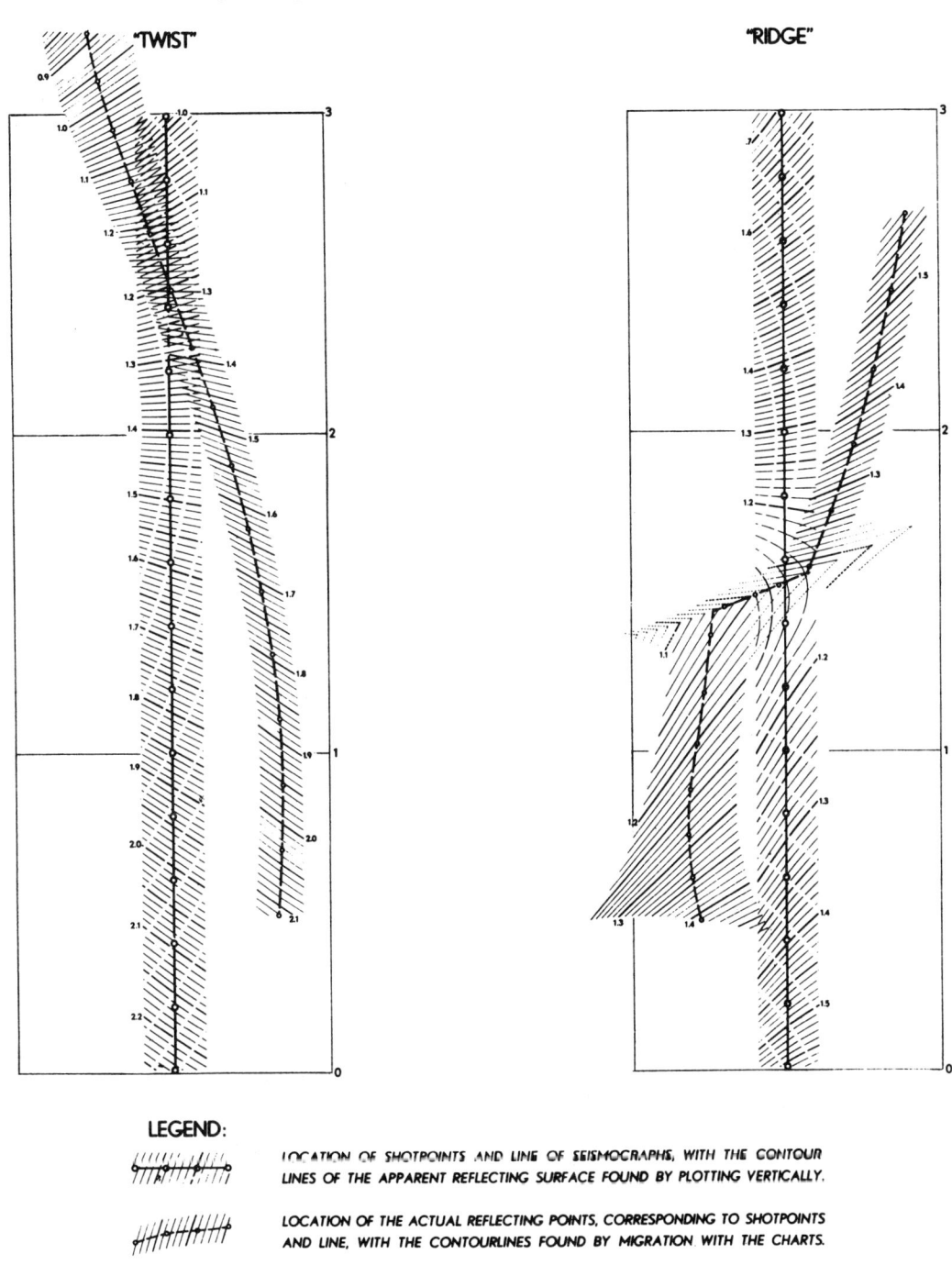

LEGEND:

///// ///// LOCATION OF SHOTPOINTS AND LINE OF SEISMOGRAPHS, WITH THE CONTOUR LINES OF THE APPARENT REFLECTING SURFACE FOUND BY PLOTTING VERTICALLY.

///// ///// LOCATION OF THE ACTUAL REFLECTING POINTS, CORRESPONDING TO SHOTPOINTS AND LINE, WITH THE CONTOURLINES FOUND BY MIGRATION WITH THE CHARTS.

///// ///// CONTOURLINES OF THE ACTUAL REFLECTING SURFACE.

Enclosure A₃

This page intentionally blank.

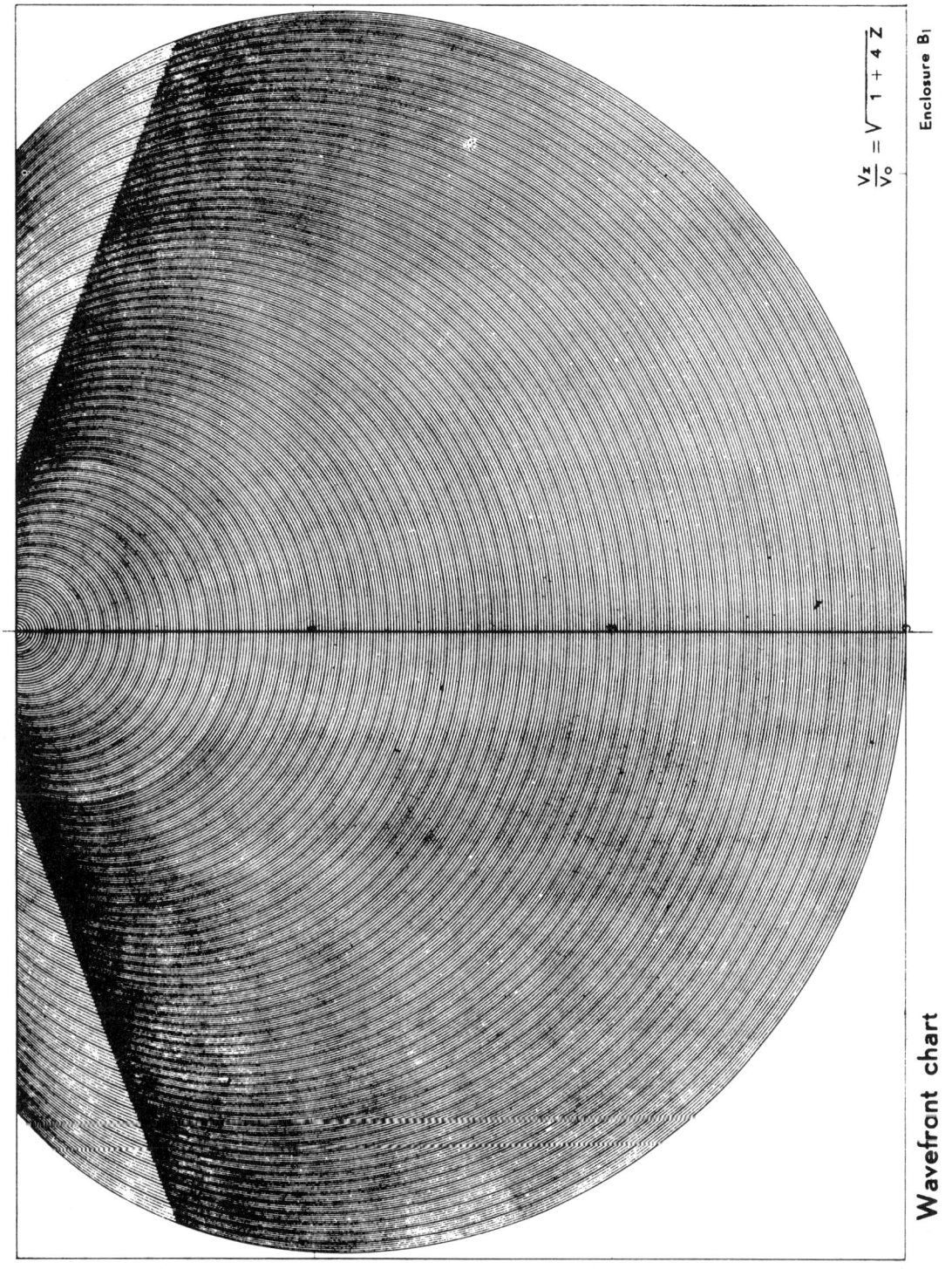

Wavefront chart

$\dfrac{V_z}{V_o} = \sqrt{1 + 4Z}$

Enclosure B₁

Chart of curves of maximum convexity

$$\frac{V_z}{V_0} = \sqrt{1 + 4z}$$

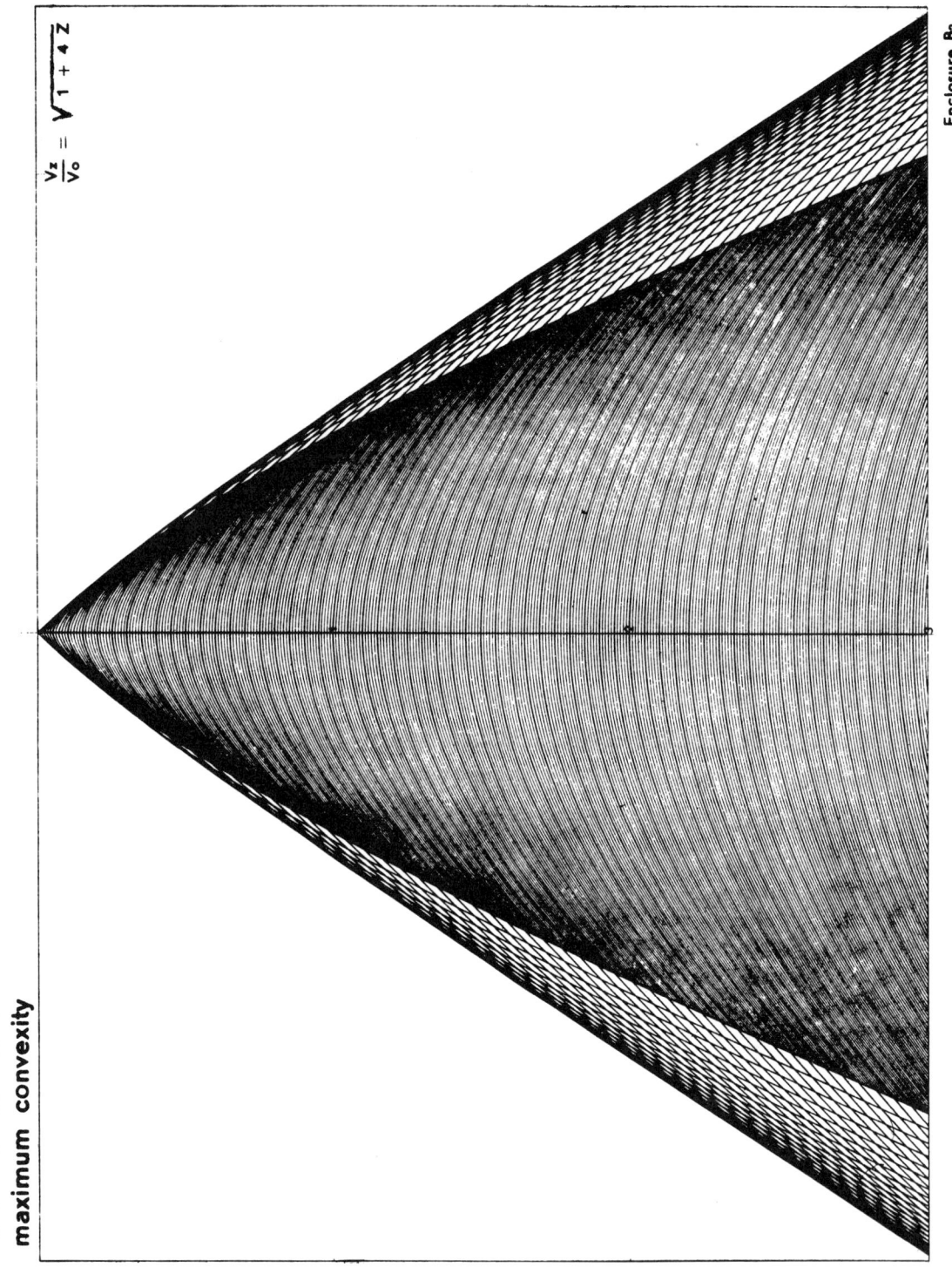

Enclosure B₂

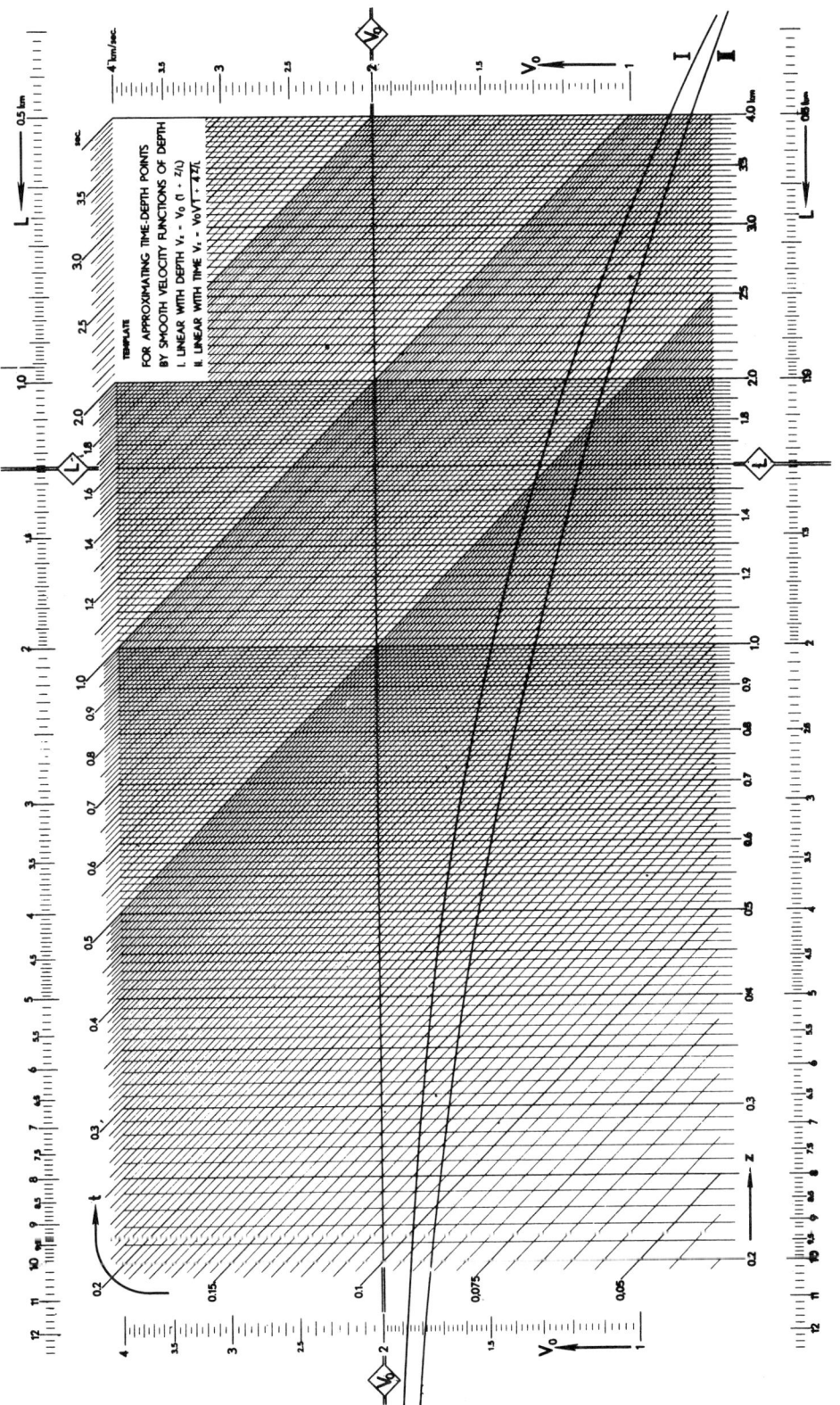

Enclosure C

This page intentionally blank.

on the accuracy required and the observational errors of the available data. It is clear that the aim is not to approximate the velocities as closely as possible but to find a function that will result in depth values that are as close as possible to the actual depths for measured time values.

This point is illustrated in fig. 23 by a simple hypothetical case of three layers of equal thickness and equal increase in velocity from one layer to the next. The velocity distribution, the stepped full line in fig. 23B, can be approximated by a linear increase with depth and the best fit is obviously by

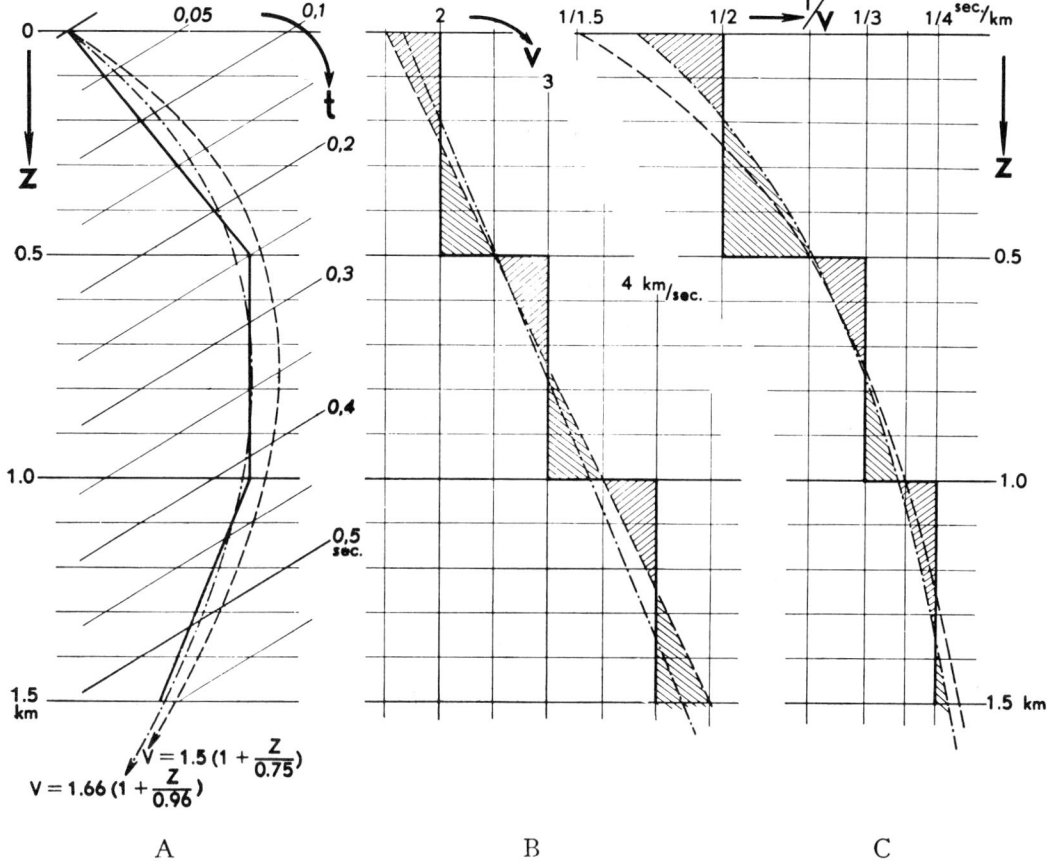

Fig. 23. The best smooth approximation to the velocity distribution is not the best approximation to the time-depth relation.

$Vz = 1.5 (1 + z/0.75)$, the dashed line in fig. 23B. The dashed line in fig. 23A is the time-depth curve corresponding to this linear velocity distribution, and is certainly not the best approximation to the actual time-depth relation given by the full line; a much better approximation is given by the dot-dash line, corresponding to the linear function $Vz = 1.66 (1 + z/0.96)$. In fig. 23B the dot-dash line also corresponds to this function but does not approximate the actual velocity function as well as the dashed line of the function $Vz = 1.5 (1 + z/0.75)$.

This apparent discrepancy is explained by realizing that, in order to arrive at a smooth velocity function Vz which leads to the best approximation of the time-depth values, the value of $t_Z = \int_0^Z dz/Vz$ should approximate the actual time value for depth Z as closely as possible. This means that it is not correct to approximate V as a function of z by a smooth curve, a straight line for a linear function. The correct procedure is to approximate the inverse value of V as a function of z by a smooth curve which is not a straight line for a linear velocity distribution. This is shown in fig. 23C in which it is seen that the best approximation is arrived at by a smooth curve for which the sum of the shaded areas on one side of the curve down to any point is equalled as closely as possible by that on the other side. Arriving at serviceable smooth velocity functions in this manner is correct in principle but would be rather laborious in practice.

2. Derivation of a serviceable smooth velocity function.

The most direct way to arrive at a serviceable, smooth velocity function is by trial and error using the actual time-depth values available. The first step in this procedure is to try to approximate the time-depth values by a linear function $Vz/Vo = 1 + z/L$ by varying Vo and L. If a close enough approximation cannot be arrived at a different function can be tried, for example the parabolic velocity distribution $Vz/Vo = \sqrt{1 + z/L}$, the velocity function that is linear with vertical traveltime and is also widely used in practice (Musgrave).

The procedure can be shortened appreciably by using a method based on the same principles as a procedure for deriving a smooth velocity function to fit a refraction time-distance curve, introduced by G. Mc G. Bruckshaw at Imperial College, London.

Velocity distributions linear with depth or vertical time can be considered as two families of all velocity functions $Vz/Vo = f(z/L)$. If F is the reverse function corresponding to f, the time-depth relation can be written as:

$$Z = L \, F(Vz/Vo)$$

$$t = \int_0^Z \frac{dz}{Vz} = \int_1^{Vz/Vo} \frac{1}{Vz} \frac{dz}{d(Vz/Vo)} d(Vz/Vo) =$$

$$= \frac{L}{Vo} \int_1^{Vz/Vo} \frac{dF(Vz/Vo)}{d(Vz/Vo)} \frac{d(Vz/Vo)}{Vz/Vo} = \frac{L}{Vo} G(Vz/Vo)$$

This is a parametric representation of the relationship between t and Z, the parameter being Vz/Vo. The functions F and G represent the time depth relationship for unit values of Vo and L. If a time-depth curve, corresponding to arbitrary values of Vo and L, is plotted on a graph with logarithmic time- and depth-scales, this curve has exactly the same shape as the curve for unit

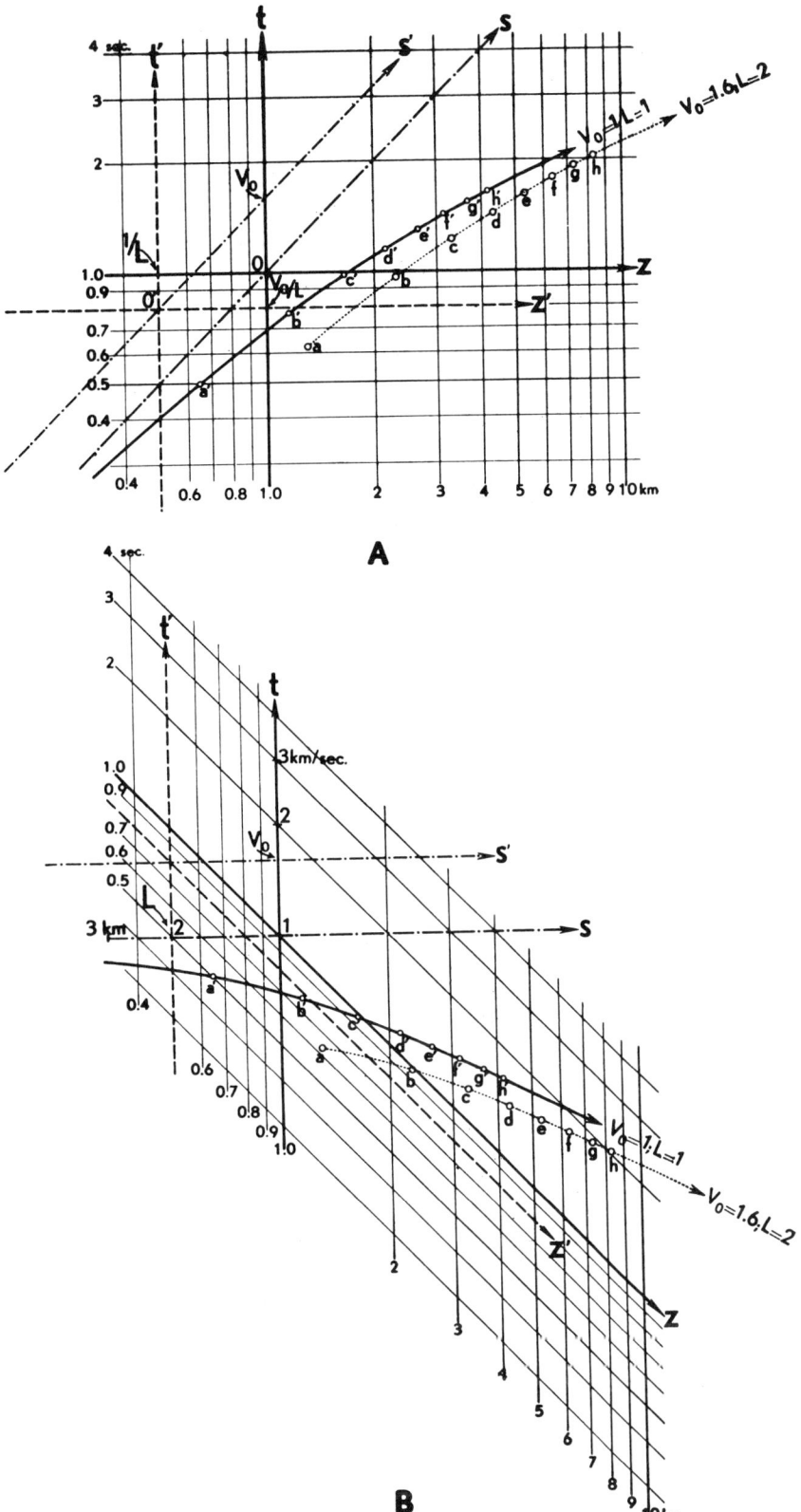

Fig. 24. Time-depth curves plotted on logarithmic graphs.

values of Vo and L. It is only shifted log L/Vo in the direction of the timescale and log L in the direction of the depth-scale.

In order to find a smooth velocity function to fit a set of observed time-depth values, these values can be plotted on a transparent sheet overlying a logarithmic graph containing a number of time depth curves corresponding to different families of functions with unit values of Vo and L. The transparent overlay can then be moved, keeping the axes parallel, and the best fit found by trial and error.

The procedure is illustrated in figure 24A, making use of a linear velocity distribution $V_z = V_o(1 + z/L)$. The indicated axes t and z are the lines of unit distance and unit time respectively. The time-depth values a, b, c, etc. are plotted on an overlay and the axes t and z are traced on this overlay. By shifting the overlay in such a way that the plotted time-distance values fall at a', b', c', etc. on the mastercurve, the traced axes t and z are moved to t' and z'. The line t' will lie on the depth scale at 1/L Km and the line z' will lie on the time scale at Vo/L seconds.

The straight line s represents the time-depth curve for constant velocity t = z. This line is moved to s' and it can be seen that it cuts the axis t at the value Vo, moving from t' to t corresponds to multiplying by L, and s is at equal angles to t and z. Thus Vo can be read directly on the axis t by tracing the line s on the overlay, omitting the tracing of the axis z.

The importance of this line s and the fact that the curves will lie within fairly narrow bands parallel to this line, leads to the conclusion that it is more logical to incline the z-axis until the line s is perpendicular to t. This transformation has been carried out in figure 24B.

3. *A template for direct determination of a smooth velocity function.*

In order to confine a template to a minimum area, the following considerations have been followed. In the first place the area of the template is fundamentally only limited by the range of t- and z-values to be expected in practice. In the second place the master curves can be drawn for any values of Vo and L if the axes to be traced on the overlay are drawn through the corresponding values on the Vo and L scales respectively. Furthermore, these reference axes can be moved together with their respective scales to any more favourable position.

These considerations have been embodied in the template on enclosure C, which can be used directly for most practical cases where the time depth relationship can be approximated either by a function linear with depth, $V_z/V_o = 1 + z/L$, or linear with vertical travel time, $V_z/V_o = \sqrt{1 + 4z/L}$. The template must be overlaid by a sheet of transparent drawing paper on which the time depth values are plotted and the axes marked

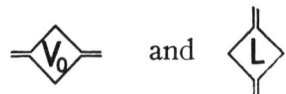

are traced. The overlay is then shifted to the best fit to one of the master curves and the values of Vo and L read off at the intersections of the traced lines with the Vo- and L-scales.

Only two master curves, corresponding to velocity distributions linear with depth and linear with vertical travel time are presented on this template. Other curves, corresponding to other functions $Vz/Vo = f(cz/L)$, can be included by computing the time-depth relation for $L = 1,25$ and $Vo = 2$. Choice of c depends on which range of the master curve is desired on the template.

The template on enclosure C is presented on a scale that is sufficiently large but enlargements can, of course, be made if required to facilitate plotting. The lines of equal time have been drawn five milliseconds apart up to the half second line, ten milliseconds apart up to one second and twenty milliseconds apart up to two seconds, so that the time-depth points can be plotted to an accuracy of one to two milliseconds, which is sufficiently accurate in most cases in practice.

It is always possible to plot unusual values within the area of this template by multiplying one or both the coordinate values by a certain factor. A change of the depth values by a certain factor will result in the same change in the values found for both Vo and L, while a change in the time values will result in only the value of Vo being different from the value read on the scale of the template.

The main difficulty when plotting on any logarithmic scale is that a certain numerical interval corresponds to different intervals at different locations. A certain error in depth or time corresponds to a diminishing interval along a time-depth curve. The most straightforward solution to this difficulty is by plotting not points but line segments; for example along a line of constant time for a depth interval of ± 20 meters. In this way the measured points can easily be fitted to a curve to within a certain error in depth.

VI. Derivation of the charts to be used for the purpose of migration

1. *Use of wavefronts instead of curves of equal reflection times.*

Up to paragraph III, 2 the concept surface of equal reflection times was used but at that point the necessity arose to use only symmetrical rotational surfaces in order to be able to evolve the two-dimensional templates for the purpose of three-dimensional migration. This will seem, on first sight, to impose a serious limitation on the proposed method for migrating contour lines.

The complications introduced by the asymmetry of the surfaces of equal reflection times at large shooting distances will, necessarily, make any strict method of three-dimensional migration cumbersome in such cases. Fundamentally, any method must be based on the determination of surfaces of equal reflection times, the intersections of which with the plane in the direction of maximum dip have a different shape for each different angle between this plane and the line shot.

In fig. 25 two wavefront charts have been centred at a shotpoint S and at a receiver R respectively, as was done in fig. 4. Three curves of equal reflection times have been drawn as full lines. The dashed curves are wavefronts that would be obtained from a shot at M and which cut the central axis through M at the same depths as the three curves of equal reflection times. These wavefronts are seen to approximate the curves of equal reflection times fairly well at depths greater than the distance between shotpoint and receiver.

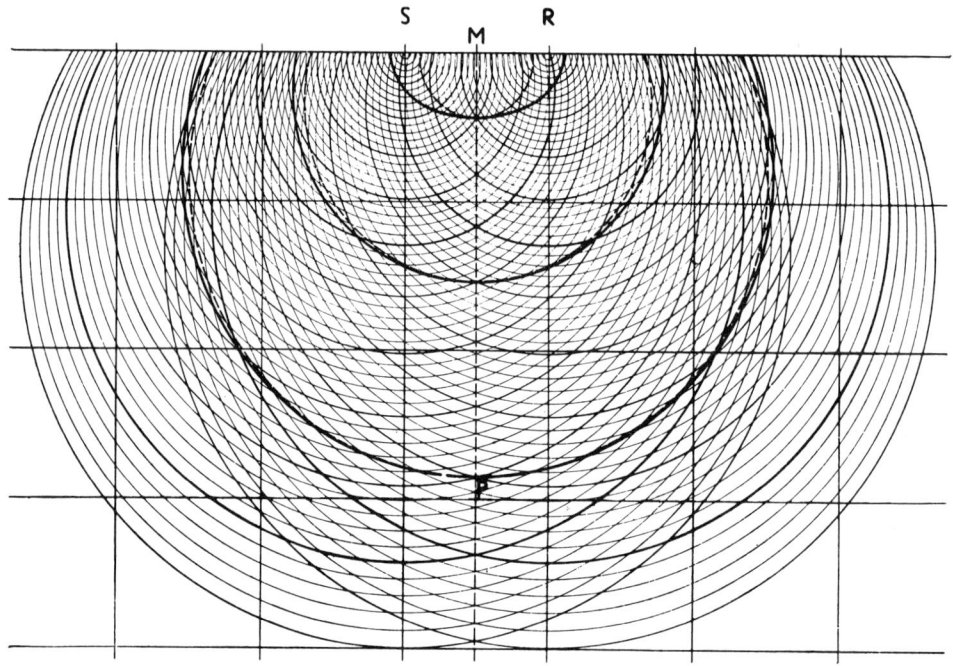

Fig. 25. Approximation of curves of equal reflection times by wovefronts.

When the widely used method of "straddle-shooting", where the shotpoint is located in the middle of a spread of geophones in the order of ½ Km long, is employed, the approximation of curves of equal reflection times by wavefronts is practically always permitted.

In some cases, however, better results are obtained by shooting to larger distances: for example, when a surface wave is too strong or when using a system of air shooting (Poulter).

The longer the distance between shotpoint and receiver, the more elongated a surface of equal reflection time becomes and the less well can it be approximated by a symmetrical wavefront surface. The closeness of this approximation is, however, not a function of depth only but also of the dip of the reflector. For larger dips the approximation is worse but, in practice, the fundamental assumptions on which the use of wavefront charts is based also become less reliable. In most cases of strongly dipping layers an appreciable change in velocity can be expected in a lateral direction and the velocity distribution cannot be evaluated accurately. Even when using fairly large shooting distances, the errors introduced by approximating curves of equal reflection times by wavefronts can, in most cases in practice, be expected to be of a smaller order of magnitude than the errors introduced by the approximation of the velocity distribution.

For large shooting distances the most straightforward procedure would be to construct a number of curves of equal reflection times corresponding to the approximated velocity distribution and the average shooting distance, both in the vertical plane through S and R in figure 25 and in the vertical plane through M which is perpendicular to the line SR. The best average fit to both sets of curves could then be sought by trying various enlargements of available wavefront charts. This is one of the purposes for which the two enclosed wavefront charts A1 and B1 may be used.

The charts which are used in this way are no longer, strictly speaking, employed as wavefront charts but only as charts of curves that can be made to fit the shape of curves of equal reflection times more or less closely. The wavefronts, shown as interrupted lines in figure 25, which originate at M and approximate curves of equal reflection times spaced at equal time intervals, are themselves not spaced at equal time intervals because, in figure 25, the difference between the traveltime from S to P to R and from M to P and back changes with the depth of P.

It is clear that this spacing of the curves is quite immaterial when used for the purpose of migration. For the purpose of three-dimensional migration the curves must be spaced along the axis to correspond with the contour lines of the reflecting surface, which means a spacing at regular depth intervals instead of a spacing at regular time intervals.

2. Charts based on velocities linear with depth or with vertical time.

Reference has already been made to smooth velocity functions in which the velocity is linear either with depth or with vertical time, and it has been explained that their fairly general popularity is due to relatively simple properties that lend themselves readily to graphical as well as analytical mathematical treatment.

The stress here is that the choice of velocity distributions of these types does not imply that the actual velocity distributions are equally smooth functions, but rather that the convenience implicit in the use of them is not offset by prohibitive errors. Because the velocity distribution linear with depth leads to circular wavefronts and circular trajectories, it is the most convenient in use. When, however, the increase of velocity with depth decreases with depth a parabolic velocity distribution, namely, one that is linear with vertical time, may have to be used and draughting of the wavefronts becomes tedious unless a suitable draughting machine (Musgrave) is available.

The discussion in paragraph II, 2 might seem to imply that for every new velocity distribution a fresh set of wavefronts has to be drawn. This procedure actually claims many adherents, with the frequent result that a chart is used beyond the range within which it agrees sufficiently with the actual velocity distribution, and sometimes where assumption of a smooth velocity function is not permissible. It is important, therefore, to consider means of making suitable charts readily available.

By first plotting vertically with the aid of a wavefront chart and subsequently migrating with another wavefront chart a simpler procedure is achieved in that, for vertical plotting, the wavefront chart is used near the axis where wavefronts can be safely approximated by arcs of circles. But migration requires a full wavefront chart and the corresponding chart of curves of maximum convexity and it may seem, therefore, that no advantage is gained. There is, however, a method of obtaining all wavefront charts and charts of curves of maximum convexity from a small number of such charts for the majority of cases likely to arise in practice. It is simply that of enlarging or reducing the relevant master charts to the required scale.

In the case of a velocity distribution $V_z/V_o = f(z/L)$ in which two wavefronts are considered at times differing by Δt, these wavefronts are separated by a distance $V_z \Delta t$ at a depth z and by $V_o \Delta t$ at the surface. Magnify this picture m times and the separation at depth mz will be $mV_z\Delta t$ and at the surface $mV_o\Delta t$. Clearly, the wavefronts remain possible wavefronts after magnification since their distance apart for a given time interval is a function of depth only. The time interval will become $\Delta t' = mV_o\Delta t/V'_o$, where V'_o is the surface velocity belonging to the case represented by the magnified picture. At the depth mz the velocity will be:

$$V'_{mz} = mV_z\Delta t/\Delta t' = V'_o V_z/V_o = V'_o f(z/L).$$

If $z' = mz$ and $L' = mL$, then $V'_z/V'_o = f(z'/L')$.

The magnified case then concerns a velocity distribution similar to the original function, only with a different value of L. If the wavefronts of the original chart were spaced at equal time intervals at integral values, those of the magnified chart will be spaced at equal time values but no longer necessarily at integral values. The time values will only remain unchanged if $V'_o/L' = V_o/L$. Thus, derivation of other wavefront charts by changing the scale of a given chart has, for plotting purposes, a limited application. But, for purposes of migration, the advantages of this method are irrefutable. For the purpose of migrating contour lines the wavefronts and the curves of maximum convexity must be spaced at equal depth intervals along the axes of the charts and this requirement remains unchanged whatever the magnification. It is only necessary, then, to ensure that the contours to be migrated are spaced at the same interval as the axial interval on the appropriate charts. Although it may be objected that this means a departure from conventional contour spacing and that interpolation is necessary after migration to restore the usual contour spacing, the overall advantages of the method should be apparent.

While several accurately drawn fundamental charts may be necessary to cover all likely practical cases, scale changing of the two enclosed pairs of charts A1,2 and B1,2 will satisfy the majority of cases to which approximations of the velocity distributions by functions linear with depth, $V_z/V_o = 1 + z/L$, or linear with vertical time, $V_z/V_o = \sqrt{1 + 4z/L}$, can be applied. The value of L on the four charts is 4 cm and the axial spacing is 0.1 L. Only in exceptional cases in which L is very large, may the contour spacing not be sufficiently close to warrant the use of these charts.

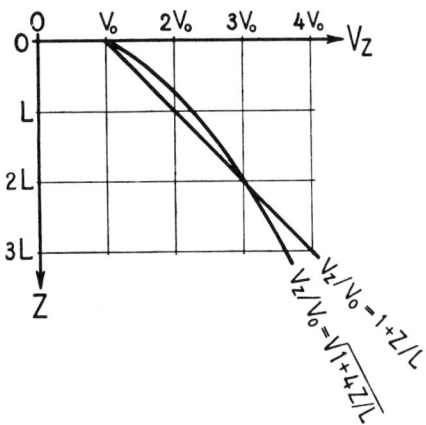

Fig. 26. Velocity functions with comparable range of velocities.

The two functions $1 + z/L$ and $\sqrt{1 + 4z/L}$ have been chosen to cover a comparable velocity range for comparable values of L. The depth range to 3L, where the velocity becomes about four times the surface velocity, should be quite sufficient for practically all cases.

Conclusion

Seismic exploration for oilbearing structures is usually carried out under circumstances where speed is as important as the results to be obtained. This is not conducive to experimental work or innovations and brings in the danger of the appearance of a gap between theorists and practicists as is so often observed in immature technical branches of physical science.

An attempt has been made, in this thesis, to avoid this pitfall by stressing the application in practice of the presented processes. Even then their use in actual practice will necessitate the acquirement of unfamiliar techniques. The drawing of a smooth line through a row of more or less scattered points is already a fairly subjective procedure, but drawing aids are available in the form of ship curves. The attainment of this same smoothness, however, in the spacing of contour lines, necessary for the purpose of three-dimensional migration, is already a much more unfamiliar technique for most seismologists.

The introduction of any new method or technique must always be justified by a marked improvement of the results obtained. A difficulty in seismic work is that it is often impossible to check the results. The value of recognizing curves of maximum convexity in order to locate faults has been clearly demonstrated in actual cases, but this alone does not warrant their introduction into an interpretational method. Their importance in this respect is due to the ease in migrating two-dimensionally with their aid, and due to their being essentially required in order to be able to migrate contour lines, whereby a simple method of three-dimensional migration becomes feasible.

Naturally it is important to be able to interpret seismic reflection data in space and not to be compelled to work only two-dimensionally because otherwise the procedures become too involved and cumbersome. The method presented in this thesis is simple enough to warrant a three-dimensional line of thought even in fast routine work. The gain in reliability, by adopting a shooting system including more cross lines or cross-spreads, will, in most cases, outweigh a possible decrease in the rate of progress. The difficulty of introducing an unfamiliar procedure into the interpretation, involving as it does a period of practice before it can compete with customary methods, is only temporary and, thus, cannot be a strong argument against acceptance.

The line of thought followed in this thesis is necessarily incomplete and must be considered as a brief outline upon which an interpretation of seismic reflection data can be based. The circumstances encountered in practice will condition the manner of application and the necessary modifications.

In some regions the assumption that the velocity distribution does not change very rapidly in a lateral direction cannot be maintained. Where a strong jump in velocity occurs at a certain lithological boundary, as, for example, to limestone, it will be necessary first to make an interpretation down to that boundary before attempting an interpretation of deeper layers. This involves, unavoidably, more cumbersome and time consuming techniques.

The advantage of plotting reflection data in two stages, first the vertical plotting and then the migration, is best appreciated in those borderline cases where plotting with the aid of a single chart for the whole depth interval is not permissible but where migration can be done with a single pair of charts, because only the approximate shapes of the curves and not their time values are of consequence. In these relatively numerous borderline cases only the procedure of vertical plotting is complicated but, for the purpose of migration, use can still be made of easily obtainable charts. Thus the process described in this thesis introduces a number of simplifications to offset the apparent complications involved by the stress which is laid on three-dimensional interpretation.

It is, of course, clear that any interpretational work can only be fruitful when reasonably clear experimental data are obtained. The limitation of the results that can be obtained is, primarily, determined by the physical circumstances encountered and by the field procedure and the recording instruments, for the particular method of interpretation employed is of secondary importance.

Acknowledgements

The author is indebted in the first place to the Bataafsche Petroleum Maatschappij for permission to publish this paper, and in the second place to all those who have assisted with their advice and criticism and especially to the draughting department for their painstaking work.

References

Baule, H.: Laufzeitmessungen an Bohrkernen und Gesteinproben mit electronischen Mitteln, Geophysical Prospecting, I, 2, June 1953.

Clewell, D. H. & R. F. Simon: Seismic Wave Propagation, Geophysics, XV, 1, January 1950.

Daly, J. W.: An Instrument for Plotting Reflection Data on the Assumption of Linear Increase of Velocity, Geophysics, XIII, 2, April 1948.

Dix, C. H. 1: On the Minimum Oscillatory Character of Seismic Pulses, Geophysics, XIV, 1, January 1949.
    2: Pulse Propagation in Two Spacial Dimensions, Geophysics, XV, 3, July 1950.
    3: Seismic Prospecting for Oil, Harper & Bros., New York, 1952.

Hagedoorn, J. G.: A Practical Example of an Anisotropic Velocity Layer, Geophysical Prospecting, II, 1, March 1954.

Krey, Th.: The Significance of Diffraction in the Investigation of Faults, Geophysics, XVII, 4, October 1952.

Melle, F. A. van: Wave-front Circles for a Linear Increase of Velocity with Depth, Geophysics, XIII, 2, April 1948.

Musgrave, A. W.: Wavefront Charts and Raypath Plotters, Quarterly of the Colorado School of Mines, Vol. 47, No. 4, October 1952.

Nettleton, L. L.: Geophysical Prospecting for Oil, McGraw-Hill, New York, 1940.

Ricker, N.: Forms and Laws of Propagation of Seismic Wavelets, Proceedings of the World Petroleum Congress, 1951. Geophysics, XVIII, 1, January 1953.

Riznichenko, G. V.: Geometrical Seismology of Stratified Media, Transactions of the Institute for Theoretical Geophysics, II, 1946.

Thornburgh, H. R.: Wave-front Diagrams in Seismic Interpretation, Bull. Am. Assoc. Petroleum Geologists, Vol. 14, No. 2, February 1930.

Interactive seismic mapping of net producible gas sand in the Gulf of Mexico

Alistair R. Brown*, Roger M. Wright‡, Keith D. Burkart*, and William L. Abriel‡

ABSTRACT

In the Garden Banks area of offshore Louisiana several gas sands have been drilled and found productive. However, the sands are laterally variable in thickness and effectiveness. An improved understanding of the spatial distribution of net producible gas sand is highly desirable for reservoir management. The bright reflections from the top and the base of each sand were tracked automatically on an interactive interpretation system. This yielded time structure maps and hence isochron maps for each gross sand interval. The horizon Seiscrop™ sections displaying amplitudes over the sand interfaces were then summed, adjusted for tuning effects, and smoothed to yield estimates of net gas/gross sand ratio over the area under study. By combining these with the corresponding isochron maps and an appropriate gas sand interval velocity, we obtained net gas sand isopach maps which tie acceptably with well data. Integration of these provided total reservoir volumes.

™Trademark of Geophysical Service Inc.

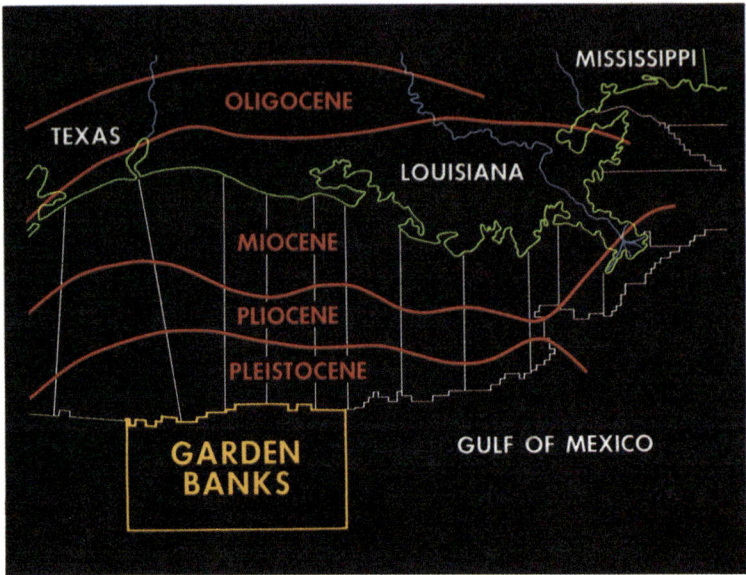

FIG. 1. Location of Garden Banks area offshore Louisiana and Texas.

Presented at the 53rd Annual International SEG Meeting September 14, 1983, in Las Vegas. Manuscript received by the Editor August 10, 1983; revised manuscript received December 21, 1983.
*Geophysical Service Inc., MS 3966, P.O. Box 225621, Dallas, TX 75265.
‡Chevron U.S.A. Inc., P.O. Box 6056, New Orleans, LA 70174.
© 1984 Society of Exploration Geophysicists. All rights reserved.

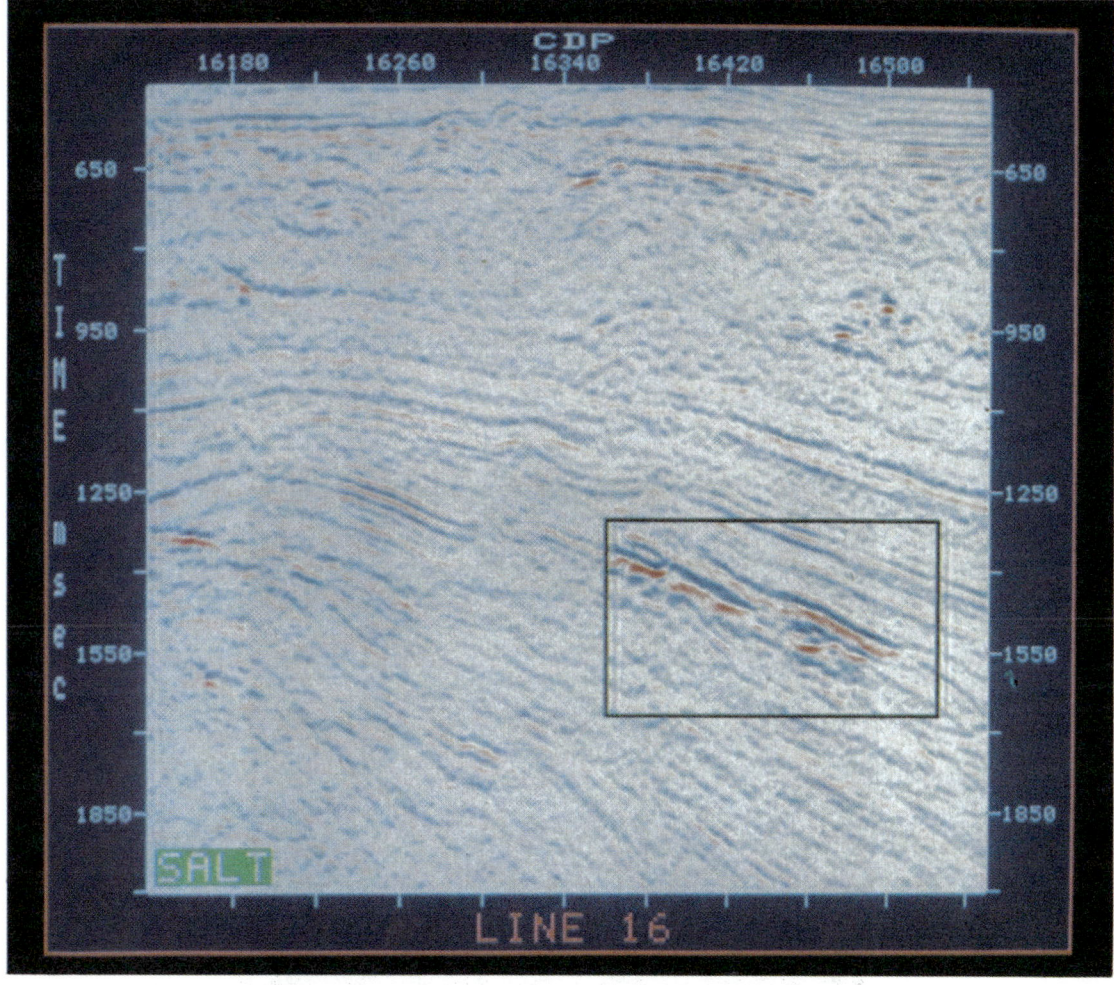

FIG. 2. Line 16 from the 3-D survey showing the setting of the bright reservoir events on the north flank of a structure probably caused by a salt swell.

INTRODUCTION

Garden Banks is a region of offshore Louisiana and Texas outside the 200 m isobath (Figure 1) where the hydrocarbon prospects are principally in Pleistocene sands. In 1977 and 1978, Chevron and partners made several gas discoveries in this area. In particular they penetrated sands at a depth of about 4500 ft (1400 m) which indicated good production. These sands are variable in thickness, porosity, and the number of lobes present, and the principal task facing the production geologist and geophysicist is to map net producible sand.

A three-dimensional (3-D) seismic survey was recorded over the gas sands in 1981. Because of the known and estimated structural and stratigraphic dips, close subsurface spacings were used: the line spacing was 50 m and the CDP spacing was 12.5 m. Figure 2 shows a north-south vertical section from the 3-D survey. The black rectangular box encloses the bright reflections associated with the gas sands. The north structural dip is assumed to be generated by a salt swell, and the irregularity of the sands is considered to result from slumping. The reservoirs are clearly stratigraphically controlled updip and the reflections from the sand interfaces are variable in amplitude. The objective of the interactive interpretive project was to utilize the seismic amplitude information to estimate the effective thickness of the gas sands.

Direct detection of gas sands on the basis of anomalous amplitude has been popular for many years. A thorough treatise on the criteria for detection was provided by Backus and Chen (1975). Most practical uses of seismic amplitude for hydrocarbon detection have been qualitative, and the difficulties of doing otherwise were explained by Domenico (1976). However, the quantitative use of amplitude to estimate the aggregate thickness of sand units encased in shale was discussed by Meckel and Nath (1977), Neidell and Poggiagliolmi (1977), and Schramm et al. (1977). 3-D migration images subsurface reflection data more accurately than 2-D migration (French, 1974), and hence the resulting amplitudes should provide more detailed and quantitative information. Brown et al. (1981) showed the ability of 3-D data to reveal subtle stratigraphic features on the basis of amplitude. An extensive reference list of 3-D case histories, both structural and stratigraphic, is provided by Brown (1983).

A small portion of the Garden Banks 3-D survey was identified as the study area (Figure 3). It measures 2.0 km north-south by 2.2 km east-west. The drilling platform to the east caused a gap in 3-D data coverage immediately to the east of the study area. Four deviated wells from the platform and one vertical well provided evaluation points for the sands under study. Figure 4 excerpts the log information for two of these wells. The interpreted producible sand zones and the resistivity logs clearly demonstrate the problem: both Upper and Lower Sand members contain many lobes of producible gas sand and there is considerable lateral variability in the sands over a short distance.

INTERACTIVE INTERPRETATION

Interactive interpretation of seismic data has been in limited use for about two years (Gerhardstein and Brown, 1984). As more and more interpreters complete interactive projects, the resulting benefits have become increasingly clear. The success of this project can, to a significant extent, be attributed to the

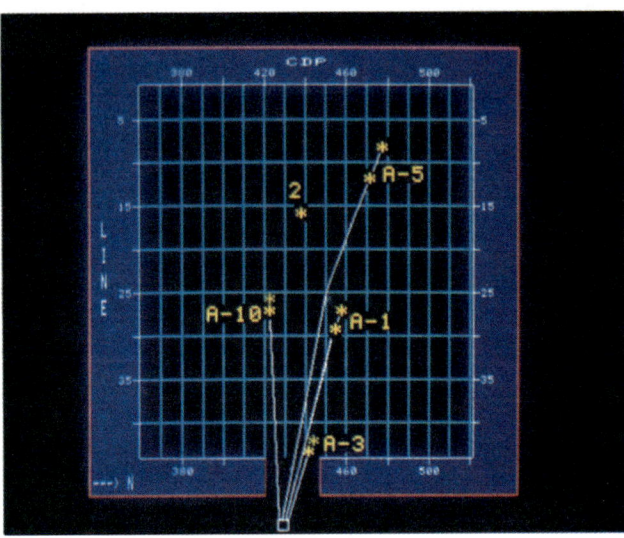

FIG. 3. Location of the five wells within the study area showing the points of intersection with the Upper and Lower Sands.

interactive system, so we will review these benefits before describing the results of the project itself.

(1) Data management. The interpreter needs little or no paper; his selected seismic data display is presented to him on the screen of a color monitor and the progressive results of his interpretation are returned to the digital data base.

(2) Color. Flexible color display provides the interpreter with maximum optical dynamic range adapted to the particular problem under study.

(3) Composition of images. Data images can be composed on the screen so that the interpreter can see just those pieces of data he needs, no more and no less, for the study of one particular issue. Pieces of data may be brought into juxtaposition for comparison or to help appreciate three-dimensionality.

(4) Idea flow. The rapid response of the system makes it easy to try new ideas. In the present project many new map and section products were generated in the course of developing the manipulative sequence which finally led to success.

(5) Time. Several automatic facilities including tracking, posting, and contouring save the interpreter significant amounts of time which he can devote instead to critical interpretive decision making.

MANIPULATIVE SEQUENCE

Figures 5 through 8 illustrate the data in north-south vertical section from every fourth line. The vertical scale is enlarged, so the north structural dip is exaggerated. Each data portion has the same CDP and time range as that outlined by the black box in Figure 2.

In general there are four bright events, and these were identified as the top and base of the Upper and Lower Sands. At the downdip limits in several places are prominent flat spots. Automatic tracking was used to follow the events wherever possible so that the track followed the true maximum amplitude of the peak or trough. Manual editing was found to be required over less than 1 percent of the total horizon distance tracked. However, this should not be taken to indicate that the horizon interpretation was easy; at times the selection of the event segment for the automatic tracker was quite difficult. Using a two-window screen display as illustrated in each of Figures 5 through 8, we were able to study the tracked horizons from the last line and the data for the next line while tracking the present line. The four horizons were only tracked where they were at least somewhat bright and the flat spots, where present, were tracked as the bases of the gas sands. The final tracks used are superimposed on the sample vertical sections in Figures 5 through 8.

The act of interactively tracking each data point caused a time and amplitude value to be stored in the prospect data base. The sequence of manipulations then applied to these times and amplitudes is charted in Figure 9. The tracked times directly yielded a structural contour map for each of the four horizons. These were then gently smoothed to remove the roughness of the automatic tracking and subtracted to yield a gross isochron map for each of the two sands.

The display of the stored horizon amplitudes directly yielded

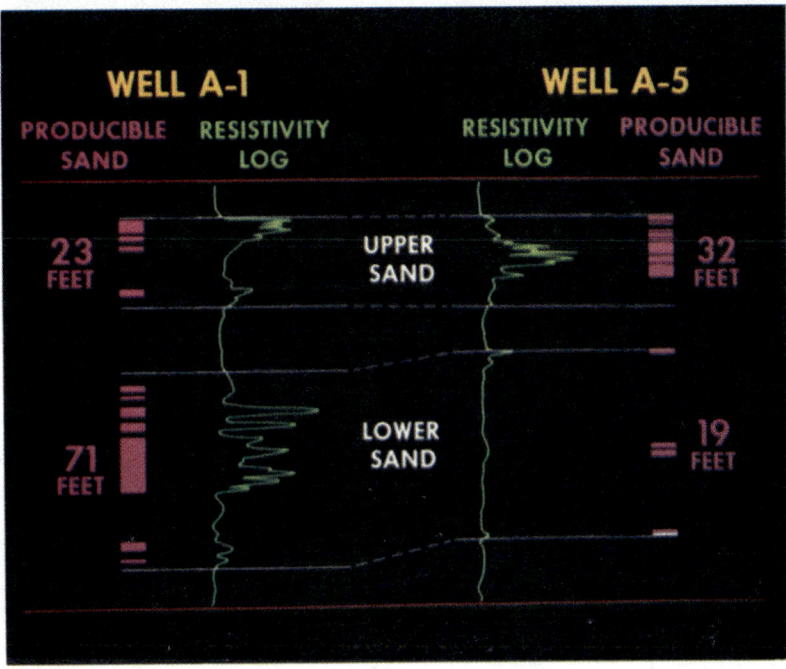

FIG. 4. The interpreted lobes of producible gas sand penetrated by wells A-1 and A-5.

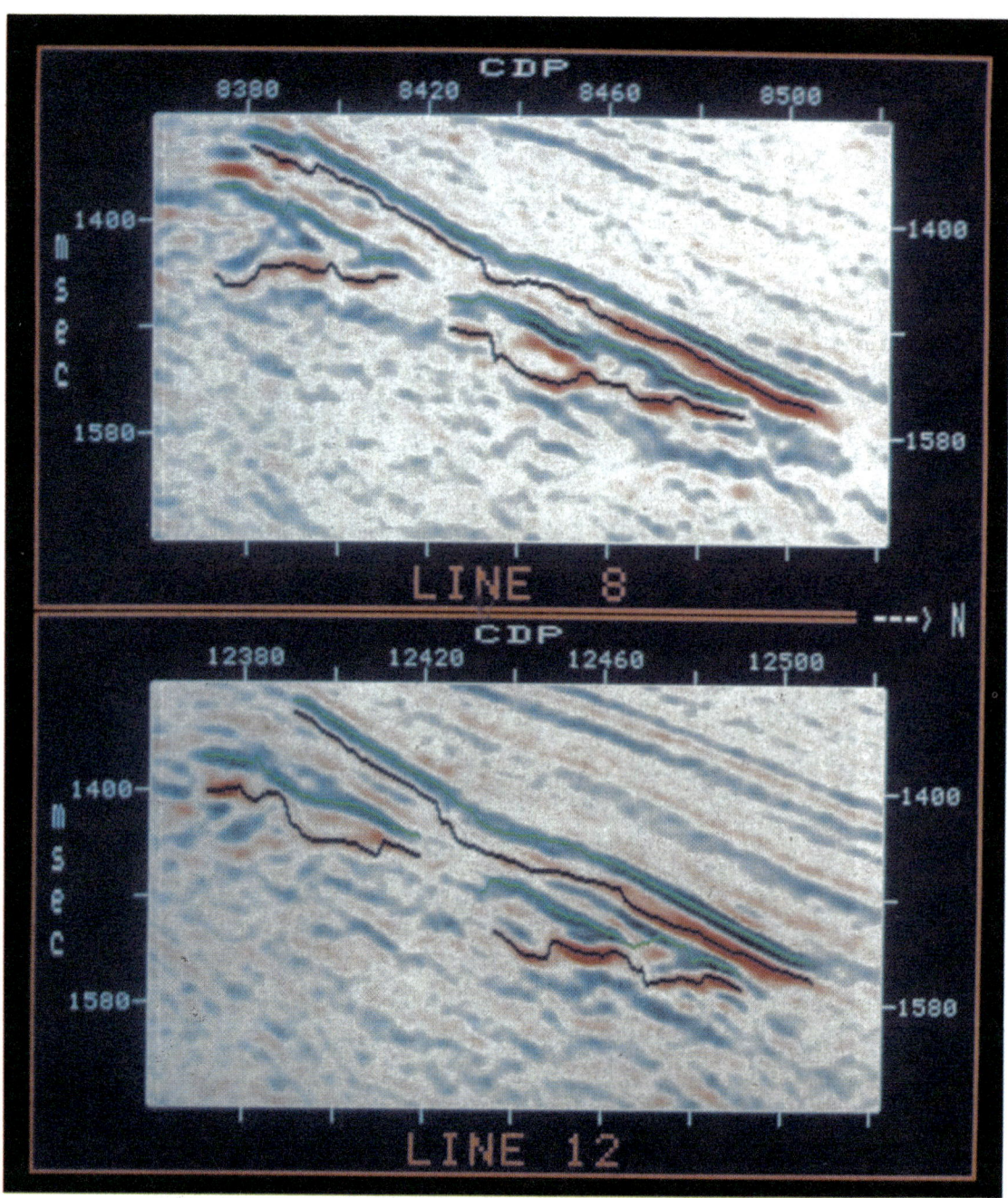

FIG. 5. Lines 8 and 12 showing tracked horizons for top and base of Upper and Lower Sands.

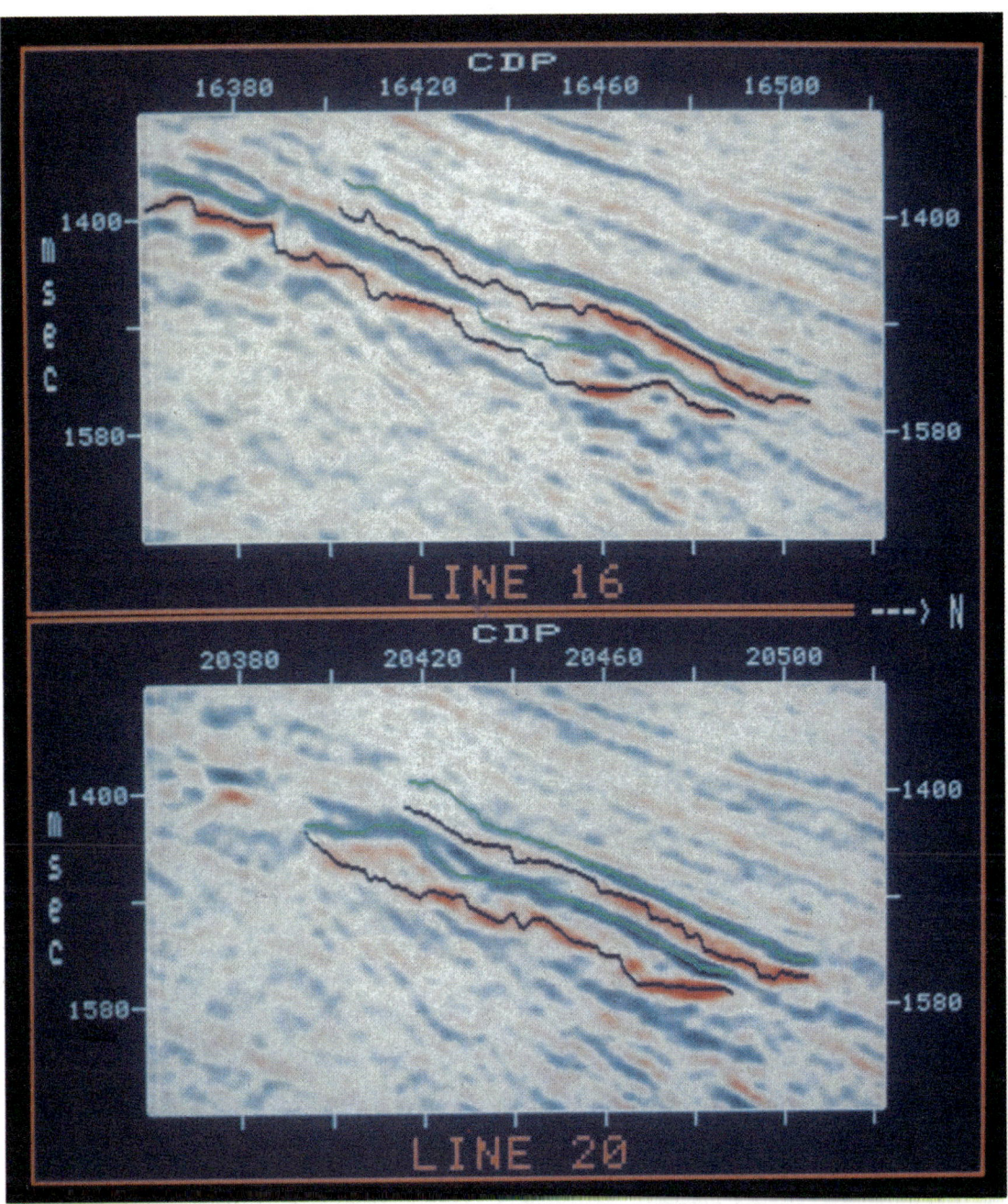

FIG. 6. Lines 16 and 20 showing tracked horizons for top and base of Upper and Lower Sands.

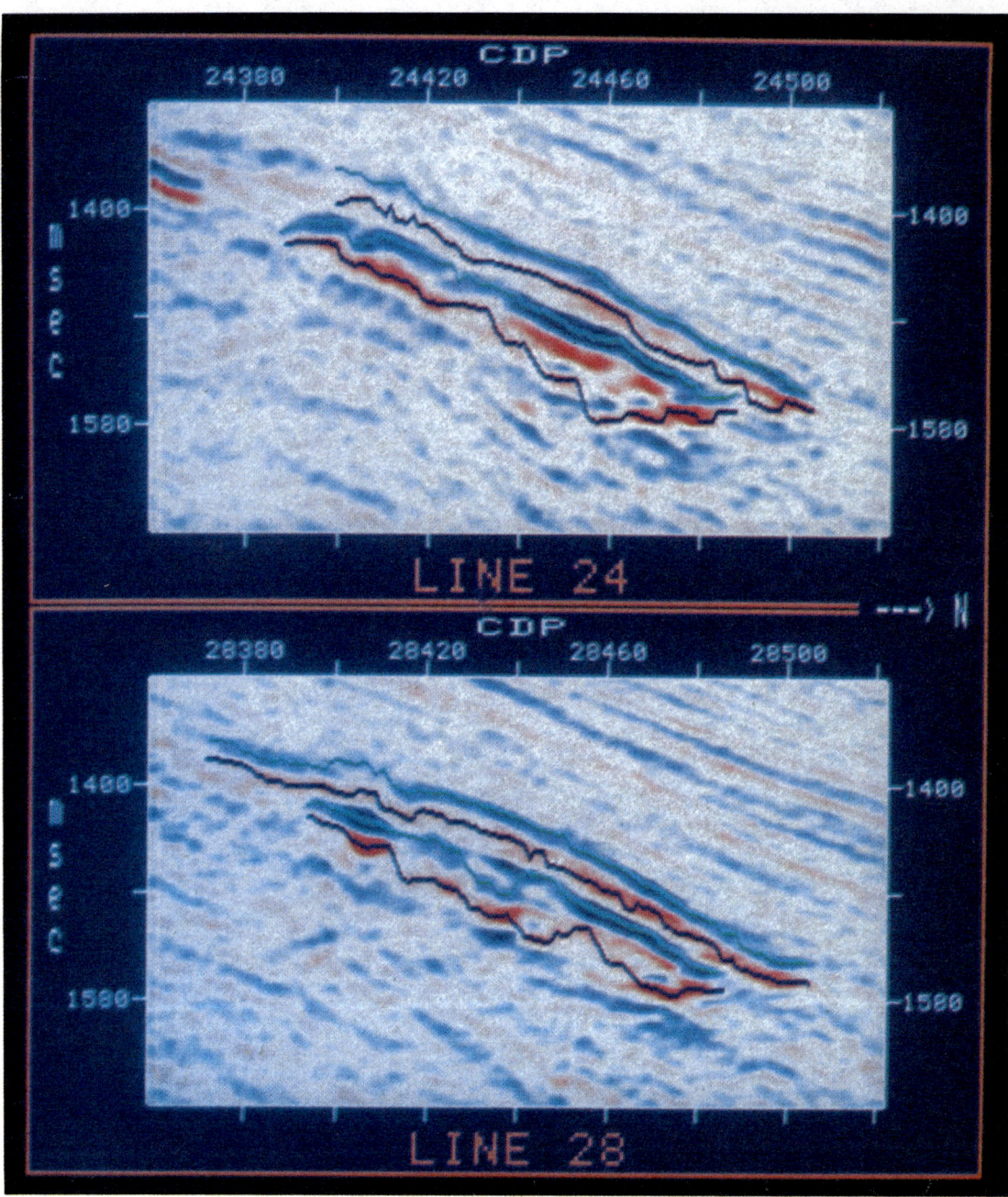

FIG. 7. Lines 24 and 28 showing tracked horizons for top and base of Upper and Lower Sands.

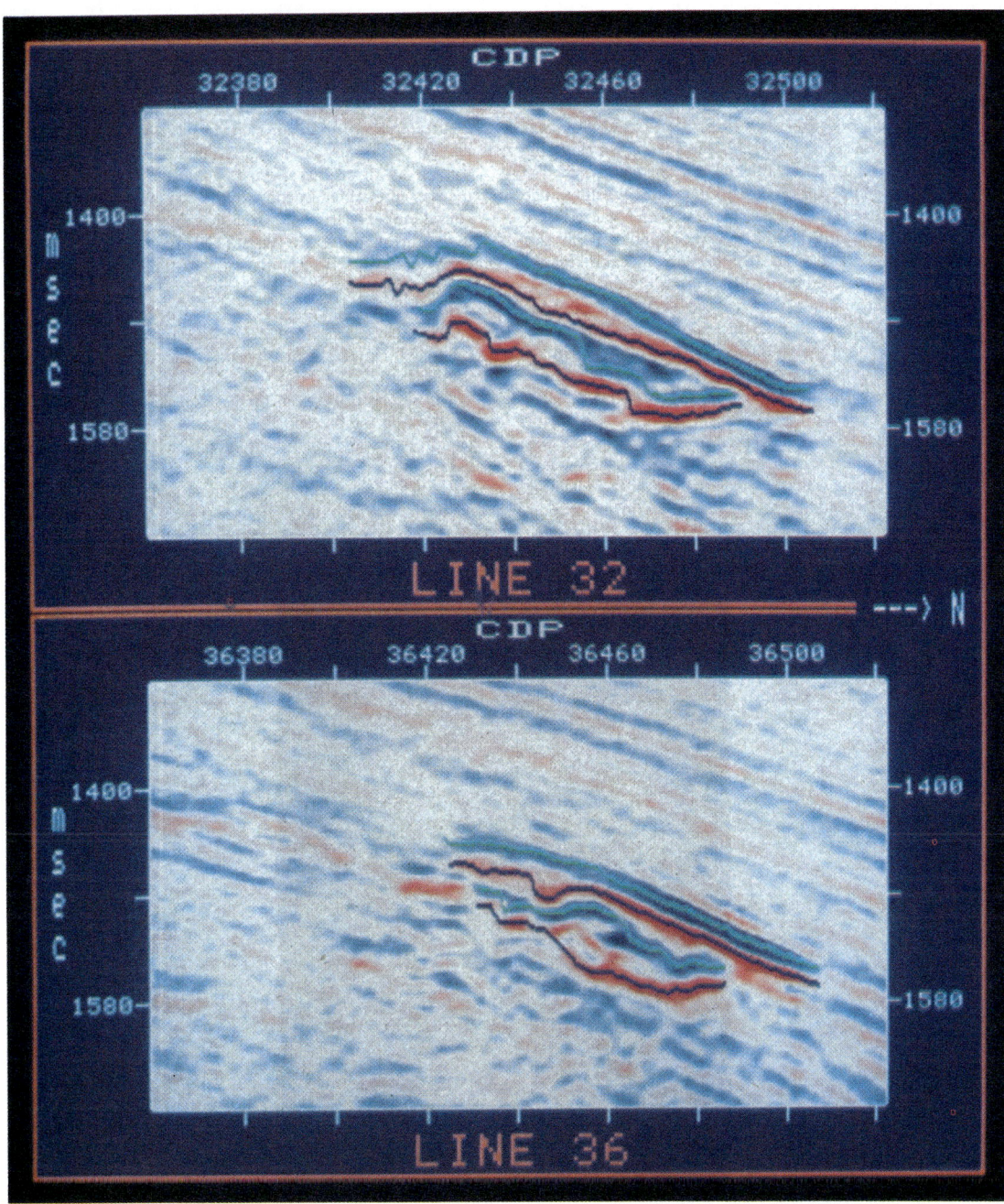

FIG. 8. Lines 32 and 36 showing tracked horizons for top and base of Upper and Lower Sands.

a horizon Seiscrop™ section for each horizon; this type of horizontal section was defined by Brown et al. (1981). For each sand the horizon Seiscrop sections from the top and base reflections were added together (in absolute value) to yield the composite amplitude response for that sand. This summation constituted a selective vertical integration, analogous to an inversion process, whereby the response of the sand was accentuated relative to that of the encasing material.

Because of constructive interference between the wavelets from the top and bottom of each sand interval, the reflection amplitudes were boosted when their time separation was around one-half seismic period; this effect is known as tuning (Widess, 1973) and the tuning thickness was here determined to be between 15 and 16 ms. Figure 10 illustrates the amplitude and measured time thickness profiles across an idealized thinning wedge and is adapted to the observed situation in this area from Meckel and Nath (1977). In order to compensate for the anomalous amplitude effects resulting from tuning, the composite amplitude response was reduced to half its previous value in the areas where the time thickness was less than 16 ms, a thickness just above tuning thickness. This edited amplitude response was then spatially smoothed to eliminate the steps in amplitude introduced by the editing, as illustrated schematically in Figure 10.

This smoothed, edited composite amplitude response was then used as an indicator of the proportion of the sand interval capable of producing gas, the higher amplitudes indicating higher proportions and the lower amplitudes lower proportions. In order to turn this amplitude response into a map of the ratio of net gas sand to gross seismic interval, we simply divided by the maximum amplitude. The multiplication of this ratio map by the gross isochron map then gave the net gas sand isochron map for each sand. Producible gas sand was assumed to have a constant velocity, and this was measured from sonic logs to be 5000 ft/s. Using this, the net gas sand isochron map was converted to a net gas sand isopach map for each sand.

RESULTS FOR UPPER SAND

The time structure maps for the top and base of the Upper Sand are shown in Figures 11 and 12. Here the maps have already been gently smoothed. A 3 point by 3 point spatial filter was used with weights selected so that the amount of smoothing was approximately equal in the two orthogonal directions, considering that the subsurface spacing was four times greater in one direction than the other. The black area surrounding the color within the blue border of the study area is where the horizons were not tracked because they were judged insufficiently bright.

The gross isochron map for the Upper Sand appears in Figure 13. Note a distinct thickness trend running northwest-southeast.

The similarity between the horizon Seiscrop sections from the top and the base of the Upper Sand (Figures 14 and 15) reinforces that the major cause of the lateral changes in amplitude is variations within the sand rather than its encasing material and hence supports the summation of the two to yield the composite amplitude response (Figure 16). Figure 17 shows the 16-ms gross isochron contour superimposed on the composite amplitude response and the amplitudes cut in half for

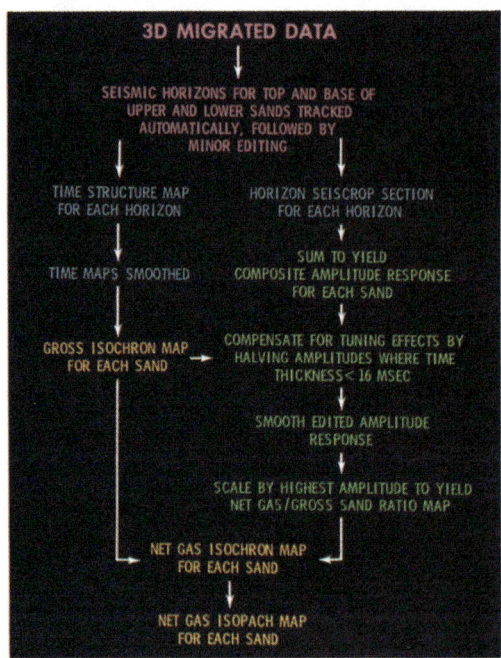

FIG. 9. Flow chart of interactive manipulative sequence used in the derivation of the net gas sand isopach maps.

thicknesses less than this value. A fairly heavy spatial smoothing of this yielded the smoothed, edited amplitude response of Figure 18. After scaling, the net gas/gross sand ratio map of Figure 19 is shown in a gradational color scheme to emphasize that this, later to be applied as a field of conversion factors for the isochron map, is indeed a smoothly varying field of numbers.

The product of the ratio map (Figure 19) and the gross isochron map (Figure 13) directly yielded the net gas sand isochron map of Figure 20. Application of the constant 5000 ft/s velocity then converted this to the net gas sand isopach map of Figure 21. Here we have removed the distinction between where the sand was not tracked, and hence assumed to be thin or absent, and where it was computed to be effectively thin; both are shown here in white.

By integration of the net gas sand isopach map of Figure 21, the total reservoir volume of the Upper Sand within the study area was computed to be 13,000 acre ft (17 million m^3).

RESULTS FOR LOWER SAND

Figures 5 through 8 demonstrate that the Lower Sand is more variable in thickness and also reaches much greater thicknesses than the Upper Sand. The Lower Sand reflections were also more difficult to track. However, the manipulative sequence used was exactly the same, namely, that charted in Figure 9; in the interest of brevity only some of the interpretive products are illustrated.

The time structure map of the base of the Lower Sand (Figure 22) shows a broad area of purple between lines 18 and 44 which indicates the extent of the flat spot reflection at the downdip limit of this reservoir. The gross isochron map (Figure 23) shows a suggestion of a thickness trend running northeast-southwest, approximately at right angles to the trend observed in the Upper Sand. The composite amplitude response for the Lower Sand (Figure 24) shows a more rapidly changing amplitude pattern than for the Upper Sand, but this is consistent with the observed greater variability of the Lower Sand. The net gas sand isochron map and the net gas sand isopach map are shown in Figures 25 and 26.

The same color assignments have been used for equivalent map products for the Lower and Upper Sands, so that comparison between the two sands is straightforward. As a consequence, the Lower Sand maps make better use of the range of colors in the legend because of its greater range of thicknesses.

By integration of the net gas sand isopach map of Figure 26, the total reservoir volume of the Lower Sand within the study area was computed to be 20,000 acre ft (26 million m^3).

THE WELL TIES

The five wells in the study area (Figure 3) each provided a control point for the Upper Sand and also for the Lower Sand against which the accuracy of the net gas sand isopach maps could be assessed. Figure 3 shows the location at which each well penetrated the top of each sand; these locations were

(*Text continued on p. 712*)

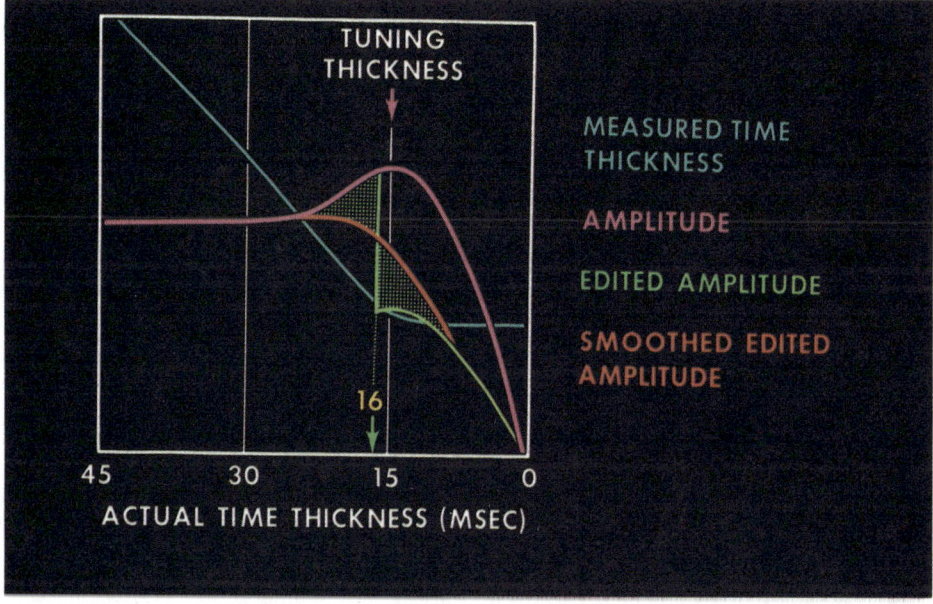

FIG. 10. Amplitude profiles across a thinning margin, adapted from Meckel and Nath (1977) to explain amplitude manipulations referred to in Figure 9.

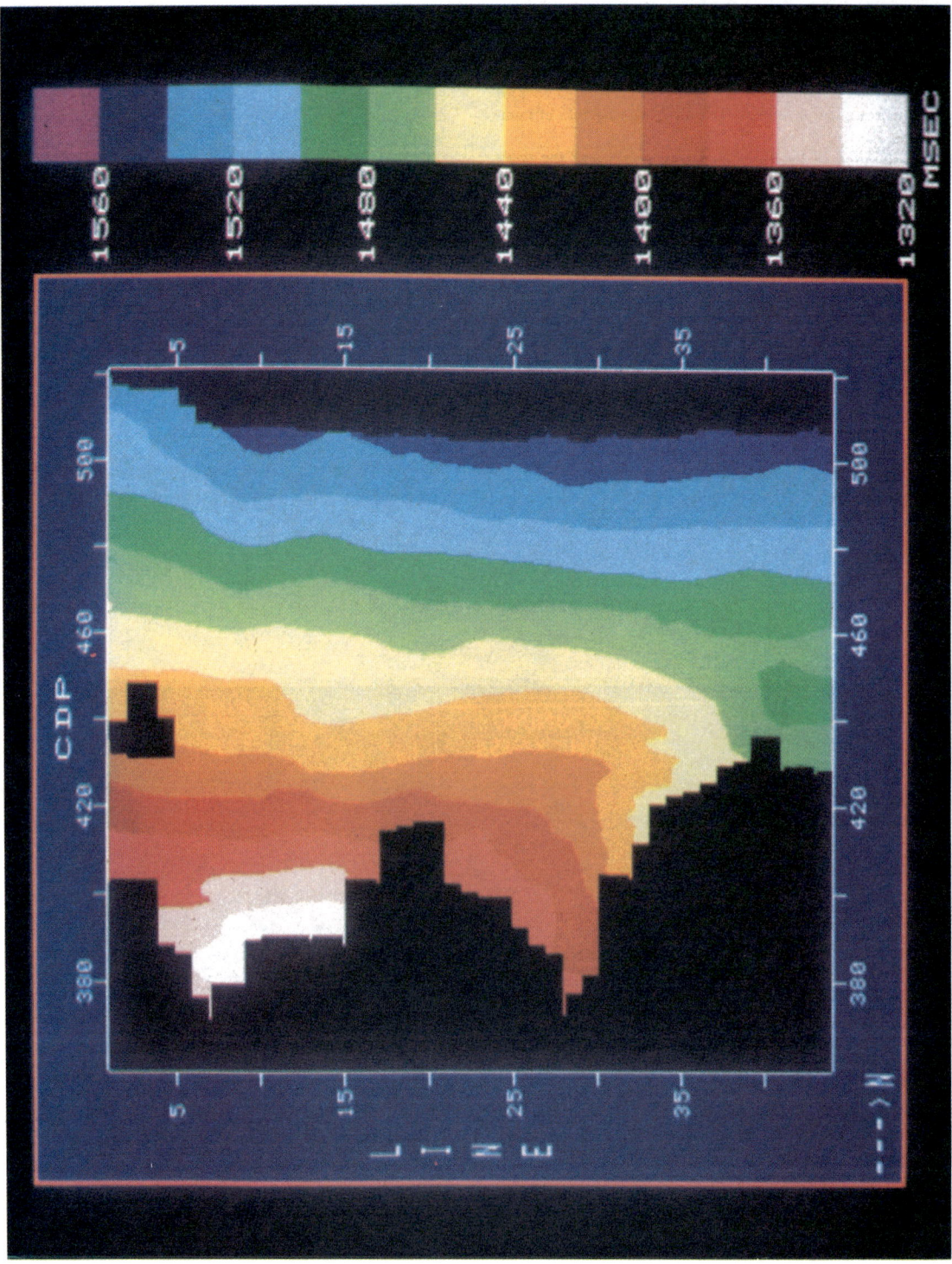

FIG. 11. Time structure map, top of Upper Sand.

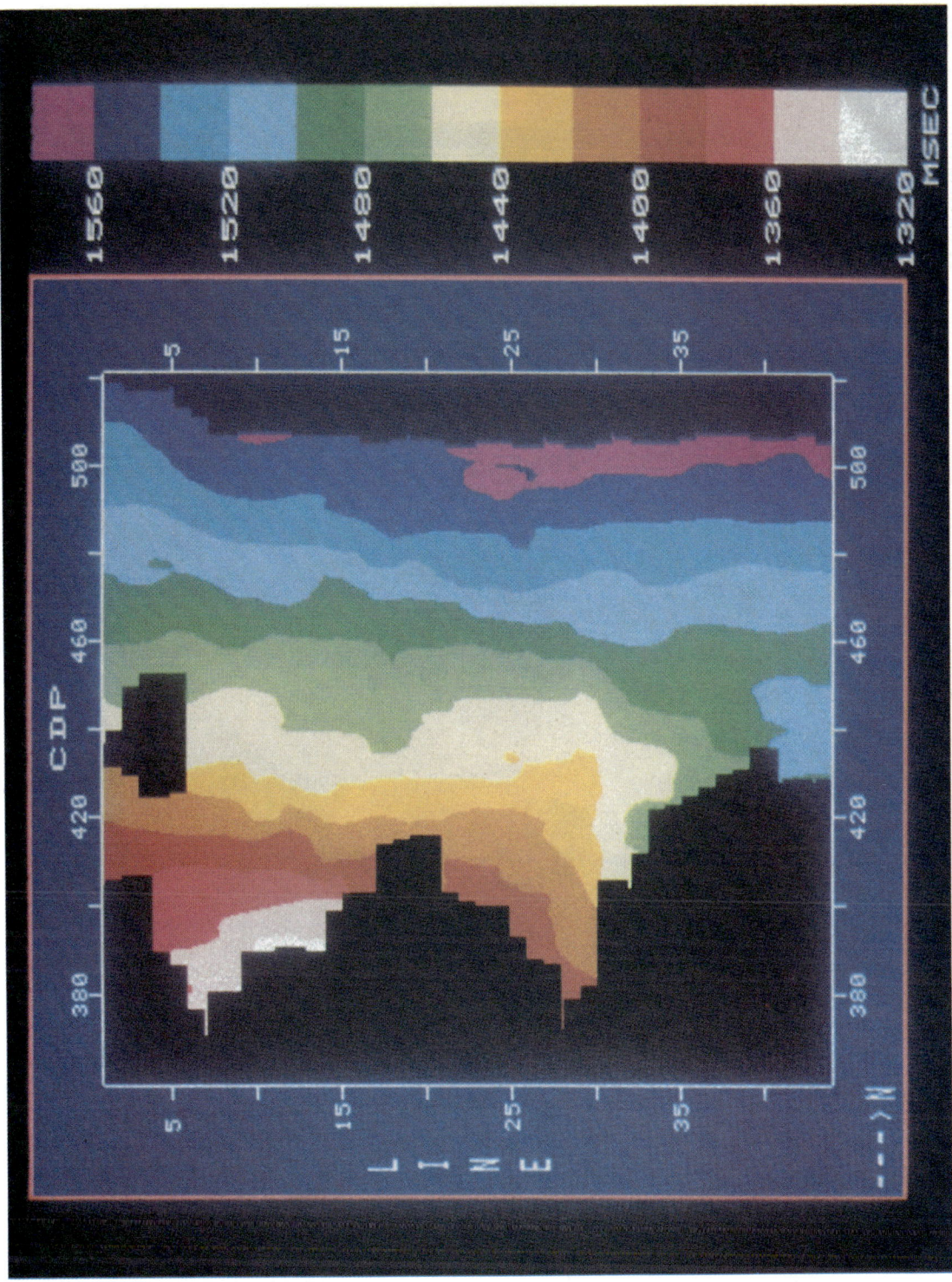

FIG. 12. Time structure map, base of Upper Sand.

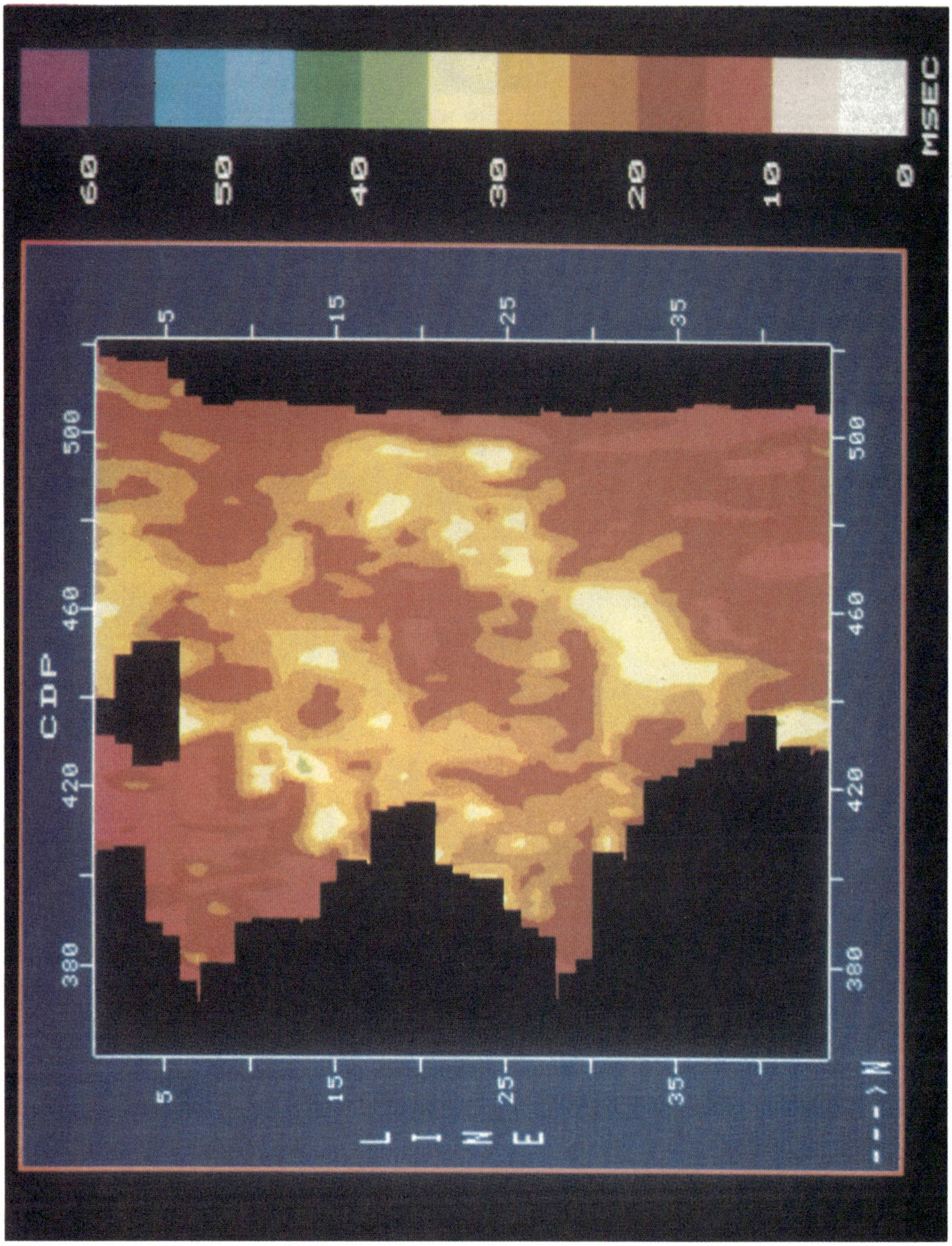

FIG. 13. Gross isochron map, Upper Sand.

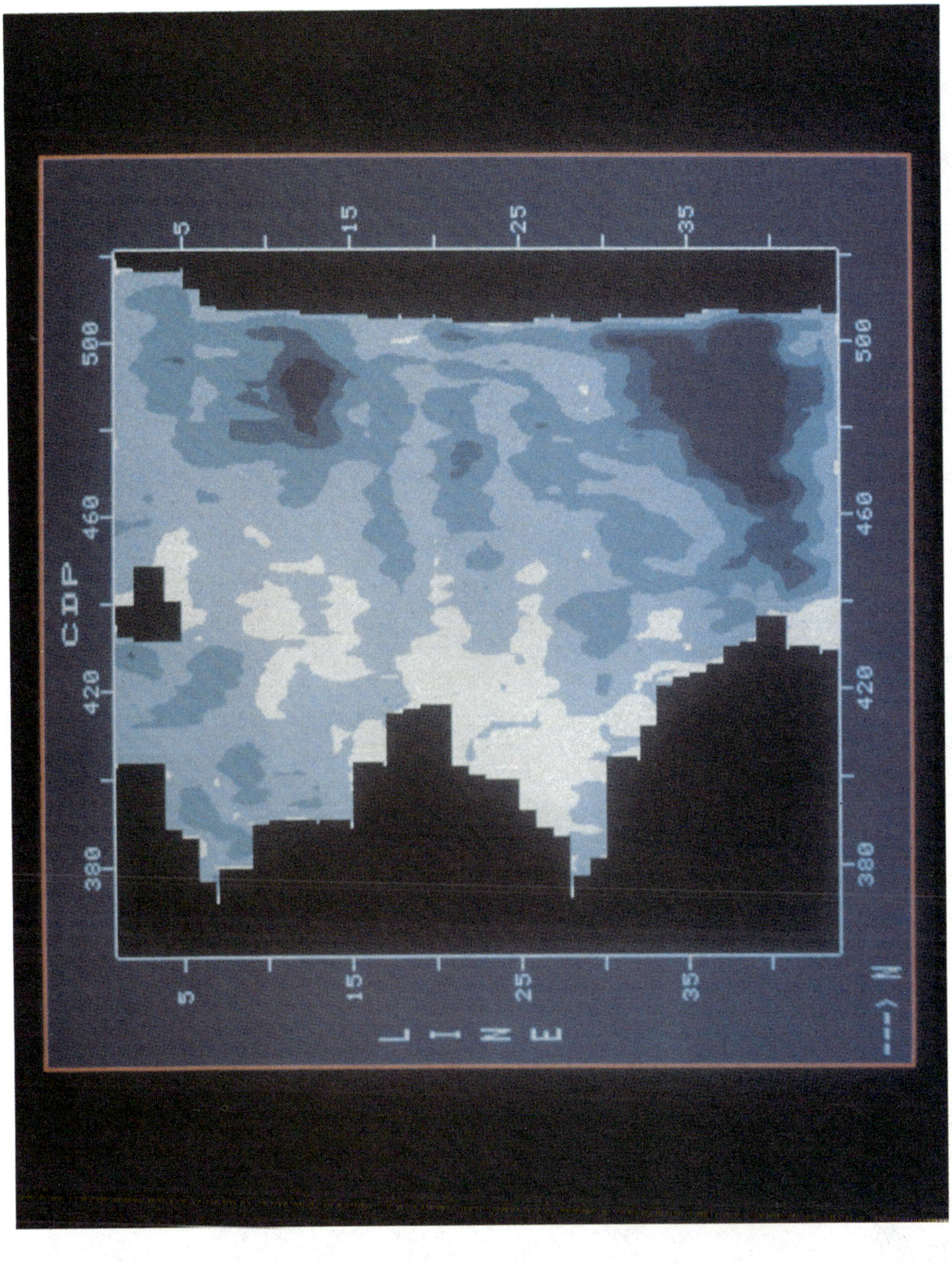

Fig. 14. Horizon Seiscrop section, top of Upper Sand.

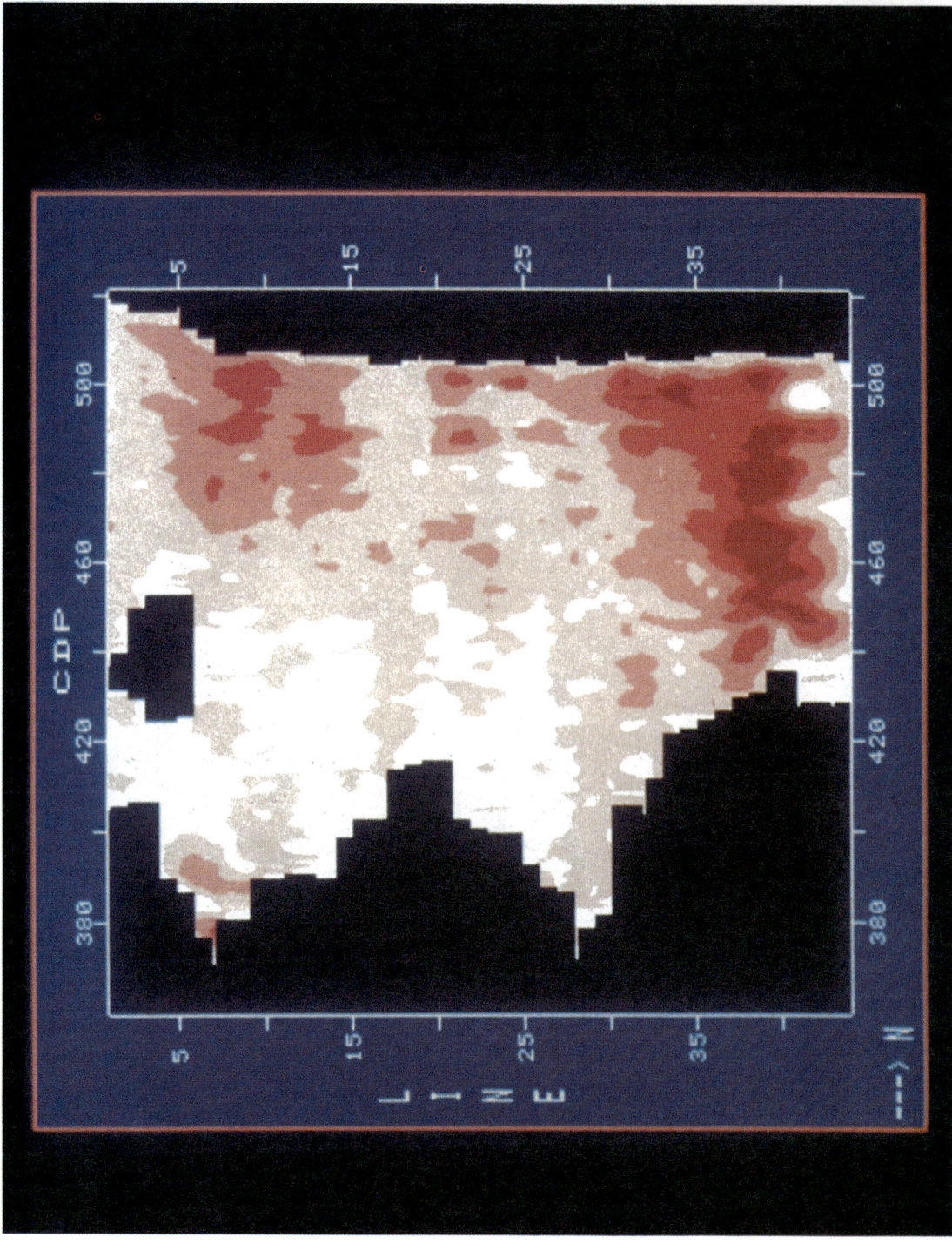

FIG. 15. Horizon Seiscrop section, base of Upper Sand.

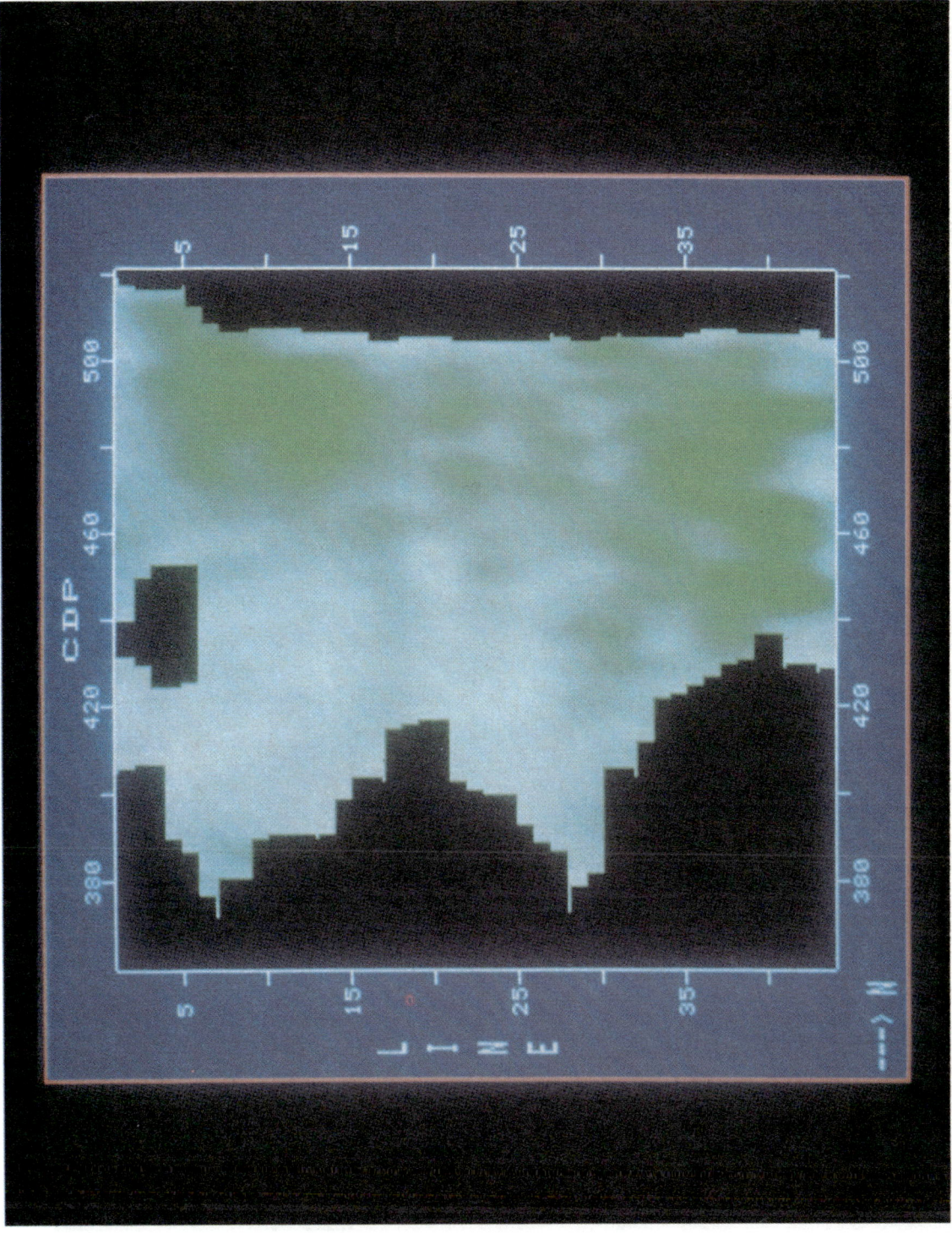

FIG. 16. Composite amplitude response, Upper Sand.

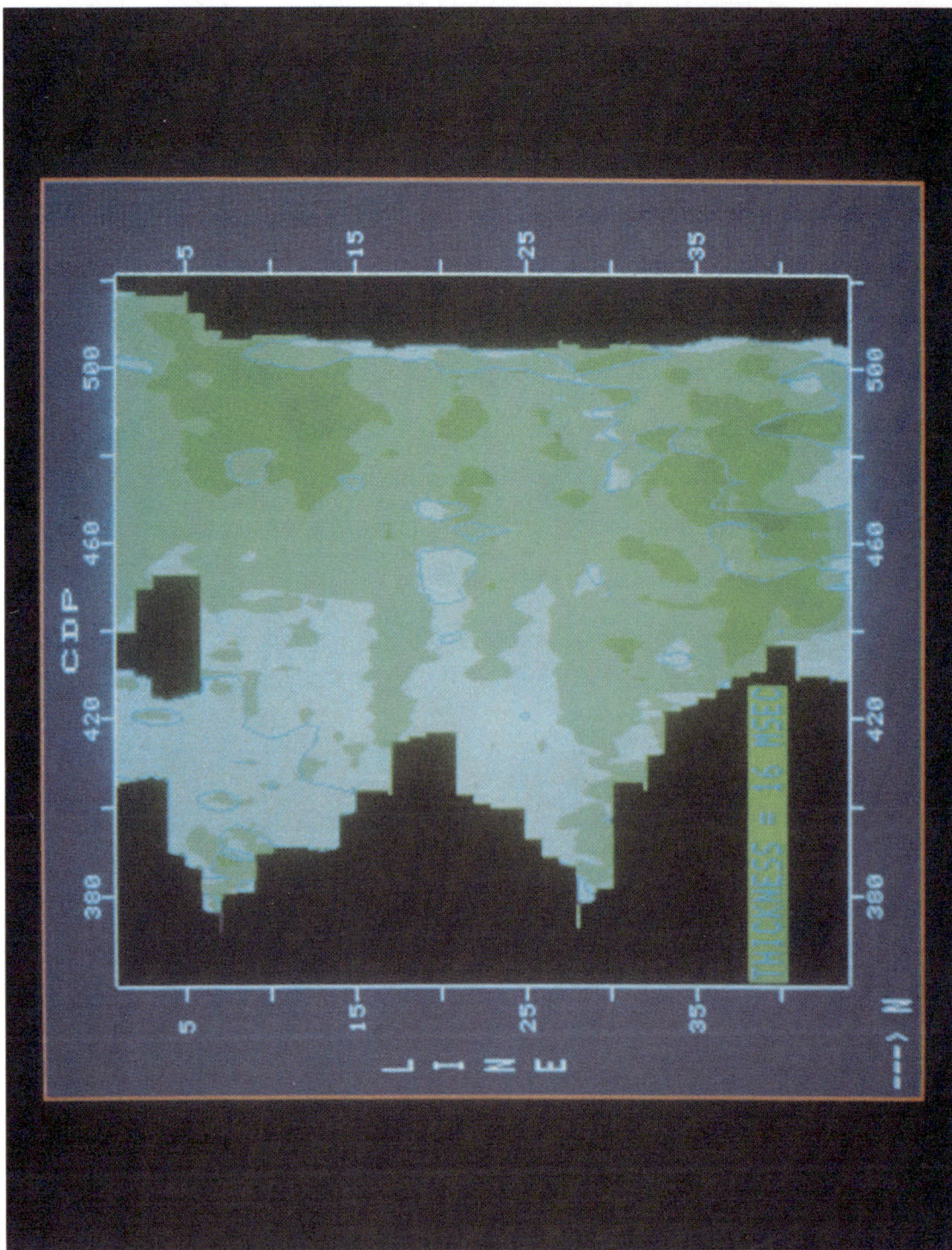

Fig. 17. Edited composite amplitude response, Upper Sand.

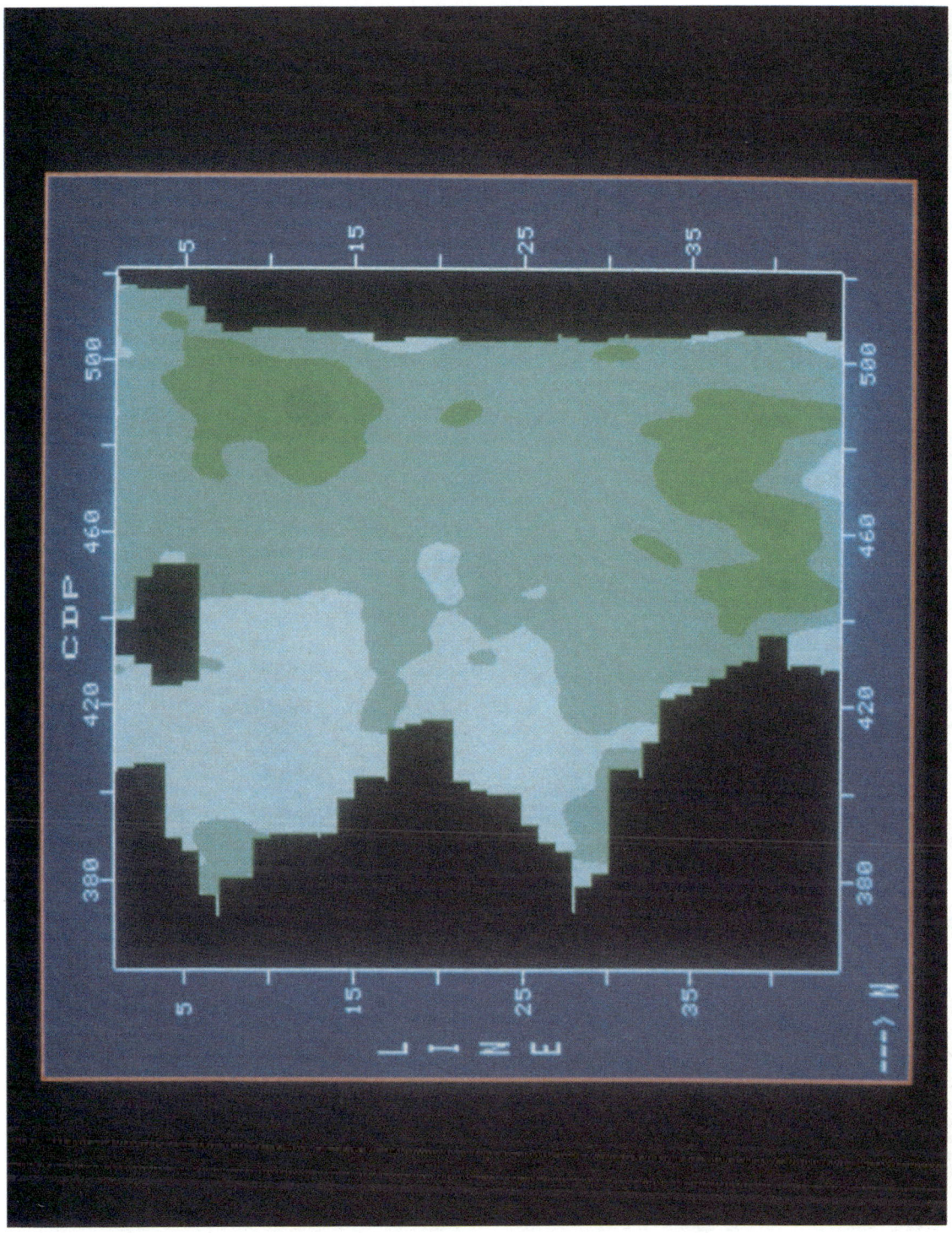

Fig. 18. Smoothed, edited composite amplitude response, Upper Sand.

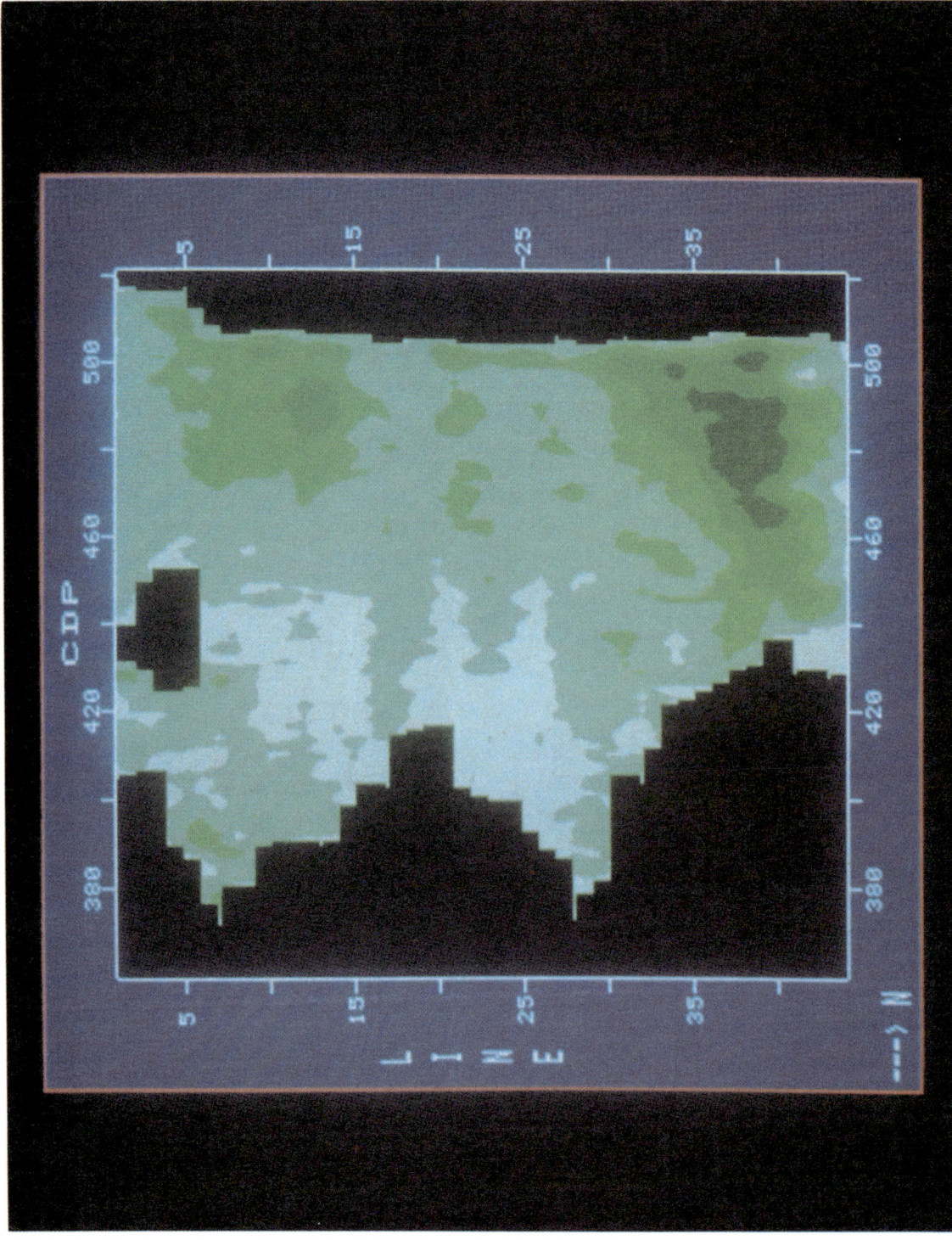

FIG. 19. Net gas/gross sand ratio map, Upper Sand.

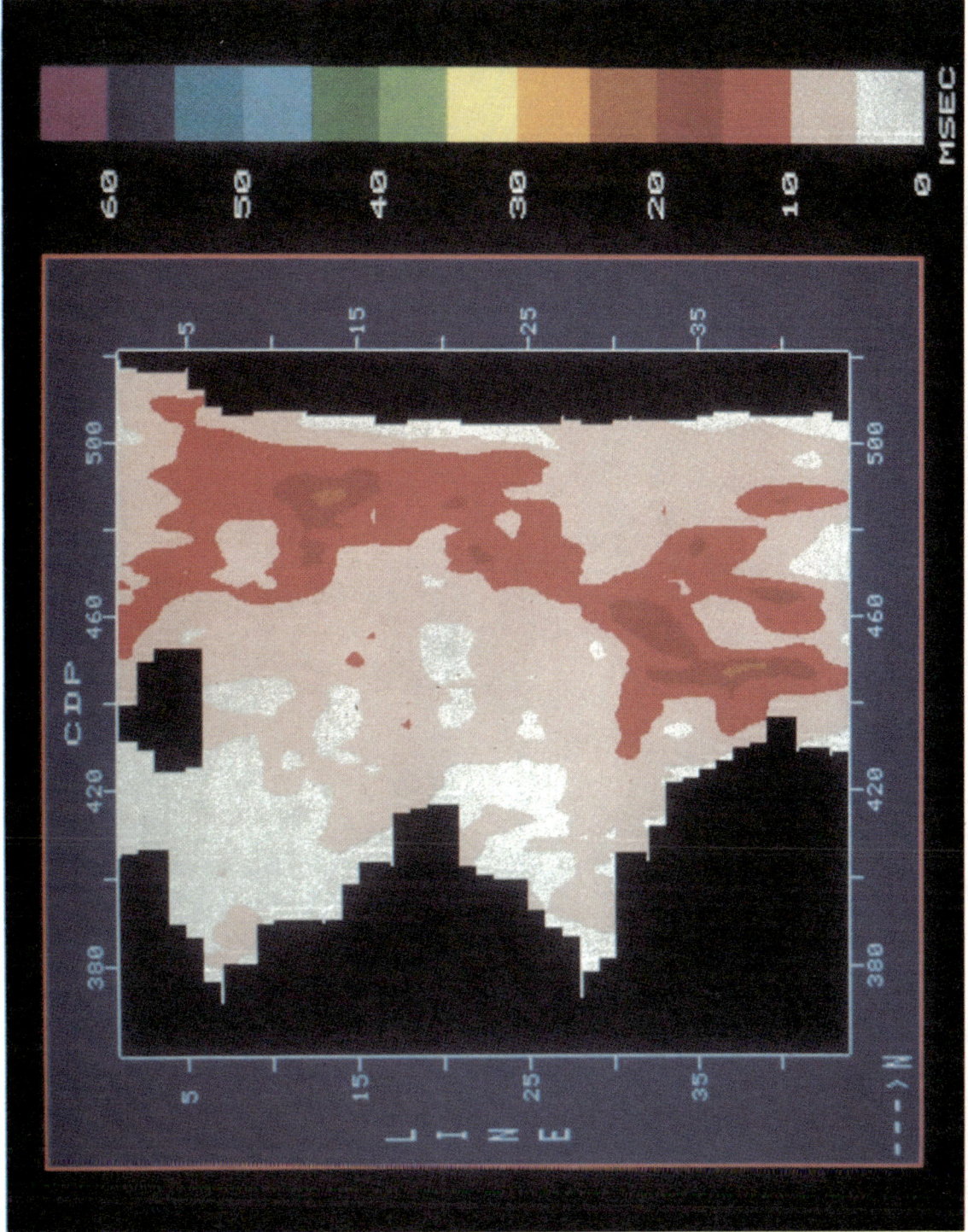

FIG. 20. Net gas isochron map, Upper Sand.

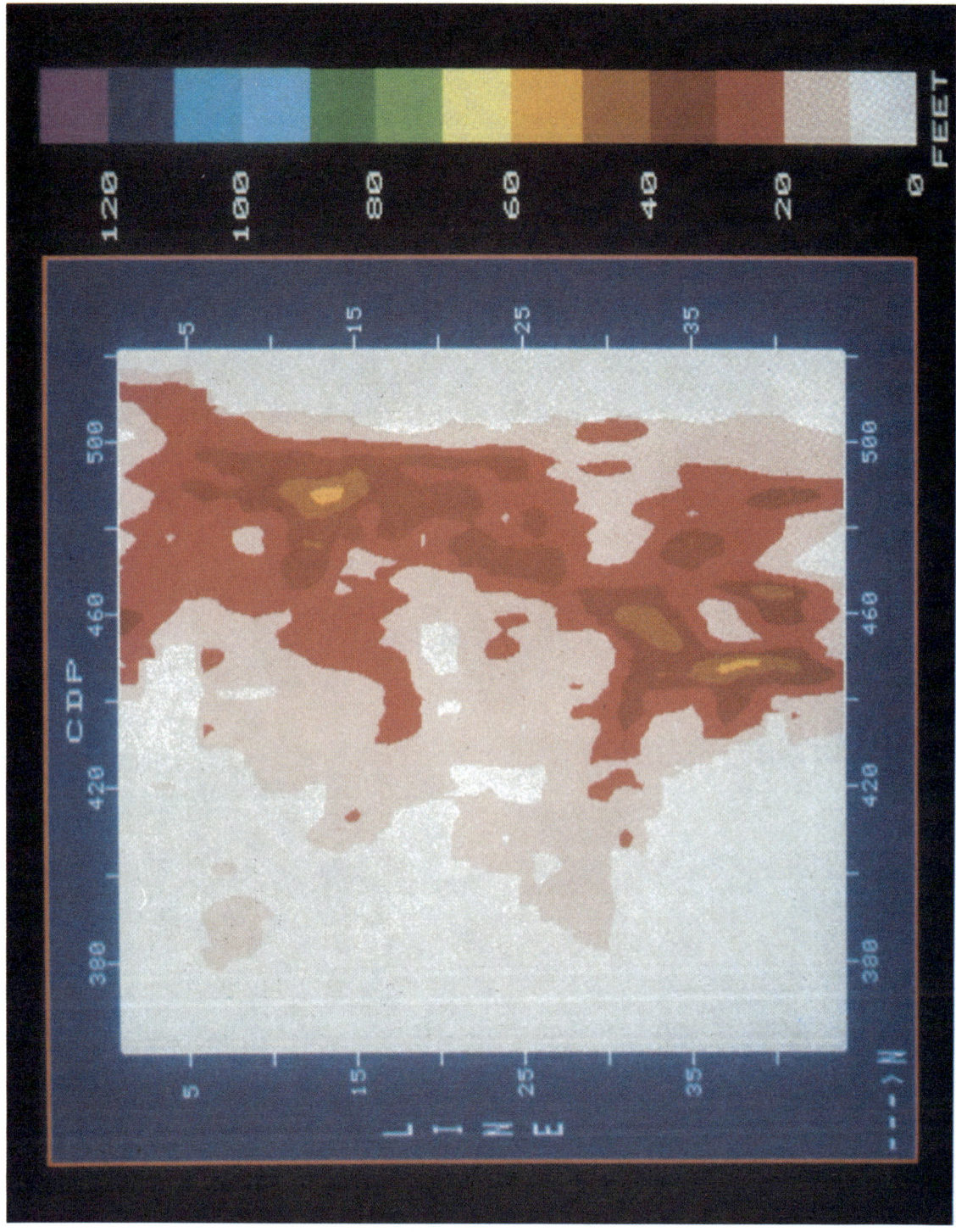

Fig. 21. Net gas isopach map, Upper Sand.

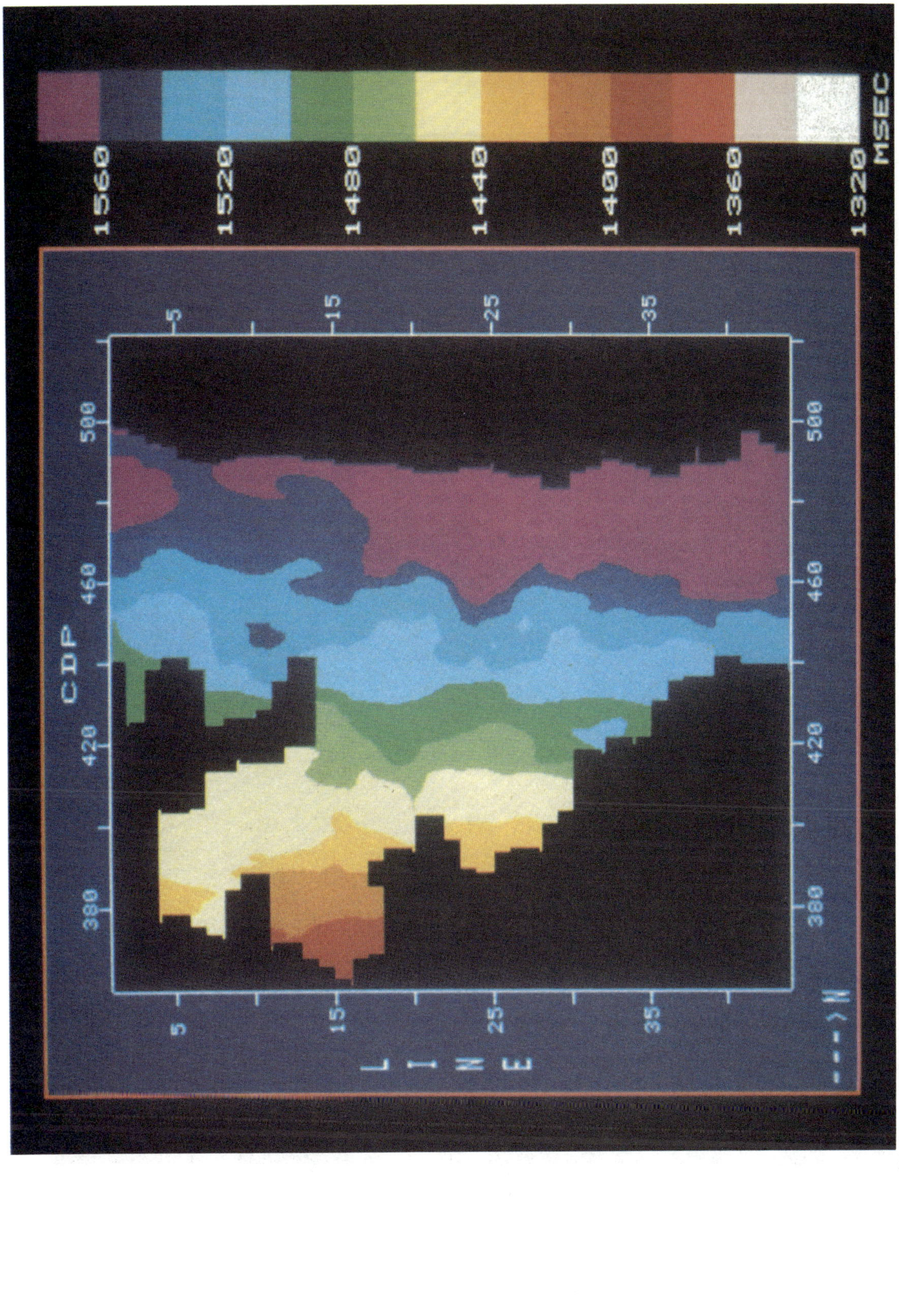

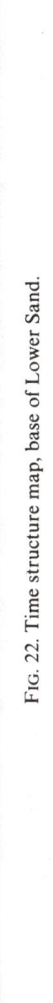

FIG. 22. Time structure map, base of Lower Sand.

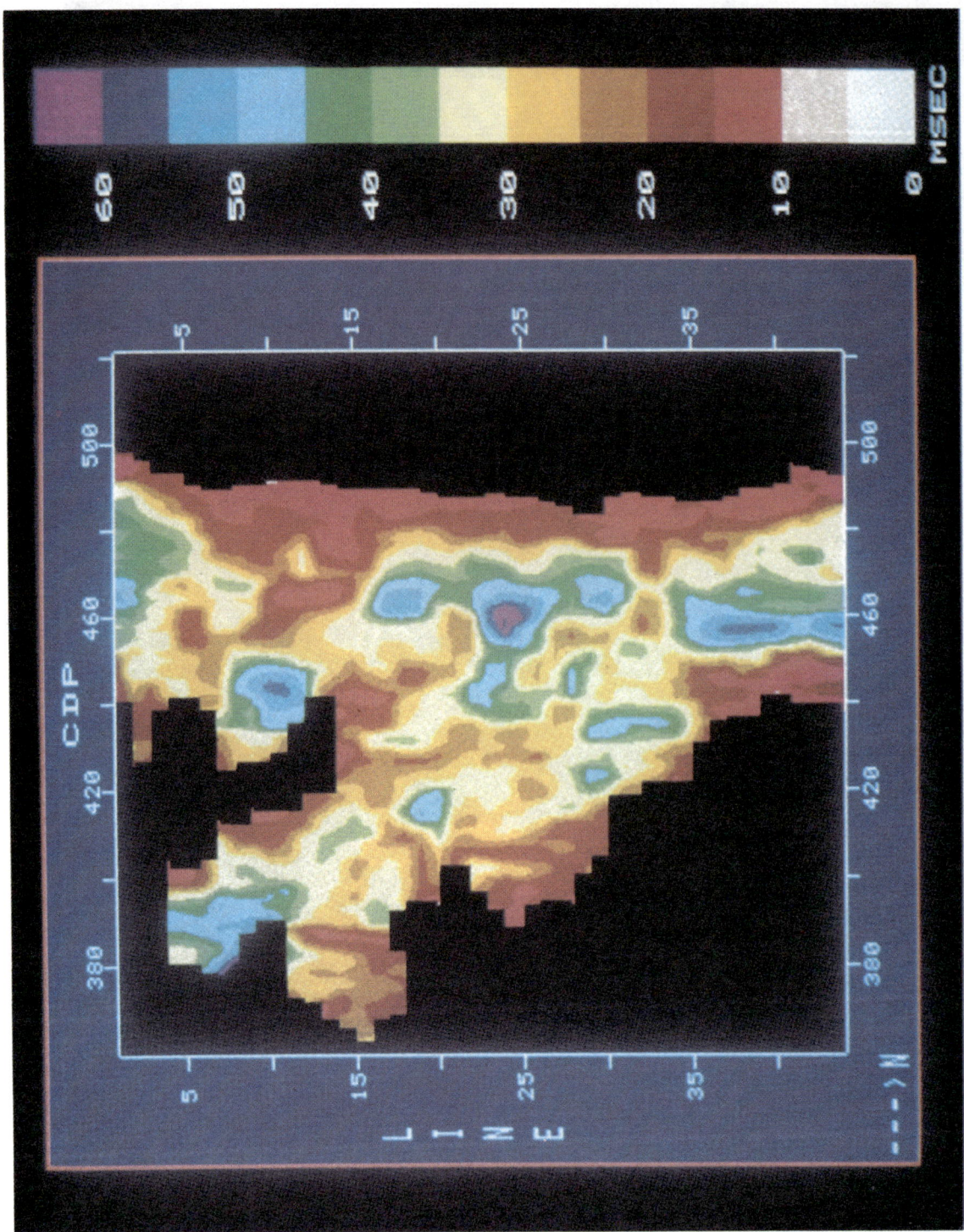

FIG. 23. Gross isochron map, Lower Sand.

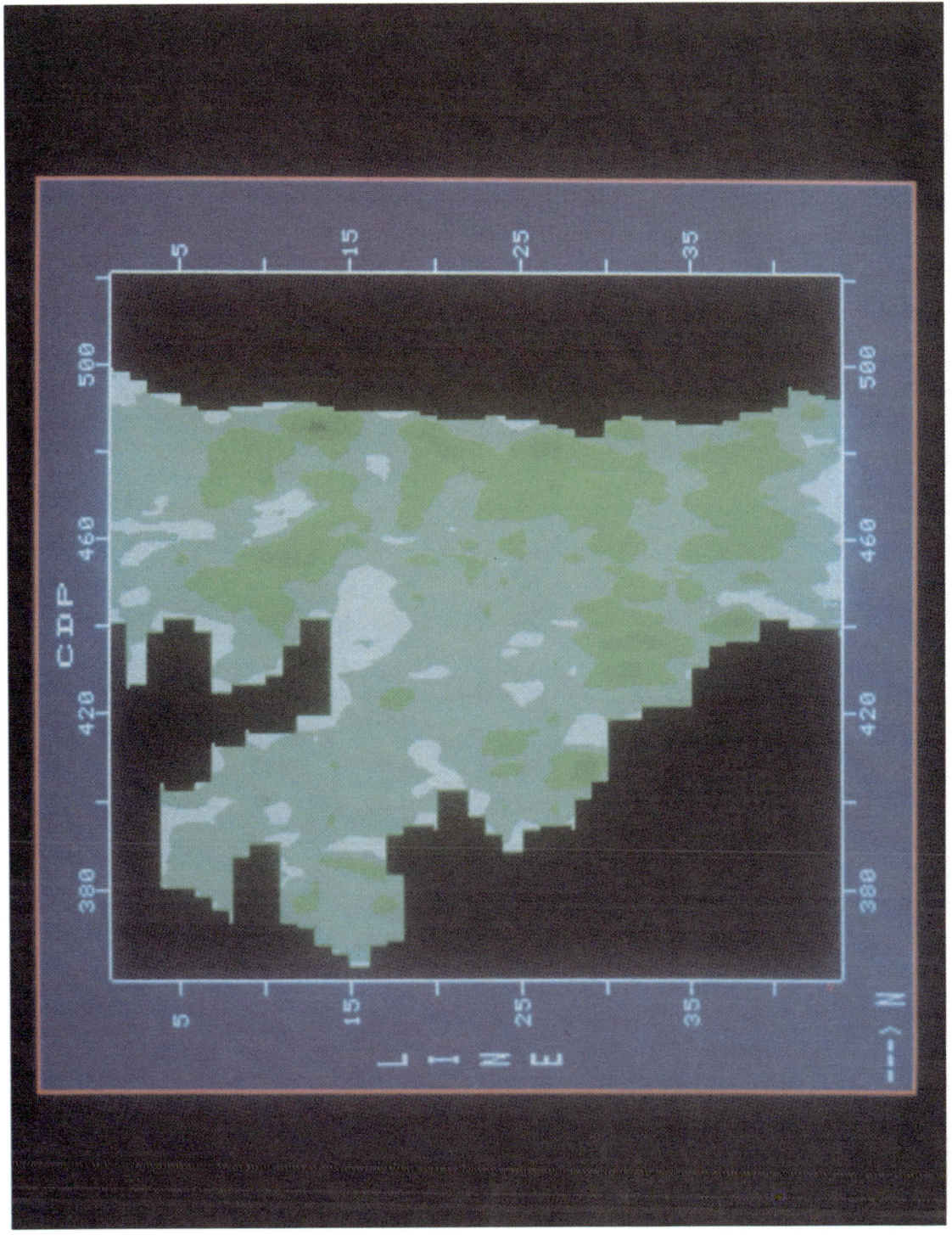

FIG. 24. Composite amplitude response, Lower Sand.

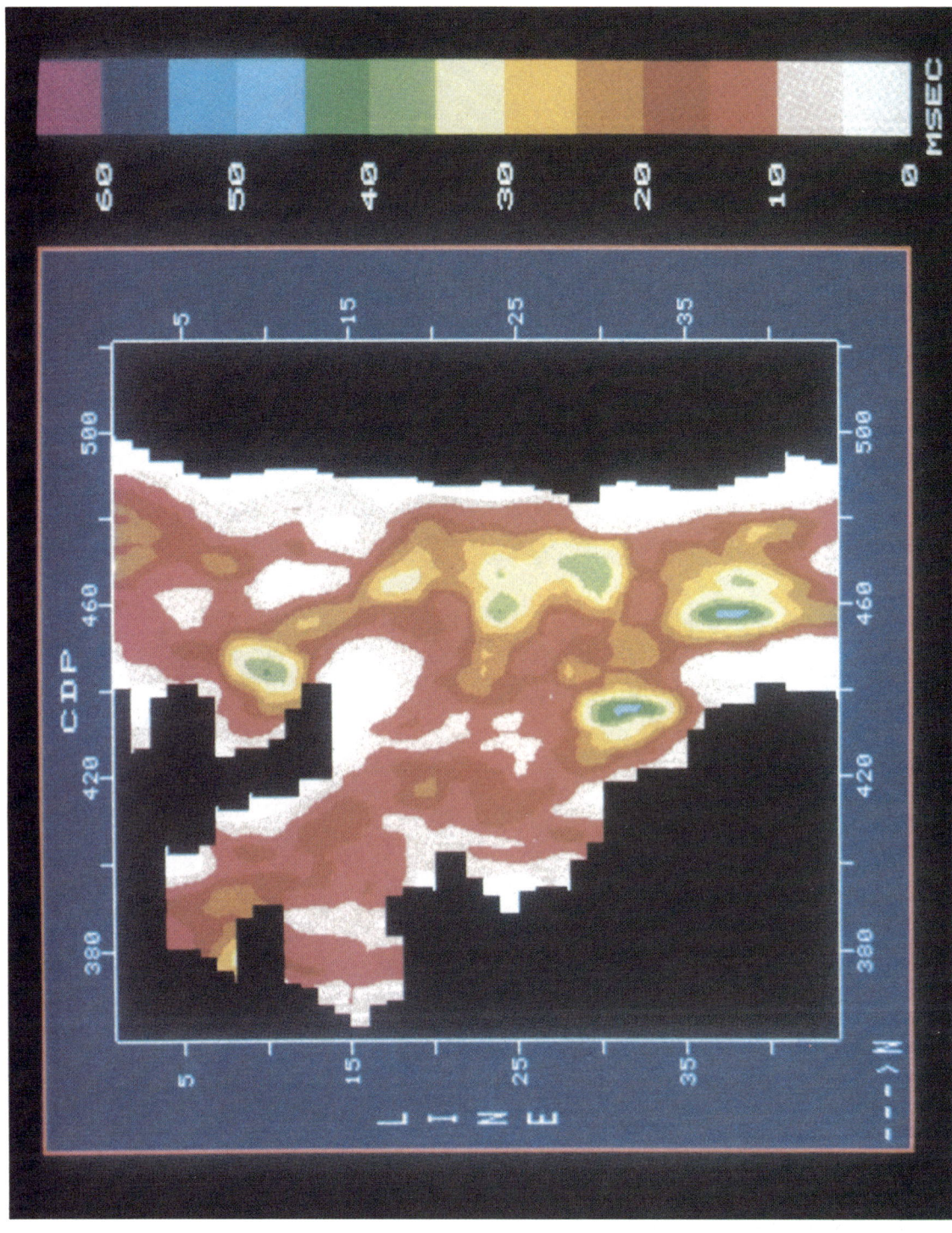

FIG. 25. Net gas isochron map, Lower Sand.

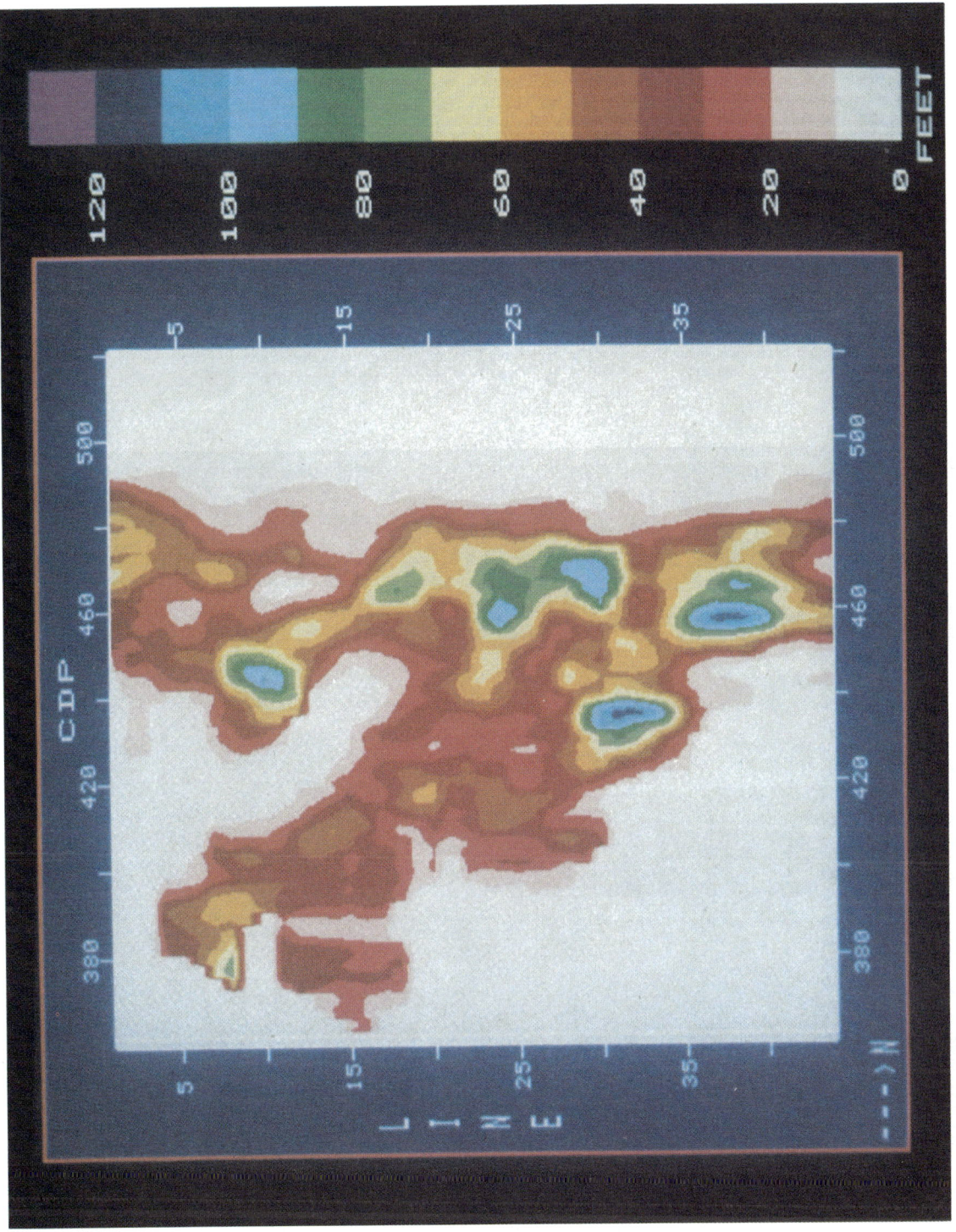

FIG. 26. Net gas isopach map, Lower Sand.

(*Continued from p. 695*)

considered accurate to within ± 100 ft (30 m). Using a version of Figure 3 as an overlay to the isopach maps of Figures 21 and 26, the seismic net gas sand values at the wells were read. These readings are tabulated in Figure 27 (except for the values at well A-3 which were later adjusted as will be discussed below) and have as a superscript an estimate of the uncertainty of the reading. This number should be added to and subtracted from the prime value in each case to give the range of thicknesses read within the 100 ft (30 m) radius location error circle.

Interpretation of the well logs provided the vertical net gas sand values at the wells, which are also tabulated in Figure 27; the sand lobes considered producible in wells A-1 and A-5 are illustrated in Figure 4. This interpretation is also subject to some uncertainty; the superscripts in Figure 27 should again be both added to and subtracted from the prime values in order to indicate the range of net gas sand thicknesses interpretable from the well logs.

The initial seismic interpretation was performed independently of the well information. The first net gas sand isopach maps indicated the need to re-examine the horizon tracks for the Lower Sand at wells A-3 and A-10 and for the Upper Sand at well 2. The first two of these were simply adjusted and the isopach maps remade. However, a local diminution in amplitude on line 26 at the location of well A-10 suggests a minor fault or other cause for further very local thinning. We made no adjustment to the tracks for the Upper Sand at well 2 because some very abrupt line-to-line changes in horizon tracks would have been required. However, line 16 suggests that a local deepening of the base Upper Sand reflection at CDP 16437 (Figure 6), the location of well 2, is quite reasonable.

Our accumulated experience from the horizon tracking in general and the well tying in particular is that the thickness of the sand bodies varies very abruptly indeed. On the crosslines, where the trace spacing is 50 m, there are occasional indications of aliasing. We therefore conclude that the subsurface sampling interval, while being adequate for all structural dips, was too

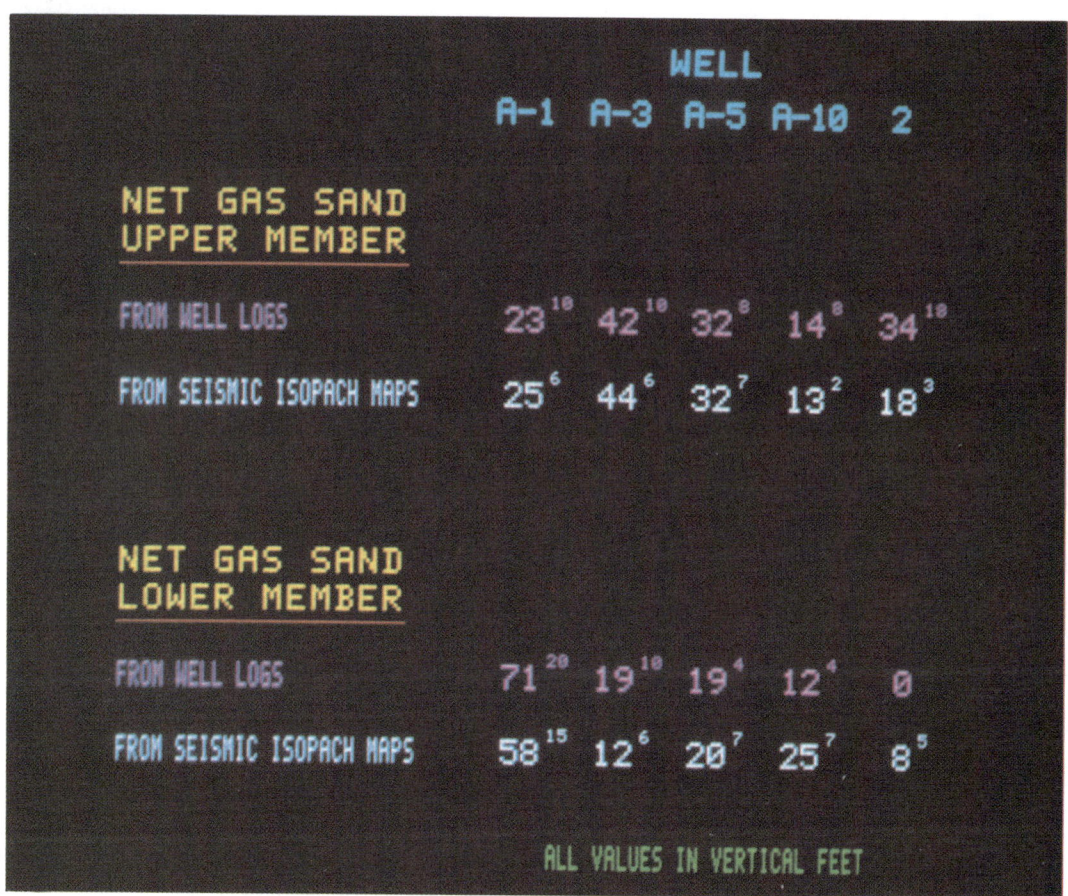

Fig. 27. Comparison of net gas sand thicknesses read from seismic isopach maps and those interpreted from well logs. Superscripts are estimates of the uncertainty, plus and minus, of the sand thickness values.

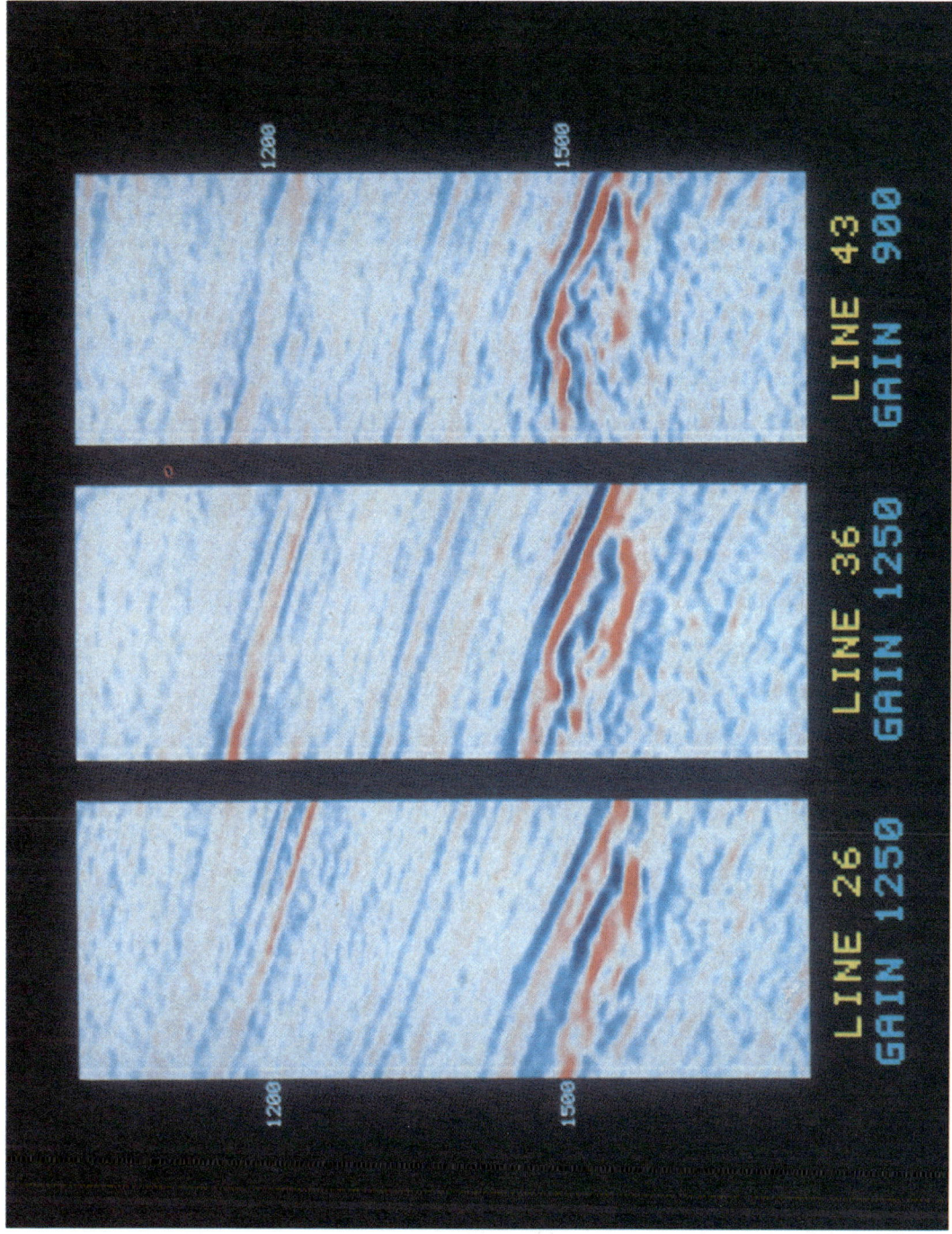

FIG. 28. Amplitude balancing exercise to estimate reduction in amplitude at well A-3 caused by a one-sided migration aperture.

coarse in the crossline direction for proper imaging of some of the depositonal dips.

The eastern boundary of the study area abuts the gap in data coverage caused by the drilling platform. Lines 41 to 44 show a progressive decrease in general amplitude level as the gap is approached, because of the effects of a one-sided migration aperture. Well A-3 intersects the Upper and Lower Sands approximately on line 43. Hence the seismic net gas sand values read from the isopach maps at the location of well A-3 are reduced accordingly. Figure 28 illustrates the way in which this amplitude reduction was estimated. Lines 26 and 36 are displayed with the gain normally used for studying this data. The gain applied to line 43 was adjusted until the general amplitude level was judged the same. The ratio of the gain settings then gave the factor by which the amplitudes on line 43 near well A-3 need to be increased in order to compensate for the migration aperture effects. This balancing exercise was repeated with other starting gains; similar factors were obtained and the average factor was determined to be 1.36. The net gas sand values read from the isopach maps were 32 ft (Upper Sand) and 9 ft (Lower Sand); hence the compensated values become 44 ft (Upper Sand) and 12 ft (Lower Sand), as tabulated in Figure 27. Note that this amplitude edge effect has been compensated only in the tabulation of Figure 27. The effect is still present in the sections and maps of Figures 14–21 and 24–26.

Comparison of the final net gas sand values from the seismic isopach maps and from the well logs (Figure 27) reveals that five of the ten sand thickness measurements agree within 2 ft and all but two agree within their estimated uncertainty ranges. The tie at well 2 is greatly improved if the values for the Upper and Lower Sands are added together.

CONCLUSIONS

In Pleistocene sands of the Gulf of Mexico, seismic amplitude has been used as a calibration factor to determine the net thickness of the producible gas sands. This led to accurate mapping of highly varying sands over two intervals.

Success of the project depended upon the maintenance of a flow of experimental ideas made possible by use of an interactive interpretation system.

ACKNOWLEDGMENTS

This project required the implementation of many new interactive capabilities in order to test the new ideas generated during the project. We thank Tony Gerhardstein of Texas Instruments Incorporated for his continual and rapid response to our needs.

We also thank the joint participants in the 3-D survey for permission to publish the results of the project. The participants were Chevron U.S.A., Union Oil Company of California, CNG Producing Company, Phillips Petroleum Company, and ICI Delaware.

REFERENCES

Backus, M. M., and Chen, R. L., 1975, Flat spot exploration: Geophys. Prosp., v. 23, p. 533–577.

Brown, A. R., 1983, Structural interpretation from horizontal seismic sections: Geophysics, v. 48, p. 1179–1194.

Brown, A. R., Dahm, C. G., and Graebner, R. J., 1981, A stratigraphic case history using three-dimensional seismic data in the Gulf of Thailand: Geophys. Prosp., v. 29, p. 327–349.

Domenico, S. N., 1976, Effect of brine-gas mixture on velocity in an unconsolidated sand reservoir: Geophysics, v. 41, p. 882–894.

French, W. S., 1974, Two-dimensional and three-dimensional migration of model-experiment reflection profiles: Geophysics, v. 39, p. 265–277.

Gerhardstein, A. C., and Brown, A. R., 1984, Interactive interpretation of seismic data: Geophysics, v. 49, p. 353–363.

Meckel, L. D., Jr., and Nath, A. K., 1977, Geologic considerations for stratigraphic modeling and interpretation, in Seismic stratigraphy—Applications to hydrocarbon exploration: Am. Assoc. of Petr. Geol. Memoir 26, p. 417–438.

Neidell, N. S., and Poggiagliolmi, E., 1977, Stratigraphic modeling and interpretation—geophysical principles and techniques, in Seismic stratigraphy—Applications to hydrocarbon exploration: Am. Assoc. of Petr. Geol. Memoir 26, p. 389–416.

Schramm, M. W., Jr., Dedman, E. V., and Lindsey, J. P., 1977, Practical stratigraphic modeling and interpretation, in Seismic stratigraphy—Applications to hydrocarbon exploration: Am. Assoc. of Petr. Geol. Memoir 26, p. 477–502.

Widess, M. B., 1973, How thin is a thin bed?: Geophysics, v. 38, p. 1176–1180.

Interactive interpretation of seismic data

Anthony C. Gerhardstein* and Alistair R. Brown‡

ABSTRACT

Interactive interpretation is urgently needed to increase the productivity of the world's hard-pressed seismic interpreters. This paper describes the use of an interactive system that displays seismic data in color on a television screen. The system is easy to use and, by automatically managing the data base for the interpreter, permits him to spend a larger portion of his time on the thoughtful process of interpretation itself. A 3-D data volume can be studied in both vertical or horizontal section form. The system can equally well handle the irregular grid from a 2-D survey.

In the course of interpreting a section on the screen, the interpreter may manipulate portions of that section in a variety of ways. He may make composite displays of multiple pieces of data, track horizons in automatic or manual modes, and zoom portions of the data to any desired extent. An interpretation made on one section can later be viewed on other sections marked at the points of intersection. Data can also be flattened to aid in structural and stratigraphic interpretations. Working maps can be produced at any time to check the progress of the interpretation. Final maps can be smoothed and manipulated to yield isochron, isopach, and other map products.

INTRODUCTION

From the advent of digital recording and signal processing methods to the development of sophisticated wavelet processing and migration techniques, the detailed subsurface information available from seismic data has improved steadily. Yet data interpretation techniques have advanced slowly during this period. This has become especially apparent with the development of three-dimensional (3-D) seismic technology where the interpreter is presented not only with vastly increased quantities of data, but also with the need to view the data from different perspectives.

The established solution to the demands of 3-D seismic data is the horizontal or Seiscrop™ section and the Seiscrop Interpretation Table (Brown, 1983). This has already increased the productivity of seismic interpreters by providing them with a means for rapidly viewing the data and extracting information from the horizontal perspective. However, further productivity lift is needed, as is the extension to 2-D data, in order to exploit fully the improved coverage and resolution available in modern seismic data.

Seismic interpretation using interactive graphics terminals has been fashionable for several years (Nelson et al, 1981; Zaccagnino et al, 1982), but few systems manipulate the seismic data itself before the eyes of the interpreter. This paper describes an approach to interactive interpretation designed initially around the requirements of 3-D seismic data but with extensive applications to 2-D data. A color graphics system is used to display seismic data in section form, either vertical or horizontal. The system also provides the tools for managing large seismic data bases.

SYSTEM DESCRIPTION

The interactive interpretation system consists of a 32-bit minicomputer, a graphics processor, auxiliary data storage media, and data input and display devices (Figure 1). The host minicomputer provides an easy-to-use interface to the user and manages the seismic data bases. It also provides all the necessary signal processing functions needed and formats the trace data for display. Each host can support multiple active work stations. The graphics processor manages the displayed images of seismic data and associated graphics, adds color to the data, and displays the results on the high-resolution color monitors. The optional color hardcopy camera generates photographic images of the displays in either 8 × 10 inch or 35 mm formats. Each work station also uses one 300 Mbyte disk for storing the seismic data bases.

User interaction with the system is provided by a video display terminal and keyboard for status information and alphanumeric input and by a graphics tablet for menu selection and for interacting with the displayed images. The interpretation tasks are implemented as a series of independent software programs that allow the user to display and manipulate the data. Each task communicates with the display and seismic data base as necessary and all communicate with each other by means of a display data base that keeps track of the current status of the system. This modular design allows the system to

™Trademark of Geophysical Service Inc.

Presented at the 52nd Annual International SEG Meeting October 17, 1982, in Dallas. Manuscript received by the Editor March 17, 1983.
*Texas Instruments Incorporated, P.O. Box 226015, MS 238, Dallas, TX 75266.
‡Geophysical Service Inc., P.O. Box 225621, MS 3966, Dallas, TX 75265.
© 1984 Society of Exploration Geophysicists. All rights reserved.

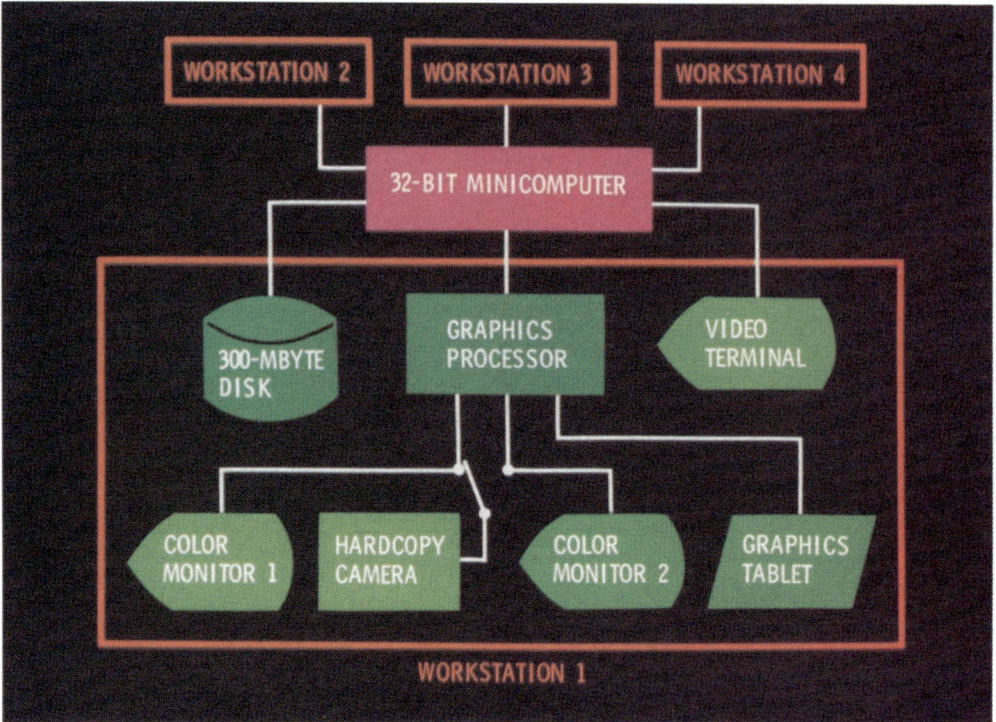

FIG. 1. Interactive interpretation system block diagram.

be extended easily by adding new interpretation tasks, and it facilitates maintenance of the system.

DISPLAY CAPABILITY

One of the major advantages of interactive interpretation is that it provides the interpreter with an extremely flexible means of viewing the seismic data. He can display each of the three principal perspectives of the data—line, crossline, and Seiscrop section—as best fits the situation. He can also create and display diagonal lines that connect wells or other locations of interest, and he can compose composite views of the data showing multiple perspectives or multiple lines for the purpose of tying loops.

Once the choice of persective has been made, the system provides the interpreter with a number of options to enhance the features of interest. He can select from multiple display formats. He can magnify all or a portion of the data to see increased detail. He can create and apply different color schemes to the data in order to highlight the features of interest. He can also fold and slide different regions of the data with respect to each other to study correlations. Many of these features are illustrated herein by data examples, all of which were photographed directly on the interactive interpretation system.

The normal format for viewing data interactively is the variable intensity format (Figure 2). In this mode the individual reflection amplitudes are color coded in the display. The standard color scheme represents the positive amplitudes in shades of blue and the negative amplitudes in shades of red. The continuous shading appearance results from the use of over 200 discrete color shades in the display. This variable intensity form of display provides a number of advantages over the more traditional methods of displaying seismic data. The greatly improved visual dynamic range over black-and-white sections is due to the eye's sensitivity to color. Also the use of different shades of color allows the peaks and troughs in the data to be represented equally on the same display. Hence faults show up more clearly due to the increased number of event terminations. Other color schemes with more contrasting colors can be used to highlight particular amplitude features in the data.

In addition to the variable intensity format, the system also provides the more traditional variable area wiggle trace mode for displaying section data (Figure 3) and also a dual polarity variable area mode. The variable area wiggle trace format is useful for studying trace character. However, since the traces are actually plotted on the screen rather than color coded, the amplitude resolution in this mode depends upon the screen width used in displaying each trace. Hence the variable area wiggle trace mode places rather severe demands on the limited

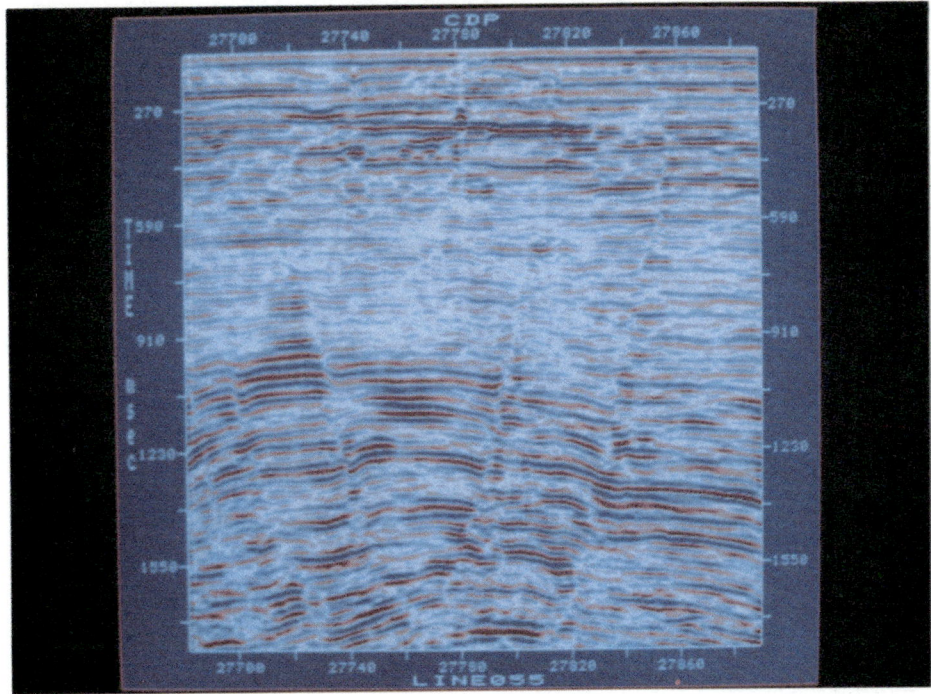

FIG. 2. Vertical section from the Gulf of Thailand displayed in the variable intensity format.

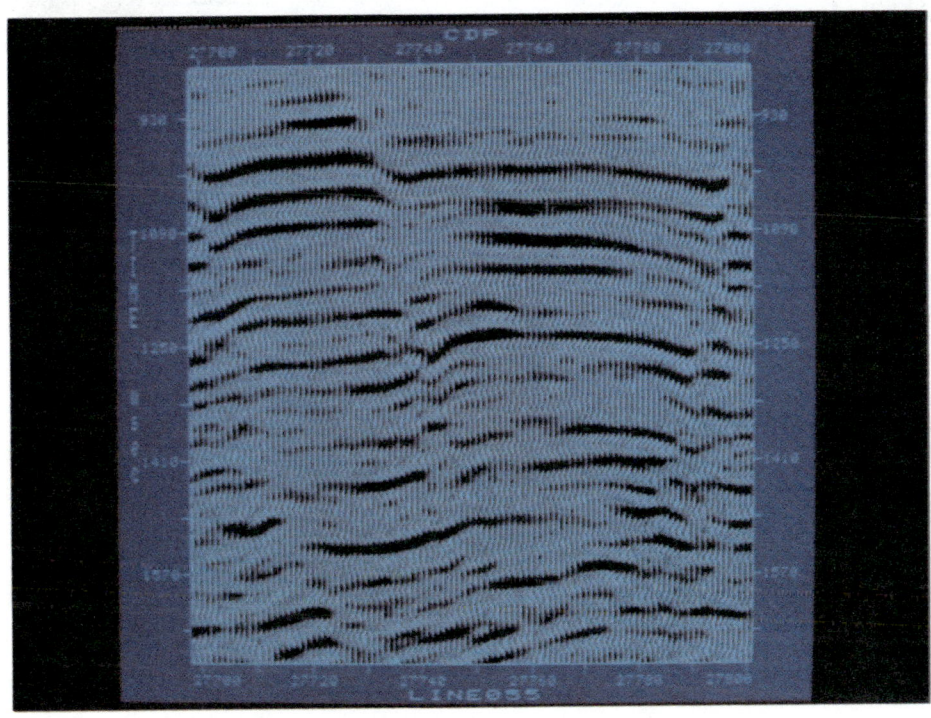

FIG. 3. Vertical section from the Gulf of Thailand displayed in the variable area wiggle trace format.

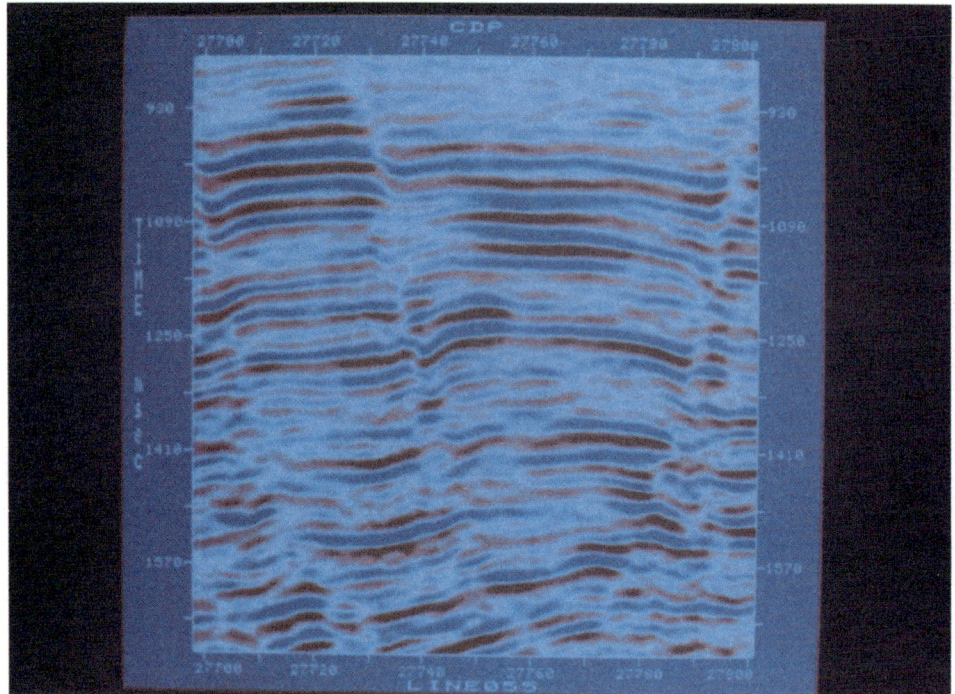

FIG. 4. Portion of vertical section shown in Figure 2 magnified by trace interpolation.

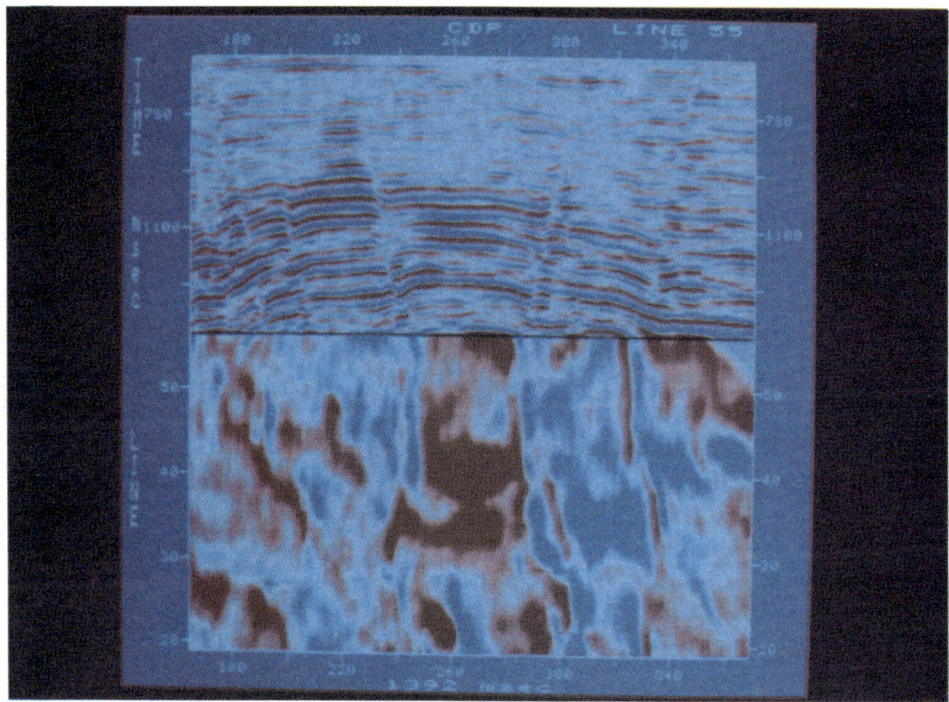

FIG. 5. Composite image of a vertical section and a Seiscrop section.

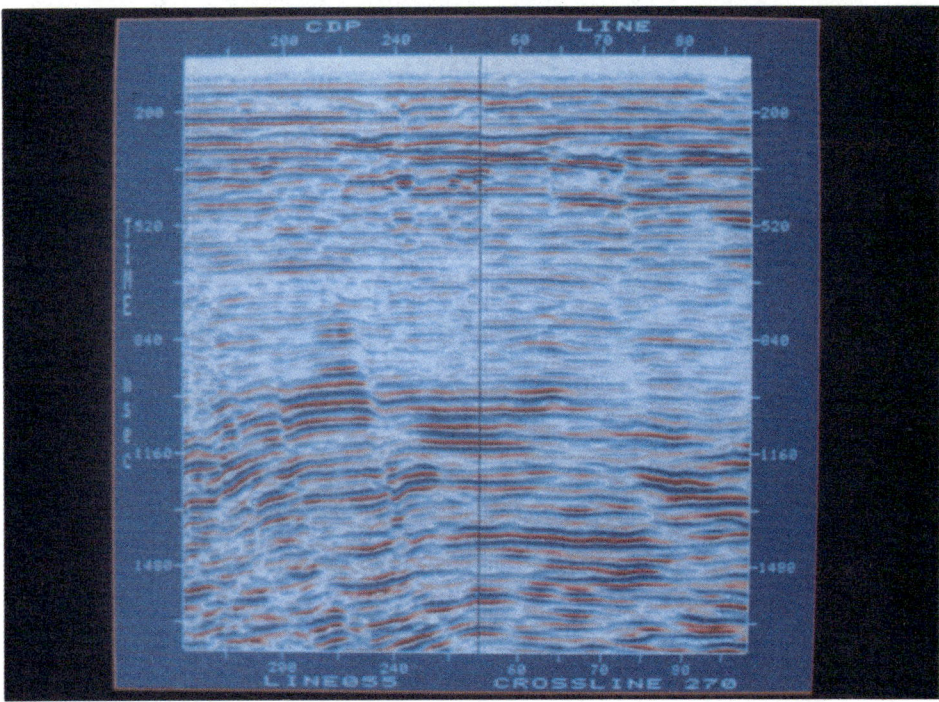

Fig. 6. Composite image of vertical sections from an intersecting line and crossline.

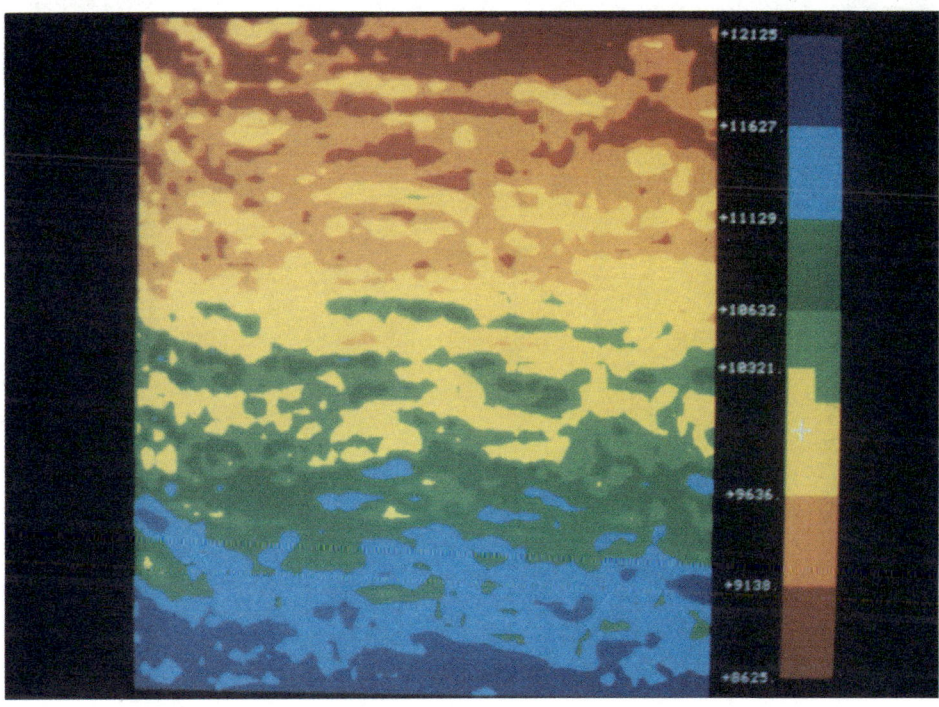

Fig. 7. G-LOG interval velocity section.

Fig. 8. Palette used for selecting color schemes.

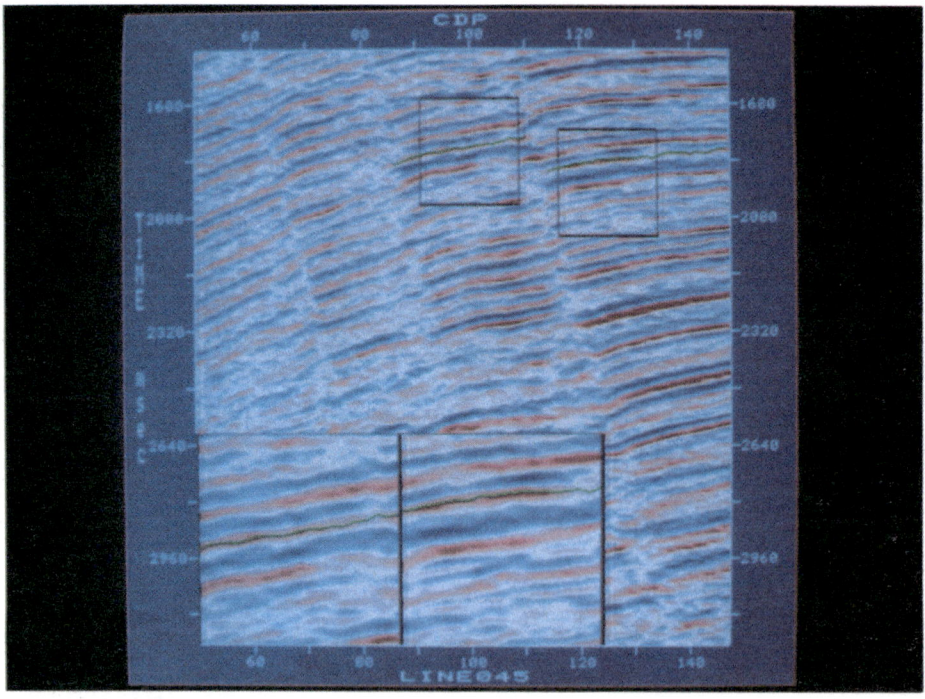

Fig. 9. Vertical section showing correlation across a fault assessed in magnified windows.

size of the graphics screen and is only useful when displaying magnified data. In the dual polarity variable area display mode, the troughs are rectified and displayed in variable area in one color while the peaks are displayed in another. This has the advantage of displaying both polarities with equal weight but suffers the same amplitude resolution restrictions as the variable area wiggle trace mode.

The displayed data can also be magnified to show more detail in a particular region (Figure 4). The magnification is accomplished by trace interpolation and does not show the blockiness typical of the zoom function on most graphic systems. Data can be magnified by any integer factor, and the magnification is independently controllable in the horizontal and vertical directions. This allows horizontal images such as Seiscrop sections and maps to be shown in correct perspective. The data can also be panned to center the region of interest.

Composite views of the seismic data can aid in visualizing the three-dimensionality of the subsurface and in determining the areal extent of features seen on an individual section. Tying a vertical section and a Seiscrop section together (Figure 5) aids in fault understanding. Tying two intersecting vertical sections (Figure 6) allows the user to see quickly the extent of features recognized on one perspective and aids in loop tying. If the intersecting lines are 2-D lines, one of them can be slid vertically with respect to the other to assess their correlation and measure the mistie. Another useful example of this technique is in correlating a seismic section with well data. Synthetic traces or logs can be inserted in the display at the well location and moved relative to the section to study the correlation.

The use of color in displaying seismic data is extremely effective in improving the perceptibility of subsurface features. However, color perception and taste vary from user to user. Therefore users may create their own color schemes, either gradational schemes (Figures 2, 4, 5, and 6) or contrasting ones (Figure 7) as suits the occasion. Colors are selected from a palette (Figure 8) using the graphics pointer and are then placed in a color bar beneath it. Once the color selection is completed, it can be named and saved for later use. Users can build entirely new color schemes or modify old ones by adding color levels, deleting color levels, or modifying them. Additional colors not shown on the palette are also available for special applications.

Once a color selection has been made, it can be applied to the displayed data and evaluated. Figure 7 shows a colored G-LOG* interval velocity section. The velocity range associated with each color is annotated next to the color bar. A contrasting set of colors is very helpful in delineating regions (in this case of interval velocity) in the data. However, the sharp boundaries between colors imply sharp amplitude contrasts that may or may not be supported by the data (Knobloch, 1982). To test this, users can select a particular color level on the color bar with the graphics pointer and adjust its boundaries either upward or downward while observing the effect on the displayed image. As the boundary is adjusted, the new value at that boundary is annotated next to the color bar as shown in Figure 7. In this way individual color levels can be expanded or contracted or simply slid through the data. Sliding the colors in effect applies a movable band-pass filter to the data. This technique has been found very useful for assigning a color to delineate precisely a particular amplitude, velocity, or other attribute range in the data for studying particular subsurface features such as bright spots. Another use is for interactively coloring contour maps.

INTERPRETATION OF A SECTION

The system provides automatic and manual tracking. Usually the interpreter starts out tracking in the automatic mode by naming the horizon, pointing to a starting location, and indicating the range of traces to be tracked. The starting and ending locations are selected by moving the location crosshair on the screen with the graphics pointer. The location of the crosshair and the reflection amplitude at that point are continuously displayed in a corner of the screen using the user's coordinate system. During tracking, the picked event is displayed on the screen overlying the data as each trace is tracked and is also stored in the data base for later use. The user can restart the tracking after a fault or obvious problem area. The manual tracking mode is used for editing the automatically picked event and for tracking faults or events in poor data areas that are difficult or impossible to track automatically. In this mode, the user picks control points on the screen with the location crosshair and the system interpolates linearly between them. Both automatic and manual tracking can be repeated until the interpretation is acceptable to the user.

Multiple views of a single section can be useful in interpreting across problem areas in the data. A tool known as a magnifying glass is used to create a magnified image of a portion of the data. The portion under study is outlined with a rectangular frame on the screen and a magnified display of the outlined region is shown in a corner of the screen (Figure 9). As the frame is moved across the screen with the graphics pointer, the magnified image is redisplayed showing the current location. Two such magnifying glasses are used to study correlations across faults or areas of poor data quality. By positioning one of the magnifying glasses to the left of a fault and another to the right, the data on each side are magnified and juxtaposed for comparison (Figure 9). Sliding one of the glasses or frames vertically with respect to the other will then bring the two portions into correlation. The shift that has been applied to the data can then be measured by the location crosshair.

Once a horizon has been tracked, the section can immediately be flattened. This provides another tool for checking the consistency of fault interpretations and for studying paleostructure. Figure 10 shows the result of flattening the faulted horizon correlated in Figure 9. Notice that the fault traces in the center which were not tracked appear unshifted in the flattening process. An alternate strategy is to track the fault as dip. Then the flattening process will apply a smooth correction across the fault.

Figure 11 shows another example of the type of fault correlations that can be performed interactively. In this case, magnified displays have been set up to correlate across the three faults in the center of the section. The four windows have been slid vertically to line up the event under interpretation. Notice that on the two outside faults, the correlation across the fault extends throughout the correlation window. However the correlation on the middle fault breaks after a short time indicating growth along the middle fault but not on the outside ones.

INTERPRETATION OF A SURVEY

Direct contouring from Seiscrop sections is a fast, convenient, and effective approach to structural interpretation of

---

*Trademark of Geophysical Service Inc.

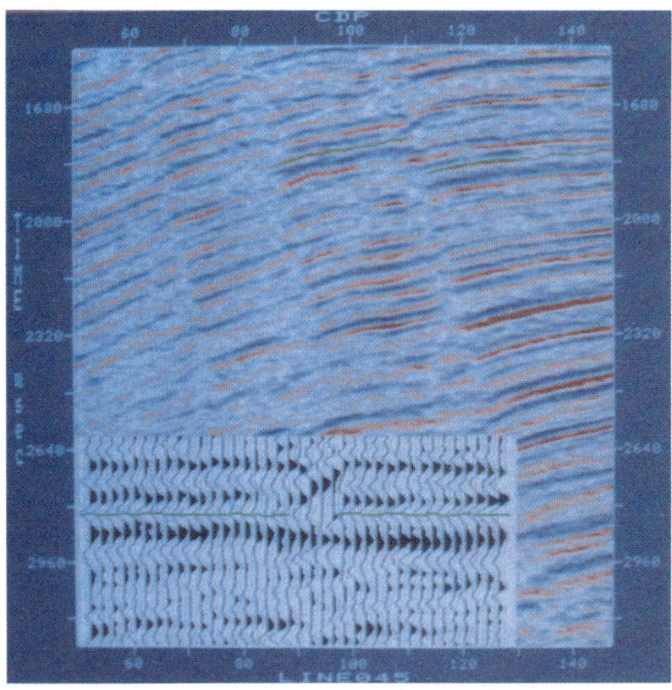

FIG. 10. Same as Figure 9 but with magnified data flattened on tracked horizon.

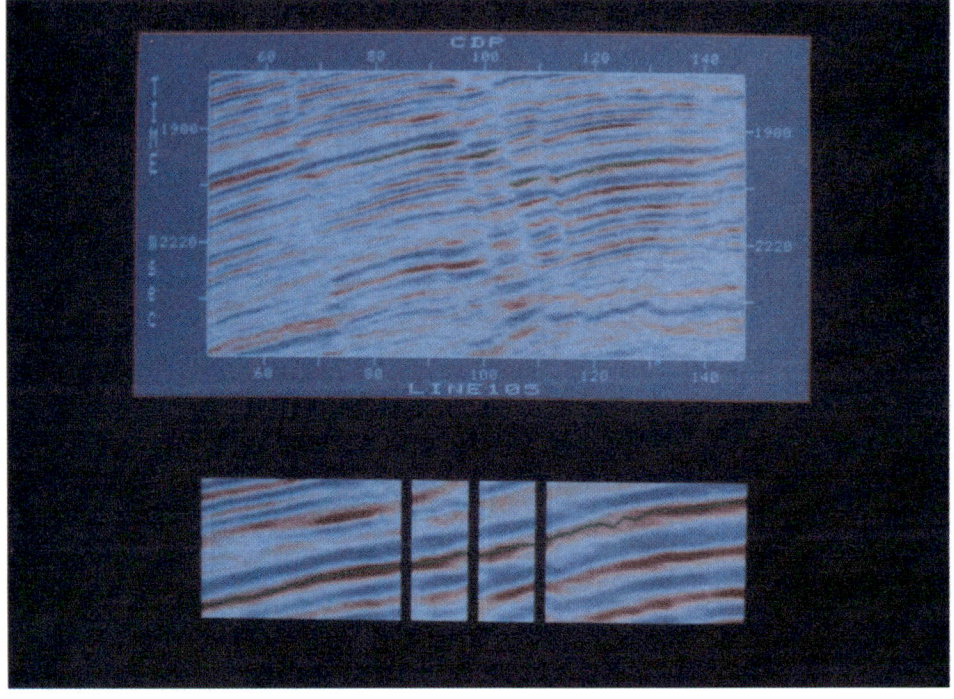

FIG. 11. Correlation across multiple faults using four magnified windows, one in each fault block.

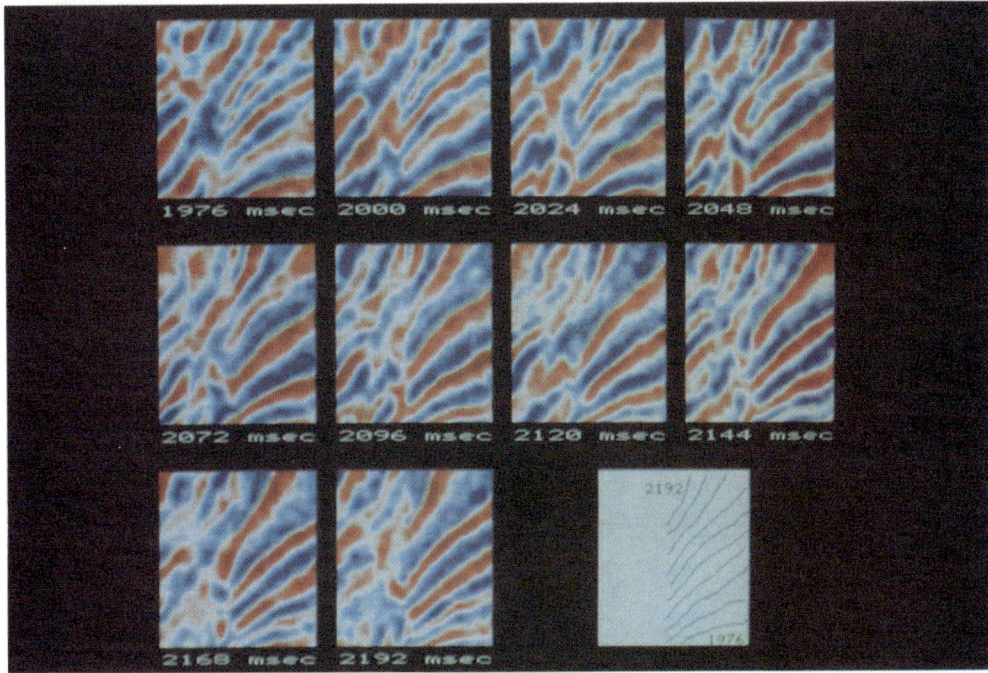

FIG. 12. Direct contouring from Seiscrop sections by automatic tracking.

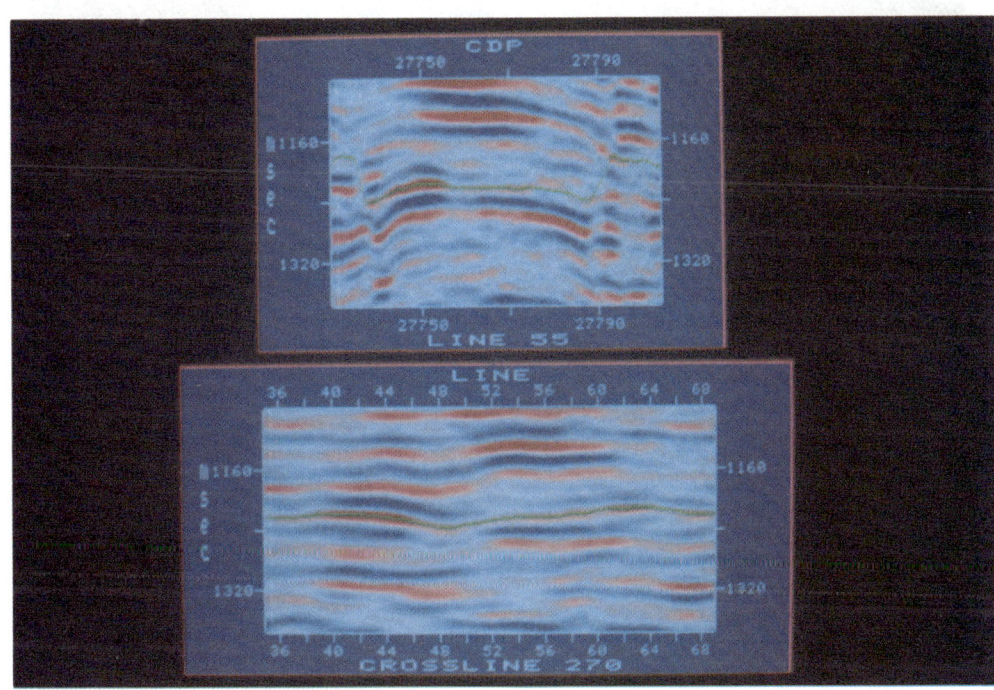

FIG. 13. Horizon tracked on a line and intersecting crossline.

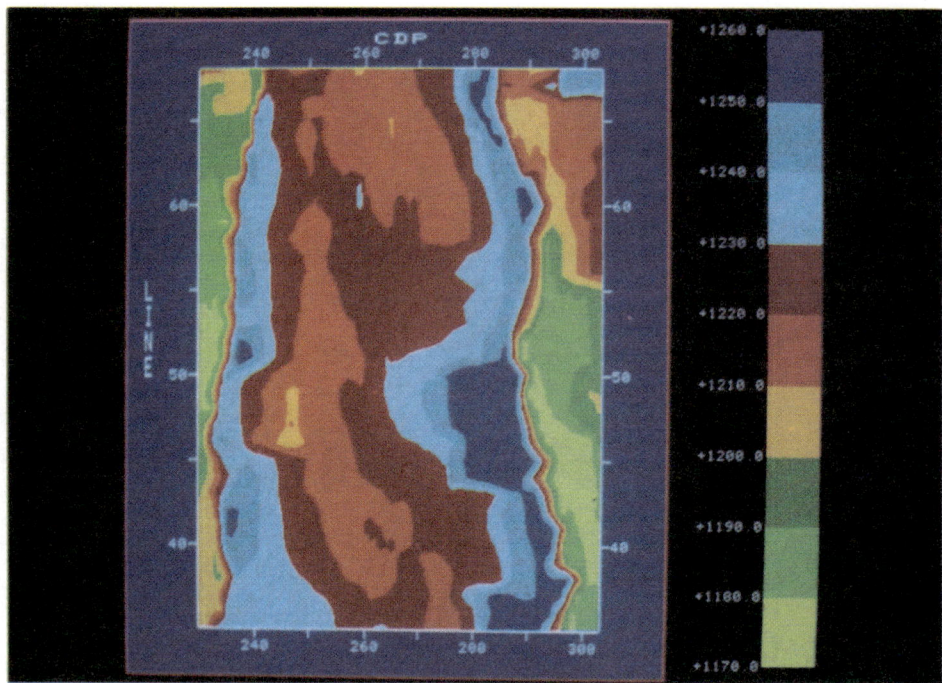

Fig. 14. Time structure map color contoured at 10 msec intervals for horizon tracked in Figure 13.

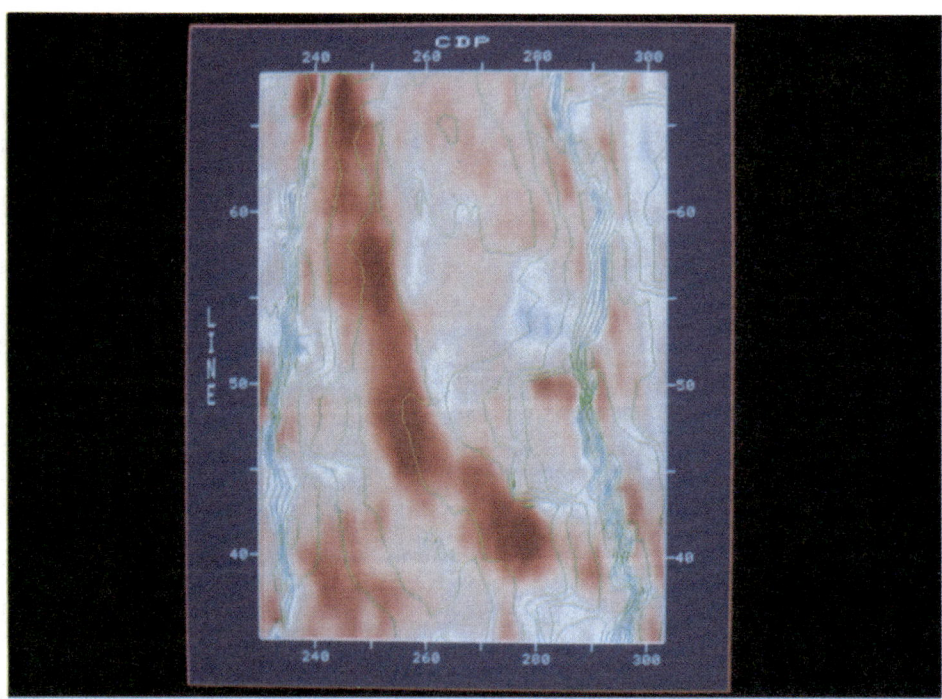

Fig. 15. Horizon Seiscrop section overlaid with 10 msec structural contours for horizon tracked in Figure 13 and mapped in Figure 14.

3-D data (Brown, 1983). Figure 12 illustrates direct contouring by automatic tracking on ten Seiscrop sections at 24 msec intervals. The track may be seen following the same event on each section, and then each track appears unaltered as part of the contour map at the lower right of the figure.

In interpreting from section to section, horizon picks from one line can be overlain on adjacent lines, helping in horizon identification. Also, once a horizon or fault has been tracked on one line, its position on intersecting sections is automatically stored in the data base. Then when one of these intersecting lines is displayed, the interpretation from the previous section is marked on the screen and serves as a starting point for the interpretation of that section. This facilitates loop ties and allows the user to build an interpretation quickly using the most advantageous perspective. Figure 13 shows a horizon that has been tracked on a portion of a line and crossline in a 3-D survey in order to interpret a particular subsurface feature.

As the event is being tracked, a time structure map of the tracked portion can be displayed as another check on the progress and consistency of the interpretation. Figure 14 shows a contour map of the time structure of the horizon that has been tracked in Figure 13. Each color band represents a 10 msec time interval using the color scheme shown on the right. As with the interval velocity section in Figure 7, the color boundaries can be interactively adjusted to vary the contouring interval. Alternatively, the time structure can be contoured with contour lines instead of the color bands, or with both. One method that has been found particularly useful is to contour with color bands at a course interval and to fill in with contour lines at a finer interval.

Another important and readily available product of the horizon interpretation and tracking is the horizon Seiscrop section (Figure 15). This section shows the amplitudes corresponding to the times in Figure 14. Brown et al (1982) defined a horizon Seiscrop section as a slice through a data volume parallel to a bedding plane and showed its utility for stratigraphic interpretation. Since the mapped horizon here is a trough, the horizon Seiscrop section is shown primarily in shades of red with regions of white and blue along the two fault lines intersecting the area. The deeper reds indicate the stronger amplitudes. The curvilinear high-amplitude feature has been interpreted as a sand-filled channel. Once the horizon Seiscrop section has been displayed, the time structure contours can be overlain on it to study the relationship between the reflection amplitude and the structure. These are shown as the green lines in Figure 15 and represent 10 msec contours.

Interpreted maps and horizon Seiscrop sections can be manipulated in a variety of ways to conclude the interpretive process. Either may be smoothed by a 2-D filter selected by the user. Two time maps may be subtracted to yield an isochron map, and depth and isopach maps may be made by multiplication using a suitable velocity field. The interrelationship of any two maps may be studied by addition, subtraction, or superposition.

CONCLUSION

The interactive system described here provides an excellent tool for managing the ever increasing volume of data in seismic surveys and serves to decrease the time required to interpret the data. The interactive display capability permits the interpreter to view the data very readily from different perspectives and in different display modes in order to represent best the features of interest for study. The mapping capability provides a means for generating subsurface maps during the interpretation process itself so that the working map becomes a useful interpretation tool. Limited use of the system described here by interpreters has already provided a significant productivity lift by permitting them to concentrate more on the thoughtful process of interpretation and less on the mechanics of data management. An interactive interpretation system will be used by most seismic interpreters in a few years.

ACKNOWLEDGMENTS

We gratefully acknowledge the substantial contributions made to the development of the interpretation system by Oscar Teoh and Avudh Ploysongsang. We also sincerely appreciate the efforts of Keith Burkart in the preparation of the interpretation examples.

REFERENCES

Brown, A. R., 1983, Structural interpretation from horizontal seismic sections: Geophysics, v. 48, p. 1179–1194.
Brown, A. R., Graebner, R. J., and Dahm, C. G., 1982, Use of horizontal seismic sections to identify subtle traps, in The deliberate search for the subtle trap: Am. Assoc. of Petr. Geol. Memoir 32, p. 47–56.
Knobloch, C., 1982, Pitfalls and merits of interpreting color displays of geophysical data: Presented at the 52nd Annual International SEG Meeting October 17, in Dallas.
Nelson, H. R., Jr., Hilterman, F. J., and Gardner, G. H. F., 1981, Introduction to interactive 3D interpretation: Oil and Gas J., v. 79, no. 40, p. 106–125.
Zaccagnino, P. A., Arens, G., and Meffre, J-F., 1982, Candice: The key for 3-D seismic interactive interpretation: Presented at the 52nd Annual International SEG Meeting October 17, in Dallas.

STRATIGRAPHIC INTERPRETATION

Inferring Stratigraphy from Seismic Data[1]

R. E. SHERIFF[2]
Houston, Texas 77036

Abstract The conventional application of seismic data to mapping depth and attitude of reflecting interfaces has been supplemented in recent years by measurements of velocity and amplitude for stratigraphic and lithologic information. Data patterns and angular relations between events provide clues to lithology and geologic history. Such data can be interpreted in terms of silica-clastic/carbonate/evaporite lithology, competent or plastic rocks, rigid or compactible sediments, depositional processes and environment, direction of sediment source, periods of subsidence/growth/erosion, etc. Special attention is given to (1) resolution of events, (2) the measurement and interpretation of seismic velocity, (3) amplitude measurements, and (4) display, so as to help an interpreter grasp the interrelations of data elements. The interpreter must bring together imagination and geologic and geophysical knowledge to discover possibilities while avoiding pitfalls of untenable geologic conclusions or of misinterpretation of seismic data.

INTRODUCTION

The traditional objective of reflection-seismic interpretation has been the mapping of geologic structure without concluding very much about the stratigraphy. Emphasis has been on finding and mapping coherent primary reflections, that is, determining their arrival times and the differences in arrival times with location, and calculating the locations and dips of the interfaces associated with them.

In some cases one event could be followed with reasonable certainty and be interpreted as representing the same geologic horizon. The first successful reflection-seismograph work near Oklahoma City identified the distinctive reflection of the contact between the thick low-velocity Sylvan Shale and the thick high-velocity Viola Limestone. However, with long seismic wave trains, multiples, and poor standout of reflections from the background noise, details often were lost and one relied on phantom horizons and the concept that bedding and reflections are generally parallel. Usually reflections were not identified specifically with formations unless they could be tied to nearby well control.

Instrumental and conceptual improvements, such as higher fidelity seismic systems and the application of the concepts of information theory, permit stabilizing reflection wavelets, shortening their duration, and improving the standout and distinctiveness of separate reflection events. Record section displays, especially in variable-area mode and reduced size and with color overlay, permit us to comprehend more data simultaneously and thus see relations which formerly were not evident. Redundant recordings, especially common-depth-point techniques, allow us to remove many of the interferences which formerly obscured primary reflections. Seismic analysis and display techniques have become more quantitative and we now are able to see and measure features (Fig. 1) in addition to arrival time and dip (which for many years constituted almost the whole of seismic interpretation):

a. The measurement of differences in arrival time with shot-to-geophone distance (offset) allows the calculation of seismic velocities and sometimes can be interpreted in terms of gross lithology.

b. The measurement of differences in reflection amplitude (reflectivity) indicates the acoustic-impedance contrasts at interfaces and often can be used to identify gas reservoirs by "bright spots."

c. Angular relations between seismic events commonly are interpreted in terms of geologic history; such relations may indicate unconformities, periods of erosion, the thinning/thickening of intervals indicating periods of structural growth, the direction to the source of the sediments, rock units which flow under pressure compared to more competent units, etc.

d. Data patterns are interpreted in terms of the depositional situation to distinguish between sediments deposited in quiet widespread seas versus those in high-energy situations, deltas, continental deposition, etc.

By combining observations in a synergetic manner, the reliability of inferences about the lithology, stratigraphy, fluid content, etc., can be improved (Marr, 1971).

As the quality of seismic data has improved, it has become more evident that there is information in the fine details of seismic traces, although extracting the geologic significance may be beyond present capabilities in many instances, especially when ties to hard control data are lacking.

© Copyright 1976. The American Association of Geologists. All rights reserved.

[1]Manuscript received, March 19, 1975; accepted, October 15, 1975. Published with the permission of Chevron Oil Co. and Seiscom Delta Inc.

[2]Seiscom Delta Inc.

The help of colleagues at Chevron Oil Co. and Seiscom Delta Inc. who assisted with this manuscript is gratefully acknowledged.

Arrival time → Depth
Differences with location → Dip
Differences with offset → Velocity
Differences in amplitude → Reflectivity
Angular relations → Geologic history
Patterns → Depositional situation
Combinations → Gross lithology / Stratigraphy / Fluid content

FIG. 1—Geophysical observations point to geologic interpretations.

The realization that detail variations may have such significance involves the temptation to interpret them without regard to the associated limitations and in inappropriate situations. "Breakthroughs" (such as the "bright-spot" technique) often are followed by disillusionment because such techniques are misused rather than because they are not valid. Attributing stratigraphic significance to seismic details constitutes a high-risk, incompletely understood technique, but it is nonetheless valuable.

Resolution imposes fundamental limitations on seismic interpretation and is considered first followed by some facets of velocity and amplitude, two important aspects from which stratigraphic conclusions are drawn. Next, the importance of display is considered. Finally the types of inferences to be drawn from seismic data and some interpretation pitfalls are discussed.

RESOLUTION

Resolution is the ability to separate two features which are very close together. Resolution is expressed quantitatively as the minimum distance between two features so that it is evident that there are two features rather than one. For seismic work which is concerned mainly with resolution in depth, the definition can be rephrased as the minimum separation between reflection events so that we may attribute separate interfaces rather than merely a single interface. (Note that this question differs from the question, "How accurately can the depth of an event be determined?")

Figure 2 shows the seismic section for a model which simulates a pinchout. The wedge must be about a quarter-wavelength thick to be seen; the actual pinchout location can be found only by extrapolation. A quarter-wavelength is also about the resolvable limit for structural features, such as shown in Figure 3 for a model which simulates faults with different amounts of throw.

To convert the threshold value, a quarter-wavelength, to thickness requires the knowledge of velocity and frequency (Fig. 4). Velocity of shallow rock is usually in the 5,000-8,000-ft/sec (1,500-2,500 m/sec) range and shallow reflections are of relatively high dominant frequency, perhaps 50-100 Hz, so that a quarter-wavelength is of the order of 15-40 ft (5-12 m). However, velocity increases and frequency decreases with depth, and hence resolution deteriorates with depth. At 10,000-15,000-ft (3,000-4,500 m) depth, velocity may be in the 12,000-20,000 ft/sec (3,700-6,000 m/sec) range and dominant frequency around 20 Hz, so that a quarter wavelength is of the order of 150-250 ft (45-75 m). Thus, deep features have to be large to be detected.

The only variable affecting wavelength over which we have any control is the frequency. Resolution can be improved (1) by using sources rich in high frequencies and (2) by not discriminating against high frequencies in recording. However, the sources richest in high frequencies commonly are limited in power and also higher frequencies are attenuated more rapidly by the earth, so that a compromise has to be reached between resolving power and depth of penetration. Profiler sources have tended to emphasize high frequencies at the expense of penetration and full-fledged seismic surveying has tended to emphasize depth of penetration at the expense of resolving power. Techniques recently have helped close the gap between these two extremes (Lucas, 1974).

Recording techniques often discriminate against high frequencies. Sampling at 4-msec intervals requires the use of alias filters which strongly attenuate frequencies above 80 Hz so that finer sampling is necessary if high frequencies are to be retained. Large geophone groups and vertical stacking also act as low-pass filters. Small time shifts among the elements of a detector group or vertical stack, such as might result from variations in offset, source, geophone plant or streamer depth, delays in weathering or seafloor sediments, and the differences in arrival of a dipping reflection across a group length, cause the rejection of high frequencies. If such shifts are random with an RMS value of only 2 msec, a modest amount, the effect is attenuation of frequencies above 90 Hz. Improved resolution thus requires less filtering because of ground mixing, shorter group lengths than usually employed, and possibly recording each geophone individually.

The length of the seismic wavelet also has an adverse effect on resolvability. The natural filtering processes in the earth lengthen the seismic wavelet and help obscure small effects. Many of these filter effects, however, can be removed at

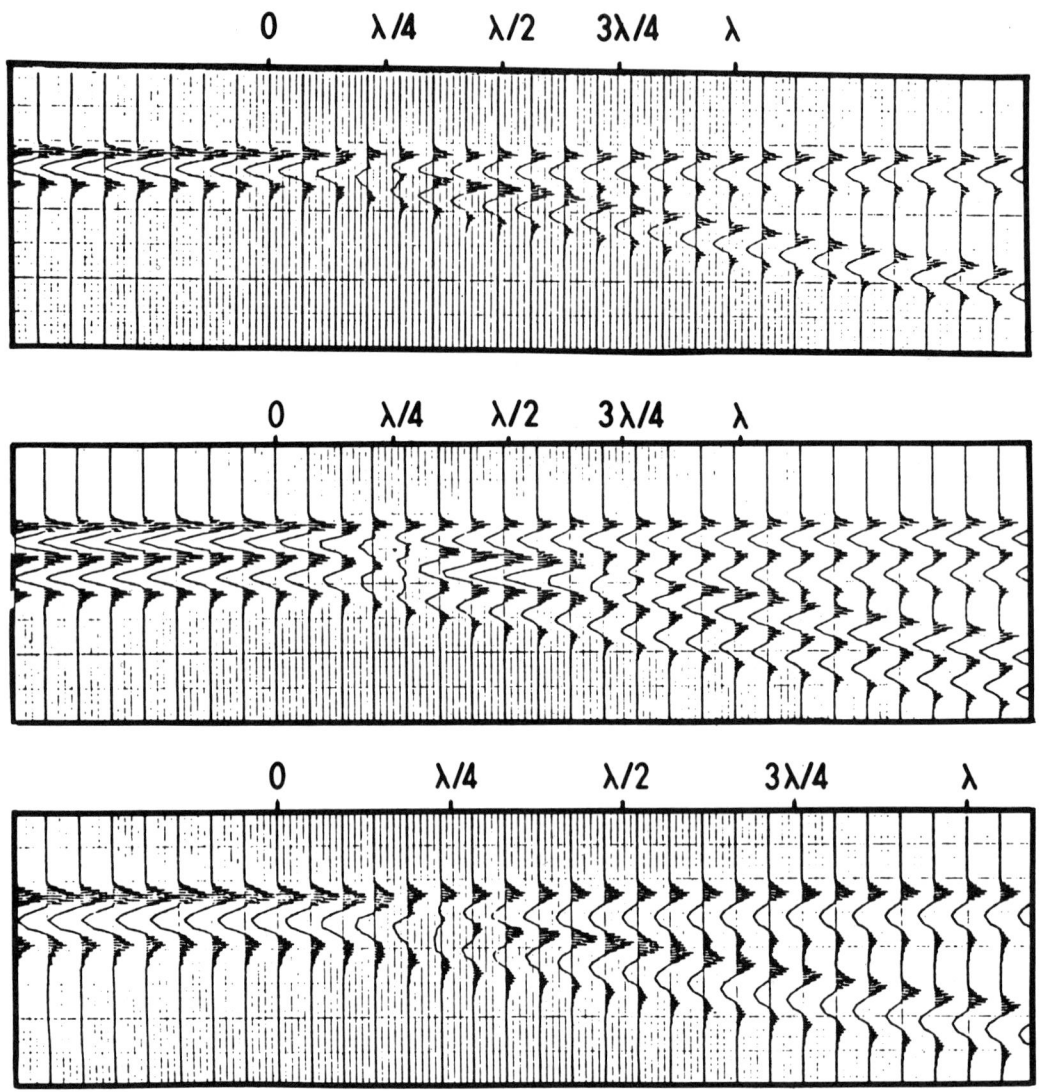

Fig. 2—Seismic evidence of pinchout. Wedge of velocity intermediate between that of rock above and below pinches out at location 0. Wedge angle is 2.7° (270 ft/mi or 51 m/km). Wedge thickness is indicated in wavelengths of the dominant frequency, f_d. (Upper) 1½-cycle wavelet, f_d = 50 Hz; (center) 2½-cycle wavelet, f_d = 50 Hz; (lower) 1½-cycle wavelet, f_d = 33 Hz.

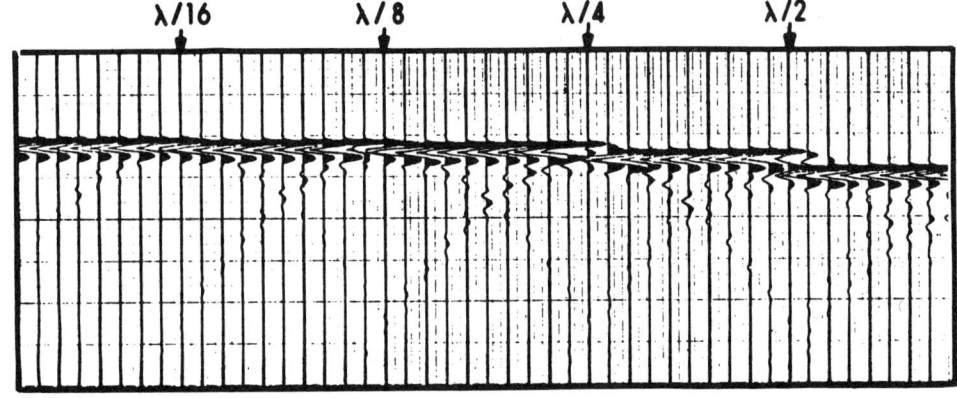

FIG. 3—Seismic evidence of faults with differing amounts of throw. Throw is measured in wavelengths of dominant frequency. At 2,000 m/sec and 50 Hz, $\lambda/4 = 10$ m at 5,000 m/sec and 20 Hz, $\lambda/4 = 62\frac{1}{2}$ m.

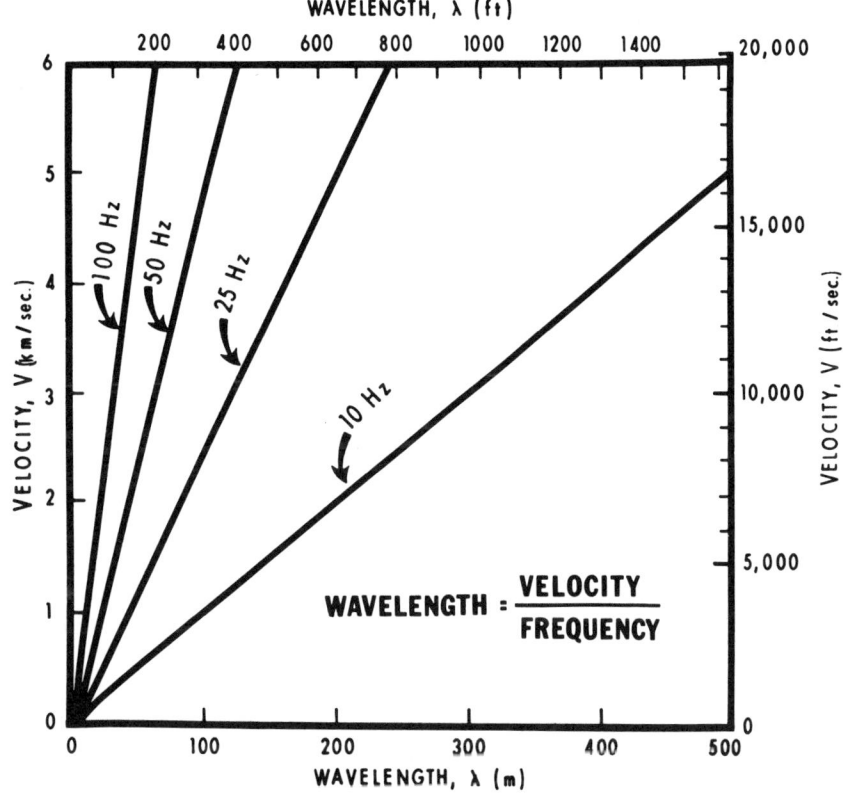

FIG. 4—Wavelength as function of velocity and frequency.

least in part by deconvolution, provided the required signal frequencies have not been attenuated so severely that they are lost irretrievably.

Velocity

Velocity values provide the basis for many attempts at lithologic identification from seismic data. The limitations on deriving velocity will be discussed before considerations as to what can be concluded from velocity measurements.

From the variation of arrival time with shot-to-geophone distance, a quantity called "stacking velocity" can be calculated (Fig. 5). Stacking velocity is used in adjusting the component traces in the common-depth-point stacking process. With good data, stacking velocities may be accurate to about 2 percent (Al-Chalabi, 1974). Decisions have to be made in the stacking process as to which stacking velocity values are to be honored, and this amounts to deciding which events on a seismic record are primary reflections. Such decisions often honor the highest velocities and assume that all events with lower stacking velocities are multiples. Although such a bias is right more than it is wrong, it is unreasonable to expect that it always is the right decision. Stacking velocity is often equated to "RMS velocity," which is nearly true if velocity layering is parallel and almost horizontal.

A quantity called "interval velocity" which is the average velocity between flat parallel interfaces can be calculated from high quality RMS velocity values for reflection events at the top and bottom of the interval by use of the Dix formula. The uncertainty in an interval-velocity calculation always is appreciably larger than in the RMS values from which it was calculated, becoming extremely large as the interval becomes very thin (Schneider, 1971). It is usually impractical to calculate interval velocity for less than 200-msec intervals because of excessive uncertainty. Assuming a velocity of 10,000 ft/sec (3,000 m/sec), this implies determining the average velocity over an interval 1,000 ft (305 m) thick. A change in velocity commonly has to be either exceptionally large or involve a large thickness to yield a definitive effect. Whereas interval velocities can be found in ways other than using the Dix formula (such as by ray tracing, Shah, 1973), this is done rarely on a routine basis because of the large amount of work involved.

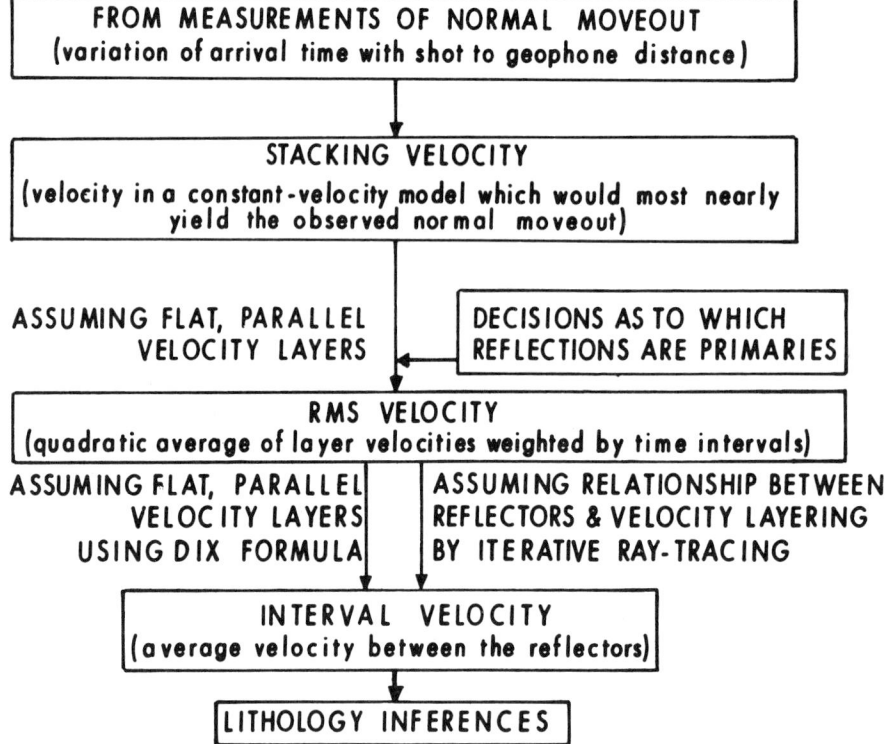

Fig. 5—Relation between different velocity terms used in seismic exploration.

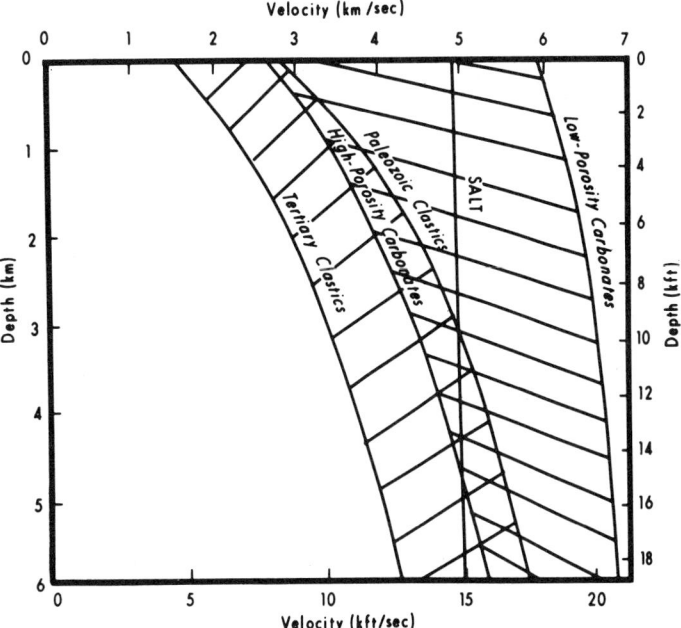

FIG. 6—Typical velocity-depth relations for clastic and carbonate rocks and salt. Younger rocks are likely to have lower velocities than older rocks because they tend to have higher porosity, be less cemented, and have undergone less deformation.

Hence, we ordinarily expect interval-velocity measurements to be meaningful only for thick beds in the case of nearly flat, parallel bedding where the data are of good quality (Taner et al, 1970; Everett, 1974). Where these conditions are not met, such calculations should be interpreted skeptically, bearing in mind that computer programs can generate values from seismic data whether they have significance in terms of actual velocities in the earth or not.

Now, assuming that we have meaningful velocity measurements, how can they be interpreted in terms of lithology? Figure 6 shows velocity-versus-depth ranges for clastic and carbonate rocks and salt. The range of values between different sandstones (or shales) at any given depth is much greater than the difference between an "average" sandstone and an "average" shale. The principal factor in determining the velocity of rocks which are matrices of particles is the porosity (Gardner et al, 1974), which can vary appreciably. The usual increase of velocity with depth is a result of the usual decrease of porosity with depth, but velocity variations are neither monotonic nor regular because porosity varies also for reasons other than depth, such as changes in the mixture of particle sizes which constitute the sand or the shale, and other reasons. Porosity also may have been lost irreversibly if the rock formerly was buried deeper (Jankowski, 1970). Variations of lithology, cementation, and other factors also are involved. The velocity in carbonate rocks also depends on the same factors. Some rock samples have velocities which fall outside the bands indicated on Figure 5. Although there is some overlap, the velocities of clastic and carbonate sections often are sufficiently different that one can use velocity to indicate gross lithology.

Although velocities in sandstone commonly are slightly higher than in shale, the differences may not be sufficient for identification of one or the other only on the basis of velocity values. The velocities in "average" sandstone and shales, usually defined by their SP-log response, are so close in some areas as to leave little hope that they can be identified from seismic-velocity measurements, even on a probabilistic basis. However, in other areas, they appear to be sufficiently different that one might be able to tell that one rock type or the other is more probable through reasonably thick intervals, that is, intervals where measurement uncertainties are not so large as to obscure the differences.

Whereas absolute-velocity values may not provide distinctive lithologic identification, systematic lateral variations which exceed the measure-

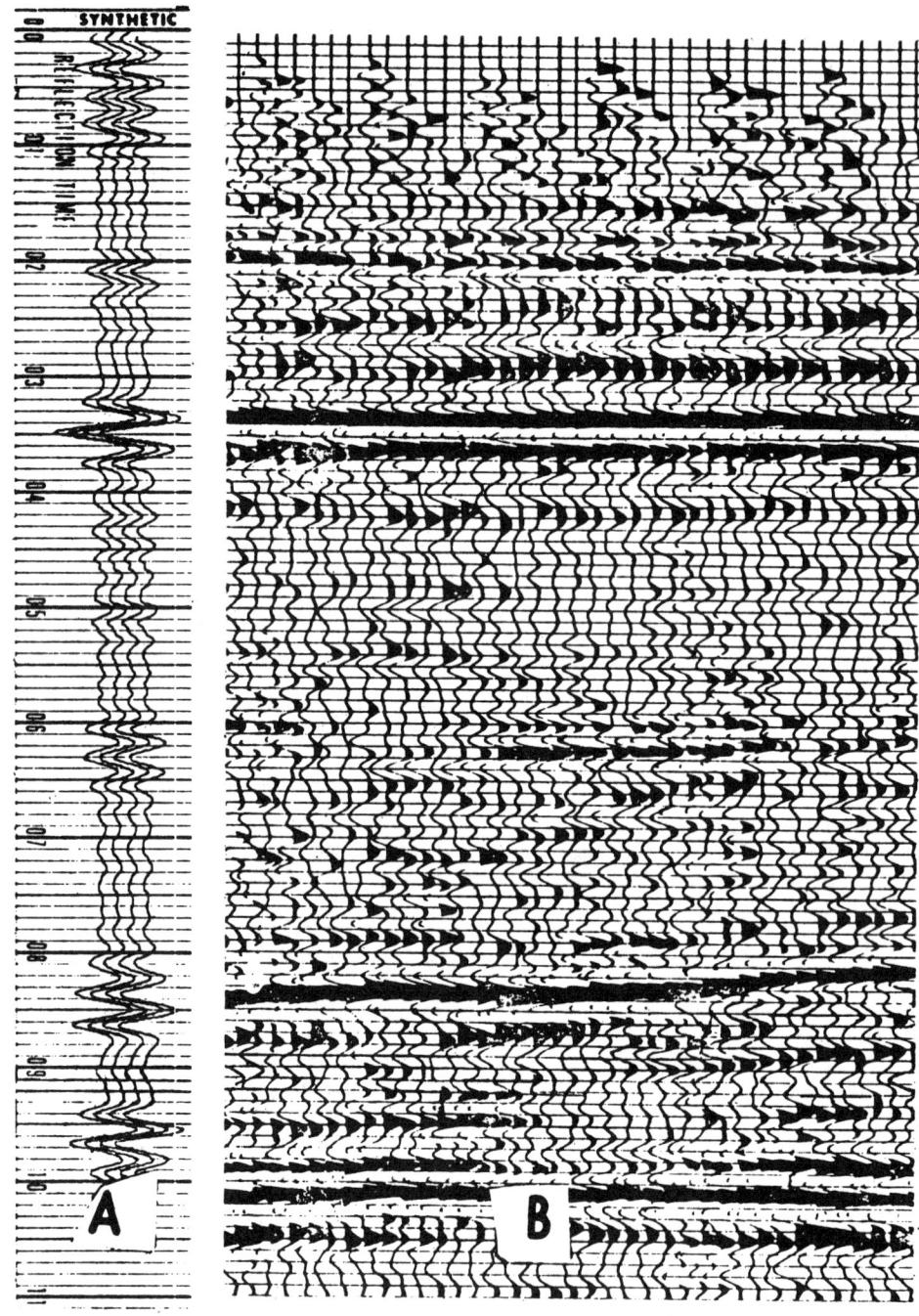

Fig. 7—Synthetic seismograms showing reef C and off-reef A records. Sonic-log data from which synthetic seismograms were made are shown in D plotted to time scale. B is actual seismic section across reef edge. Courtesy Digitech.

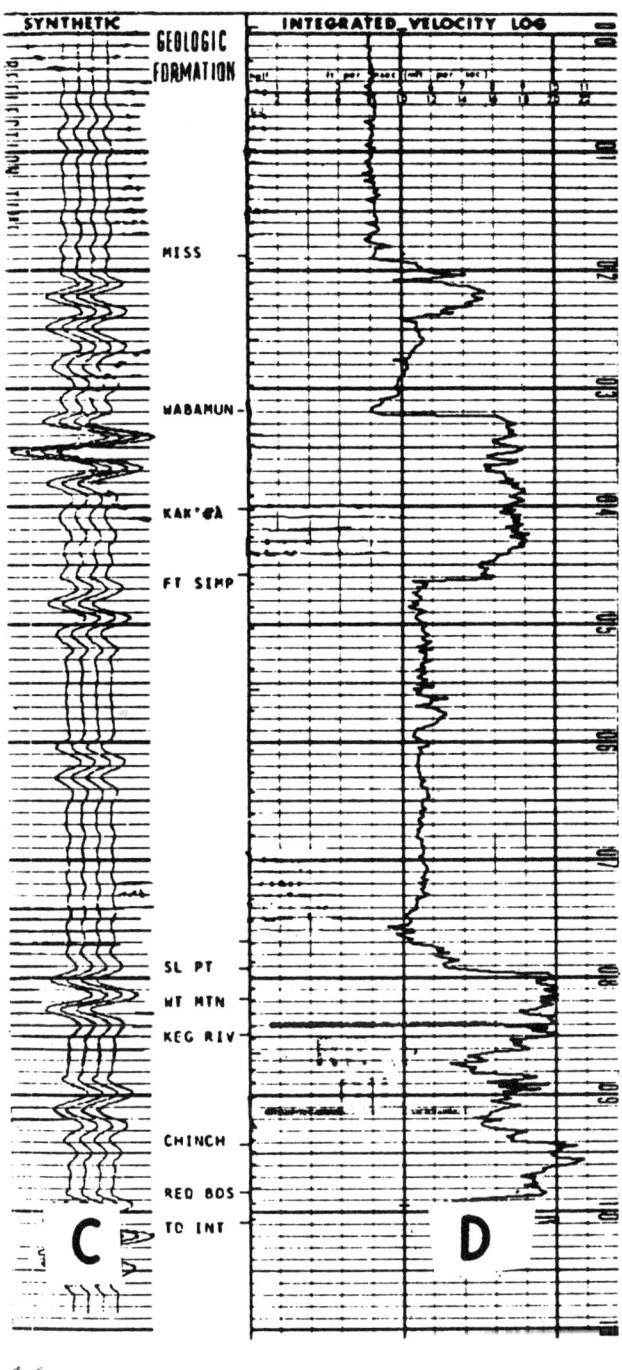

ment uncertainty may indicate systematic changes in lithology, hydrocarbons in the pore spaces, and other information. Thus, a lateral increase in the lime content of a sandstone might show as a systematic increase of velocity, or gas in a rock's pore space as a decrease of velocity (Domenico, 1974). The interpreter must have excluded nonsignificant causes, such as refraction at shallower velocity irregularities, of course.

One special case of importance is the overpressured formation. The effective pressure which determines the velocity in a rock is the difference between the overburden pressure (which is trying to squeeze out the rock's porosity) and the interstitial-fluid pressure (which is trying to hold the pore space open). Hence, an overpressured rock "feels" an effective pressure which is appropriate for a shallower depth and hence the rock has a velocity which is abnormally low for its actual depth of burial. If reflectors are present within or below the overpressured rock, it may be possible to measure the interval velocity accurately enough to predict the overpressured situation with confidence. Inasmuch as such situations usually involve impermeable shale, it is common to identify velocity inversions as massive shale despite the possibility of prospective sandstone within or below the shale.

Amplitudes

Although amplitude of a seismic reflection is a function of many factors other than the reflection coefficient of the reflecting interface (O'Doherty and Anstey, 1971; Sheriff, 1975), careful processing attempts to remove the effects of such factors. Assuming that this has been accomplished successfully, the amplitude of an isolated reflection can be related to the acoustic impedance (the product of velocity and density) of the rocks on either side of the interface from which the reflection occurs. The relation for geophones near the shotpoint is

$$\text{reflection coefficient} = \frac{\text{difference in acoustic impedance}}{2 \text{ (average acoustic impedance)}}$$

Strong reflections indicate large changes in acoustic impedance, such as at the boundaries of free gas in the pore spaces of an unconsolidated sand, clastic-carbonate rock interfaces or unconsolidated sediments lying on basement. Lateral amplitude variations convey information about changes in the acoustic impedance which may have stratigraphic significance. Where amplitude anomalies indicate gas accumulations, there is no assurance that gas is present in commercial quantities. Conditions other than the presence of gas also can cause amplitude anomalies, and gas can be present without giving distinctive amplitude anomalies.

An aspiration of geophysicists is the calculation of an "acoustic-impedance log" from measurements of reflection amplitude (Lindseth, 1975). To the extent that it may be possible to remove nonprimary reflection events and extract the true amplitude of primary reflections, it may be possible to achieve this goal. Conceptually, such information may be combined with accurate velocity information and thus separate out density variations. However, this cannot yet be achieved except under ideal circumstances.

Display

Interpretation requires bringing together different kinds of data and concepts so that interrelations can be seen and the ways in which data are displayed can help or hinder the perception of interrelations. The thickening/thinning of the interval between reflections can have vastly different meaning, whether it occurs on a regional scale or is associated with a structural or stratigraphic feature. A gradual thinning on a regional scale may indicate starvation, that less material was available for deposition as the sediment source became more distant. Thinning associated with a local structure may indicate that the structure was growing during the deposition of the sediments; such growth may have affected the sorting of the sediment components or provided a local source for the sediments and, thus, indicate facies variations. Thinning over a reef may indicate differential compaction and thus be an indication of the deeper feature.

The horizontal compression of a long line of regional seismic data often helps in revealing reflection patterns and systematic thickness variations which are not evident on sections at a more conventional scale. A section of small size thus may display data over a large distance, often with 5 to 20 times vertical exaggeration. Regional seismic lines selected so as to avoid structural features are apt to be more suitable for analysis for stratigraphic features. Of course sections involving tremendous distortion of structural relations may not be suitable for many interpretation purposes. A complete interpretation is thus apt to require several displays of the available data.

Combining different data or different aspects of data on the same display helps make interrelations apparent. Thus, it often is useful to plot a gravity profile on the seismic section, overlay well logs at the scale of the seismic section, or overlay seismic-velocity calculations. The display in color of measurements such as reflection strength, interval velocity, frequency, reflection polarity, and changes in reflection dip superimposed on the reflection pattern helps an interpreter realize the significance of interrelations. Isometric displays showing similar features on adjacent seismic lines assist in attributing a cause to the features. Looking at data in different ways is likely to make otherwise obscure interrelations readily apparent.

INFERRING STRATIGRAPHIC VARIATIONS NEAR SUBSURFACE CONTROL

A geophysicist familiar with the geology of an area sometimes can make remarkably accurate predictions about the lithology and stratigraphy, given high-quality seismic data and well control. Sometimes a "change in the character of an event" at one point on a record will indicate to an experienced geophysicist the presence of a reef, channel sand, salt solution, or some other feature, whereas a similar change at another arrival time will be dismissed as accidental. The prediction from the one and dismissal of the other is possible because the geophysicist knows that geologic conditions were right in the one instance for reef growth, channel sand deposition, or whatever, whereas they were not right in the other instance. He thus finds the subtle evidences (Langstroth, 1971) because he knows where to look and what to look for.

Synthetic seismograms are valuable in helping the geophysicist know what seismic evidence to look for as indicating a stratigraphic change (Harms and Tackenberg, 1972). Synthetic seismograms involve calculating from well-log data how a seismic record at that location should look (Fig. 7). If the synthetic is close enough in appearance to an actual seismic record, confidence develops in the assumptions involved and hence in calculations of the changes that variations in the rocks would produce. If another nearby well is present which has such changes and a connecting seismic line, it may be possible to determine exactly where the change occurs. Tying of seismic data with well information is probably the best method for finding stratigraphic traps.

INFERRING STRATIGRAPHY IN UNKNOWN AREAS

An entirely different type of problem is involved in inferring stratigraphy where geologic control is lacking. Under these conditions it is not possible to obtain the detail nor the accuracy achievable when using seismic data to extend well control. However, in such areas it is helpful to have even imprecise suggestions as to the stratigraphy of the area.

Some of the seismic elements useful in stratigraphic interpretation are listed in Figure 8. The

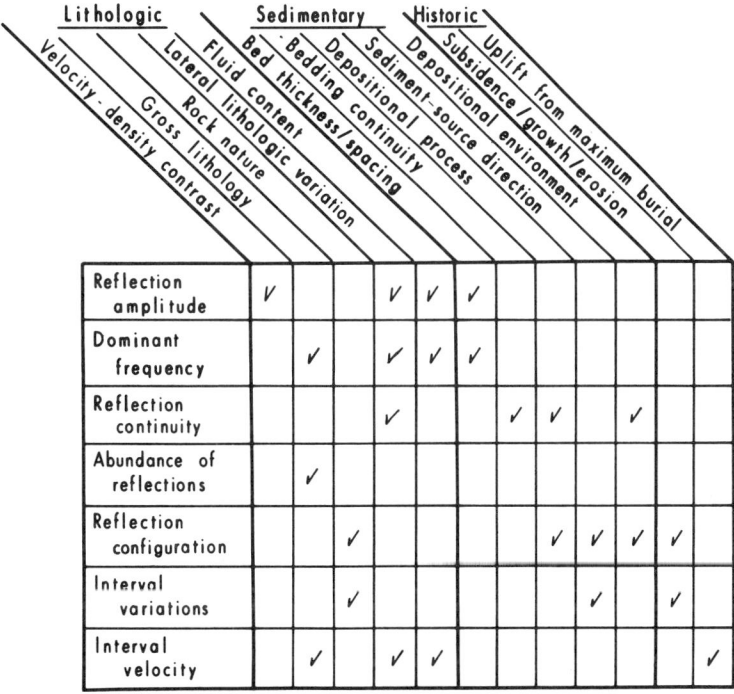

FIG. 8—Seismic elements useful in stratigraphic interpretation.

first and last of the seismic features in this figure, reflection amplitude and interval velocity, already have been discussed briefly.

The interference between successive reflections, which depends on the thickness of rock units, will affect the amplitude and character (or frequency content) of the composite reflection, but commonly in unobvious ways. Hence, if the amplitude or frequency changes laterally, it might be inferred that the bedding has changed, but it will be difficult to tell how it has changed.

Significant lateral frequency variations sometimes occur where reefs are present (Davis, 1972) or where other major lateral changes in the section occur (Balch, 1971), such as at a major fault. Marked vertical frequency variations may indicate major lithologic breaks as shown in Figure 9H or at the base of the topset beds in Figure 9D. Major abrupt lithology changes often give rise to characteristically low frequency reflections. A gas accumulation may attenuate high frequencies strongly and thus be indicated by changes in frequency content. However, displays of frequency have not proved very definitive by themselves except where tied to well control.

Continuous reflections commonly indicate widespread uniformity of depositional environment. Continuous reflections may be indicative of marine deposition, although a gentle unconformity also can result in a continuous reflection. Abundant reflections indicate frequent changes of lithology with depth. In clastic sections they may indicate interbedding of sandstone and shales. A general lack of distinctive reflections may indicate massive beds such as thick, relatively uniform shale, a thick carbonate sequence or basement.

Reflection configuration concerns the pattern of nonparallel reflections. Long regional (Fig. 10) lines that are not complicated by local structure are best for detecting many of the features from which stratigraphic conclusions can be drawn, but the behavior in structural areas also gives lithologic information. Assuming that appropriate allowances or corrections have been made for reflection migration, velocity variations, and nonprimary reflections, nonparallel reflections indicate the thickening/thinning or disappearance of lithologic units. These patterns may indicate the depositional process or environment. Thickening/thinning may indicate the direction of the source of the sediments. Thickening-thinning which is not systematic with location may indicate the presence of sediments capable of flowing, such as shale or salt, or possibly soluble sediments such as salt. Thickness variations in folds may distinguish competent from plastic rock units. Thickness variations also may indicate the presence of rigid rocks such as reef limestone in comparison with more compactible shale surrounding the reef. Reflections from gas-oil-water contacts ("flat spots") where the reflections from nearby lithologic interfaces are not flat are evidences of hydrocarbon accumulations. The interpretation of some reflection configuration patterns is shown in Figure 9.

PITFALLS

An interpreter needs a thorough understanding of geophysical principles and a knowledge of how the data have been recorded and processed because selecting these procedures and determining the parameters to be used prejudice subsequent interpretation. The more data that can be in-

FIG. 9—Examples of seismic-reflection configurations.

A, Migration of syncline with depth indicates movement of salt during deposition of sediments. Salt is now in dome at right end of section. Courtesy Seiscom-Delta.

B, Above dashed line all units thin to left, below it, to right indicating reversal in direction to source of sediments. Courtesy Seiscom-Delta.

C, Pattern of oblique reflections indicates progradation in ancient delta. Chances for clean sandstones are greater than in more uniform marine deposition above or below. Courtesy Chevron Oil Co.

D, Oblique pattern indicates outbuilding and upbuilding of shelf, culminating in present shelf edge. Undulations in data deeper than this prograding unit which are not present at shallower depths (pseudoweathering anomalies) suggest large lateral-velocity variations in prograding part and appreciable carbonate content. Courtesy United Geophysical.

E, Response of competent versus incompetent rock units to folding stresses shows in this migrated section. Courtesy Geocom.

F, Velocity pull-up under syncline between two salt lenses shows high-velocity rocks in syncline. Courtesy Western Geophysical.

G, Variable thickness and diffraction commonly are patterns characteristic of salt. Courtesy C.G.G.

H, Strong low-frequency reflection from massive carbonate rock overlain by clastic strata.

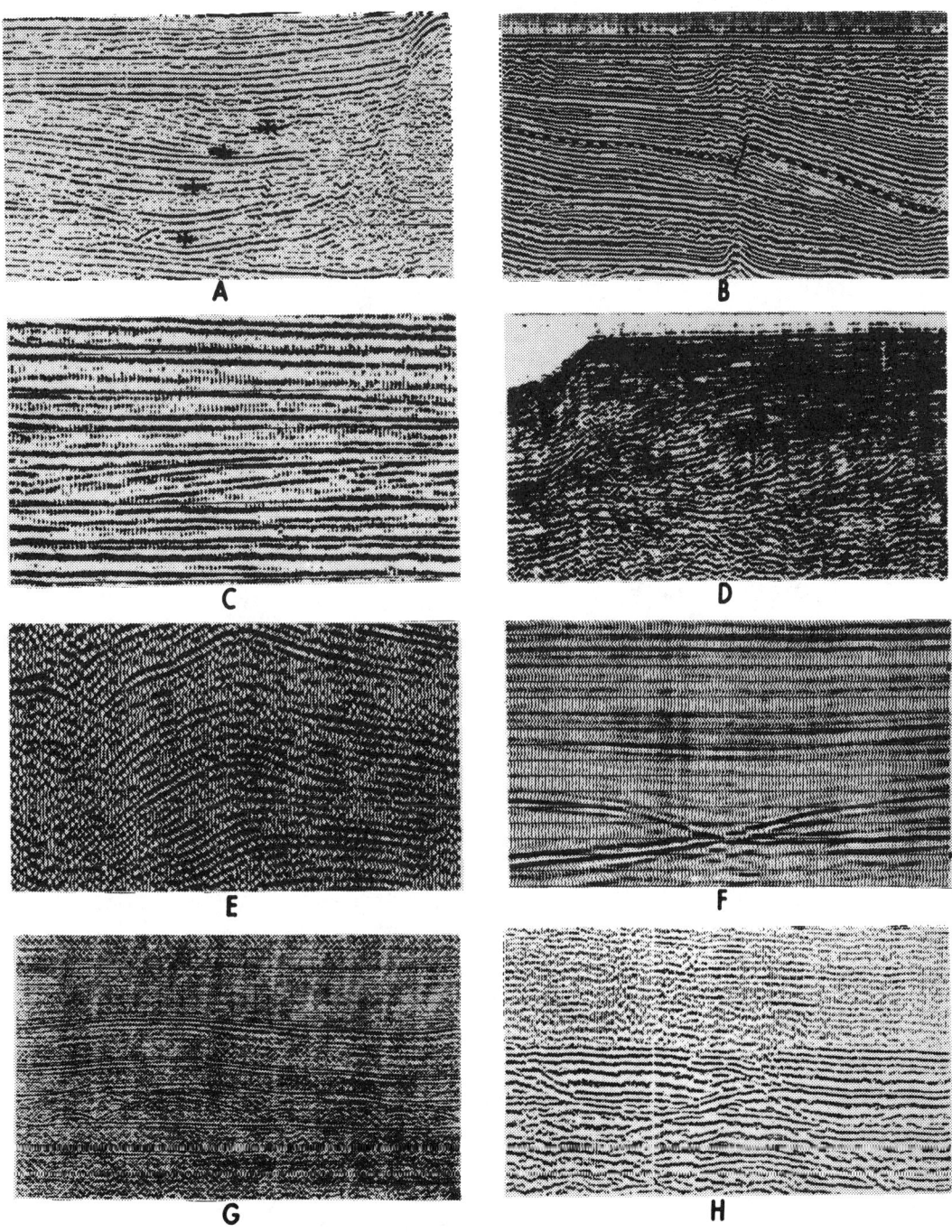

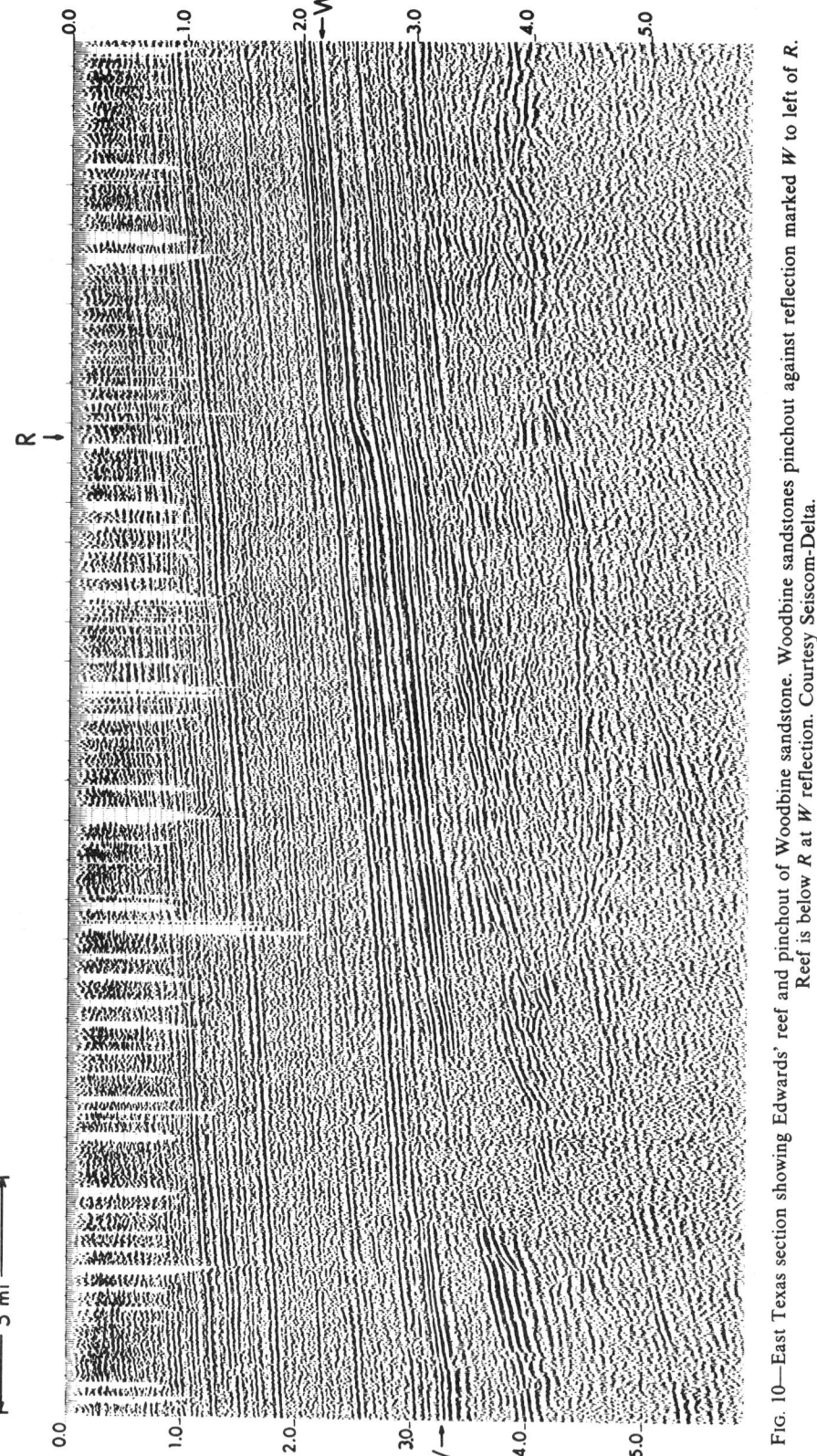

Fig. 10—East Texas section showing Edwards' reef and pinchout of Woodbine sandstone. Woodbine sandstones pinchout against reflection marked W to left of R. Reef is below R at W reflection. Courtesy Seiscom-Delta.

volved in interpretation, the less likelihood of misinterpretation.

Failure to allow correctly for reflection migration is probably the most serious cause of misinterpretation. Dipping reflections represent the attitude of reflectors which are somewhere else rather than under the observation point. Migration attempts to position reflection events under the surface location above the reflectors involved. Without migration, data may be superimposed to give the illusion of pinchouts or structure which are not actually present. Most migration procedures assume that strike is perpendicular to the line and thus simple automatic migration may not produce correct results. Most migration procedures also require a knowledge of the actual distribution of seismic velocity, which may not be obtainable in a simple way from velocity analyses (Dohr and Stiller, 1975), especially in structurally complex areas where migration is most important.

Near-surface anomalies, such as weathering channels, cause distortion throughout the section. These distortions are most evident when the same type of effect appears as a vertical alignment on the seismic section, but other, less obvious effects may be produced by refraction at dipping velocity interfaces, lenses of material of different velocity within the section, high-velocity material near the surface (such as permafrost), and other situations. Incorrect choices of stacking velocity may result in the misinterpretation of nonprimary reflections. Poor choice of processing parameters such as of the position of design gate or of filter length, may build up erroneous alignments or noise. Amplitude recovery may be biased by failing to allow for amplitude effects other than reflection-coefficient variations. Paging of profiler data may superimpose data from another shot or another arrival time. Multiple energy may obscure and confuse primary reflections. Many other opportunities for misinterpretation may be present. An interpreter who does not understand geophysics thoroughly may be tempted strongly to think of a seismic section as a geologic section without allowing for the fact that seismic energy can be of many types and can arrive from many directions.

The detection and prevention of errors in recording or processing data require care in the recording and processing operations, the careful examination of intermediate processing products, and a knowledge of the objectives. An interpreter should be suspicious of any conclusions which are not confirmed by adjacent seismic lines or other data. The main criterion for best interpretation of geophysical data is overall consistency.

Conclusions

Much stratigraphic information is contained in seismic data but extracting it requires knowledge and information on the part of the interpreter. Several "morals" can be drawn:

1. The more the interpreter knows about the geology, the more complete and accurate interpretations can be. A corollary is that a geophysicist cannot do a meaningful interpretation without knowledge of the geology.

2. Aberrations inherent in geophysical data or resulting from the methods of data acquisition or processing can be misinterpreted as having geologic significance. A corollary is that a geologist cannot do accurate interpretation without knowledge of geophysics.

3. Geologic and geophysical knowledge constrains an interpreter's imagination; he cannot postulate features that either geologic or geophysical data contradict.

References Cited

Al-Chalabi, M., 1974, An analysis of stacking, RMS, average, and interval velocities over a horizontally layered ground: Geophys. Prospecting, v. 22, p. 458-475.

Balch, A. H., 1971, Color sonagrams: a new dimension in seismic data interpretation: Geophysics, v. 36, p. 1074-1098.

Davis, T. L., 1972, Velocity variations around Leduc reefs, Alberta: Geophysics, v. 37, p. 584-604.

Dohr, G. P., and P. K. Stiller, 1975, Migration velocity determination. Pt. 2, Applications: Geophsyics, v. 40, p. 6-16.

Domenico, S. N., 1974, Effect of water saturation on seismic reflectivity of sand reservoirs encased in shale: Geophysics, v. 39, p. 759-769.

Everett, J. E., 1974, Obtaining interval velocities from stacking velocities when dipping horizons are included: Geophys. Prospecting, v. 22, p. 122-142.

Gardner, G. H. F., L. W. Gardner, and A. R. Gregory, 1974, Formation velocity and density—the diagnostic basics for stratigraphic traps: Geophysics, v. 39, p. 770-780.

Harms, J. C., and P. Tackenberg, 1972, Seismic signatures of sedimentation models: Geophysics, v. 37, p. 45-58.

Jankowski, W., 1970, Empirical investigation of some factors affecting elastic wave velocities in carbonate rocks, 1970: Geophys. Prospecting, v. 18, p. 103-118.

Langstroth, W. T., 1971, Seismic study along a portion of the Devonian salt front in North Dakota: Geophysics, v. 36, p. 330-338.

Larner, K. L., 1974, An overview of continuous velocity analysis: Paper presented at Continuing Education Seminar of Houston Geophys. Soc., December 1974.

Lindseth, R., 1975, Matching interpretation techniques to modern seismic reflection data: Paper presented at "New trends in seismic interpretation" seminar, Denver Geophys. Soc., April 1975.

Lucas, A. L., 1974, A high resolution marine seismic survey: Geophys. Prospecting, v. 22, p. 665-682.

Marr, J. D., 1971, Seismic stratigraphic exploration, pt. 1: Geophysics, v. 36, p. 311-329; pt. 2: Geophysics, v. 36, p. 533-553; pt. 3: Geophysics, v. 36, p. 676-689.

O'Doherty, R. F., and N. A. Anstey, 1971, Reflection on amplitudes: Geophys. Prospecting, v. 19, p. 430-458.

Sangree, J. B., and J. M. Widmier, 1974, Interpretation of depositional facies from seismic data: Paper presented at Ann. Mtg. Soc. Exploration Geophysicists, Dallas, November 1974.

Schneider, W. A., 1971, Developments in seismic data processing and analysis (1968-1970): Geophysics, v. 36, p. 1043-1073.

Shah, P. M., 1973, Use of wavefront curvature to relate seismic data with subsurface parameters: Geophysics, v. 38, p. 812-825.

Sheriff, R. E., 1975, Factors affecting seismic amplitudes: Geophys. Prospecting, v. 23, p. 125-138.

Taner, T., E. E. Cook, and N. S. Neidell, 1970, Limitations of the reflection seismic method; lessons from computer simulations: Geophysics, v. 35, p. 551-573.

OTC 3568

SEISMIC STRATIGRAPHY, A FUNDAMENTAL EXPLORATION TOOL

by G.R. Ramsayer, Exxon Production Research Co.

© Copyright 1979, Offshore Technology Conference

This paper was presented at the 11th Annual OTC in Houston, Tex., April 30-May 3, 1979. The material is subject to correction by the author. Permission to copy is restricted to an abstract of not more than 300 words.

ABSTRACT

The basic principle of seismic stratigraphy is that primary seismic reflections parallel stratal surfaces (bedding planes) rather than gross lithostratigraphic boundaries. This concept coupled with the Law of Superposition can be used to answer a number of fundamental stratigraphic questions which are critical to exploration. For example, unconformity surfaces can be located on the seismic section by the identification of reflection terminations. This permits the interpreter to recognize the major shifts and changes in deposition which have occurred during a basin's evolution and provides a geologically meaningful basis for the subdivision of the geologic section into genetically significant stratal units (depositional sequences). Because reflectors follow bedding planes, reflection geometry parallels depositional geometry. Therefore after palinspastic reconstruction, the internal reflection patterns of a sequence and their areal distribution provide valuable information for paleoenvironmental reconstruction. Once the distribution of depositional environments within an area is determined, facies predictions and estimates of possible reservoir rock content can be made. Finally, the seismic section can be used as a true chronostratigraphic correlation tool. This attribute is particularly helpful in areas where reliable well log correlations are difficult due to rapid lateral facies changes, sparse well control and/or insufficient paleontologic control. Ideally these correlations are made using synthetic seismograms carefully tied to the seismic section. However, in the absence of sonic and/or density logs, the reflection geometry of the seismic section can often give the interpreter a correlation style or concept that will permit proper correlations.

INTRODUCTION

The concept that primary seismic reflectors parallel bedding planes (time lines) and unconformities forms the basic principle of seismic stratigraphy and may be used to answer some fundamental

References and illustrations at end of paper

stratigraphic questions critical to exploration. This paper will demonstrate that applying this principle in conjunction with the Law of Superposition permits 1) the recognition of unconformity surfaces and the subdivision of the stratigraphic sections into genetically meaningful units, 2) the reconstruction of paleogeometry and therefore paleogeography and paleoenvironments, 3) true chronostratigraphic correlations, and 4) the recognition of a stratigraphic trap.

FUNDAMENTALS

Primary seismic reflectors are generated at physical surfaces having a velocity and/or density contrast. In a sedimentary section only two types of reflection-generating interfaces are present at the time of deposition - stratal surfaces and unconformities. Each of these has chronostratigraphic significance. Stratal surfaces are the depositional bedding planes which separate the episodes of sedimentation. They represent ancient depositional surfaces and are therefore geologically synchronous over the area of their occurrence (Vail et al. 1977). Although unconformities are not time synchronous, and in fact are usually quite variable in terms of the hiatus they represent, they also have chronostratigraphic significance in that strata above the unconformity are younger than strata below, i.e. an unconformity separates older rocks from younger rocks.

As pointed out by Vail et al. (1977), there is no physically continuous surface that follows the top of a time-transgressive lithostratigraphic boundary and therefore, within the resolution of the tool, no seismic reflector will follow such a transition. The lateral continuity of a seismic reflector cuts across time-transgressive lithostratigraphic boundaries with lateral changes in reflection amplitude and frequency occurring in response to the changing reflection coefficients and bed spacing.

Like all exploration tools, however, the seismic section has its resolution limits. The accuracy of chronostratigraphic correlations using the seismic section is dependent upon data quality and is, in practice, limited to ± 1/2 a wavelength. Non-primary seismic events such as multiples and coherent noise, and events not migrated to their proper position such as defractions and steeply dipping events, can mislead the interpreter if not properly identified.

SEISMIC RECOGNITION OF UNCONFORMITIES AND DEPOSITIONAL SEQUENCES

Given that primary seismic reflectors parallel stratal surfaces (time lines) and unconformities, we can begin attaching a chronostratigraphic significance to the four basic reflection termination patterns we commonly see on a seismic section: onlap, downlap, toplap and truncation (Figure 1.).

Onlap (Figure 1-A) occurs where "an initially horizontal stratum laps out against an initially inclined surface, or in which an initially inclined stratum laps out against a surface of greater inclination" (Mitchum et al., 1977, p. 57-58). This pattern is schematically represented in Figure 2-A which shows strata onlapping a preexisting surface. Applying the Law of Superposition, the chronostratigraphic relation depicted in Figure 2-A results. The onlapping pattern signifies a nondepositional hiatus of increasing duration as successively younger strata lap out against the preexisting surface.

Downlap (Figures 1-B and 2-B) occurs when "an initially inclined stratum terminates downdip against an initially horizontal or inclined surface" (Mitchum et al. 1977, p. 55). Again applying the Law of Superposition and our concepts of reflection chronostratigraphy, the chronostratigraphic significance of downlap is shown in Figure 2-B. Like onlap, downlap signifies a nondepositional hiatus of increasing duration as successively younger strata lap out against a preexisting surface.

Toplap (Figures 1-C and 2-C) is the lapout pattern in which "initially inclined stratum terminate against an overlying surface mainly as a result of nondeposition" (Mitchum 1977, p. 211). Typically this pattern results from sedimentary bypassing under conditions where local base level is too low to allow strata to deposit further updip. Minor erosion commonly is associated with toplap. As pointed out by Mitchum et al. (1977, p. 58), toplap is most often associated with shallow marine progradational environments such as deltas, but it can also be found in deep marine deposits where deep sea currents define a submarine depositional base level. Figure 2-C depicts the chronostratigraphic relationships typical of toplap.

Erosional truncation (Figure 1-D and 2-D) represents the lateral termination of strata by erosion. Where these strata are inclined, this pattern can often be difficult to distinguish from toplap.

Using these discordant reflection termination patterns, unconformity surfaces can be recognized on a seismic section. These can be carried into areas where the reflectors become concordant because strata become conformable or the unconformity becomes a paraconformity and is therefore not seismically recognizable. This is a powerful concept for it allows us to recognize the major shifts and changes in deposition which occur during the evolution of a basin. Looking at it slightly differently, we can say that the ability to seismically recognize unconformities allows us to define depositional sequences.

A depositional sequence is a stratigraphic unit composed of a succession of relatively conformable and genetically related strata which are bounded at their top and base by unconformities or their correlative conformities (Mitchum et al. 1977, p. 53). The advantage of interpreting a stratigraphic section within a sequence framework is that it objectively divides the section into units which represent periods of essentially continuous sedimentation and may therefore be interpreted in terms of the naturally occurring episodes of sedimentation which have filled a basin.

An example of unconformity recognition and sequence definition is shown on seismic line 1-A (Figures 3 and 4). This is a north-south trending dip line from the Polk County area of East Texas. Two unconformity surfaces can be recognized within the zone of interest between 1.9 and 2.1 seconds at the north end of line 1-A. The lower unconformity is characterized by downlap between shot points 2205 and 2280 and by onlap between shot point 2205 and the north end of the line. The upper unconformity surface is characterized by toplap beneath it between shot points 2205 and 2250 and onlap onto it from shot points 2250 to the south end of the line. From shot point 2180 to the north end of line 1-A the upper unconformity is concordant. Even without well or paleontologic control a generalized chronostratigraphic sketch along line 1-A can be drawn to show the relative age relationships within this sequence (Figure 5). This is accomplished by simply plotting the seismic reflectors (time lines) as horizontal lines of appropriate lateral extent in stratigraphic succession. No vertical scale (time scale) can be assigned to the chart without more data.

Using tie lines these unconformities and the sequence they define can be carried to lines 1 and 4 (Figures 6 and 7) where their character is similar.

SEISMIC FACIES ANALYSIS AND MAPPING

Seismic sequences define natural depositional units within which the primary reflectors parallel bedding planes and therefore parallel actual paleo-depositional surfaces. Thus the physiographic characteristics of the depositional environment under which sediments were deposited are displayed on the seismic section and can be used to reconstruct paleo-environments, predict facies and infer reservoir rock content. The process of describing and interpreting the seismic reflection characteristics within a sequence is seismic facies analysis.

A seismic facies unit is a mappable, three-dimensional seismic reflection pattern whose characteristics differ from those of adjacent patterns (Mitchum et al. 1977, p. 122). Like lithofacies, a seismic facies is a rather informally defined unit whose internal characteristics are unique enough relative to the units around it, to justify separating it from those adjacent units. The basic parameters used to define seismic facies units are those dealing with the reflection geometry within a sequence and the external form of the sequence itself. Reflection continuity, amplitude, frequency and seismically derived interval velocity should be incorporated in seismic facies definitions if they appear variable and significant.

1860

Mitchum et al. (1977) discuss in detail a number of the more common reflection configuration patterns and discuss their interpretive significance in terms of depositional energy. The reader is referred to their article for a complete presentation of this topic. However, because of their general importance and their relevance to the example used in this paper, the progradational reflection patterns will be discussed here briefly.

Progradational patterns result from the lateral advance of sedimentation by outbuilding or progradation. Two end member types are recognized: oblique progradation and sigmoid progradation (Figure 8). An oblique progradational pattern is ideally characterized by "a number of relatively steep-dipping strata terminating updip by toplap at or near a nearly flat upper surface, and downdip by downlap against the lower surface of the facies unit" (Mitchum et al. 1977, p. 125). The nearly flat, toplapping upper surface indicates that sedimentation was able to build up to local base level (commonly sea level) and then prograde into the basin. It can be inferred from this pattern that sediments were deposited under conditions of abundant sediment supply at or near base level where sediment reworking can take place. This pattern indicates a high depositional energy level and is likely to contain significant quantities of sand.

A sigmoid progradational pattern is characterized by a series of stacked sigmoid or "S"-shaped reflectors which parallel "strata with thin, gently dipping upper and lower segments and thicker, more steeply dipping middle segments" (Mitchum et al. 1977, p. 125). Unlike the oblique progradational pattern, the sigmoid pattern builds upwards as well as outwards thus indicating that depositional base level is never reached and sediment reworking is unlikely. A relatively low depositional energy level is indicated and the likelihood of significant sand content within this unit is small.

A combination of sigmoid and oblique prograding patterns is known as complex sigmoid-oblique (Figure 8c). This pattern reflects alternating episodes of progradation and aggradation and, as might be expected, represents variable energy levels. It is typical of large deltas prograding into relatively deep waters.

With this background we can now define the two-dimensional seismic facies character of the sequence we defined earlier on line 1-A. Between shot point 2205 and 2250 the sequence is characterized by toplap at its upper boundary, downlap at its base and an oblique progradation reflection pattern internally. North of shot point 2205 a reflector onlapping onto the preexisting shelf margin and a long, gently onlapping reflector characterize the base of the sequence. Here the upper surface is concordant and shows no terminations. Since the entire sequence is only one reflector thick, no internal reflection pattern is identifiable. South of shot point 2250, toplap ceases and the upper boundary of the sequence dips basinward parallel to the sequence's internal reflections. In this portion of the line the sequence is characterized by concordant strata at the top, downlapping at the base and a parallel internal configuration. These three seismic facies patterns describe the sequence on line 1-A.

In order to understand the three-dimensional character of the sequence we must however map the distribution of the identified seismic facies. A shorthand notation has been devised to accomplish this (Figure 9) in which the upper boundary character (A), the lower boundary character (B), and the internal reflection pattern (C) of a sequence are abbreviated and placed in the fashion indicated. The seismic facies unit between shot points 2205 and 2250 on Line 1-A would be characterized:

$$\frac{\text{Top - Dwn}}{\text{Ob}}$$

and the seismic facies unit south of shot point 2250 would be characterized:

$$\frac{\text{C - Dwn}}{\text{P}}$$

the system is flexible and can be adjusted to accommodate various descriptions. In this case of the northern seismic facies unit on line 1-A:

$$\frac{\text{C - ON}}{\text{thin}}$$

is an appropriate notation.

These seismic facies notations can easily be placed on a work map (Figure 10) along with arrows indicating the direction of onlap (solid arrows) and downlap (open arrows). The seismic facies units of lines 1, 2, 3 and 4 have also been placed on Figure 10 along with other significant features including the sequence's depositional edge and the prominent pre-sequence shelf margin. Note that the sequence is not present on line 3. Integrating these data, a seismic facies map (Figure 11) showing the areal distribution of the seismic facies units was constructed.

The distribution of seismic facies in both cross-section and map view strongly indicates a progradational wedge of sediment building out and over a pre-existing shelf margin. The sediment source is strongly focused producing a tear-drop form and the toplap character suggests that sediments built up to sea level. These stratal patterns fit nicely into a classical deltaic model with the distribution of depositional environments shown in Figure 12. In the terms of Mitchum et al. (1977) this would be a sand-prone system with the coarsest clastics in the fluvial and the upper part of the delta plain depositional environments.

WELL CORRELATION USING SEISMIC STRATIGRAPHY

Since seismic reflections parallel bedding planes, the seismic section is an excellent chronostratigraphic correlation tool, particularly in areas where well log correlations are difficult due to rapid lateral facies change, sparse well control, and/or insufficient paleontologic control. Ideally these correlations are made using synthetic seismograms carefully tied to the seismic section. Lacking sonic and/or density logs, or in cases where the units to be correlated are too thin to be resolved, the reflection geometry of the seismic section can often give the interpreter a correlation style or concept that will permit proper correlations.

Synthetic seismograms were constructed for the Shell 1 S. P. Mills, Shell 2 S. P. Mills and Placid 1 D. D. (Figure 3). Each of these wells penetrated the mapped sequence and show it to be the Woodbine, a gas producing unit of Cenomanian age. Above the Woodbine is the Eagle Ford Shale. This is in turn overlain by the Austin Chalk. Below the Woodbine lies a non-porous carbonate, the Buda Limestone. It was the Buda shelf margin that was mapped earlier as the "preexisting shelf margin." Two wells, the Phillips 1 Coke and the Exxon 1 Ogletree, were drilled beyond the Woodbine sequence's depositional limit as mapped during seismic facies analysis. These encountered no sands, penetrating only a thick Eagle Ford Shale section between the Austin Chalk and Buda Limestone.

A cross section (Figure 13) from the Shell 2 S. P. Mills, through the Shell 1 S. P. Mills and on northward parallel to seismic line 1 shows how the Woodbine sands shale out towards the south at just about the point predicted by our seismic facies maps. The G. M. A. S. P. Mills 1B and 2B were drilled prior to the discovery of Hortense Field. The upper three sands of the Shell 1 S. P. Mills are water wet as are the upper two sands of the G. M. A. 1B S. P. Mills well. The presence of Hortense Field may or may not have been predicted depending upon how these two wells were correlated. However, although the individual Woodbine sands are too thin to be resolved seismically, the geometric configuration of the seismic allows us to recognize that the sands are inclined toplapping units and should be correlated as shown in Figure 13. With this correlation the presence of a stratigraphic trap at Hortense Field would have been predicted.

CONCLUSIONS

Using reflection termination patterns, unconformity surfaces can be seismically recognized and used to define unconformity-bound stratal units known as depositional sequences. A depositional sequence represents a period of essentially continuous sedimentation and as such may be interpreted as a single episode of sedimentation. Having defined a sequence, interpretation proceeds with seismic facies analysis; the process of describing, mapping and interpreting the seismic reflection characteristics within a sequence. Reflection geometry is the basic characteristic used. Because the seismic reflectors within a sequence parallel bedding planes, the geomorphologic character of the sediment's depositional environment is displayed on the seismic section. From this geometry paleoenvironment can be inferred and facies predicted. Where well data are available, they can be integrated with the seismic using synthetic seismograms. The chrono-stratigraphic properties of seismic reflectors make the seismic section a powerful correlation tool which can be used to aid well log correlations in areas of facies change, sparse well control, or limited paleontological data.

Thus applying only the Law of Superposition and the concepts of reflection chronostratigraphy, the seismic stratigrapher can make fundamental stratigraphic interpretations on the seismic section. In addition, the depositional sequence framework developed during seismic stratigraphic analysis provides an excellent time-stratigraphic framework for the integration of all available well log, lithologic and paleontologic data. Armed with this complete geologic package, the explorationists can locate stratigraphic traps and define the facies framework of structural plays.

ACKNOWLEDGEMENTS

The author wishes to acknowledge the contributions of his colleagues at Exxon Production Research Company, particularly P. R. Vail and R. G. Todd whose discussions and suggestions helped greatly in the preparation of this paper. The permission of Seiscom Delta Inc. to publish seismic lines 1 and 4 (Figures 4 and 7) is gratefully acknowledged.

REFERENCES

1. Mitchum, R. M. Jr., 1977, Seismic Stratigraphy and Global Changes of Sea Level, Part 11: Glossary of Terms Used in Seismic Stratigraphy, In: Payton, C. E. ed., Seismic Stratigraphy - Applications to Hydrocarbon Exploration: AAPG Mem. 26, 205-212.
2. Mitchum, R. M. Jr., P. R. Vail and J. B. Sangree, 1977, Seismic Stratigraphy and Global Changes of Sea Level, Part 6: Stratigraphic Interpretation of Seismic Reflection Patterns in Depositional Sequences, In: Payton, C. E. ed., Seismic Stratigraphy - Applications to Hydrocarbon Exploration: AAPG Mem. 26, 117-133.
3. Mitchum, R. M. Jr., P. R. Vail and S. Thompson III, 1977, Seismic Stratigraphy and Global Changes of Sea Level, Part 2: The Depositional Sequence as a Basic Unit for Stratigraphic Analysis, In: Payton, C. E. ed., Seismic Stratigraphy - Applications to Hydrocarbon Exploration: AAPG Mem. 26, 53-62.
4. Vail, P. R., R. G. Todd, J. B. Sangree, 1977, Seismic Stratigraphy and Global Changes of Sea Level, Part 5: Chronostratigraphic Significance of Seismic Reflections, In: Payton, C. E. ed., Seismic Stratigraphy - Applications to Hydrocarbon Exploration: AAPG Mem. 26, 99-116.

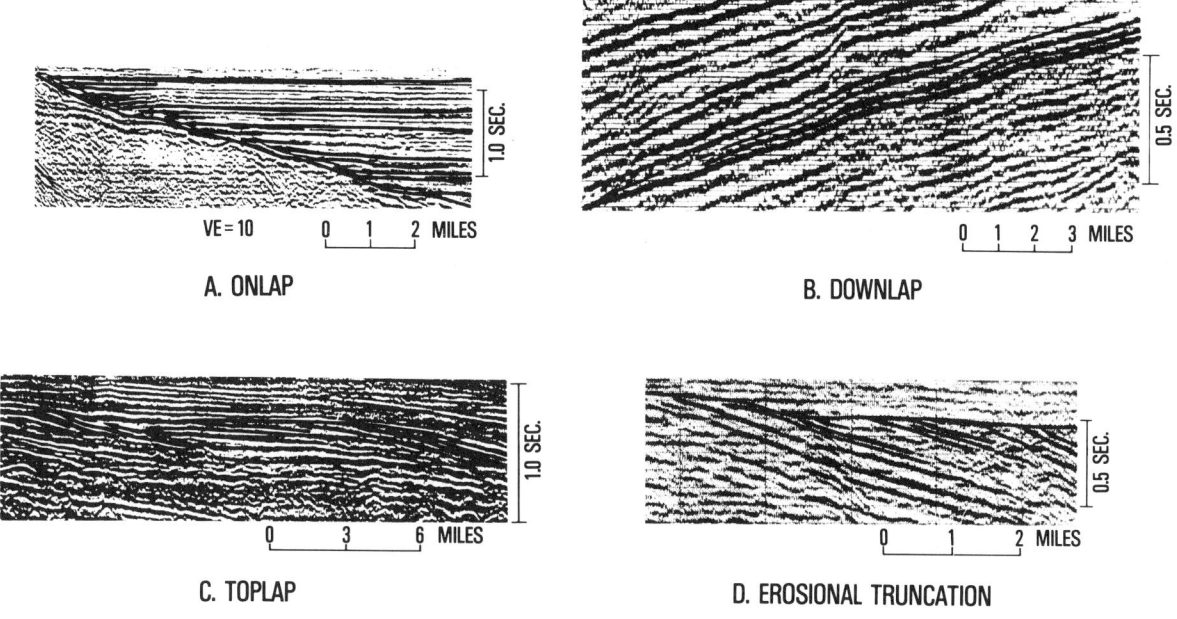

Fig. 1 - Basic seismic reflection termination patterns.

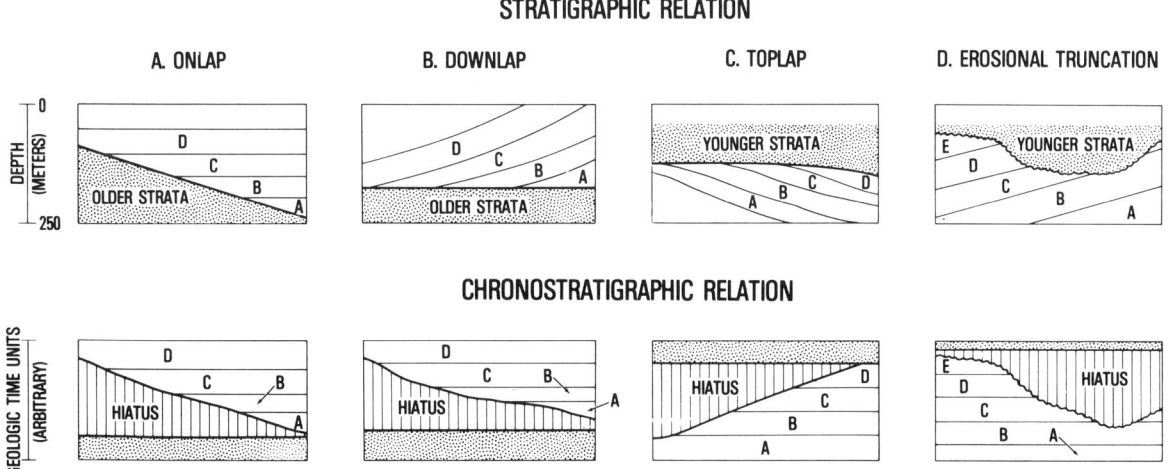

Fig. 2 - Stratigraphic and chronostratigraphic relations of onlap, downlap, toplap and erosional truncation.

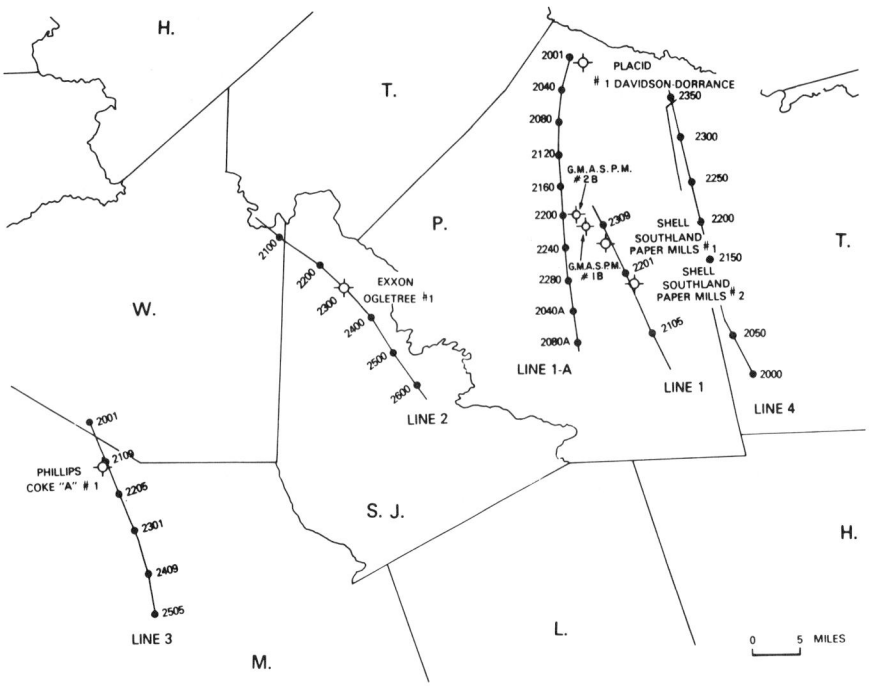

Fig. 3 - Distribution of seismic and well data in the East Texas example.

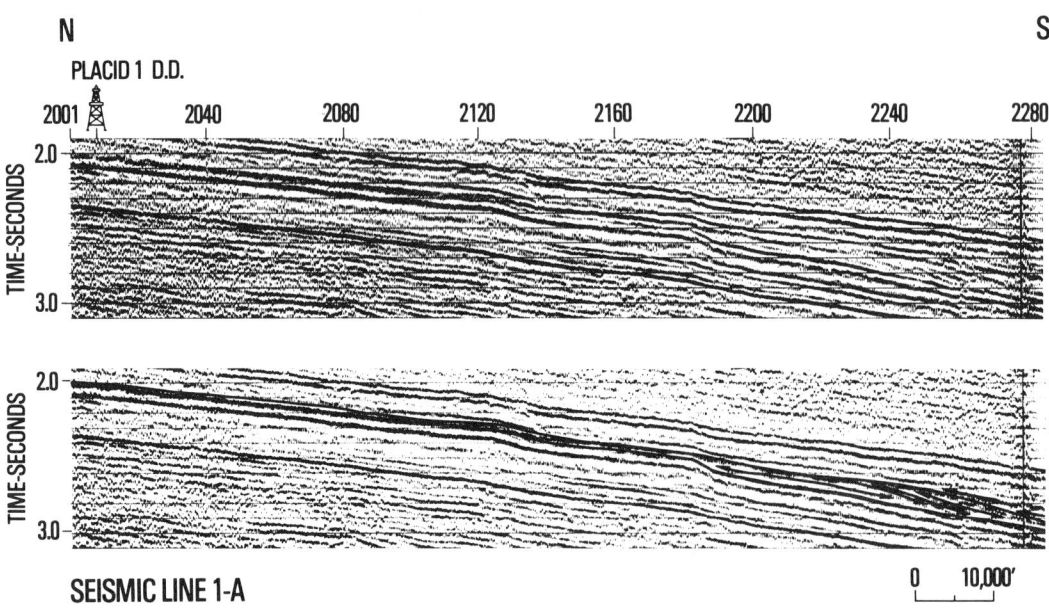

Fig. 4 - Uninterpreted and interpreted presentations of a portion of seismic line 1-A (24 channel, 6 fold). Seismic line courtesy of Seiscom Delta Inc.

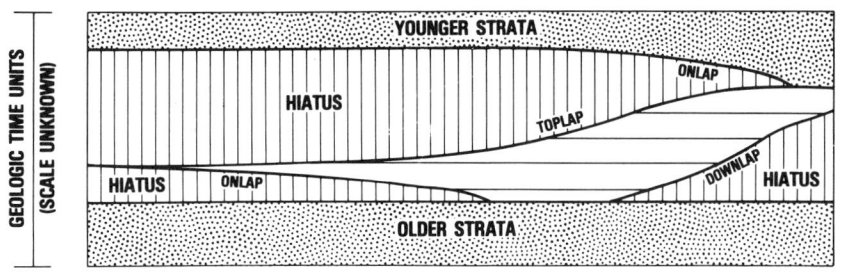

Fig. 5 - Relative chronostratigraphic relationships for the sequence indentified along seismic line 1-A as indicated by seismic reflection patterns.

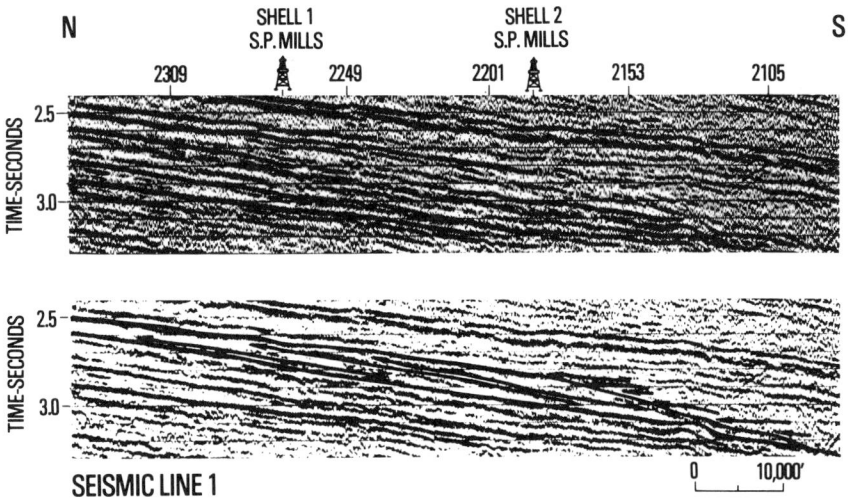

Fig. 6 - Uninterpreted and interpreted presentations of a portion of seismic line 1 (24 channel, 12 fold).

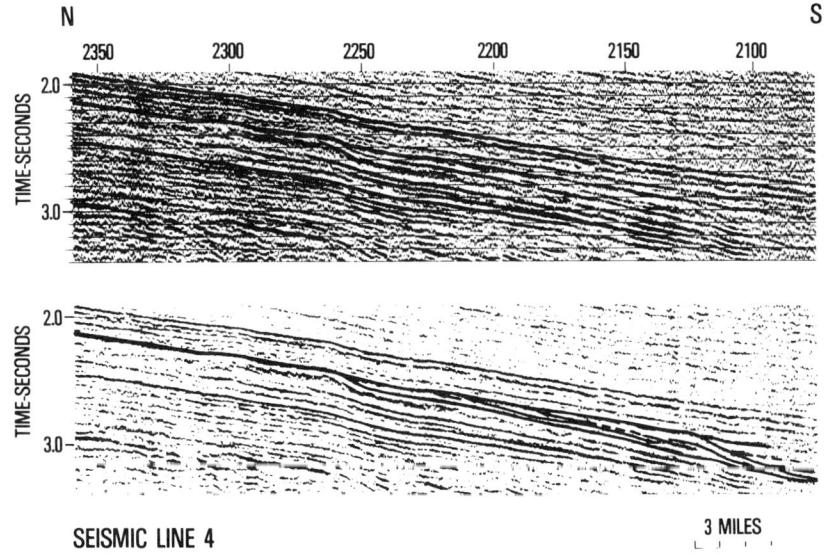

Fig. 7 - Uninterpreted and interpreted presentations of a portion of seismic line 4 (48 channel, 6 fold). Seismic line courtesy of Seiscom Delta Inc.

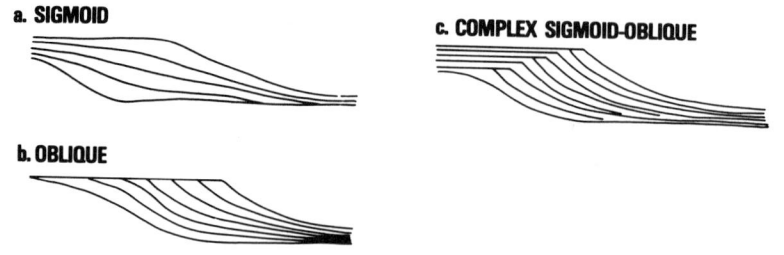

FIG. 8 - PROGRADATIONAL REFLECTION PATTERNS.

CODE SYSTEM

UPPER SEQUENCE BOUNDARY

T_e — EROSIONAL TRUNCATION
Top — TOPLAP
C — CONCORDANT

LOWER SEQUENCE BOUNDARY

On — ONLAP
Dwn — DOWNLAP
C — CONCORDANT

$$\frac{A-B}{C}$$

INTERNAL CYCLE CONFIGURATION

P — PARALLEL
D — DIVERGENT
C — CHAOTIC
W — WAVY
DM — DIVERGENT MOUNDY

M — MOUNDED
Ob — OBLIQUE PROGRADATIONAL
Sig — SIGMOID PROGRADATIONAL
Rf — REFLECTION FREE
Sh — SHINGLING

FIG. 9 - SHORT HAND NOTATION FOR SEISMIC FACIES MAPPING.

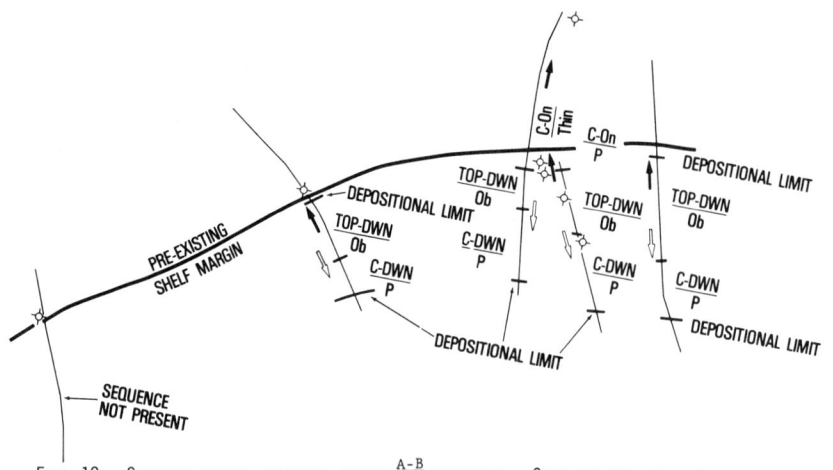

FIG. 10 - SEISMIC FACIES WORKMAP USING $\frac{A-B}{C}$ NOTATION. OPEN ARROWS INDICATE THE DIRECTION OF DOWNLAP, SOLID ARROWS INDICATE THE DIRECTION OF ONLAP.

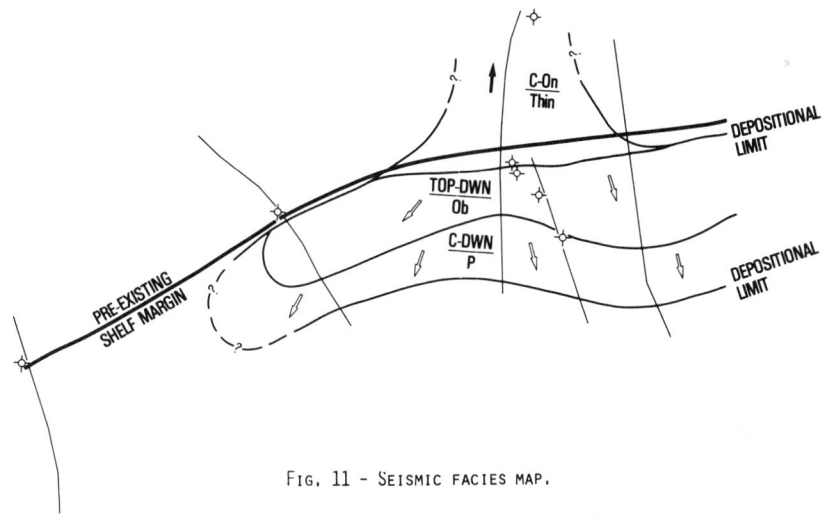

Fig. 11 - Seismic facies map.

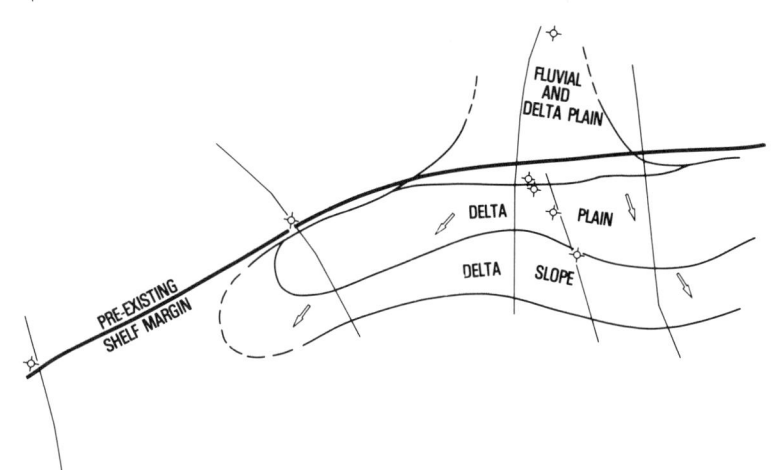

Fig. 12 - Geologic interpretation of the seismic facies patterns. Coarsest would be predicted in the fluvial and the upper part of the delta plain environments.

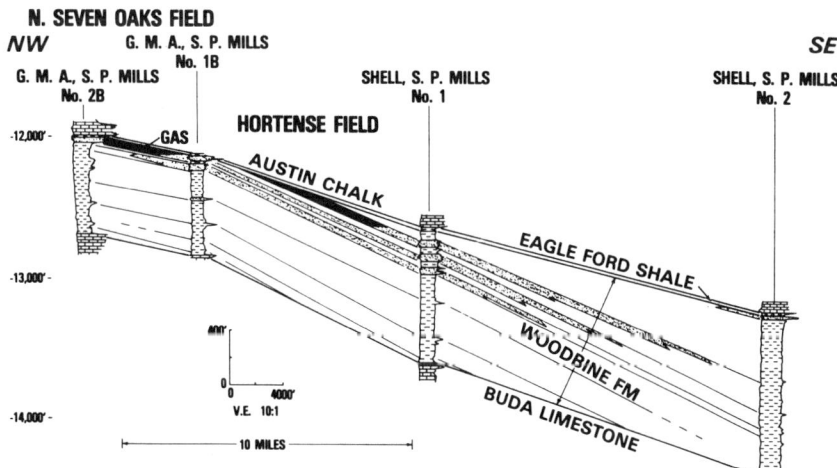

Fig. 13 - Electric-log cross section showing distribution and geometry of Woodbine deltaic sandstone beds, Polk County, Texas (Vail et al., 1977, p. 114).

SEISMIC FACIES ANALYSIS CONCEPTS *

BY

M. M. ROKSANDIĆ **

Abstract

ROKSANDIĆ, M. M., 1978, Seismic Facies Analysis Concepts, Geophysical Prospecting 26, 383-398.

Seismic facies analysis makes use of different seismic parameters in order to get other than structural information. A review is given of possibilities and usefulness of seismic facies analysis in oil exploration.

A seismic facies unit can be defined as a sedimentary unit which is different from adjacent units in its seismic characteristics. Parameters that should be taken into consideration in the seismic facies analysis are as follows: reflection amplitude, dominant reflection frequency, reflection polarity, interval velocity, reflection continuity, reflection configuration, abundance of reflections, geometry of seismic facies unit, and relationship with other units.

Interpretation of seismic facies data may be either direct or indirect. The purpose of the direct interpretation is to find out geological causes responsible for the seismic signature of a seismic facies unit. So, the direct interpretation may be aimed at predicting lithology, fluid content, porosity, relative age, overpressured shales, type of stratification, geometry of the geological body corresponding to the seismic facies unit and its geological setting. The indirect interpretation is intended to reach some conclusions on depositional processes and environments, sediment transport direction, and some aspects of geological evolution (transgression, regression, subsidence, uplift, erosion).

The results of the seismic facies analysis may be shown on seismic facies cross-sections and seismic facies maps. Depending on the available seismic data and geological conditions in the area under consideration, the seismic facies maps may be of different types such as general seismic facies maps showing distribution of different seismic facies units, sand-shale ratio maps, direction of cross-bedding and paleo-transport maps etc.

Several kinds of seismic facies units and their geological interpretation are discussed as examples of seismic facies analysis.

Introduction

The term "seismic facies" was introduced in exploration geophysics only a few years ago, but has not been used widely. However, the procedure, which we can call "seismic facies analysis", had been used before the term came into use, although only occasionally, specially in exploration of stratigraphic traps.

---

* Paper read at the Silver Anniversary Meeting of the European Association of Exploration Geophysicists in The Hague, June 1976.
** Present address: SOQUIP, 3340 de la Pérade, Ste-Foy, Québec G1X 2L7, Canada.

Dramatic improvement of seismic data quality during the last years has enabled seismic facies analysis to be used in many cases and for different purposes. The present paper is intended to bring into sharper focus some of the basic notions of seismic facies analysis.

Seismic facies analysis makes use of different seismic parameters in order to get other than structural information.

This procedure can be applied in all phases of exploration. However, the more geological control there is, the better is the interpretation. It goes without saying that the interpreter must take into consideration the recording and processing procedures and all restrictions that are consequences of those procedures. Seismic facies analysis may be based on routine processed seismic sections. However, special processing and modelling will make it possible to use more parameters and, therefore, to increase the scope and improve the quality of the interpretation.

In geology, many definitions and classifications of facies have been proposed, but the term "facies" has been chiefly used to denote differences in appearance, composition, and biologic content between different rock or lithostratigraphic units or their parts. Similarly, one part of a sedimentary sequence can be distinguished from others according to general seismic appearance. So, a seismic facies can be defined as a lithostratigraphic or seismostratigraphic unit (or part of such a unit) which has appropriate seismic characteristics distinguishable from those of other units (or other parts of a particular unit). In this context, a seismostratigraphic unit can be defined as a subdivision of sedimentary sequence between two distinctive seismic markers mappable over a considerable area. The notion of seismostratigraphic unit is useful when lithostratigraphic units cannot be separated on seismic sections because of their physical properties or/and the resolution of the seismic method.

Seismic facies may sometimes correspond to geological facies. However, in some cases a seismic facies does not correspond to a geological facies owing to different factors:

1. Resolution provided by the seismic method is much less than that provided by geological methods.
2. Seismic data do not include much information which is sometimes essential for definition of a geological facies.
3. Some seismic parameters are influenced by secondary processes which are not strictly connected with a geological facies. For example, the presence of gas may essentially change the seismic signature.

On the other hand, the characteristic seismic signature of some geological bodies essentially composed of one geological facies may be caused by features originating from post-depositional processes. Such is the case with diapir cores.

Seismic Facies Elements

Sangree and Widmier (1974) and Sheriff (1975) pointed out that the following seismic data (i.e. seismic facies elements) should be taken into consideration in seismic facies analysis: reflection amplitude, dominant frequency, interval velocity, reflection configuration, reflection continuity, external form of seismic facies unit (i.e. the geometry of the seismic facies unit) and areal association. Reflection polarity and abundance of reflections should be added to the list. The presence of diffractions can also be an important diagnostic element.

On fig. 1 the seismic facies elements, together with the more important properties of the geological environment which are responsible for them, are listed. The list of geological-physical causes is not quite consistent, and the listed parameters are not independent (e.g. absorption depends upon lithology, porosity, and fluid content). However, the list has been made in such way in order to emphasize the geological and physical properties responsible for given seismic parameters.

One geological or physical factor may affect more than one seismic facies element. On the other hand, most seismic facies elements are caused by several geological or physical factors. This ambiguity makes an interpretation rather difficult.

Seismic facies data can be interpreted in two ways (fig. 1). Firstly, the purpose of the interpretation may be to find out geological causes responsible for the seismic signature of a seismic facies. Such an interpretation can be called the direct interpretation. The direct interpretation may be aimed at predicting lithology, fluid content, porosity, relative age, overpressured shales, type of stratification, geometry of the geological body corresponding to the seismic facies unit, and its geological setting.

From results of the direct interpretation and seismic facies data we can deduce some aspects of depositional processes and environments, sediment transport direction, and geological evolution (transgression, regression, subsidence, uplift, erosion). This kind of interpretation can be called the indirect interpretation. In a short paper it is not possible to discuss all seismic facies elements and all aspects of seismic facies analysis. The present discussion will be restricted to a few problems.

The reflection configuration may contain information about type of stratification, lithology, depositional process and environment, sediment transport direction, and geological history. On the other hand, the reflection configuration can be studied, at least to a certain degree, on every reasonably good seismic section. This is why this seismic facies element deserves our attention.

Sangree and Widmier (1974) proposed a classification of typical reflection configurations. An expanded version of the classification of reflection con-

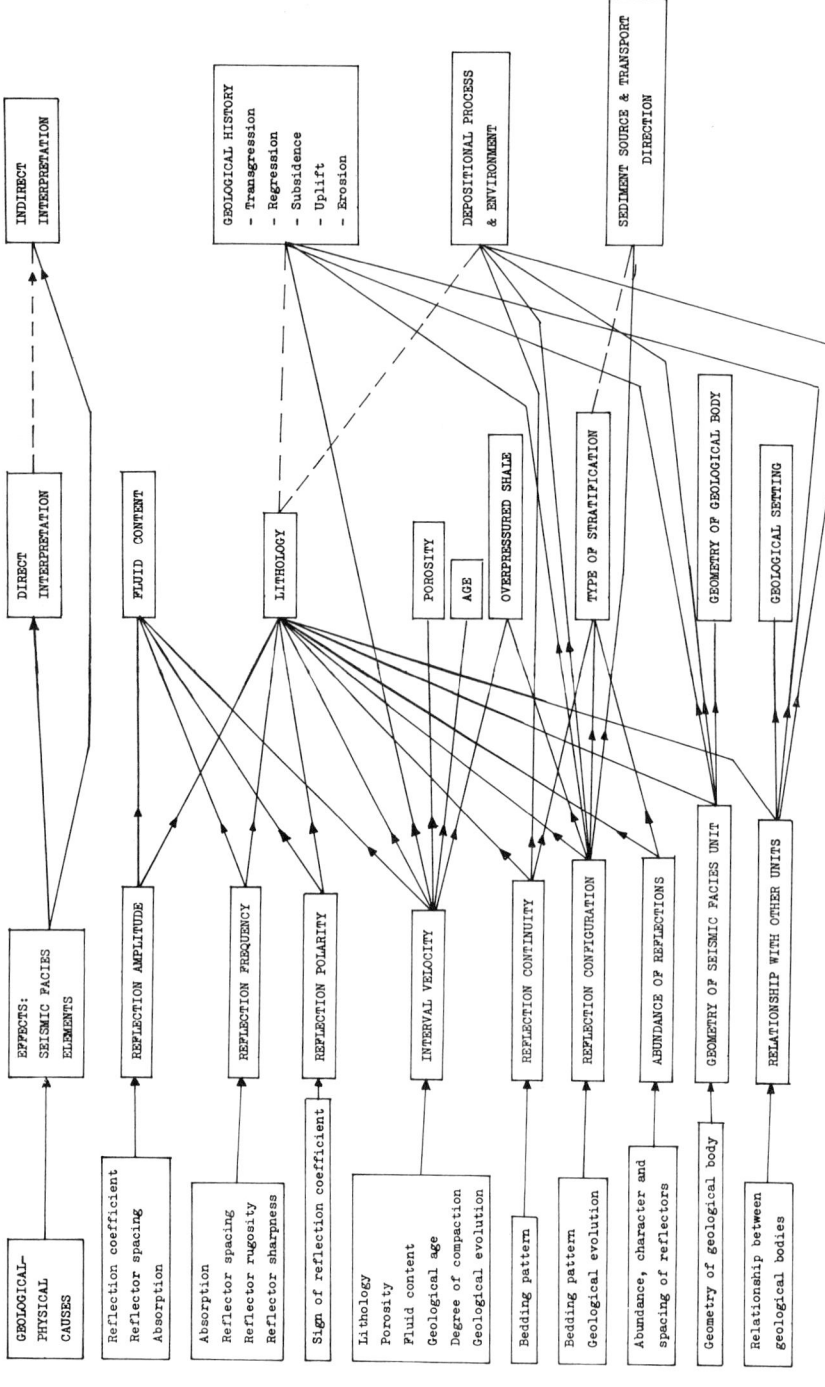

Fig. 1. The relationship between geological-physical causes, seismic facies elements and interpretation.

figurations usable in both shallow penetration and deep penetration seismics is shown on fig. 2 with the corresponding diagrammatic representations. Three basic types of reflection configurations are: reflection free, layered, and chaotic reflection configurations. The reflection free configuration may be with or without diffractions. The layered reflection configuration can be subdivided into simple layered and complex layered configurations. Simple layered configuration can be either parallel or divergent, whereas complex layered configuration is cross-layered, oblique, or sigmoid. Chaotic reflection configuration can exist with or without diffractions. It should also be mentioned that the reflection configuration in some cases may depend upon orientation of seismic lines relative to depositional strike. Because of that the reflection configuration should be studied on seismic sections of different orientation before making any deduction.

The study of reflection configuration should take into consideration continuity of reflections, their amplitudes and frequencies as well as variations of those parameters.

BASIC TYPES OF REFLECTION CONFIGURATIONS

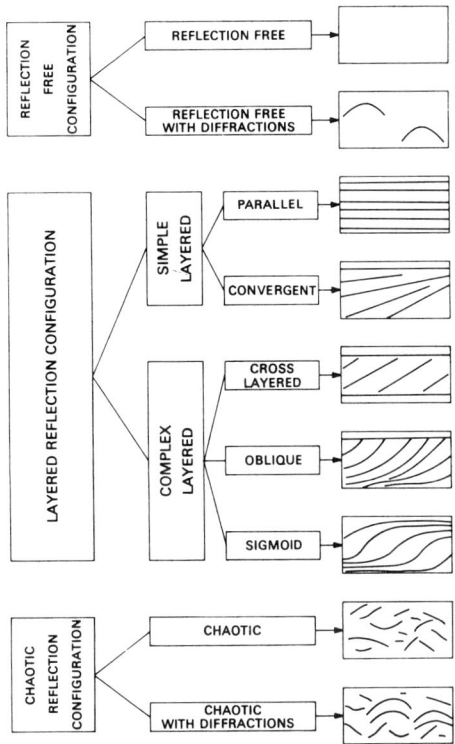

Fig. 2. Basic types of reflection configuration.

Several examples of some reflection configurations are presented on fig. 3. Cross-layered and sigmoid configurations can be seen on fig. 9.

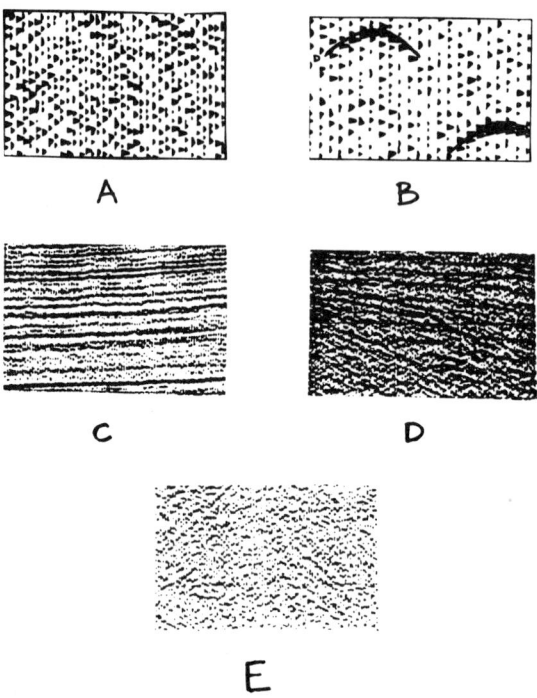

Fig. 3. Several examples of reflection configurations. A. Reflection free configuration. B. Reflection free configuration with diffractions. C. Parallel reflection configuration. D. Divergent reflection configuration. E. Chaotic reflection configuration.

In the following a few examples both from deep- and shallow penetration seismics are given. The reflection free configuration is characteristic of reefs. On shallow penetration seismic sections reflection free configurations with or without diffractions have been found in glacial tills. The parallel configuration is naturally the most widespread configuration in sedimentary rocks. The convergent configuration may be caused either by primary, depositional factors (for example, pinching out), or by secondary factors (for example, differential compaction). The chaotic configuration is characteristic of diapiric cores, whose internal structure is very complex. If a diapiric core, mostly composed of plastic rocks (such as salt and shale), contains more competent layers (such as anhydrite or limestone), the more competent layers are broken in pieces of different size. Such pieces or their edges are responsible for many diffractions that can be seen on seismic sections of diapiric cores. The cross-layered reflection configuration corresponds to cross-bedding. Cross-bedding is characteristic of psammites. Thus a cross-layered reflection pattern is a diagnostic criterion for sandy rocks (sands, sandstones, or calcarenites). On the other hand, the cross-layered reflection configuration allows us to study paleo-currents (paleo-transport of sedimentary material). To do this, it is necessary (see fig. 4):

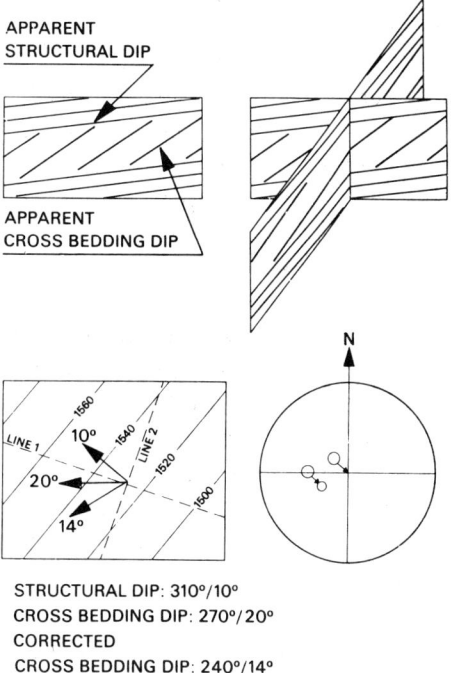

STRUCTURAL DIP: 310°/10°
CROSS BEDDING DIP: 270°/20°
CORRECTED
CROSS BEDDING DIP: 240°/14°

Fig. 4. Determination of paleo-current from the cross-layered reflection configuration.

1. to calculate real cross-bedding dip from apparent dips at the intersection of two seismic sections;
2. to rotate cross-bedding planes by the structural dip value. The structural dip can be easily found from a structural contour map or apparent structural dips at the intersection of two seismic sections. The rotation can be done by means of stereographic nets, using the procedure known from structural geology and interpretation of dipmeter data.

Cross-bedded sequences may be deposited in different environments (alluvial, deltaic, eolian, shoal). However, the paleo-current patterns and the relationship of paleo-currents to the geometry of lithostratigraphic bodies and to the paleo-slope is different for different environments, and this fact can be the base for deduction of the depositional process and the depositional environment.

Currents in alluvial and deltaic sediments are controlled by the existing slope. With this fact in mind, it is possible to find out the relative age of some structures (Selley 1976). Paleo-currents are independent of the post-depositional structures. On the other hand, they diverge from axes of syndepositional uplifts and converge on centers of syndepositional lows.

Sigmoid and oblique reflection configurations occur in connection with progradational patterns on the shelf margin. According to Sangree and Widmier (1974), the sigmoid pattern tends to be of low depositional energy, while the oblique pattern is characteristic of high-energy depositional conditions. We shall see later that those configurations may correspond to delta slopes.

The three-dimensional approach by means of structural contour and isochore (isopach, isochron) maps corrected for post-depositional deformations is the best way of studying the geometry of a seismic facies unit. However, in some cases properly oriented seismic sections can give very good insight into the shape of a geological body (e.g. channels).

While studying the geometry of seismic facies units and relationships of different seismic facies units it is essential to find out whether they are primary or secondary features, because they may be very valuable diagnostic criteria in both direct and indirect interpretation.

To find out the relationship between two seismic facies units lying one above the other is not always an easy task. The difference between a conformity and a disconformity cannot be deduced from geometrical relationships on a seismic section. Also, a low angular unconformity may be taken for a conformity. Sometimes the knowledge of interval velocities can be very helpful in distinguishing a conformity from a disconformity or a low angular unconformity. A non-conformity between a sedimentary sequence and the basement is often easily recognized due to the strong low frequency reflection. The loss of higher frequencies may be explained by two factors:

a) scattering of higher frequencies waves due to the rugosity of the interface;

b) presence of a paleo-weathered zone which acts as a transitional zone.

An onlap may be recognized both by geometrical relationship and facies changes. However, an offlap usually cannot be distinguished purely on the basis of geometrical relationship from consequences of an erosional phase. An offlap could be deduced from lateral facies changes, if such changes are clear on seismic sections.

The interpretation of a seismic facies unit requires that not only properties of the facies under consideration be taken into account, but also those of neighbouring facies. For example, the diagnostic criteria of a reef are not only the reflection free configuration and high velocity of the reef, but also the presence of a basinal sedimentary sequence with a parallel reflection configuration off the reef, a polarity reversal of the reflection indicating the top of the reef at its edges, the draping in the overlying sediments, the higher frequency of reflections above the reef in comparison with the frequency off the reef due to the differential compaction etc. (Dobrin 1960; Morrison ?).

A FEW EXAMPLES FROM SHALLOW PENETRATION SEISMICS

A few examples from shallow penetration seismics are briefly discussed in this section (see fig. 5).

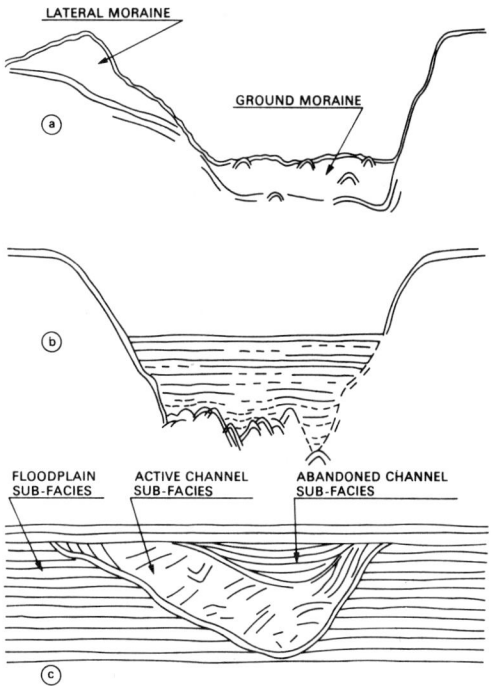

Fig. 5. Diagrammatic presentation of some facies on shallow penetration seismic sections: a. Moraines; b. Channel fill; c. Meandering river sub-facies.

The reflection free configuration with or without diffractions is characteristic of moraines. Diffractions are caused by the presence of boulders and larger blocks. From the geometry of moraines and the relationship to elements of glacial morphology, it is possible to recognize the different types of moraine, such as the lateral, terminal, and ground moraines.

In arctic areas some submarine channels and fjords (or parts of some fjords) have flat bottoms unusual for glaciated valleys. Sparker seismic data show that they are glaciated valleys partly filled with a well-stratified sedimentary sequence. The bottom of such a valley has all the properties of a glacial trough. The sedimentary sequence has a parallel reflection configuration. Some reflections have high amplitudes, the others moderate and low ones. Lateral variations of amplitudes are also noticeable. The prediction is that the sequence is composed of relatively fine-grained sediments: sands, silts, and shale. This is, in fact, outwash brought into fjords and submarine channels by meltwater streams.

Shantser (1951, cited after Selley 1970) found that the deposits of a meandering river can be sub-divided into three main sub-facies due to deposition in three different sub-environments: floodplain sub-facies, active channel sub-facies, and abandoned channel sub-facies. Those three sub-facies can be recognized on sparker seismic sections (King 1973; Roksandić 1976). The active channel sub-facies is characterized by a chaotic reflection configuration. The floodplain and abandoned channel sub-facies have the same parallel reflection configurations. However, thanks to their different geometry, they can be distinguished: the former sub-facies has a sheet geometry while the latter has a channel geometry.

A few Examples from Deep Penetration Seismics

A few depositional environments together with corresponding lithology and some seismic facies elements are listed very briefly on table 1. Only three examples will be discussed in order to illustrate the seismic facies analysis approach in the interpretation of routine processed seismic data.

Fig. 6 shows an example of the vertical seismic facies distribution deduced from a deep penetration seismic section. The non-conformity between the basement and sedimentary cover is characterized by strong, low-frequency reflection. In the sedimentary cover it has been possible to distinguish seven velocity layers, the interval velocities of which vary from 1900 m/s to 3280 m/s. The interval velocity curve and specially the curve showing interval velocity plotted versus the middle point depths of the velocity layers show that there are at least two different sequences, the first one including the layers A, B, and C, while the second one comprises the layers E, F, and G. The layer D has too high an interval velocity for the upper sequence and too low an interval velocity for the lower sequence. It might be an independent sequence.

Interval velocities of the layers A, B, and C show that this is a Neogene (perhaps Quaternary-Neogene) clastic sequence. The layers A and B have similar, parallel reflection configurations, although the continuity of reflections is better in the layer B than in the layer A. The prediction is that these layers are composed of an alternation of sands and shales. The reflection configuration of the layer C indicates that it is chiefly made up of a fairly uniform fine-grained sequence.

The interval velocity of the layer D also belongs to the range of values for the Tertiary clastics. The parallel, discontinuous, variable amplitude reflection

TABLE I

Relationship of some seismic facies to sedimentary environments and facies

Environment	Lithology	Seismic facies elements		Other important diagnostic features
		Reflection pattern	Geometry	
Reef				
1) reef facies	limestones (biolithites) sometimes dolomitized and recrystallized	reflection free	like a linear ridge, like a brachyanticline, domelike, irregular	a) interval velocity of reef limestone is often higher than that of basinal facies.
2) basinal facies (back-reef and/or fore-reef)	calcarenites, calcilutites, shales, evaporites	parallel layered	sheet	b) polarity reversal may exist on the reef edges c) draping of overlying sediments is common feature.
Neritic with rapid succession of transgressions and regressions	clastics, sometimes carbonates; frequent lateral and vertical lithological changes	parallel, discontinuous, variable amplitude	sheet geometry of the unit as a whole	
Deltaic				
1) subaerial delta platform	sands (distributary channels), silts (levees), silts and clay (flood basins and lagoons), peat (swamps)	parallel discontinuous or continuous, cross-layered, chaotic	prismatic or fan-shaped geometry as a whole	
2) subaqueous delta platform	mostly sands	parallel, sometimes cross-layered		
3) delta slope	silts, sometimes sands	sigmoid or oblique		
4) prodelta	shale	parallel		
Low energy marine	mostly shale	parallel, continuous, low amplitude, usually low frequency	sheet	

pattern indicates a shaly-sandy sequence deposited in an environment of considerable and fast changes of energy and sediment influx in time and space.

According to the interval velocities, the layers E, F, and G are mostly composed of clastics, probably of Lower Tertiary-Upper Mesozoic (Cretaceous) age. However, the layers have very different reflection configuration. The layer E is in this aspect similar to the layer D. The layer F has a chaotic configuration which suggests a high energy sequence, if it is composed of clastics. The reflection pattern of the layer G is characteristic of a fairly uniform sequence composed mostly of fine-grained rocks.

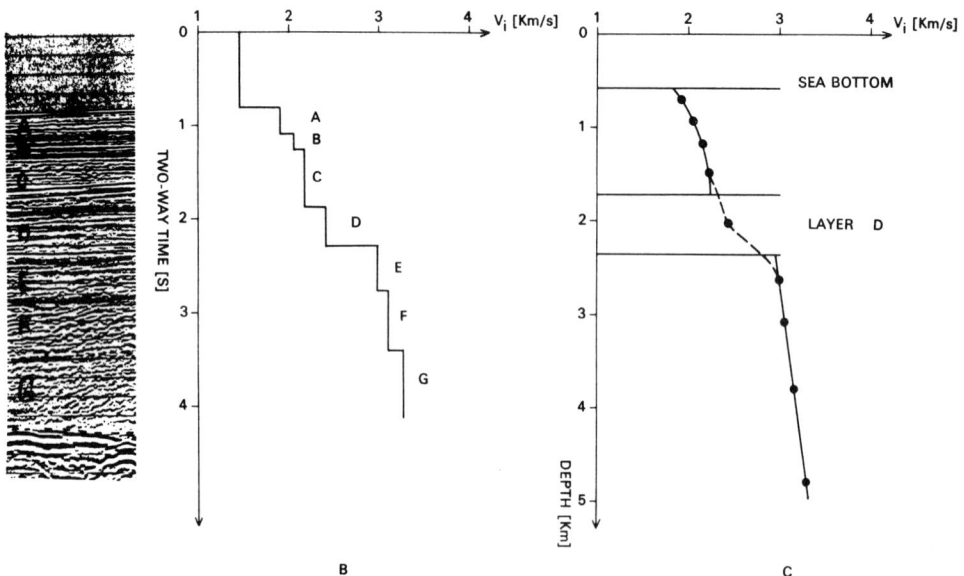

Fig. 6. Vertical seismic facies distribution: A. Seismic section; B. Interval velocities, C. Interval velocities vs. middle point depths of velocity layers.

Fig. 7 shows an example of lateral seismic facies changes between two markers A and B. The change of the reflection configuration indicates that a relatively homogeneous, mostly fine-grained sequence on the left passes into an alternation of sandstones and shales on the right.

The next example concerns a delta sequence. On the top of fig. 8 the geomorphology of a delta is shown. Three main geomorphological elements of a delta and three different deposition environments are: delta platform (or topset) with the subaerial and subaqueous parts, delta slope (or foreset) and prodelta (or bottomset).

In the middle of the figure a theoretical model of the mechanism of formation of a delta sequence and distribution of facies is presented. The model, made

Fig. 7. Lateral seismic facies changes.

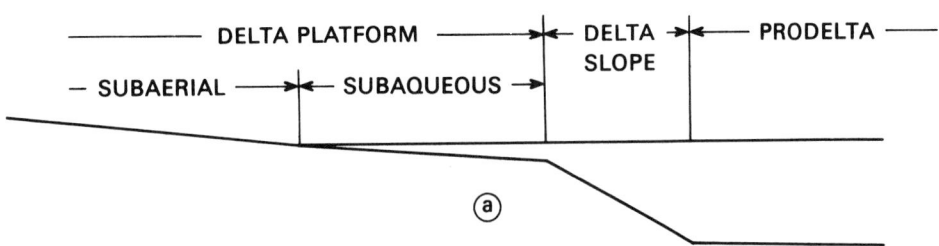

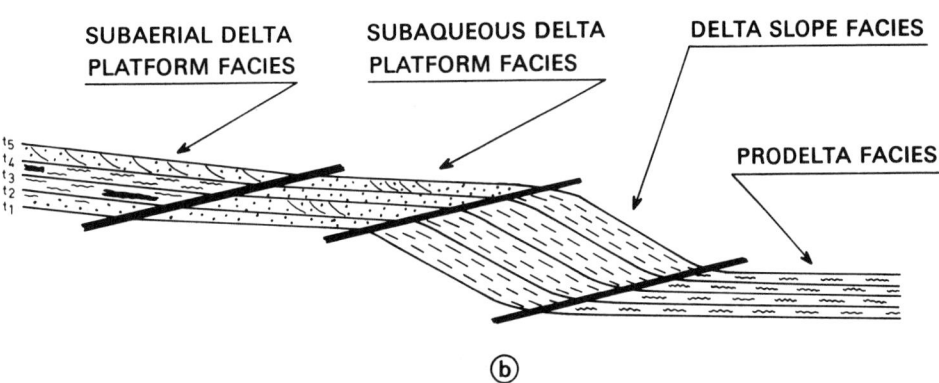

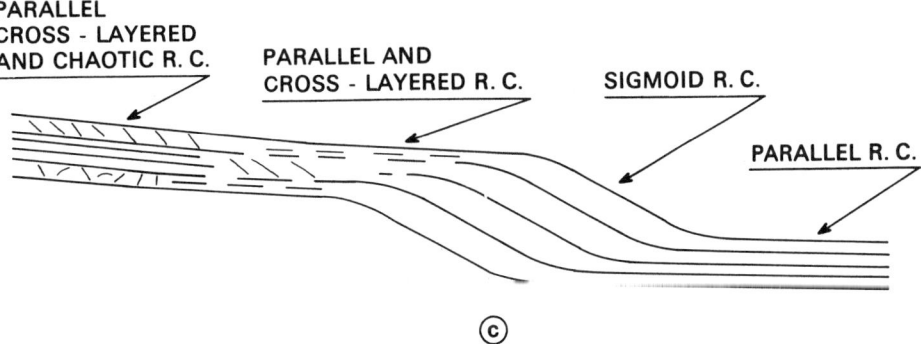

Fig. 8. Theoretical model of a delta: a. geomorphology of a delta; b. theoretical model of the mechanism of formation of a delta; c. expected reflection configuration of a deltaic sequence.

Fig. 9. Seismic section of a delta sequence.

in a way similar to that followed by Dailly (1975), is based on the assumption that progradation is constant in time i.e. that the vertical subsidence and sediment influx do not vary essentially with time. The lines t_1-t_5 are time lines.

Prodelta clay passes up transitionally into delta slope facies consisting mostly of silt. The subaqueous delta platform facies is composed mostly of sands. On the subaerial part of a delta platform different deposits can be formed simultaneously (Selley 1970): sands in distributary channels, silts on the levees which flank distributary channels, silts and clays in flood basins and lagoons, and peat in swamps.

The third diagram of the same figure shows the expected reflection configuration in different parts of a deltaic sequence. The reflection configuration of the subaerial delta platform facies may be different depending on the subfacies along the seismic line. The distributary channel sands will have an oblique reflection configuration, but sometimes also may have chaotic or reflection free configurations. The flood basin and lagoon facies will have a parallel reflection configuration. Subaqueous delta platform sands may have a discontinuous parallel configuration, sometimes also an oblique reflection configuration. The sigmoid reflection configuration will correspond to delta slope facies, while prodelta clay will have a parallel reflection configuration.

Fig. 9 is an example that shows the close analogy of the described theoretical model with an actual seismic section.

It should be emphasized that the described model cannot explain every delta sequence. Evolution of many deltas is much more complex than assumed by the model. Therefore the distribution of sedimentary facies will be different and often more complicated. Consequently, the distribution of seismic facies will be somewhat different.

Seismic Facies Maps

The areal variation of seismic properties can be studied and presented on seismic facies maps. Different types of such maps have already been used or can be proposed for future use:

1. Maps showing some properties of a reflector. In fact, these are structure contour maps (time or depth contour maps) with the presentation of the variations of some seismic facies elements (e.g. amplitude, polarity).

2. The second group of seismic facies maps is designed to show seismic facies distribution of a lithostratigraphic or seismostratigraphic unit. The maps can be either qualitative (e.g. isopach or structural contour maps showing the distribution of seismic facies) or quantitative (i.e. contour maps showing the variation of a seismic facies element within a lithostratigraphic or seismostratigraphic unit, e.g. velocity variation, or a property of the units derived from interpretation of seismic data, e.g. sand-shale ratio).

3. As the third group of seismic facies maps we can regard paleo-current maps based on the study of cross-bedding, maps of progradation derived from the mapping of several reflectors from a zone with sigmoid or oblique reflection configuration, etc.

If the geological data are available, the results of seismic facies analysis may be integrated with the geological data and incorporated in (geological) facies maps.

Acknowledgements

I wish to thank the Director of the Geological Survey of Greenland for permission to release this paper and Mr. G. Henderson (Geol. Survey of Greenland) for improvements in the English text.

References

Dailly, G., 1975, Some remarks on regression and transgression in deltaic sediments, in "Canada's continental margins and offshore petroleum exploration", Canadian Society of Petroleum Geologists, Calgary, 791-820.

Dobrin, M. B., 1960, Introduction to Geophysical Prospecting, McGraw-Hill Book Co., Inc., 163-165.

King, V. L., 1973, Sea bed geology from sparker profiles, Vermilion block 321, offshore Louisiana, 1973 Offshore Technology Conference, Preprints I, 657-666.

Morrison, O J., ?, A discussion on the features on the AIMS™ modeling system, Seismograph Service Corporation (without year, presumably 1974 or 1975).

Roksandić, M. M., 1976, Physiographic and geological mapping of the sea-floor off West Greenland, The Geological Survey of Greenland, Report of Activities 1975, 31-35.

Sangree, J. B., and Widmier, J. M., 1974, Interpretation of depositional facies from seismic data, Paper presented at National Convention SEG, November 1974, Dallas, Texas.

Selley, R. C., 1970, Ancient Sedimentary Environments—A brief survey, Chapman & Hall Ltd., London.

Selley, R. C., 1976, An Introduction to Sedimentology, Academic Press, London—New York—San Francisco.

Sheriff, R. E., 1975, Inferring stratigraphy from seismic data, 1975 Offshore Technology Conference, Preprints II, 253-263.

SEISMIC SIGNATURES OF SEDIMENTATION MODELS†

J. C. HARMS* AND P. TACKENBERG*

Seismic techniques have been used mainly for structural interpretation, but mounting interest in stratigraphic applications is evident. Estimation of sand-shale ratios from seismically derived average velocities is a recent example of a stratigraphic application. Except in the case of tall pinnacle reefs, today direct location of stratigraphic traps by reflection methods is restricted, at best, to areas of very high quality data and abundant well control. However, it may be possible to interpret some useful stratigraphic characteristics from seismic reflections, the interpretation being based upon the concept of sedimentation models.

Most stratigraphic sequences are not random stacks of various lithologies. Commonly, they are well organized and have units with characteristic contacts, thicknesses, lateral extents, lateral facies changes, and vertical sequence. These orderly characteristics are summarized in sedimentation models, where the control of lithologic distribution by dominant depositional processes is emphasized.

Three sedimentation models for sandstone and shale sequences are presented. For each, one example is described and converted to a synthetic reflection seismic cross-section. These cross-sections are each distinct in terms of reflection polarities, areal changes in reflection amplitudes, continuity of events, and lateral interval velocity changes.

The simplified models, although limited in their scope, suggest that additional stratigraphic information can be gleaned from reflection seismic data. To exploit this promise, record processing techniques that emphasize recognition of reflection polarities, amplitudes, continuity, and interval velocities must be developed or improved. It is also necessary to improve our knowledge of seismic boundaries in a variety of stratigraphic sequences. Though difficult, these valuable goals appear attainable.

INTRODUCTION

Reflection seismic surveys have undergone an astonishing renaissance in the past decade. Success and increased activity have been fostered by digital recording, new seismic energy sources, burgeoning computer technology, and lower unit costs of computation. Although these techniques and improvements have greatly advanced the capabilities of the reflection seismograph, the basic objective has remained the location of structural traps.

During this same period, increasing attention has been devoted to stratigraphic trap exploration methods. Most authorities agree that an increasingly larger proportion of future reserves must be found in stratigraphic traps. Some believe that the volume will equal or exceed that found in structural traps (Lyons, 1968). All agree that stratigraphic traps are difficult to find by any known method, and a large proportion of geologic research has been devoted to developing or improving stratigraphic trap-finding methods.

With the promise of large future reserves and the acknowledged difficulty of detection, it is only natural that seismic methods be turned toward stratigraphic trap exploration. Although there has been some industry effort in this direction in the past few years, very few results were reported until two symposia in 1968 sponsored by the SEG (21st Annual Midwestern Exploration

---

† Manuscript received by the Editor December 12, 1969; revised manuscript received July 8, 1971.
* Marathon Oil Company, Littleton, Colorado 80120.
© 1972 by the Society of Exploration Geophysicists. All rights reserved.

Meeting; 38th Annual International Meeting). Attempts to identify stratigraphic traps specifically by seismic techniques have been disappointing or only marginally successful, with the single major exception of exploration for tall pinnacle reefs. Most other types of stratigraphic trap reservoirs are simply too thin to be identified within the limitations of data quality, frequency, and velocity contrast imposed upon the reflection seismic method by geologic sections.

There appears to be little expectation among authorities that the reflection seismograph can be used as a tool for the *direct* location of stratigraphic traps at the prospect level. Although this is an important goal, it is probably unattainable in most sandstone and shale sequences with current reflection seismic technology. A more general but still valuable goal is the detection of other stratigraphic information, such as sand-shale ratios, the thickness, continuity, and spacing of sandstone beds, and estimates of reservoir properties. It is this second goal which we will emphasize in our discussion.

The step from seismic field records to stratigraphic interpretations can be made only, we believe, by using the concept of sedimentation models. Sedimentation models summarize the characteristics of strata deposited in a certain environment and predict the anticipated thickness and continuity of various lithologies, the type and rate of lateral change, and the nature of contacts. These sedimentation models indicate that stratigraphic sequences are not random stacks of the various lithologies, but commonly are systematically organized. If lithologic changes are systematic, then so too may be the rock velocities and densities which govern reflections. It is the use of reflection seismic data for recognizing and deciphering such systematic stratigraphic organization in areas of sparse well control that holds most promise.

SEDIMENTATION MODELS

A sedimentation model is a simplified summary of the properties of strata deposited in a general environment. Within any environment, only certain processes of erosion, transport, and deposition of sediment operate. These processes impose restrictions on the thickness, continuity, and spacing of beds, the nature of contacts, and the type and rate of vertical or lateral change. The concept of models and comparisons of various depositional environments were discussed by Potter (1967), Shelton (1967), and Visher (1965). The geometric comparison of sandstone bodies of different depositional types is illustrated in the volume assembled by Peterson and Osmond (1961).

Only three of the better known and contrasting sedimentation models are described in this report. Important oil reservoirs are found in all three. Other models exist, but the attempt here is to illustrate a principle rather than to present a complete catalog of sedimentation models.

Shoreline and shallow marine model

Shoreline and shallow marine environments represent one of the most studied and best understood models, and many oil reservoirs occur in such sequences. An example of shoreline and shallow marine environment where sand is supplied mainly along shore, that is, not on an active delta fringe, is illustrated in Figure 1. Modern shorelines of this type exist along the central Texas Gulf Coast, the southern Atlantic coast of the U. S., the German North Sea, and many other places. Waves in the sea are the principal energy source in this setting. The shoreline profile develops in response to the typical waves, and sand is moved offshore to a depth dictated by the depth of significant wave motion. Tides and nearshore currents also influence the distribution of sand. But beyond a certain depth, between 50 and 100 ft off most modern coasts, sand grains are seldom moved; and deposition is mainly silt and clay grains settling from suspension out of slightly turbid water masses moved by tidal or nearshore currents. With time, if sand is supplied along the shoreline faster than the rate of subsidence, deposition drives the shoreline profile seaward.

These depositional processes dictate many of the characteristics of strata that will be encountered in such an environment. The sandstone bodies are generally extensive sheets or elliptical plates. Their maximum thickness, governed by the depth of wave action, rarely exceeds 100 ft. The width of such bodies is at least several miles, and they can be broadly continuous at some stratigraphic levels if sea level was constant over long periods of time. These sandstone units commonly have sharp upper contacts where they are overlain by lagoonal or marine muds. Because high energies are concentrated in the beach and surf zones, the cleaner and coarser sands also occur near the top of the bodies. The bases and margins of these sandstones are transitional, with marine deposits on the one side and lagoonal deposits on the other. The bodies thin rather gradually from their maximum development toward the seaward margin. As for the mudrock units that enclose the sandstones, bedding on the marine side tends to be very continuous and parallel. In many cases, the content of silt decreases in a seaward direction offshore. On the lagoon side, bedding is much less continuous and the deposits are silty, coaly, and in some places contain tidal or stream channels.

Some of the details of this sedimentation model depend upon the relative supplies of sand and mud, rate of sediment supply, and rate of basin subsidence, as pointed out by Sears et al (1941, Pl. 25), Spieker (1949), and Young (1955, Figure 4). The broadly continuous sandstone units illustrated in Figure 1, which suggest large cyclic shifts in shoreline position, are common in the Tertiary and Mesozoic rocks of many areas. Outstanding examples of Mesozoic transgressive-regressive shoreline deposits, based upon detailed surface or subsurface control, were illustrated by Sears et al (1941, Figure 22), Spieker (1949, Figure 2), Young (1955, Pl. 3), Weimer (1961, Figures 5 and 13), and Asquith (1970, Figures 1, 19, 20, and 34). Very thick bodies of sandstone unbroken by shale beds would be expected only where the rate of sediment supply and basin subsidence were nearly balanced; in such a case, the position of the shoreline would remain stable over long periods of time.

Meandering stream model

Deposition by meandering streams would result in sediment units that are very different from those found in a shoreline and shallow marine setting. Streams meander and shift constantly in time and space. The transporting energy is exerted by the flow of the stream itself, so that sandstone

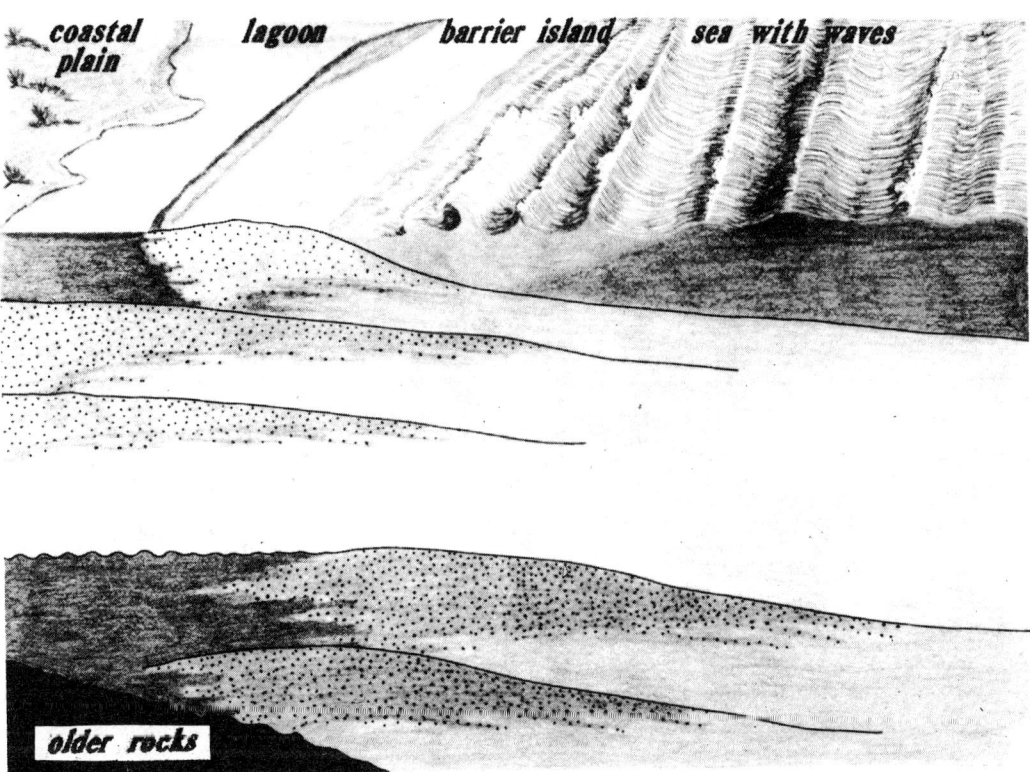

FIG. 1. Shoreline—shallow marine model. Sandstone bodies are commonly extensive. Wave energy concentrates the cleanest sands along beach and very nearshore zones; sandstone bodies have sharp tops and transitionally increasing silt and clay content laterally and downward.

bodies are localized in or near main channels (Figure 2). Those sandstone bodies, like the streams, tend to be sinuous ribbons. Some major rivers scour to depths of perhaps 200 ft during floods, so that the maximum thickness of individual channel fills can be as much as 200 ft, although they are commonly considerably less. Width of such bodies can range from a few hundred feet to several thousand feet, depending on the size of the stream and the length of time that the active channel occupied a specific meander belt. Streams erode along their bottoms and at their banks, so that the sand bodies tend to have sharp bases and margins. Cleaner and coarser sands occur near the base of such deposits because the greatest energy is exerted in the deepest parts of the channels, especially during floods. As a stream channel is abandoned in this continuing process of shifting, silt and clay may be mixed with sand, so that the upper parts of the channel bodies are commonly transitional with overlying mudrock units. Bedding within the shales and siltstones that enclose such channels is commonly complex and relatively discontinuous. For example, the channels are often enclosed in an envelope of silts deposited along natural levees that grade outward onto the flood plains into more clayey and peaty deposits in short distances. These mudrock bodies are themselves commonly later cut by younger channels and partly destroyed. An extensive review and bibliography of river sediments was compiled by Allen (1965).

Deposits built by streams, shown schematically in Figure 2, have extreme stratigraphic variability. Ancient examples are well known in continental facies of many ages. A variety of surface and subsurface examples were illustrated by Nanz (1954, Figure 4), Fisher and McGowen (1969, Figure 3), Koesoemadinata (1970, Pl. 23 and 24), and Hilpert (1969, Pl. 4). Such deposits conform most closely perhaps with the concept of randomly stacked lithologic units. Although broadly continuous sandstone units do occur in such sequences, they commonly represent an

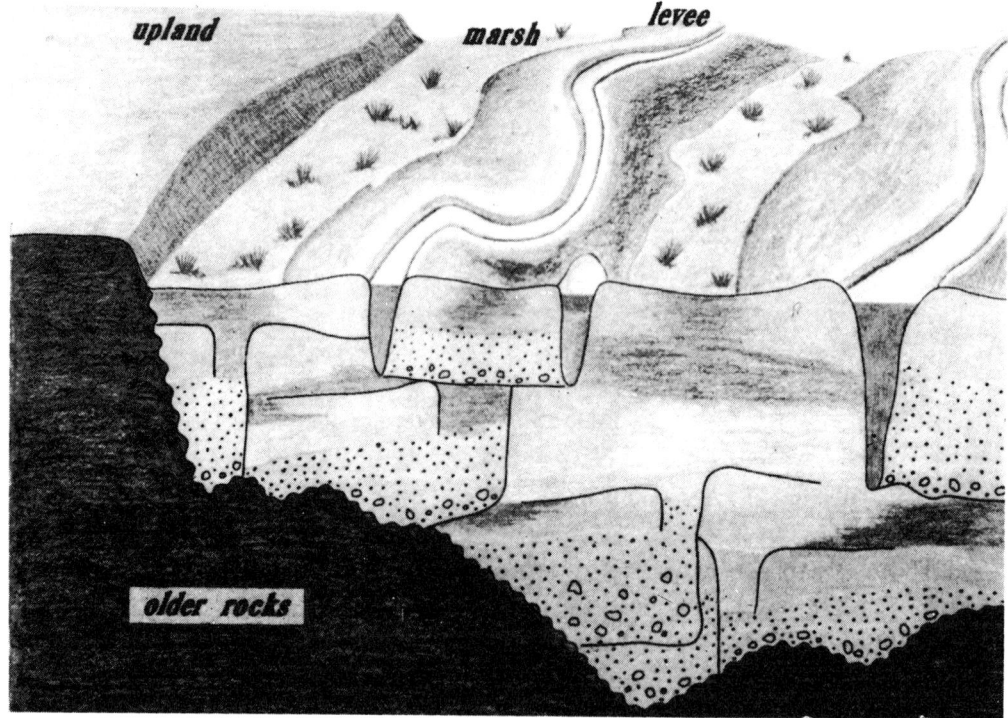

FIG. 2. Stream model. Sandstone bodies are commonly shaped like sinuous ribbons, and discontinuity is increased by repeated scour and channel shifts. Scouring currents concentrate the cleanest sands in the deepest parts of channels; sandstone bodies have sharp bases and margins and transitionally increasing silt and clay content upward.

FIG. 3. Deep-sea fan model. Sandstone bodies are commonly extensive bundles of individually thin beds in the lower fan. Turbidity currents deposit numerous thin beds of sand repeatedly in low areas of the fan; bundles of sandstone beds may have sharp bases and tops and wedge out laterally against bathymetric highs existing at the time of deposition.

episode of notable tectonic activity. Regional unconformities would be expected at the bases of such units.

Deep sea fan model

Coarse detritus is moved into deep ocean basins by turbidity currents, dense suspensions of sediments created at river mouths, shorelines, or the heads of submarine canyons. These dense suspensions flow under the clearer, lighter marine water, moving through canyons, down channels across the upper parts of fans, and then spreading as thin, turbid sheets over the lower fan and abyssal plains. In the lower fan areas, illustrated in Figure 3, sandstone bodies are extensive sheets that are formed by numerous thin beds deposited individually by the turbid flows. Because the water is deep and the process operates as long as a slope is maintained, there is no limit to the maximum thickness of such sand bodies (although individual beds are only a few inches or a few feet thick). The fan builds upward by the shifting of turbidity currents through time from one lower area to another.

The sandstone bodies deposited by this process can be continuous over broad areas, and their maximum width may be controlled only by the dimensions of the fan itself, commonly many miles. Both the tops and the bases of the larger sandstone units can be sharp, in contrast to the situation characteristic of the shoreline and shallow marine and stream models previously described. The sandstone units, composed of bundles of individually thin beds, wedge out at their margins against bathymetric highs, which are either uplifts of the older seafloor or previously developed fans. In the direction down the slope, the sandstone units thin gradually across abyssal plains as coarser material is lost because of diminished gradient.

Shales and siltstones are deposited across the lower fan and abyssal plain by settling of suspended material from turbid water masses. The clay and silt-bearing water masses may move slowly downslope because of their slightly greater density or parallel to bottom contours following currents generated by oceanic circulation. The bedding in the fine-grained units is very continuous and parallel. Beds formed from muddy sediment mantle many of the existing bathymetric highs.

The proportions of terrigenous sand and mud supplied to the basin determine the proportions of sandstones and shales. During periods of slack land supply, calcareous or siliceous pelagic material can be deposited and form beds of appreciable volume. On the lower parts of deep sea fans and on abyssal plains, strata are very continuous and lateral rates of change of thickness or texture are gradual compared to those of most other sedimentation models. Near the heads of deep sea fans or within submarine canyons, channeling is common; irregularities in stratification may be great; and the sediments may be dominantly coarse textured. At those sites, persistent beds may form only during periods when solely fine terrigenous or pelagic material is supplied. Figure 3 emphasizes stratification only in the lower part of a deep sea fan complex.

There are many well-known ancient and modern examples of turbidity current deposits from almost every geologic period, summarized in part in a volume edited by Bouma and Brouwer (1964). The topography of modern deep sea fans and abyssal plains was illustrated by Menard (1960, Figure 1), Hand and Emery (1964, Figures 3 and 8), and Normark (1970, Figure 22). The distribution of turbidite layers and resulting seismic reflections in the deep sea have been reported in many publications, a notable example being that of Emery et al (1970). Ancient turbidite sequences with persistent bundles of sandstone beds have been illustrated by Natland and Kuenen (1951, Figures 4 and 5), Sullwold (1961, Figures 1, 2, and 7), Walker and Sutton (1967, Figures 10 and 11), Rad (1968, Figure 7), and Enos (1969, Figures 2, 4, and 25).

SEDIMENTATION MODELS AND SEISMIC RESPONSES

Sedimentation models summarize a large amount of geologic data. The three examples cited are both simple and distinct. There can be a great deal of variation and complexity superimposed on these simple themes. Although oversimplified, these models do provide a basis for considering, and illustrations of, the interplay between stratigraphy and seismic responses. Each model is different in terms of the nature of contacts between lithologies, the thickness, continuity, and spacing of lithologies, and the nature and rate of lateral changes. These differences influence reflection seismic traces in predictable ways.

Influence of lithologic boundaries on reflections

The boundaries between sandstone bodies and enclosing mudrocks are generally quite different in each of the three models illustrated in Figures 1 to 3 (Visher, 1965; Shelton, 1967). This persistent and predictable difference of contacts can influence the reflections associated with these interfaces. Figure 4 illustrates how these contacts modify synthetic reflection traces for wavelets of three different frequencies.

Velocity distributions for individual units within each of the three models have the forms illustrated along the left column of Figure 4, if the simplifying assumptions are made that velocity increases with sand content, mudrocks have a velocity of about 9000 ft/sec, and clean sandstones a velocity of 12,000 ft/sec. Many observations support the notion that velocity in detrital rocks increases with sand content and decreases as shale laminae or clay matrix become more abundant (Sarmiento, 1961, Figure 5; Pickett, 1960, Figures 9 and 10; Tegland, 1970, Figures 5 and 14). In shallow marine and nearshore sandstones, the upper contact tends to be sharp, the coarsest and cleanest sandstone occurs near the top of a unit, and the base is transitional with underlying mudrocks. Therefore, the velocity distribution would be wedge shaped with the highest velocities near the top of the unit. The second velocity distribution shown in Figure 4 is also wedge shaped, but the higher velocities occur near the base of a sandstone unit. This velocity distribution would be typical of channel sands deposited by a stream, because such sandstones tend to have sharp bases, the coarsest and cleanest sandstone concentrated near the base, and finer-grained, transitional upper contacts. The third velocity distribution shown on Figure 4 has both a sharp top and a sharp base, like the turbidity current sandstones on deep sea fans. To simplify the illustration, each sandstone unit was considered to be 100 ft thick. This is approximately the maximum thickness anticipated for most shoreline and stream sands.

Synthetic reflections for three Ricker wavelets of 25, 45, and 60 hz are shown for each of these velocity distributions. The contrast between reflections caused by these units is clearest for the higher frequencies. The wavelets reflected from the upper and lower contacts are shown to the right of each trio of synthetic traces. These two wavelets are added together to form the traces shown in the solid dark lines.

The nature and sequence of acoustic boundaries can be interpreted from the higher frequency reflections shown on Figure 4. The polarity and amplitude of the synthetic seismogram deflections are different for a sharp-topped, transitional-based reflector, as compared with those from reflectors which have transitional tops and sharp bases or from reflectors with both sharp tops and bases. In these idealized examples, we can reconstruct vertical lithologic variations within reflecting sequences from the seismograms themselves, using the polarity and amplitude of reflections and the average velocity of the section. Reconstruction is one of the important goals of interpretation of seismic records for stratigraphic purposes. Additional examples of the influence of sharp and transitional contacts and velocity distributions were given by Sengbush et al (1961, Figures 13, 16, and 19). The complicated factors influencing reflection amplitudes were reviewed by O'Doherty and Anstey (1970).

On field records, it may not be possible to establish absolute polarity of reflections. However, polarity may be determined relative to a prominent and consistent reflection whose polarity can be estimated from acoustic logs, interval velocity changes, or stratigraphic knowledge. These relative polarities can be compared in the same manner illustrated in Figure 4.

Although the goal of detailed stratigraphic interpretation appears possible, several very restrictive simplifying assumptions have been made. First, the synthetic records shown on Figure 4 are noise free. There are no multiples or reverberations in the records to complicate the reflection pattern. Also, the characteristic wavelets are generated without interference of reflection events associated with directly overlying beds. Second, differences in velocity distribution within the reflectors are discernible only for the higher frequencies. Notice that the reflection pattern for all three beds is nearly identical for frequencies of 25 hz, using the velocities and thickness of the units selected for this example. Higher frequency signals may be too weak relative to noise to use on many field records, particularly at greater depths. Third, velocity has been made directly proportional to sand content in these examples. Other controls of velocity, such as burial depth, differential cementation, fluid content, or fluid pressure, have been excluded. Some of these points are reiterated in the last section.

Distribution of reflecting beds

The nature of the contacts is not the only characteristic of sedimentation models that might be expressed seismically. Continuity, spacing, and lateral changes of reflectors are also important. The distribution of sandstone and shale units is shown in a stratigraphic cross-section of a Cretaceous marine regressive deposit in Figure 5a. The cross-section is oriented approximately perpendicular to the ancient shorelines. Each of the sandstone units was deposited at or near a shoreline during periods of relatively stable sea level stand. The sandstones represent periods of local regression; the intervening marine shales represent temporary transgressions. Through the time recorded by 1000 ft of deposition, sand bodies built farther and farther seaward (to the right), indicating that the rate of sediment supply somewhat exceeded the rate of basin subsidence. This example is based upon Upper Cretaceous rocks exposed in the Book Cliffs (Young, 1955, Pl. 3; Young, 1957, Figure 3).

Each sandstone unit, 40 to 100 ft thick, has a sharp top and transitional base and margins. The character of spontaneous potential and resistivity logs shows that the sand content increases upward in each unit (Young, 1957, Figure 7; Munger, 1965, Figures 6 to 9). Seaward of the extreme ends of the sandstone units, spontaneous potential and resistivity logs have only small excursions because the marine shales are relatively homogeneous. Beds of higher resistivity and greater silt content within the marine shale section can commonly be traced laterally to the distal edges of sandstone units.

The stratigraphic section (Figure 5a) can be converted to a synthetic seismic section (Figure 5b) by assuming that velocity depends directly on sand content and therefore follows a configuration like the resistivity logs. The light gray curves shown at 5-mile intervals on Figure 5b indicate wedge-shaped velocity distributions, abrupt at the top and tapering toward the base. Synthetic seismograms derived from those velocity distributions are shown in heavy lines, also plotted at 5-mile spacings. The vertical scale of Figure 5b has been converted to time, using a velocity of 9000 ft/sec for shale and 12,000 ft/sec for sandstone.

The first reflections from the top of the mass and reflections from beds spaced widely enough to avoid interference have polarity and form that indicate reflectors with sharp tops and transitional bases. But the reflections also have great lateral continuity, gradually decreasing amplitudes in the seaward direction, and spacings that change only gradually over distances of miles. Even where reflecting units are closely spaced and cause interference for 60 hz frequencies, the interference patterns change slowly. In the part of the section with the greatest stratigraphic variability, where sandstone bodies are closely stacked in the upper central part of the stratigraphic diagram, the seismograms also show the greatest variability. Although the configuration of individual units cannot be determined using even these high frequencies, the sites of greatest stratigraphic variability could be selected from the synthetic seismic section.

Stream deposits can be illustrated on similar stratigraphic and seismic cross-sections (Figures 6a and 6b). This section shows some Cretaceous fluvial sandstones and mudrocks in central Mississippi, cut in a direction about perpendicular to the trend of the sinuous channels. Note that the thickness and lateral extent of this section are much smaller than for the shoreline and shallow marine example of Figure 5.

Sandstones in the channels have sharp bases, the cleanest sand near the base, and merge transitionally upward into mudrocks, as shown by spontaneous potential and resistivity logs; these features are common characteristics of many channel deposits (Hopkins, 1958, Figure 7; Harms, 1966, Figure 12; Shelton, 1967, Figures 2a and 2b). The cross-section shown in Figure 6a is based on fairly closely spaced well control within a field, so that the margins of channel sandstones must be very abrupt. The mudrocks and thin

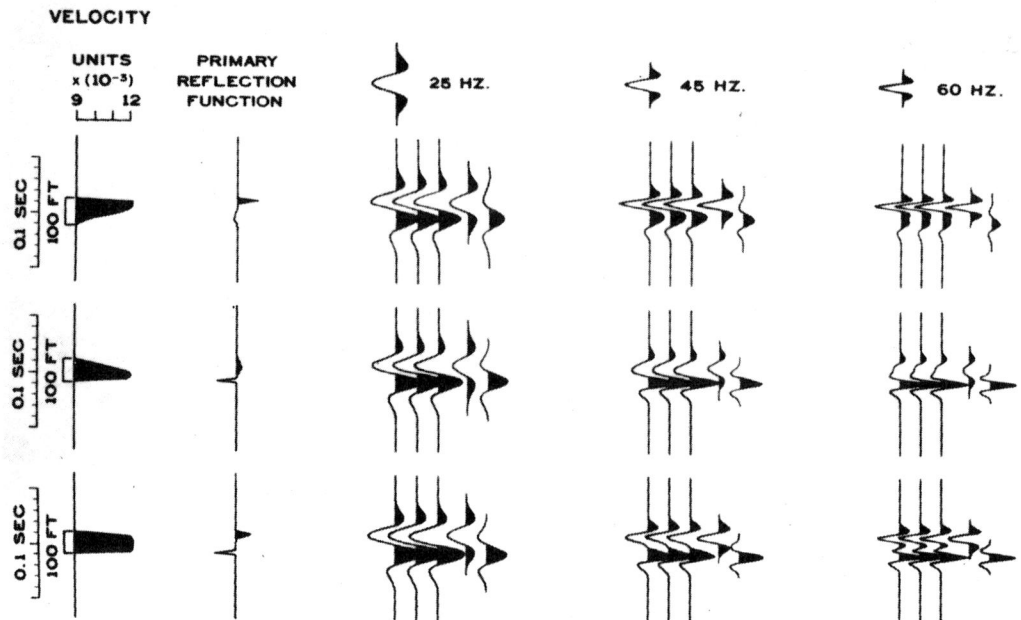

FIG. 4. Influence of contacts on reflections for three models. Velocity distributions are shown at the left and Ricker wavelets of three frequencies across the top. Individual wavelets associated with upper and lower contacts are shown to the right of each trio of synthetic traces.

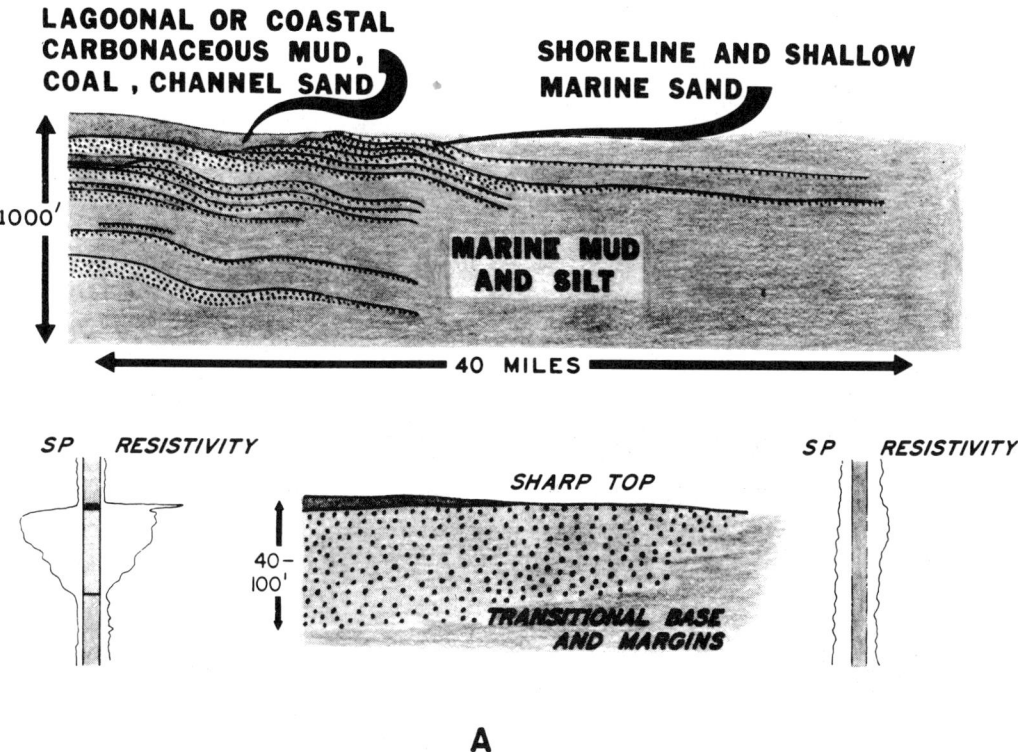

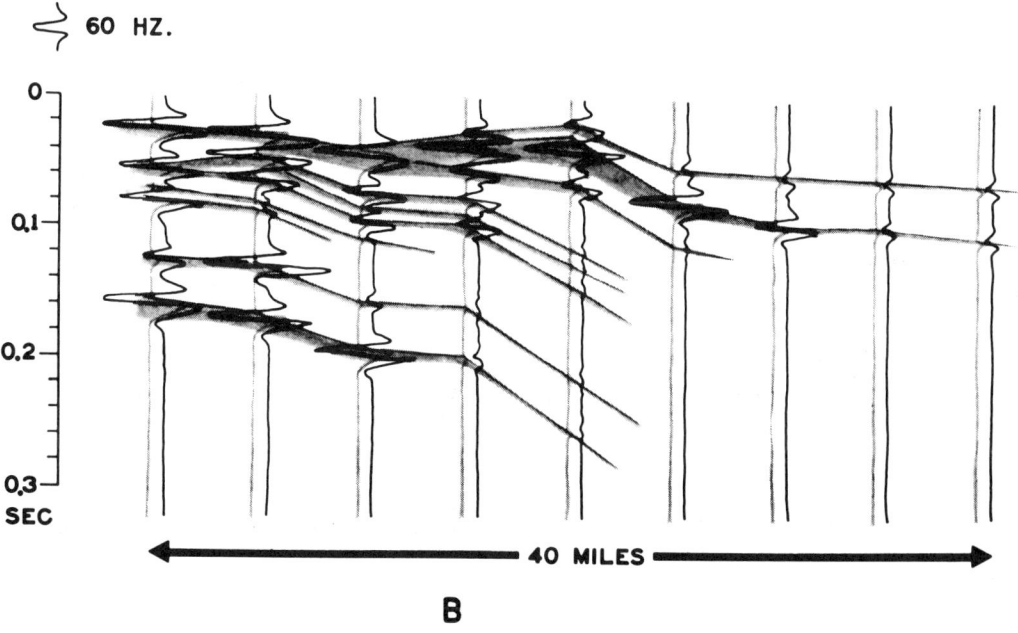

FIG. 5. Reflectors in regressive marine deposits. (A) Distribution of lithologies on a large scale and mechanical logs typical of sandy and shaly facies of an individual sandstone unit. (B) Synthetic seismic section of the stratigraphic section shown in A. The synthetic traces are heavy lines; the velocity values from which each trace was derived are shown to the left as lighter lines, using the same assumptions as in Figure 4. Correlative sandstone units are marked as shaded bands.

sandstones outside of channels have extremely variable spontaneous potential and resistivity logs. Stratigraphic variability in these finer-grained facies is also great, and few units can be correlated over distances exceeding 2 or 3 miles.

Reflections from such a section (Figure 6b) are

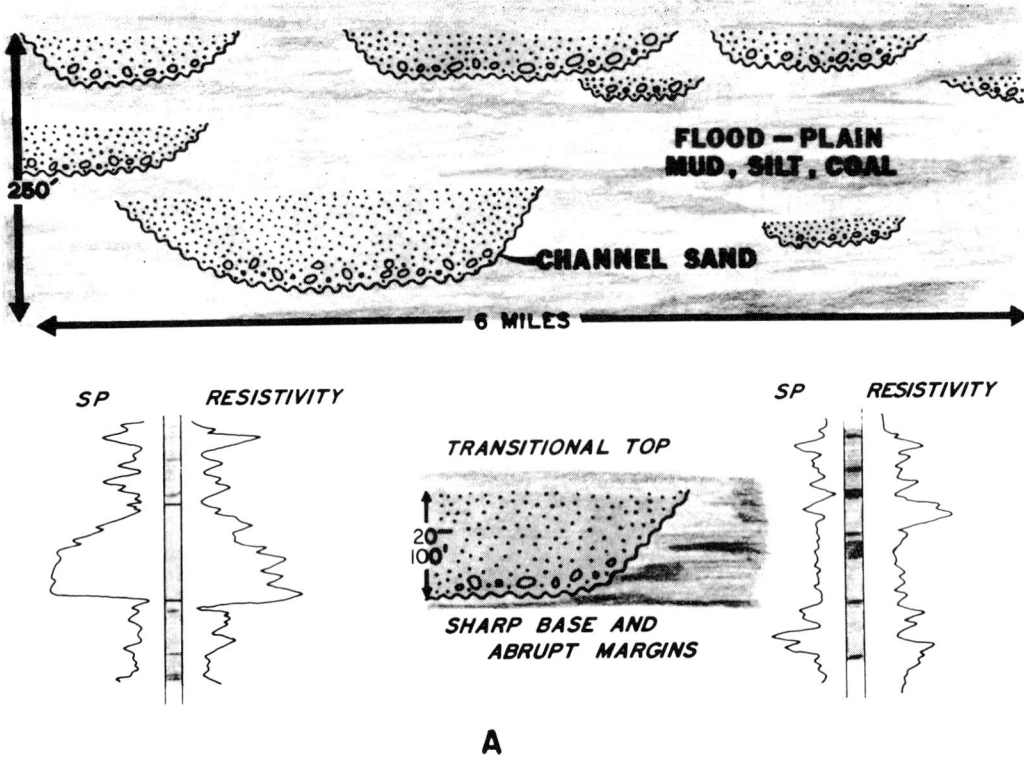

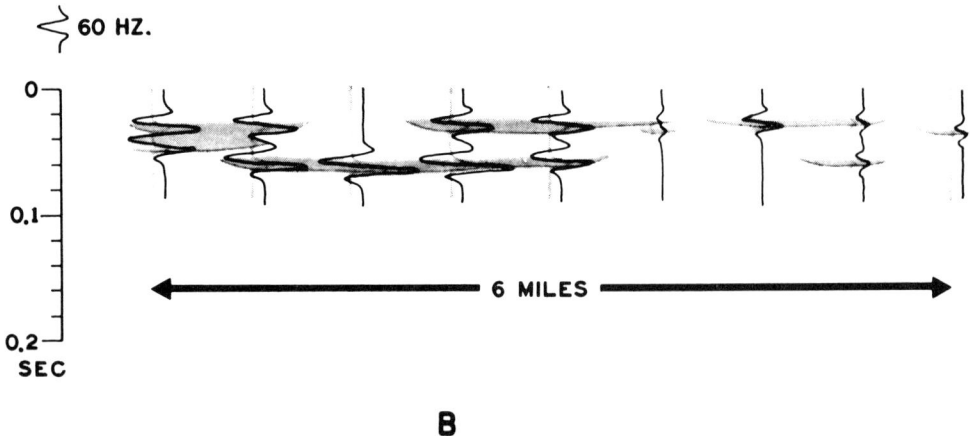

Fig. 6. Reflectors in stream deposits. A and B are explained by the caption of Figure 5.

also discontinuous and variable in character. The gray lines at three-quarter-mile intervals show velocity distributions dependent on sand content alone. The velocity distributions are wedge shaped with sharp bases and transitional tops. The first reflections from the top of the section or from within the section where interference is absent show polarities that indicate a sharp-based reflector. Reflection amplitude is small for thin and less sandy channels. The extreme discontinuity of reflections and variable interference indicate the discontinuous nature of sandstone bodies within this sedimentation model.

The third sedimentation model, deep-sea fan deposits, is shown in Figures 7a and 7b. This example was taken from Tertiary turbidity current deposits in the Ventura basin, California, in a section approximately perpendicular to dominant transport direction. Note that the dimensions of this example are similar to those of Figure 5, but much larger than those of Figure 6.

The dominantly sandstone units can be very thick in this sedimentation model, ranging up to 500 ft. These units are composed of bundles of individually thin graded beds, each commonly less than a few feet thick, as illustrated very schematically in the lower part of Figure 7a. These bundles of sandstone beds tend to have sharp bases and sharp tops. They wedge out along their margins by onlap against slopes representing ancient bathymetric highs. Mudrock sections contain vertically variable amounts of silt and clay, but beds within mudrock units are very continuous and parallel.

The synthetic seismic section for these deep-sea fan deposits is shown in Figure 7b. The distinct synthetic reflections caused by the box-like velocity distributions indicate by their polarity that the reflectors have both sharp tops and bases. The onlap thinning of sandstone units causes reflections to converge laterally, but without an attendant loss of amplitude or change in character until thinning is sufficient to cause interference. Seismic sections of Cenozoic deposits under modern deep seas show reflections from extensive turbidite bundles (Emery et al, 1970, p. 99) and onlap against sea floor highs (Scholl et al, 1968, Figures 2a and 2b).

Average velocity

Average velocities, computed from reflection moveout times for record stacking purposes, have already been pressed into service to aid in stratigraphic interpretation. Computer approaches required to make velocity analyses have improved markedly in the past few years. If sand-shale sequences are sufficiently thick (1500 to 2000 ft), continuous velocity analysis along the sections shown in Figures 5, 6, and 7 would add another facet to interpretation.

Interval velocities for the section containing regressive marine deposits (Figure 5) would decrease systematically in a seaward direction. Interval velocities for stream deposits (Figure 6) could be irregular and unsystematic, like all other aspects of the seismic response. Interval velocities for the deep-sea fan deposits (Figure 7) would decrease, perhaps fairly abruptly, toward ancient bathymetric highs. Note, however, that an analysis of reflection waveforms (Figure 4) could be applied to stratigraphic intervals that are too thin for detection by velocity analysis methods.

REQUIREMENTS FOR IMPROVED STRATIGRAPHIC INTERPRETATION

Synthetic seismic sections are distinctly different for the three sedimentation models selected for this discussion. Certain simplifications are incumbent upon this synthetic approach; they include close control of velocity by sand content, relatively high-frequency signals, and noise-free records. If differences could not be observed under these ideal conditions, there would be no point in considering much less ideal field conditions. But because some promise is suggested by these simplified models, it is pertinent to discuss what knowledge or techniques must be developed to use field records successfully for stratigraphic interpretation.

More data must be gathered on the relationship of lithology to acoustic velocity. Velocity appears in general to be controlled by sand content in many stratigraphic sequences. This observation suggests that the depositional controls are transferred fairly explicitly into acoustic properties. However, exceptions are noted where compaction or cementation is variable, gas saturation is high, or pore fluid pressures exceed hydrostatic.

The most useful aspects of seismic records for stratigraphic interpretation are relative polarity of reflections, areal change in reflection ampli-

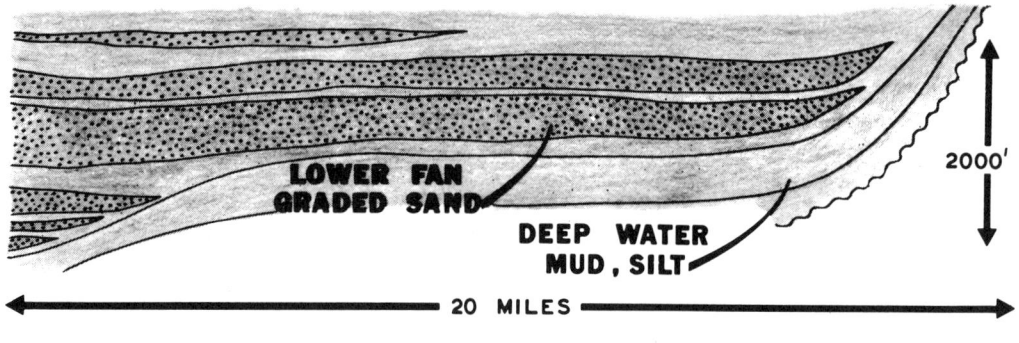
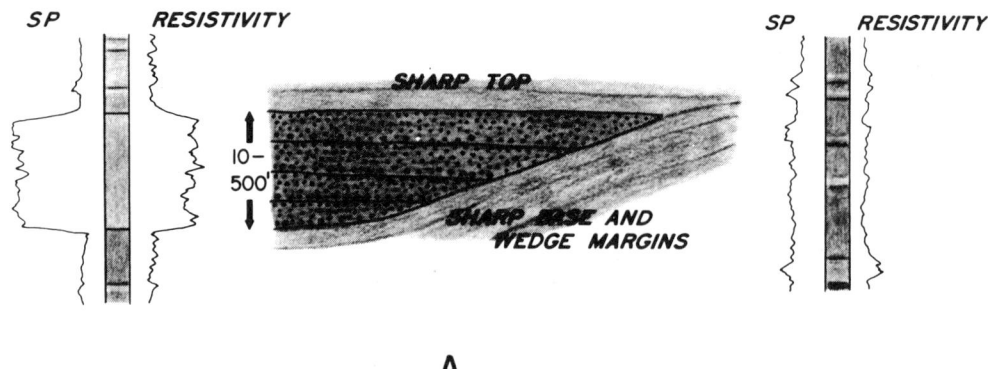

A

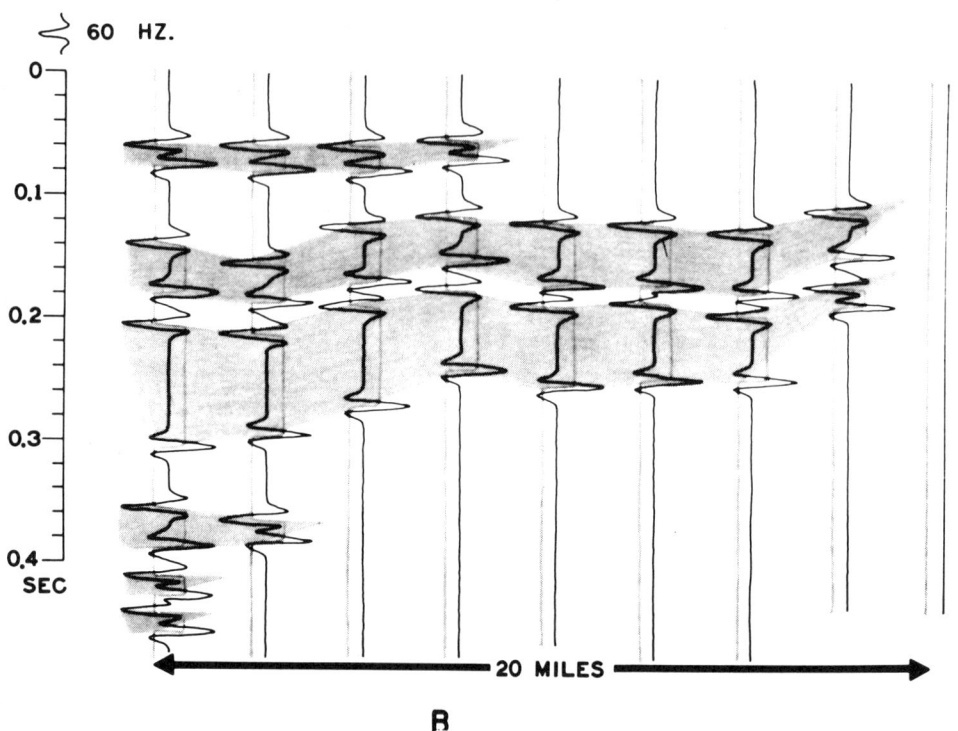

B

FIG. 7. Reflectors in deep-sea fan deposits. A and B are explained by the caption of Figure 5.

tudes, continuity of individual reflection events, and interval velocities. Of these, only continuity of prominent events and interval velocity determinations are emphasized in current record-processing techniques. Additional work must be done on evaluating the polarity of reflection events and comparing quantitatively areal changes in reflection amplitudes (O'Doherty and Anstey, 1970). Attempts to improve methods for extracting high-frequency, noise-free components of seismic signals must be continued.

Although these are difficult goals, partial attainment is possible. The strides made in record processing for other purposes suggest that concerted efforts can produce valuable results.

CONCLUSIONS

1. Sedimentation models summarize much geologic information. Lithologic units have characteristic contacts, thickness, spacing, lateral extent, and lateral changes that depend directly on depositional processes. These characteristics can be usefully applied to exploration problems, especially in sparsely drilled areas.

2. If the lithologic variations summarized by sedimentation models control velocities, different stratigraphic sequences should yield recognizable and systematically different reflection seismic cross-sections.

3. Synthetic seismic sections are markedly different for the three different sedimentation models selected for study. Incumbent simplifications include close control of velocity by sand content, relatively high-frequency signals, and noise-free conditions.

4. The most useful aspects of seismic records for stratigraphic interpretation are polarity (or relative polarity) of reflection events, areal change in reflection amplitudes, continuity of individual events, and interval velocities. Of these, only continuity of prominent events and interval velocity determinations are emphasized in current record processing.

5. If greater stratigraphic use of seismic information is to be made, record processing techniques that emphasize recognition of reflection polarity and amplitude, particularly for the higher frequency signal components, must be developed. Understanding of seismic boundaries in a variety of stratigraphic sequences must also be improved.

REFERENCES

Allen, J. R. L., 1965, A review of the origin and characteristics of Recent alluvial sediments: Sedimentology, v. 5, p. 89–191.

Asquith, D. O., 1970, Depositional topography and major marine environments, Late Cretaceous, Wyoming: AAPG Bull., v. 54, p. 1184–1224.

Bouma, A. H., and Brouwer, A., eds., 1964, Developments in sedimentology 3, Turbidites: Amsterdam, Elsevier Pub. Co.

Emery, K. O., Uchupi, E., Phillips, J. D., Bowin, C. O., Bunce, E. T., and Knott, S. T., 1970, Continental rise off eastern North America: AAPG Bull., v. 54, p. 44–108.

Enos, P., 1969, Anatomy of a flysch: J. Sed. Petrol., v. 39, p. 680–723.

Fisher, W. L., and McGowen, J. H., 1969, Depositional systems in Wilcox Group (Eocene) of Texas and their relation to occurrence of oil and gas: AAPG Bull., v. 53, p. 30–54.

Hand, B. M., and Emery, K. O., 1964, Turbidites and topography of north end of San Diego trough, California: J. Geol., v. 72, p. 526–542.

Harms, J. C., 1966, Stratigraphic traps in a valley fill, western Nebraska: AAPG Bull., v. 50, p. 2119–2149.

Hilpert, L. S., 1969, Uranium resources of northwestern New Mexico: USGS Prof. Paper 603.

Hopkins, M. E., 1958, Geology and petrology of the Anvil Rock Sandstone of southern Illinois: Ill. State Geol. Survey Circ. 256.

Koesoemadinata, R. P., 1970, Stratigraphy and petroleum occurrence, Green River Formation, Redwash field, Utah: Quart. Colorado Sch. Mines, v. 65, p. 1–77.

Lyons, P. L., 1968, The potential and the challenge—stratigraphic traps (abstract): Geophysics, v. 33, p. 1036.

Menard, H. W., 1960, Possible pre-Pleistocene deep-sea fans off central California: Geol. Soc. Amer. Bull., v. 71, p. 1271–1278.

Munger, R. D., 1965, Subsurface exploration mapping, southern Uinta basin, Castlegate and Dakota-Cedar Mountain formations: The Mountain Geologist, v. 2, p. 141–166.

Nanz, R. H., 1954, Genesis of Oligocene sandstone reservoir, Seeligson field, Jim Wells and Kleberg Counties, Texas: AAPG Bull., v. 38, p. 96–117.

Natland, M. L., and Kuenen, Ph. H., 1951, Sedimentary history of the Ventura basin, California, and the action of turbidity currents: Soc. Econ. Paleontologists and Mineralogists Spec. Pub. 2, p. 76–107.

Normark, W. R., 1970, Growth patterns of deep-sea fans: AAPG Bull., v. 54, p. 2170–2195.

O'Doherty, R. F., and Anstey, N. A., 1970, Reflections on amplitudes: presented at the Edinburgh meeting of the EAEG, May 20–22, 1970.

Peterson, J. A., and Osmond, J. C., eds., Geometry of sandstone bodies: Tulsa, AAPG.

Pickett, G. R., 1960, The use of acoustic logs in the evaluation of sandstone reservoirs: Geophysics, v. 25, p. 250–274.

Potter, P. E., 1967, Sand bodies and sedimentary environments, A review: AAPG Bull., v. 51, p. 337–365.

Sarmiento, Roberto, 1961, Geological factors influencing porosity estimates from velocity logs: AAPG Bull., v. 45, p. 633–644.

Scholl, D. W., von Huene, R., and Ridlon, J. B., 1968, Spreading of the ocean floor, undeformed sediments in the Peru-Chile trench: Sci., v. 159, no. 3817, p. 869–871.

Sears, J. D., Hunt, C. B., and Hendricks, T. A., 1941, Transgressive and regressive Cretaceous deposits in the southern San Juan basin, New Mexico: USGS Prof. Paper 193F, p. 101–121.

Sengbush, R. L., Lawrence, P. L., and McDonal, F. J., 1961, Interpretation of synthetic seismograms: Geophysics, v. 26, p. 138–157.

Shelton, J. W., 1967, Stratigraphic models and general criteria for recognition of alluvial, barrier-bar, and turbidity current sand deposits: AAPG Bull., v. 51, p. 2441–2461.

Spieker, E. M., 1949, Sedimentary facies and associated diastrophism in the Upper Cretaceous of central and eastern Utah: Geol. Soc. Amer., Mem. 39, p. 55–82.

Sullwold, H. H., Jr., 1961, Turbidites in oil exploration, *in* Geometry of sandstone bodies: Tulsa, AAPG.

Tegland, E. R., 1970, Sand-shale ratio determination from seismic interval velocity: 23rd Annual Midwestern Regional Mtg., SEG and AAPG, Dallas, Texas, March 8, 1970.

Visher, G. S., 1965, Use of vertical profile in environmental reconstruction: AAPG Bull., v. 49, p. 41–61.

von Rad, Ulrich, 1968, Comparison of sedimentation in the Bavarian flysch (Cretaceous) and Recent San Diego trough (California): J. Sed. Petrol., v. 38, p. 1120–1154.

Walker, R. G., and Sutton, R. G., 1967, Quantitative analysis of turbidites in the Upper Devonian Sonyea Group, New York: J. Sed. Petrol., v. 37, p. 1012–1022.

Weimer, R. J., 1961, Spatial dimensions of Upper Cretaceous sandstones, Rocky Mountain area, *in* Geometry of sandstone bodies: Tulsa, AAPG.

Young, R. G., 1955, Sedimentary facies and intertonguing in the Upper Cretaceous of the Book Cliffs, Utah-Colorado: Geol. Soc. Amer. Bull., v. 66, p. 177–202.

Young, R. G., 1957, Late Cretaceous cyclic deposits, Book Cliffs, eastern Utah: AAPG Bull., v. 41, p. 1760–1774.

INTEGRATION OF BIOSTRATIGRAPHY AND SEISMIC STRATIGRAPHY: PLIOCENE - PLEISTOCENE, GULF OF MEXICO

JOHN M. ARMENTROUT
Mobil Oil Corporation
P.O. Box 650232
Dallas, Texas 75265

INTRODUCTION

The sequential procedure for seismic stratigraphic analysis has been clearly outlined by Mitchum *et al.* (1974), Vail *et al.* (1977), and Sangree *et al.* (1978). In working the Gulf of Mexico Pliocene and Pleistocene, I have found that careful integration of paleobathymetric interpretations with seismically defined depositional architecture results in geologically reasonable map patterns. By using multiple data sets, anomalous patterns existing in any one data set can be analyzed and usually explained.

This paper focuses on procedures used and general interpretations made from a study of 100 exploration wells integrated with a regional seismic grid for the eastern Texas offshore. The study area encompasses the High Island, Galveston South and East Breaks areas (Fig. 1).

Schematic diagrams of two seismic lines (Lines A and B) are used as examples of the integrated approach. The first is characterized by relatively uniform cycles of sediment accumulation across a prograding shelf-slope system. This system is interrupted by growth-fault and salt-withdrawal-basin architecture. The second example consists of multiple shelf-slope clinoforms prograding across a tectonically stable shelf margin. Discussions include: (1) identification of time-significant *versus* ecologically controlled biohorizons; (2) recognition of inplace *versus* displaced faunal assemblages; (3) construction of biofacies models for ancient oceans; and (4) the impact of eustatic fluctuations on faunal distribution.

Documentation of the biostratigraphy and reproduction of the seismic Lines A and B (see Figs. 1, 4, and 8) will be published separately, but examples will be shown at the poster session of this meeting. The original study was done at Mobil Exploration and Producing Services Inc., and provides the basis for this discussion. The documentation of detailed biostratigraphy suitable for publication is being prepared by personnel of MICRO-STRAT, Inc. (please see Acknowledgments).

METHODOLOGY

Detailed analysis of stratigraphic architecture and seismic facies distribution begins with definition of depositional sequences. In the Texas offshore study area, the Pliocene-Pleistocene sequence boundaries are most clearly defined by high-amplitude continuous reflectors. These reflectors represent pelagic and hemipelagic drape

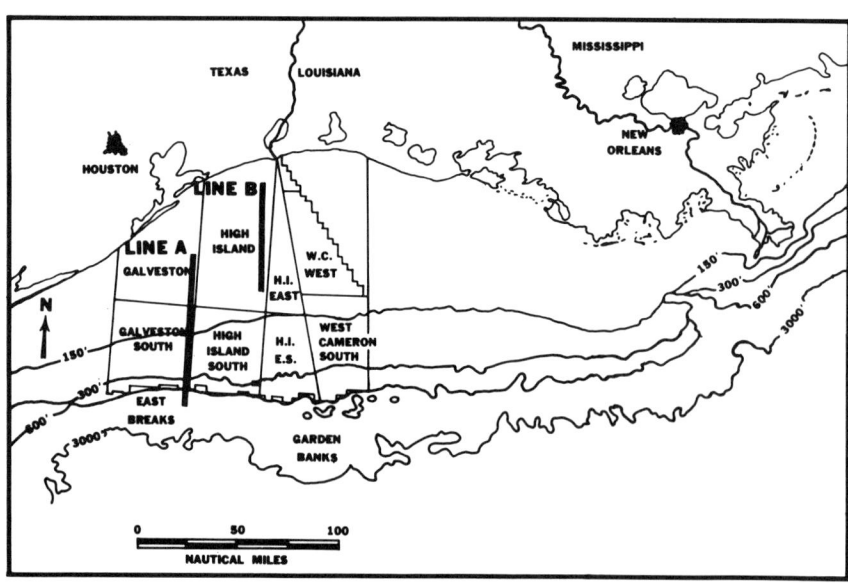

Figure 1. Map of study area showing location of seismic Lines A and B. The East Breaks 160 No. 1 well (Figs. 2, 5, and 6) is located on Line A immediately south of the 600-foot isobath.

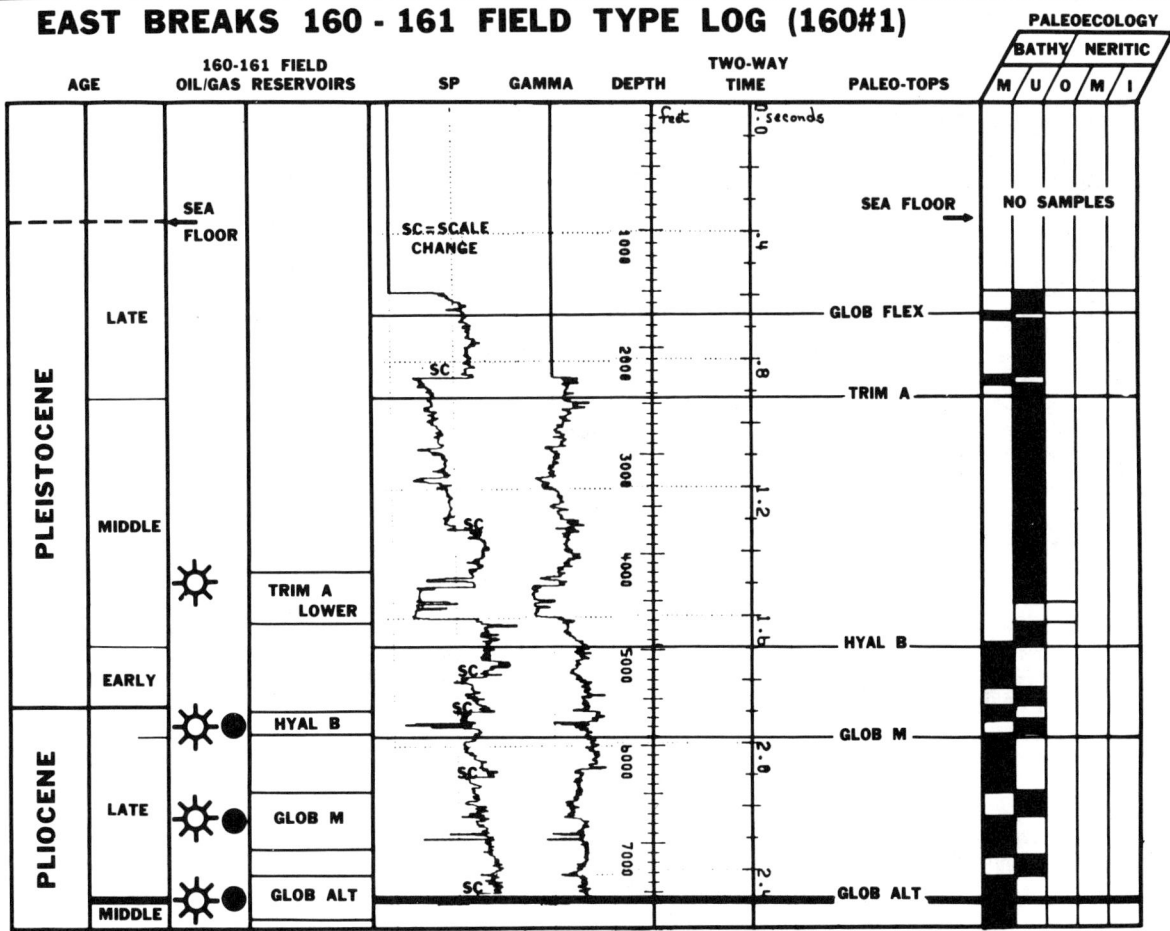

Figure 2. Example of well-data set used to calibrate seismic lines. Abbreviations for bioevents (paleo-tops) are: *Glob flex* = *Globorotalia flexuosa*; *Trim A* = *Trimosina denticulata*; *Hyal B* = *Hyalinea balthica*; *Glob M* = *Globorotalia miocenica*; *Glob alt* = *Globoquadrina altispira*. Paleoecologic abbreviations are: Bathy = Bathyal; M = Middle; U = Upper; O = Other; I = Inner.

across the basin, slope and outer shelf, and transgressive facies across the eroded inner shelf and coastal plain. Each depositional sequence contains those sediments accumulated between transgressions.

Correlation of sequence boundaries across faults is done initially by matching seismic packages having similar reflection character. This is difficult where faulting is coincident with facies boundaries resulting in age equivalent seismic packages having different reflection configuration and attributes. Age significant bioevents provide chronostratigraphic control points which permit correlation between wells. Figure 2 shows five bioevents (paleo-tops) for the Late Pliocene through Pleistocene strata penetrated by the East Breaks 160 No. 1 well. These five bioevents are defined by the first downhole occurrence of the microfossil listed, in this case planktonic and benthic foraminifers. Calcareous nannoplankton fossils are also useful for correlation in the Gulf of Mexico Neogene strata. [See Stainforth *et al.* (1975) and Poag and Valentine (1976)].

The bioevents used in subsurface work are identified from detailed checklists of species found in each 30-foot sample. The first downhole occurrence of a fossil represents the last or geologically youngest occurrence of that fossil within the stratigraphic section drilled. The last occurrence may be due to: (1) extinction of the species; (2) local extermination of the species due to environmental factors; (3) preservation or reworking artifacts; or (4) sample problems. Careful analysis of species occurrence patterns from several adjacent wells is required to optimize the identification of useful bioevents. Figure 3 shows the correlation of bioevents across a major growth fault where seismic reflection patterns do not provide a unique correlation solution.

The bioevents listed on Figure 2 were identified in a series of wells along seismic Line A (Figs. 1 and 4) and were used to calibrate the depositional sequences across a growth-fault salt-withdrawal-basin architecture. The correlation procedure involved: (1) identification of bioevents in each well; (2) annotation of those bioevents on the electric log; (3) correlation of the electric log to the seismic data using two-way time formatted logs and syn-

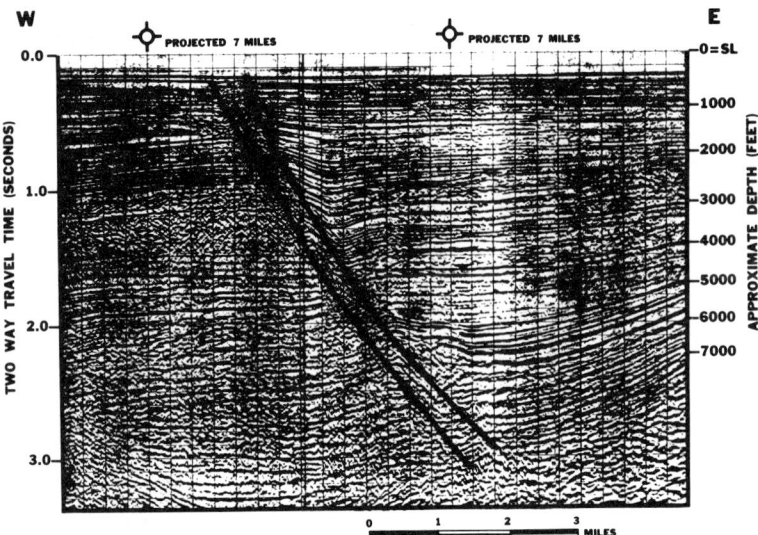

Figure 3. Correlation across a growth fault in the High Island area. Bioevent tops from the two wells were correlated into this section through a regional seismic grid. Extinction bioevents indicate that correlative datums intersect the figure margins for *Trimosina denticulata* at 1.0 seconds (left) and 4000 feet (right); for *Hyalinea balthica* at 1.5 seconds (left) and 7000 feet (right). This correlation suggests an offset of 3000 feet at *Trimosina denticulata*, and 6000 feet at *Hyalinea balthica*.

thetic seismic displays derived from the logs (Fig. 5). This procedure permits correlation of the depositional sequences across rapidly changing stratigraphic thicknesses and facies. In Figure 4, the sequence boundaries are shown by the heavy lines and are numbered 1 to 4. Each sequence shows a seaward thinning except where local thickening occurs downthrown to coeval growth faults. The growth faults result from differential sediment accumulation which loads and displaces salt or shale. The resulting depositional architecture reflects a retrograde pattern of progressive seaward development of growth-fault salt-withdrawal-basins. Each basin accommodates sediment accumulation until salt or shale can no longer be displaced and the depocenter steps seaward. Electric log and sample analysis provide local information on rock type. Rock type can be integrated with seismic facies analysis to define a probable lithofacies distribution pattern within the depositional architecture. Such a schematic facies distribution is shown for Line A (Fig. 4) with each depositional cycle containing prograding shelf facies and basinal aggradational mounds.

Distribution patterns of environmentally controlled organisms also provide information on the depositional environment. Benthic Foraminifera are especially useful in such work because they provide a tool for estimating water depth [see Tipsword *et al.* (1966) and Poag (1981)]. Paleobathymetric curves can be constructed from checklists of species occurrence and abundance in conjunction with rock type, and where conventional cores are available, analysis of sedimentary structures. Comparison of paleobathymetric curves between wells permits identification of local and regional relative sea level fluctuation events. Regionally significant sea level fluctuations provide an additional correlation tool for comparison with both seismic and bioevent correlations.

Figure 2 shows the paleoecologic interpretation of benthic foraminiferal assemblages from the East Breaks 160 No. 1 well. The entire well is interpreted to have been deposited in outer neritic to bathyal water depths. Depositional environments interpreted at the well location

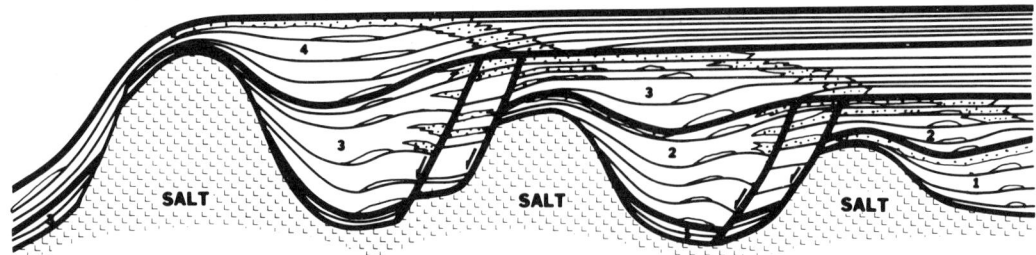

Figure 4. Schematic diagram of offlap depositional architecture based on interpretation of seismic Line A, approximately 50 mi north-south. The four cycles are seismically defined depositional sequences calibrated by bioevents from several wells along the section. Stippled pattern suggests sandy facies. Key to cycles shown: 1 = Early Pliocene, 2 = Late Pliocene, 3 = Early Pleistocene, 4 = Late Pleistocene

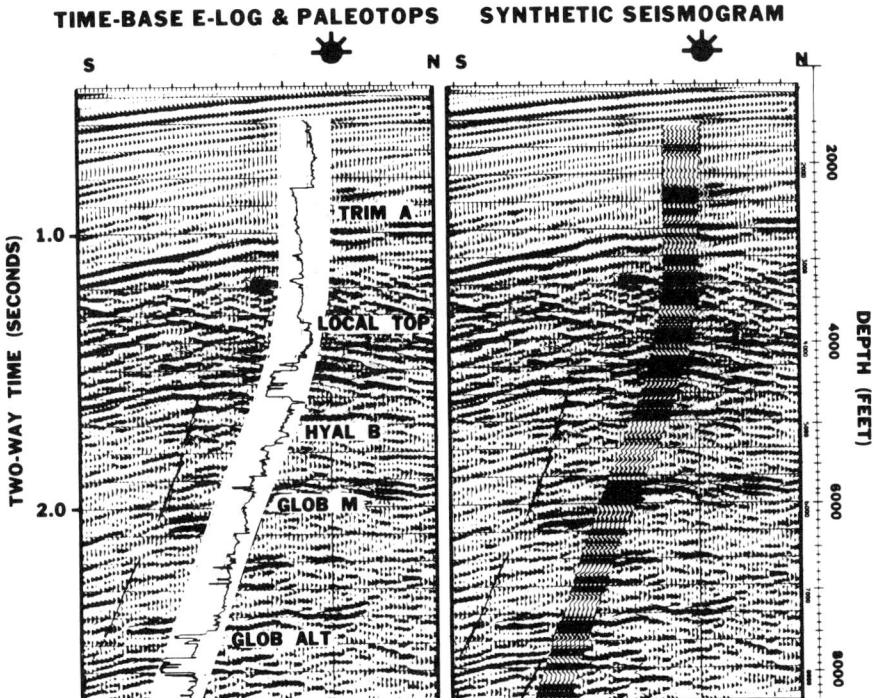

Figure 5. Calibration of seismic record section with two-way time formatted well data, including spontaneous potential electric log and synthetic seismogram; East Breaks 160 No. 1 Well.

cesses active far below normal wave-base. Paleoecologic interpretation of several wells along Line A is shown schematically in Figure 6. Calibration of seismic facies with biofacies data at each well permits extrapolation of biofacies patterns between wells.

Interpretation of paleobathymetry usually relies on models for foraminiferal distribution based on the occurrence of living forms. Extrapolation of modern models into the geologic past becomes increasingly less precise as the fossil assemblages become less similar to the modern and as ocean water-mass structures become less like the present. Analysis at the higher levels of taxonomy (family and genus rather than species) is one technique used to accommodate for the increasing number of extinct taxa (see Poag and Valentine, 1976). Another approach is to redefine the faunal assemblage characteristic of each biofacies for specific geologic intervals. This involves the reconstruction of the ocean floor topography and identification of the faunal assemblages present along the physiographic provinces of that sea floor.

Figure 7 shows a schematic reconstruction of the Early Pleistocene *Hyalinea balthica* datum along seismic Line A. The equilibrium sea-floor profile is reconstructed from seismic record sections and is based on the depositional architecture of the sequence being studied. Careful backstripping is required to remove tectonic and depositional overprinting of younger events, such as growth-faults and salt-withdrawal. Seismic facies analysis permits recognition of shelf/slope clinoforms between updip platform facies and downdip slope and basinal facies. The faunal assemblages recovered from samples along the reconstructed *Hyalinea balthica* profile provide a data base to identify biofacies assemblages. These assemblages can be used to test existing biofacies models or for the definition of new models that are age and basin specific. Assignment of absolute water depths to faunal assemblages defined from such paleoecologic reconstructions is dependent on the precision of the reconstructed sea-floor equilibrium profile [see McDougall (1980) and Olson (1987) for similar reconstructions].

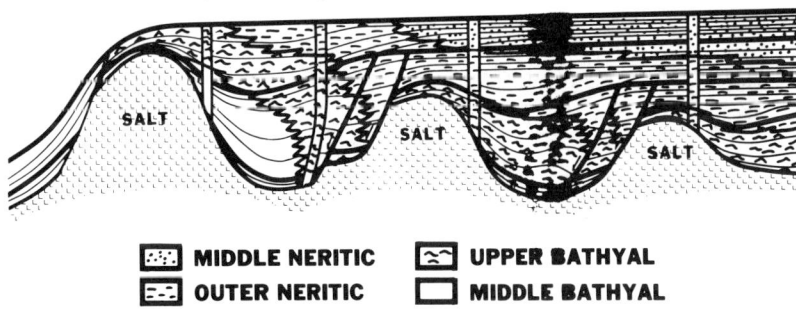

Figure 6. Schematic diagram for distribution of benthic foraminiferal biofacies based on well data (vertical columns) and extrapolated using the depositional fabric interpreted for Line A.

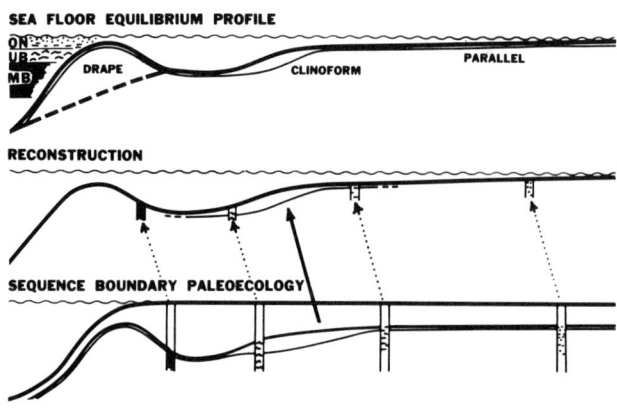

Figure 7. Paleoecologic reconstruction for Line A at *Hyalinea balthica* time. Biofacies patterns are indexed on Figure 6. Abbreviations: ON = Outer Neritic; UB = Upper Bathyal; MB = Middle Bathyal. Drape, Clinoform, and Parallel refer to seismic facies types interpreted as basinal, slope, and platform facies, respectively, and are used to construct the physiographic profile. The foraminiferal assemblages recovered from samples immediately below the sequence boundary (Sequence Boundary Paleoecology) are placed at the appropriate physiographic position (Reconstruction) along a sea-floor profile (Sea Floor Equilibrium Profile). Note that the reconstruction has not been corrected for the upper slope high. As a consequence, the middle bathyal biofacies of the leftmost well appears to reconstruct at the same apparent depth as the upper bathyal biofacies of the second well. Backstripping of the salt/shale tectonic high to a normal equilibrium slope profile (dashed line) would reconstruct the biofacies to a more typical position along the physiographic profile.

(1980) and Olson (1987) for similar reconstructions].

BIOSTRATIGRAPHIC PITFALLS

In analysis of seismic data over several tens of miles, bioevents often do not consistently correlate with seismically defined sequence boundaries. Along Line A, the bioevents consistently correlated with the seismic sequence boundaries, but along Line B the first downhole occurrences of *Globoquadrina altispira* (*Glob alt*) cut across the sigmoidal pattern of reflectors between the *Globorotalia miocenica* (*Glob M*) and *Globigerina nepenthes* (*Glob N*) datums (Fig. 8). In most of the Line A wells, the biofacies reflect relatively deep, normal marine waters whereas Line B wells reflect a much shallower inner to middle neritic depositional environment for most of the section above the *Globoquadrina altispira* occurrences. *Globoquadrina altispira* is a planktonic foraminifer that lives in open oceanic water masses, and is seldom found preserved in inner shelf sediments. Therefore, the first occurrence of *Globoquadrina altispira* in updip wells (A 104-1, A 72-1 and A 33-1) along Line B, where the first occurrence is immediately below the middle neritic biofacies, are interpreted as ecologically-depressed bioevents, not time-significant first occurrences. True extinctions are present in the downdip wells (A 446-1 and A 267-1) where the first occurrence is within outer neritic and deeper biofacies.

Recognition of the ecologically depressed bioevents permits selection of true extinction bioevents for seismic calibration. The seismic reflector can then be used to extrapolate the chronostratigraphic datum updip. An analogous ecologic problem exists in correlation of shallow water bioevents downdip into deep water facies. Resolution of these problems requires an integrated approach using several fossil groups carefully tied to seismic lines across the regional depositional dip. The integrated data set allows identification of the depth range within which a specific species is most useful as a chronologic datum, and which other species can be used as coeval or nearly coeval datums in shallower or deeper environments. Regional mapping datums are named for widely distributed, time-significant fossil taxa but in fact may be identified by coeval taxa in areas outside the optimal preservational environment of the index species.

Note in Figure 8 that the middle to outer neritic biofacies boundary extends down the face of the clinoform seaward of the seismically defined physiographic shelf/slope break. This may occur as a consequence of three factors:

(1) Lowering of sea level shifts the water mass – substrate-linked biofacies assemblages seaward, possibly far enough to place the inner neritic biofacies at the shelf/slope break; this shifts the middle to outer neritic bio-

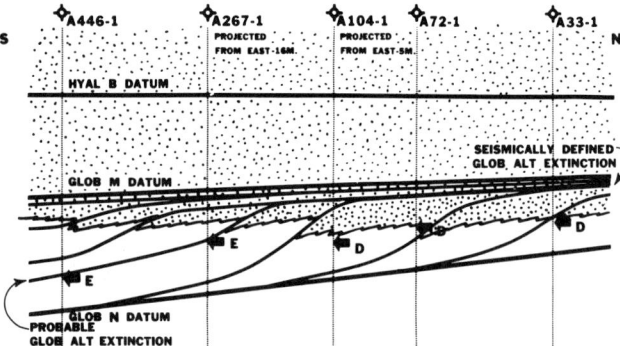

Figure 8. Integration of biostratigraphy and seismic data along Line B (approximately 50 mi north-south, see Figure 1 for location). See text under Biostratigraphic Pitfalls for explanation.

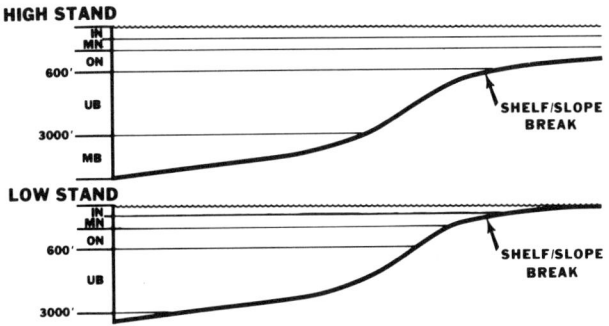

Figure 9. Impact of sea level drop on the distribution of watermass biofacies. As sea level drops, the biofacies boundaries shift down the physiographic profile of the basin margin. The independence of biofacies from physiography may result in correlation of the physiographic shelf/slope break with the outer neritic/upper bathyal biofacies boundary at highstand of sea level and with the inner neritic/middle neritic biofacies boundary at lowstand of sea level, depending on how much change in sea level occurs.

facies down onto the upper slope of the clinoform (Fig. 9).

(2) The middle neritic fauna may reflect a seaward extension of a major sediment input area such as a delta. High rates of sedimentation, possibly coarser-grained and containing abundant terrigenous organic matter, and modifications of salinity and temperature, alter the local environment. Biofacies distribution responds to these environmental modifications (see Pflum and Frerichs, 1976, for a discussion of the "delta-depressed fauna").

(3) The down-slope transport of shallow water faunas by gravity processes which mix or juxtapose biofacies assemblages from different environments (Fig. 10) (see Woodbury et al., 1978). The further mixing of stratigraphically separate assemblages by rotary drilling makes the identification of mixed assemblages difficult, but careful sample examination and the use of closely spaced sidewall core samples helps.

Resolution of these problems requires an integrated approach by the geologist, the geophysicist, *and* the biostratigrapher working in cooperation. Figure 8 is an example of integrating multiple data sets to define the stratigraphic framework of both bioevent and biofacies distribution. This data set allowed identification of the ecologic control on the diachronous first-occurrence of *Globigerina quadrina* between depositionally upslope and downslope wells.

MAP PATTERNS

Seismic facies interpretations provides definition of

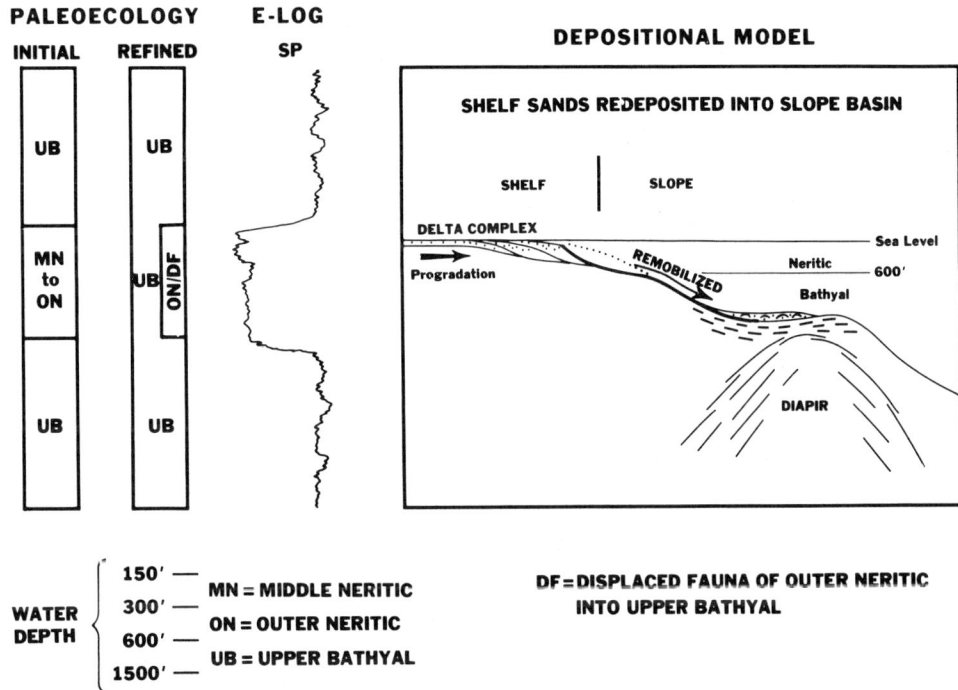

Figure 10. Paleoecologic interpretation must take into consideration the displacement of faunal assemblages into deeper water by gravity processes, such as slumps or turbidity flow events. Apparent paleoecologically defined shallowing events based on initial sample analysis may in fact reflect a mixed fauna resulting from remobilized neritic assemblages being redeposited into bathyal water depths.

facies tracts (see Sangree et al., 1978, and Haq et al., 1987). Calibrations of rock type from cuttings and electric log interpretations, and water depth from paleoecologic analysis of microfossils, provide additional facies information. Mapping different types of facies patterns for genetically related seismic sequences defines an integrated facies pattern suitable for assignment of depositional environments.

Figure 11 is a paleoecologic map for the Late Pliocene *Globoquadrina altispira* interval. It is based on paleoecologic analysis of benthic Foraminifera in 48 *Glob alt* Sand Interval samples from five wells. These wells are on depositional strike with wells having typical middle bathyal assemblages. The southward biofacies excursion could be due to: (1) a local topographic high, elevating the sea floor into upper bathyal water depths; (2) the downslope resedimentation of upper bathyal sediments masking the presence of inplace middle bathyal assemblages; or (3) an ecologic shift in environment excluding the middle bathyal species from this area. Inspection of seismic record sections gives no indication of a *Globoquadrina altispira* age paleotopographic high in

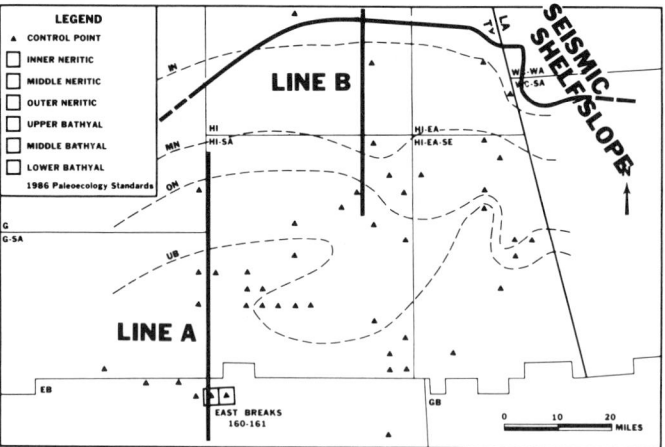

Figure 11. Paleoecologic map of benthic foraminiferal biofacies based on analysis of forty-eight wells, Offshore Gulf of Mexico. The interval mapped is the Late Pliocene *Globoquadrina altispira* sequence encompassing deposits between the *G. altispira* bioevent and the older *Globorotalia nepenthes* bioevent. See text under Map Patterns for further explanation.

wells. The map pattern is defined by the relative distribution of benthic foraminiferal biofacies (Tipsword et al., 1966). Two significant patterns are indicated: (1) the relationship of the seismically defined shelf/slope break to the biofacies distribution; and (2) the southwestward downslope excursion of the upper bathyal biofacies.

The seismic shelf/slope break is identified by the seismic facies transition from large-scale clinoform to upslope parallel reflectors. The clinoforms are interpreted as upper slope progradational deposits. The parallel reflectors, usually of moderate to high amplitude and continuous, are interpreted as shelf deposits. On Figure 11, the seismic shelf/slope break occurs within the inner neritic biofacies. This is because the *Glob alt* Sand Interval Paleoecologic Map represents an interval of deposition at lowstand sea level. The *Globoquadrina altispira* datum occurs at approximately 2.8 Ma (Stainforth et al., 1975). This correlates to the eustatic sea level low of the Tejas B 3.7 third order cycle (Haq et al., 1987) (P.R. Vail, personal communication, 1987). During lowering of sea level the biofacies shift seaward (Fig. 9). The pattern of Figure 11 suggests a shift of inner neritic biofacies to the seismically defined upper slope, approximating a shift of about 300 feet water depth from current biofacies/bathymetric relationships in the Gulf of Mexico. The Global Cycle Chart (Haq et al., 1987) suggests a similar 300-foot lowering of relative water depth for cycle Tejas B 3.7.

The southward excursion of the upper bathyal biofacies reflects the absence of middle bathyal biofacies taxa in the area of the biofacies excursion. Examination of the cutting samples and sidewall core samples does not yield any typically middle bathyal taxa, suggesting that their absence is not a sample artifact. Thus, the probable cause of the biofacies excursion is an excursion in the optimal environment for the benthic Foraminifera, perhaps similar to that described by Pflum and Frerichs (1976) for the "delta-depressed fauna." Modification of environments due to areas of major sediment input at times of lower sea level would probably impact facies distribution well down onto the slope-face, as the biofacies pattern of Figure 11 suggests. The coeval pattern of upper-slope clinoform progradation on Line B suggests a major sediment input area north and upslope from the biofacies excursion. This further supports the interpretation of the biofacies excursion as an environmental modification and neither a local topographic high nor displaced assemblages. Again, the integration of multiple data sets helps reduce the number of possible working hypotheses for resolution of apparently anomalous data patterns.

DEPOSITIONAL PATTERN

Analysis of the seismic facies and integration of biofacies patterns allows definition of a series of depositional events. Figure 12 shows the seismic facies of Line B clinoform packages between the *Globigerina nepenthes druryi* (*Glob N*) and *Globorotalia miocenica* (*Glob M*) datums. Each clinoform has a subpackage of rotated

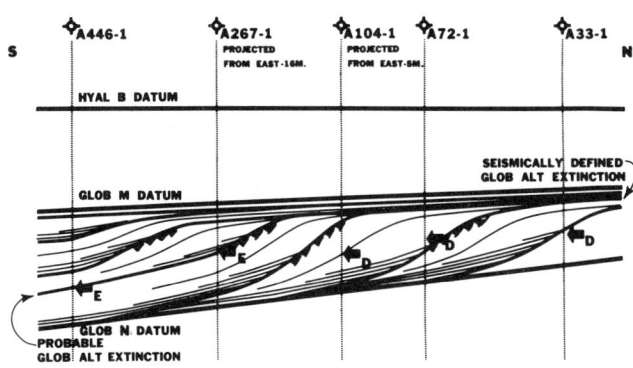

Figure 12. Schematic seismic facies pattern for Line B (approximately 50 mi north-south); specifically, the prograding sigmoidal clinoforms between the *Globorotalia nepenthes* (*Glob N*) and *Globorotalia miocenica* (*Glob M*) datums. The clinoform packages have subpackages of rotated reflectors within the upper part (triangular pattern) and are onlapped by parallel reflectors. See text under Depositional Pattern for further discussion.

reflectors in the upper part of the clinoform surface. The rotated subpackage maps out as a seismic facies common to the upper clinoform and sometimes extending down into basinal areas. The rotated-reflector seismic facies is interpreted as slumped sediments. Additionally, the upper surface of each clinoform package is onlapped by low to moderate amplitude, parallel reflectors. These onlaps are interpreted as aggradational basin filling.

The clinoforms of Line B are coeval with Global Cycle Chart eustatic cycles TB 3.6 and 3.7 (Haq *et al.*, 1987). The occurrence of multiple subcycles of clinoform progradation-slump-onlap packages within the third order eustatic cycles suggests the packages are fourth order events, possibly fourth order eustatic events or the consequence of shifts in the sediment input area. Shifts in sediment input area would result in abandonment of one clinoform and development of another with the bottomset reflectors onlapping the previous clinoform surface. Preliminary mapping of these seismic facies has not provided a clear definition of probable origin, mostly due to the slumping which makes difficult the mapping of the original depositional architecture of each subcycle.

Detailed analyses of a regional seismic grid, represented here by Lines A and B, permits the construction of a general facies sequence model for High Island-East Breaks Pliocene-Pleistocene deposition. The pattern of events is shown on Figure 13 and can be summarized as four principal phases:

(1) Progradation of muddy clinoforms occurs during lowering of sea level and is followed by the regressive coastal plain facies. With a very high rate of sediment input, the clinoforms oversteepen and slump. Multiple clinoform-slump-onlap cycles form due to switching of the axis of sediment input at the top of the slope-ramp, or due to fourth-order eustatic fluctuations.

(2) At some point in the sea-level lowering process, the delta-front coastal facies arrives at the physiographic shelf/slope break. Clinoforms continue aggrading and prograding, and river-fed turbidites bypass the slope-ramp depositing as lower-slope and basinal aggradation-

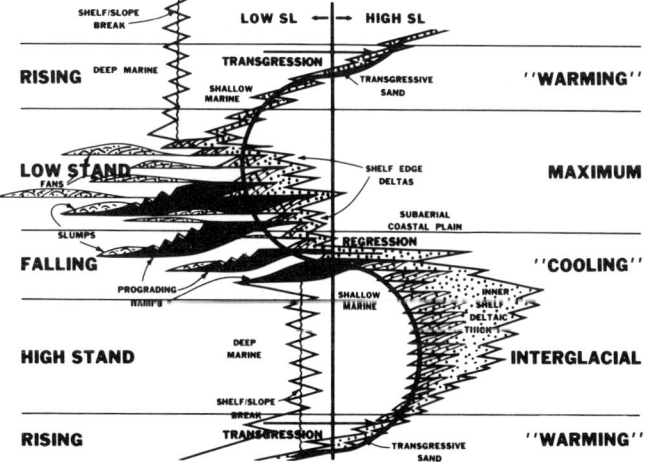

Figure 13. Preliminary facies model for Pliocene-Pleistocene sedimentation in the High Island-East Breaks area during one glacially driven eustatic sea-level rise and fall. See text under Depositional Pattern for further discussion.

al mounds.

(3) Rise in sea level initiates transgression of the shelf. The basin becomes progressively more sediment starved with dominance of hemipelagic sedimentation. Transgressive sands are deposited on the shelf as the coastal plain facies of the lowstand sea level system are reworked.

(4) During highstand sea level, most sediment is deposited within inner shelf and deltaic coastal-plain facies. Depocenter thicks form if coastline subsidence accommodates the volume of sediment. If not, a thin, widely distributed sediment layer accumulates due to delta-lobe switching and wave induced sediment redistribution.

CONCLUDING REMARKS

The above summary is inadequate to convey the complexity of the interpreted depositional pattern. Copies of Lines A and B with enlargements of specific facies packages are being prepared for publication elsewhere. Both Lines A and B will be displayed at the Poster Session at this meeting along with examples of the detailed biostratigraphic data.

Even this brief review of integrating biostratigraphic and seismic stratigraphic data suggests the type of information available for reconstructing depositional environments. The integrated data sets help resolve apparent discrepancies between the independent data sets rather than attributing discrepancies to "bad data."

Acknowledgments

This paper is part of an ongoing project documenting the biostratigraphic control available from wells and its utility in analysis of seismic facies. The original study was done at Mobil Exploration and Producing Services Inc. and is now being documented for public release with the biostratigraphic support of MICRO-STRAT. Inc., including J.F. Clement, T.C. Huang, and W.W. Wornardt, Jr. The author acknowledges his appreciation to Mobil Oil Corporation for permission to publish, and to his co-workers at Mobil for their input to this project, most especially G.M. Ragan, G.D. Taylor, R.C. Becker, T.D. Lee, R.W. Morin and D.W. Ford for biostratigraphic support, and P. Braithwaite, C.E. Beeman, S.J. Malecek, T.D. Crutcher, J.W. Wheeler, and E.A. Pennick for help in facies interpretation.

REFERENCES

Haq, B.U., J. Hardenbol, and P.R. Vail, 1987, Chronology of fluctuating sea levels since the Triassic (250 million years ago to present): Science, v. 235, p. 1156-1166.

McDougall, K.A., 1980, Paleoecological evaluation of late Eocene biostratigraphic zonations of the Pacific coast of North America: Journal of Paleontology, v. 54, no. 4, supplement (Paleontological Monograph 2), 75 p.

Mitchum, R.M., Jr., P.R. Vail, and J.B. Sangree, 1977, Stratigraphic interpretation of seismic reflection patterns in depositional sequences, *in* Seismic Stratigraphy — Applications to Hydrocarbon Exploration: Amer. Assoc. Petroleum Geologists, Memoir 26, p. 117-143.

Olson, H.C., 1987, Biofacies zonation of middle Miocene benthic Foraminifera, Southeastern San Joaquin Basin, California (Abst.): Amer. Assoc. Petroleum Geologists Bull., v. 71, no. 5, p. 599.

Pflum, C.E., and W.E. Frerichs, 1976, Gulf of Mexico deep water foraminifers: Cushman Found. Foram. Res., Spec. Pub. 14, 125 p.

Poag, C.W., 1981, Ecologic Atlas of Benthic Foraminifera of the Gulf of Mexico: Marine Science International, Woods Hole, Massachusetts, 174 p.

Poag, C.W., and P.C. Valentine, 1976, Biostratigraphy and ecostratigraphy of the Pleistocene Basin, Texas-Louisiana continental shelf: Gulf Coast Asso. Geol. Soc., Trans., v. 26, p. 185-256.

Sangree, J.B., D.C. Waylett, D.E. Frazier, G.B. Amery, and W.J. Fennessy, 1978, Recognition of continental-slope seismic facies, offshore Texas-Louisiana: Amer. Assoc. Petroleum Geologists, Studies in Geology no. 7, p. 87-116.

Stainforth, R.M., J.L. Lamb, H. Luterbacher, J.H. Beard, and R.M. Jeffords, 1975, Cenozoic planktonic foraminiferal zonation and characteristics of index forms: The University of Kansas Paleontological Contributions, Article 62, 425 p.

Tipsword, H.L., F.M. Setzer, and F.L. Smith, Jr., 1966, Interpretation of depositional environment in Gulf Coast exploration from paleoecology and related stratigraphy: Gulf Coast Assoc. Geol. Soc., Trans., v. 16, p. 119-130.

Vail, P.R., R.G. Todd, and J.B. Sangree, 1977, Chronostratigraphic significance of seismic reflections, *in* Seismic Stratigraphy — Applications to Hydrocarbon Exploration: Amer. Assoc. Petroleum Geologists, Memoir 26, p. 99-116.

Woodbury, H.O., J.H. Spotts, and W.H. Atkers, 1978, Gulf of Mexico continental-slope sediments and sedimentation: Amer. Assoc. Petroleum Geologists, Studies in Geology, no. 7, p. 117-137.

The Role of Horizontal Seismic Sections in Stratigraphic Interpretation

Alistair R. Brown
*Geophysical Service Inc.
Dallas, Texas*

Horizontal (or Seiscrop™) sections sliced from a three-dimensional seismic data volume normally intersect structural horizons and are effective for structural interpretation. A feature not interpretable as structure may have stratigraphic significance and is often recognizable by characteristic shape. Sink-holes, for example, appear on individual horizontal sections as discrete circular objects.

Seiscrop sections have even greater value for stratigraphic interpretation if sliced along individual seismic horizons, thus removing the effects of structure. Such a section, defined as a horizon Seiscrop section, displays the spatial distribution of seismic amplitude, or other attribute, over a single bed. Horizon Seiscrop sections have been shown to have a powerful ability to reveal subtle depositional features such as bars and channels, particularly in Tertiary clastic rocks like those of the Gulf of Mexico. A horizon Seiscrop section can be regarded effectively as the reconstitution of a depositional surface.

The extent of a gas reservoir may be revealed directly by the shape of the bright spot on a horizon Seiscrop section. A study of reflection amplitude as a function of recording offset can help validate a bright spot as resulting from hydrocarbon content. A horizon offset section displays the amplitude of one horizon as a function of offset and horizontal position along a seismic line.

A seismic data volume may be inverted to a volume of seismic logs displaying the acoustic velocity in the subsurface. A slice following one structurally interpreted horizon through this volume is a horizon Seiscrop velocity section and has further application for delineating depositional features and the extent of hydrocarbon reservoirs.

Amplitude, displayed in the form of horizon Seiscrop sections over the top and base reflections of a known gas reservoir, has been used to assess the proportion of producible gas sand within a mapped seismic interval. This has led to net gas sand isopach maps which tie the thickness of producible gas sand in the wells. Integration of these maps has yielded total net reservoir volumes.

The interpretive approaches discussed above involve the manipulation of large quantities of seismic data. Much of this is now done interactively, as this helps the interpreter greatly with his data management. Also flexible color display increases visual dynamic range, and the construction of composite views helps in the appreciation of data complexities. Most important of all, the interactive interpreter can maintain a rapid flow of ideas because of the short response time of the system.

INTRODUCTION

Horizontal seismic sections provide a new dimension, figuratively as well as literally, to the stratigraphic interpreter. Their most direct application is to reveal characteristic shapes interpretable as depositional, erosional, lithologic or reservoir features; they can also synthesize horizon properties for improved stratigraphic interpretability. Seiscrop* sections are a normal product of three-dimensional seismic surveys and are simply horizontal slices through a three-dimensional (3D) data volume. Proper planning is critical to the success of a 3D survey (Brown and McBeath, 1980), and the value of 3D migration in clarifying subsurface structure and stratigraphy was vividly demonstrated by French (1974) and Hilterman (1982). The utility of horizontal sections was first presented in 1975 by Bone, Giles, and Tegland (1983) and their use for structural interpretation was discussed by Brown (1983a).

Several authors have discussed the value of horizontal seismic sections for the identification and study of subtle features. Johnson and Bone (1980) demonstrated their importance in the recognition of subtle faults. Brown, Dahm, and Graebner (1981), and Brown, Graebner, and Dahm (1982) demonstrated their importance for the identification and delineation of sand bars and channels. Where beds are flat-lying, a normal Seiscrop section parallels bedding planes and is directly the section required for stratigraphic interpretation. In the presence of structure, it may be necessary to slice through the data volume along or parallel to a structurally interpreted horizon surface, in which case the resulting product is a horizon Seiscrop section. If one of these slices displays an attribute other than seismic amplitude, this will appear in the name; for example, a horizon Seiscrop velocity section displays spatial variations in velocity over one seismic horizon. The generation of a horizon Seiscrop section is, in effect, the reconstitution of a depositional surface.

RECOGNITION OF STRATIGRAPHIC FEATURES

Figures 1 and 2 are Seiscrop sections from a 3D survey recorded in the Gippsland basin, offshore southeastern

---
*Trademark of Geophysical Service Inc.

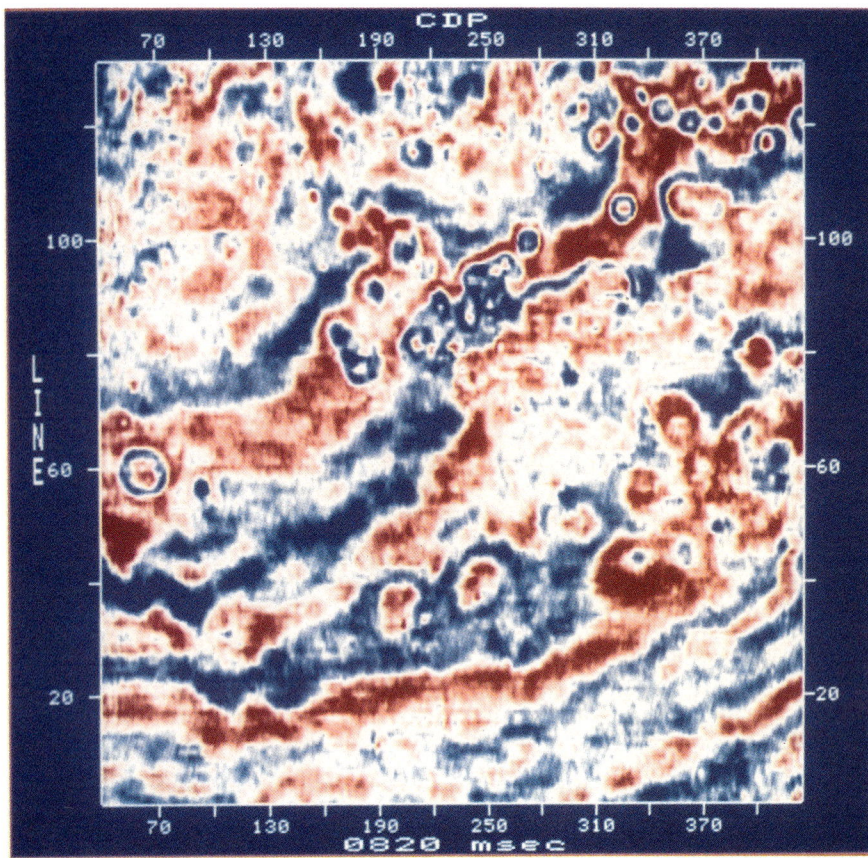

Figure 1. Seiscrop section at 820 msec from 3D survey over Mackerel field in offshore Gippsland basin, southeastern Australia. Circular objects are interpreted as sinkholes in karst topography. (Courtesy Esso Australia, Ltd.)

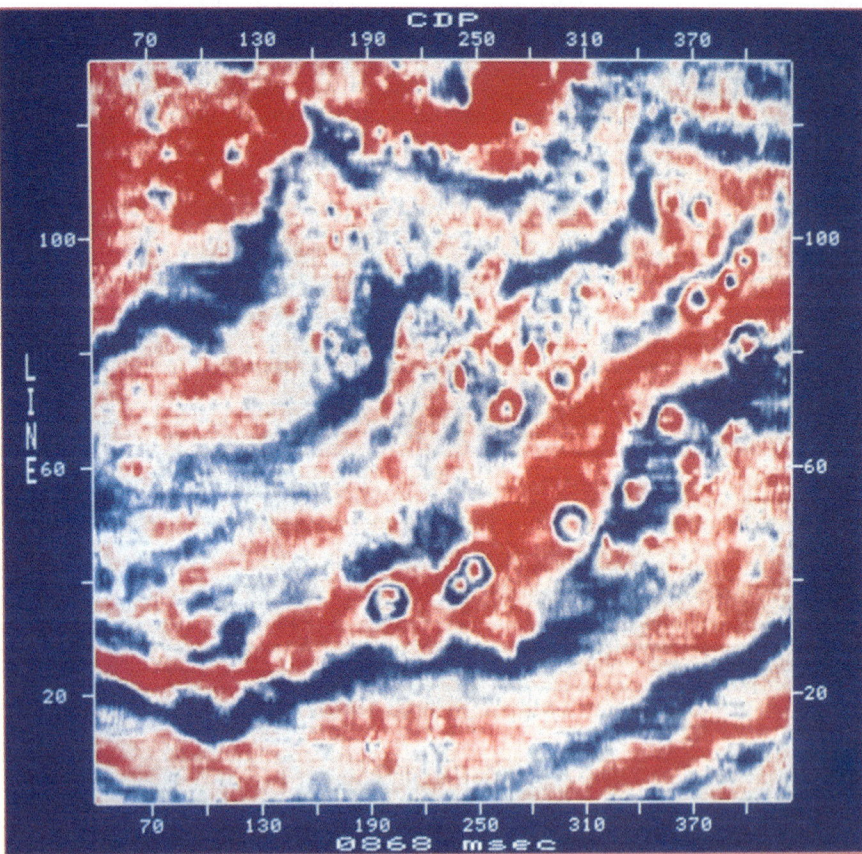

Figure 2. Seiscrop section at 868 msec from 3D survey over Mackerel field in offshore Gippsland basin, southeastern Australia. Circular objects are interpreted as sinkholes in karst topography. (Courtesy Esso Australia, Ltd.)

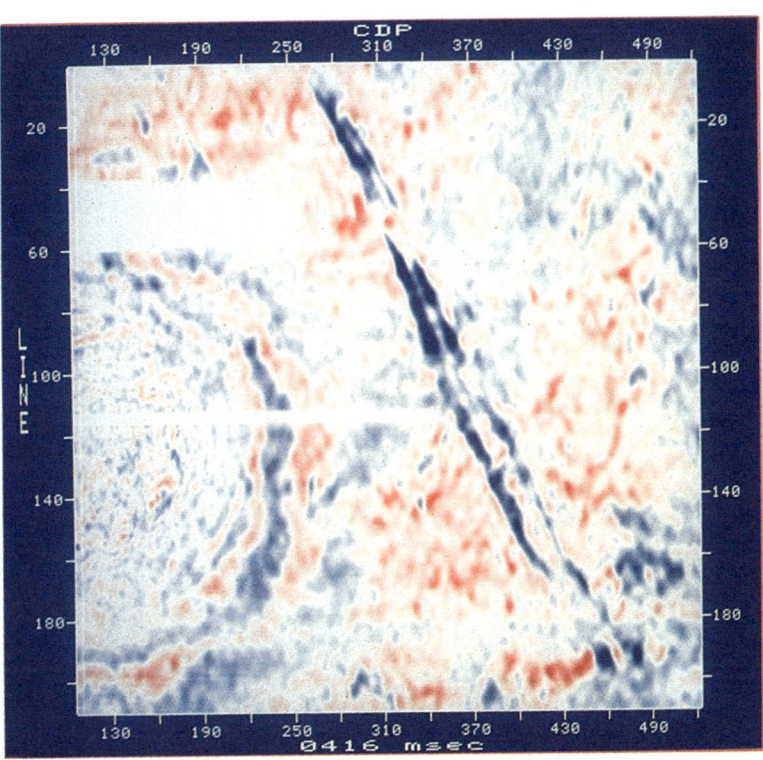

Figure 3. Seiscrop section at 416 msec from 3D survey in the Gulf of Mexico. The bifurcating channel is seen close to the edge of a salt dome. (Courtesy Chevron U.S.A., Inc.)

Australia (Sanders and Steel, 1982). Several small circular features are strikingly evident. They measure 200 to 500 m (656 to 1,640 ft) in diameter and are tentatively interpreted as sinkholes in a karst topography of Miocene age. The beds in which these features exist are dipping from upper left to lower right (eastward) in Figures 1 and 2. The width of the reflection is a function of the seismic frequency and the structural dip (Brown, 1983a). The prominence of the sinkholes results because their diameters are less than this reflection width.

Another type of circular carbonate feature was discussed by Brown (1978). In flat-lying beds of the Williston basin in North Dakota very subtle draping of seismic reflections was barely discernible on vertical sections, whereas Seiscrop sections showed a distinct, almost circular outline.

Figures 3 through 6 show channels in two different areas of the Gulf of Mexico. The bifurcating channel in Figure 3 lies in sufficiently flat-lying beds that the one Seiscrop section at 416 msec shows it clearly. Some parts of the same channel were more clearly seen on the adjacent section at 412 msec; simple addition of these two Seiscrop sections provided the most complete view of this channel system. Adding together of Seiscrop sections is a useful approach to the enhancement of stratigraphic features if, but only if, the structural variation across the feature is less than half a period of the appropriate seismic signal. The circular object on the left of Figure 3 is a salt dome.

Figure 4 shows another channel system deeper in the same data volume. A narrow channel branches at Line 70, CDP 470. The western branch is visible on the sections from 812 and 816 msec and is seen most clearly on a summation of the two. The eastern branch is seen most clearly on the Seiscrop section at 820 msec indicating that it is structurally slightly downdip. The seismic period does not permit the addition of more than two adjacent sections, so in Figure 4 both frames are required to view the full extent of this channel.

Figures 5 and 6 present a Gulf of Mexico channel in the presence of a moderate amount of structure. Here it was necessary to interpret that structure and then slice through the data volume according to the interpretation. The high amplitude red "blobs" under study at about 830 msec just right of center on both Lines 57 and 60 (Figure 5) are not part of readily definable structural continuity. Thus an event one-and-one-half periods above was tracked. A slice parallel to the resulting structurally interpreted surface was taken through the data volume at the level of the red "blobs" to yield the horizon Seiscrop section of Figure 6.

The channel is evident in different shades of red with the darker reds indicating higher seismic amplitudes. In order to study these amplitudes in relation to structural position, the time contours of the structurally interpreted horizon were superimposed. They reveal that the higher amplitudes in the channel are all structurally higher than the lower amplitudes. We thus conclude that the channel is probably composed of porous sand throughout, which is water-filled to the southwest and gas-filled to the northeast.

DELINEATION AND VALIDATION OF HYDROCARBON RESERVES

The extent of a gas reservoir may be revealed directly by the shape of the bright spot on a horizon Seiscrop section. A probable gas reservoir is described in the composite display of Figure 7. The horizon Seiscrop section (lower left) shows the approximately circular outline of the probable gas reserves. Two of the vertical sections show the tracked reservoir horizon which provided the amplitudes and the times displayed in the lower panels. Comparison of these lower panels shows that the probable reserves are structurally high (the orange and red areas, lower right) but not in total conformity with structure, particularly on

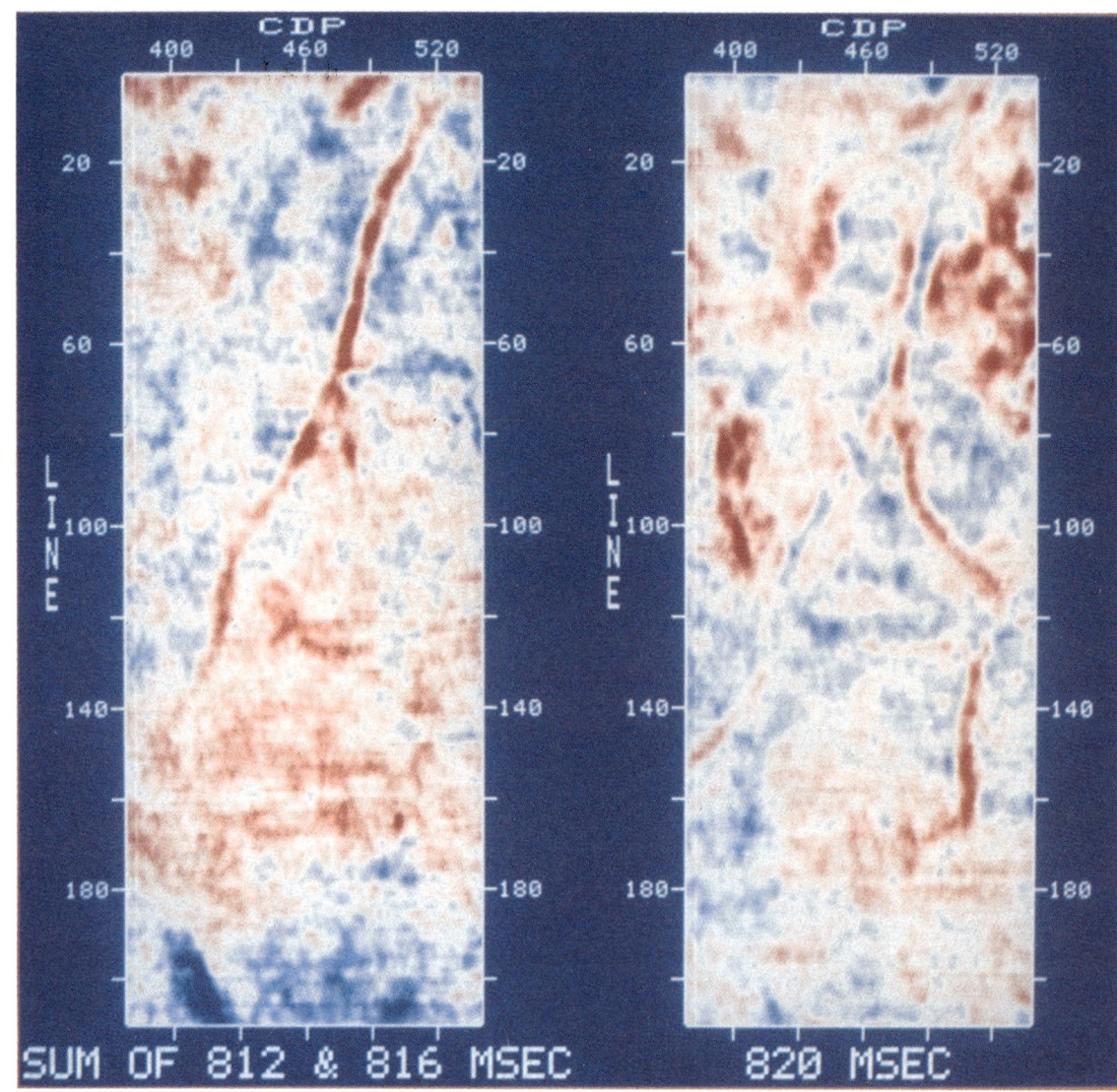

Figure 4. Composite display of Seiscrop sections at 812, 816 and 820 msec from same survey as Figure 3 showing branching channel. (Courtesy Chevron U.S.A., Inc.)

the north, which suggests some stratigraphic control.

The change in seismic reflectivity with recording offset, that is with angle of incidence, has recently become a subject of widespread interest, as it has potential for determining several rock properties (Backus and Goins, 1984). Specifically, Ostrander (1984) has shown that an increase in reflection amplitude with offset is often indicative of gas and so can be used to validate gas bright spots. Ostrander presents vertical sections with traces from different offsets or groups of offsets for one common depth point.

Figure 8 shows a horizon offset section in which the variation in amplitude with offset can be studied for many common depth points for one selected horizon. In this case we chose a line of unstacked data over a known gas reservoir in the Gulf of Mexico, in which the thickness of gas sand varied laterally. CDP gathers from a suite of adjacent depth points were treated as a 3D data volume where one dimension was offset. By tracking one of the reservoir reflections and extracting the corresponding amplitudes, we obtained a horizontal section displaying the peak amplitude of this reflection as a function of offset and depth point along the line. This horizon offset section (Figure 8), with the darker reds indicating higher amplitudes, shows a general increase in amplitude with offset for most depth points as expected for gas.

Kurfess, Giles, and Bone (1977) discussed the development of a field with 3D seismic methods in the jungles of Peru. Many successful development wells have since been drilled and the field is now producing oil from the Vivian Sandstone. The 3D seismic data was recently reprocessed and reinterpreted to investigate whether the limits of production could be further understood.

A line from the reprocessed data volume is shown in the lower panel of Figure 9. This was then inverted to a section of seismic logs using the G-LOG* process (Graebner, Steel, and Wason, 1980). The

---

*Trademark of Geophysical Service Inc.

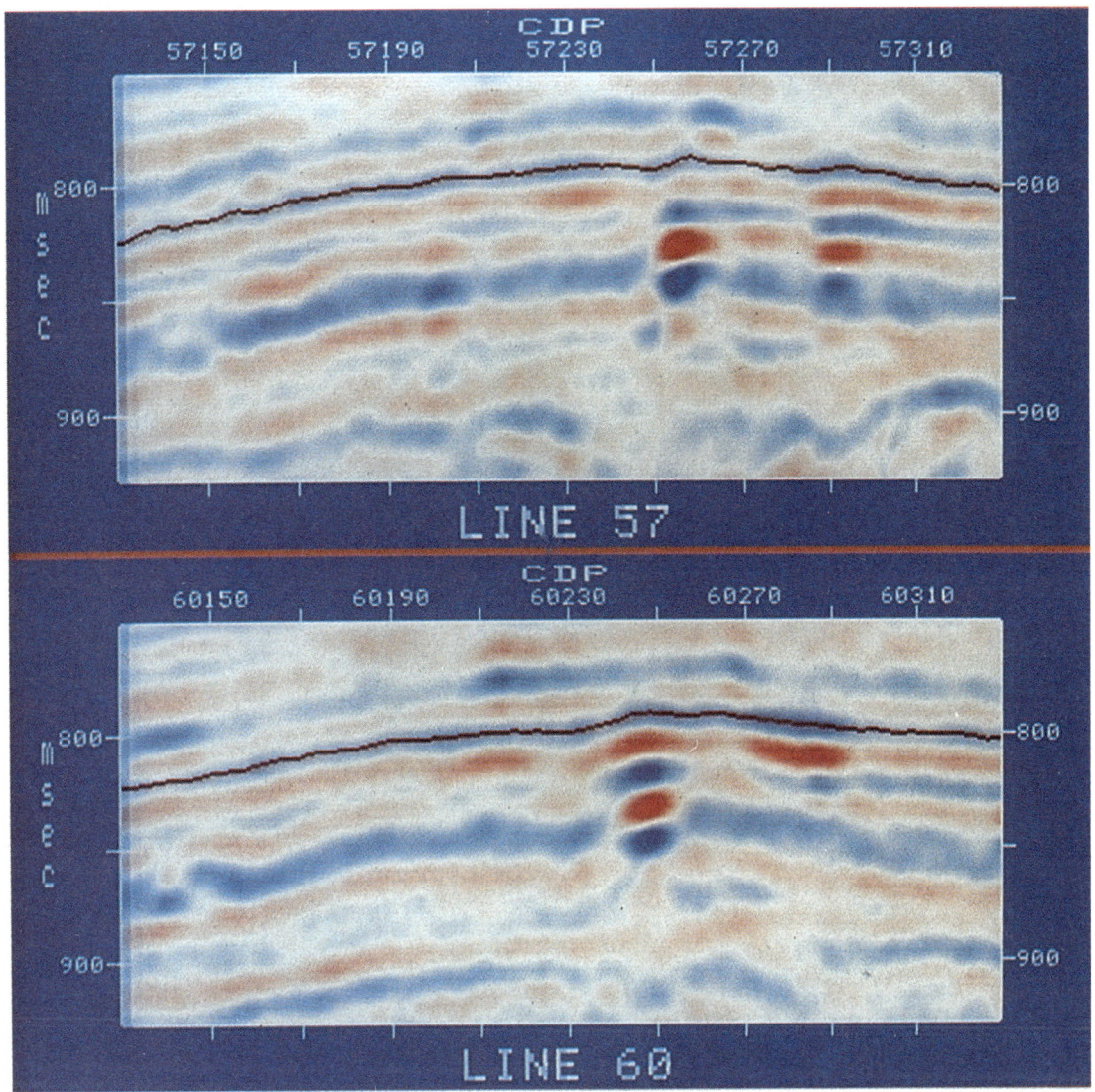

Figure 5. Lines 57 and 60 from 3D survey in the Gulf of Mexico showing tracked horizon above bright events indicating channel intersections. (Courtesy Chevron U.S.A., Inc.)

resulting G-LOG section (Figure 9) displays velocity according to the color legend appearing in Figure 10. The Vivian Pay sandstone shows lower velocity over the structure. The whole data volume was similarly treated and a horizon Seiscrop velocity section extracted through the Vivian Sandstone (Figure 10). This displays the lateral variation in velocity over the reservoir and demonstrates that the region of lower velocity, approximately defined by the outer limit of the orange color, contains the producing wells and excludes the dry ones.

MAPPING OF RESERVOIR PROPERTIES

Seismic amplitude in the form of horizon Seiscrop sections over the top and base reflections of a known gas reservoir has been used to assess the proportion of net producible gas sand within a mapped seismic interval. Figure 11 shows four bright events on two vertical sections which indicate the top and base of two gas sand members in this area of Garden Banks, offshore Louisiana in the Gulf of Mexico.

Figure 12 describes the interpretive sequences applied to this data and presents the intermediate products for one sand over a small portion of the area. Automatic tracking (Figure 11) yielded times and amplitudes of the reservoir reflections. Subtracting of the resulting time structure maps gave an isochron map of the gross sand interval. Summing of horizon Seiscrop sections yielded the composite amplitude response for the reservoir sand. The lateral amplitude changes reflect lithologic changes in the reservoir but they also contain some spurious effects due to tuning (Meckel and Nath, 1977). Various approaches to editing the tuning effects have been investigated, the simplest of which was discussed by Brown et al (1984). After editing the amplitude response, it was smoothed and scaled to yield an estimate of net gas/gross sand ratio within the reservoir. By combining this with the gross isochron map, a net gas isochron map was obtained and hence a net gas isopach map.

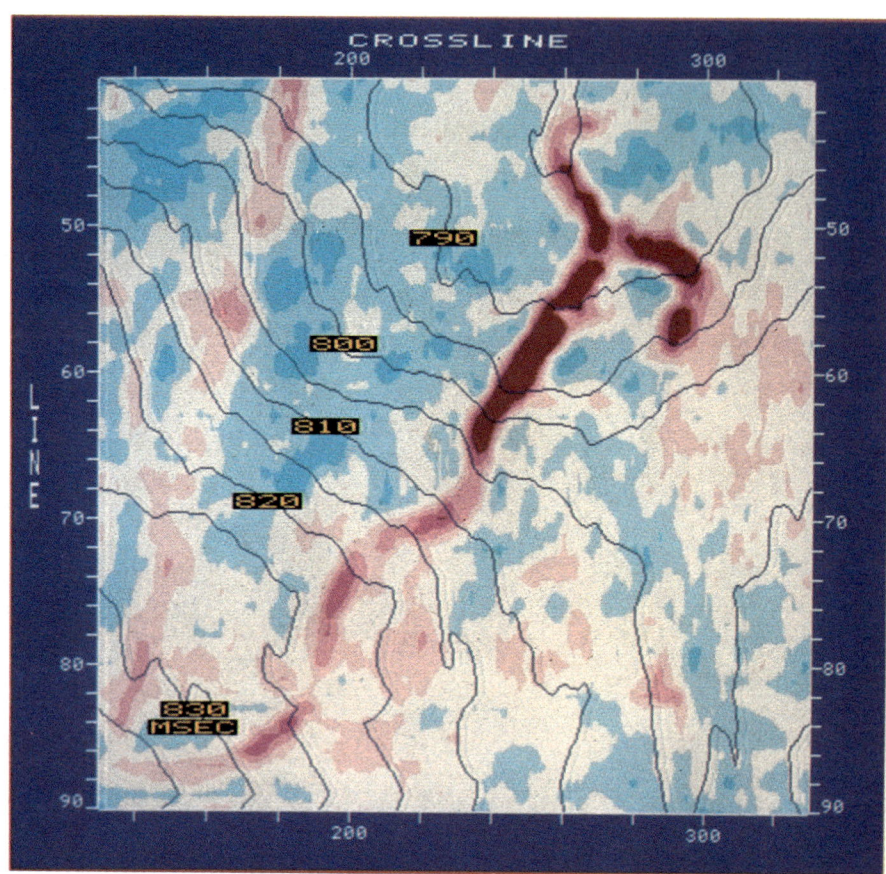

Figure 6. Horizon Seiscrop section showing channel intersected in Figure 5. Superimposed structural contours show that brightest portion of channel is structurally high. (Courtesy Chevron U.S.A., Inc.)

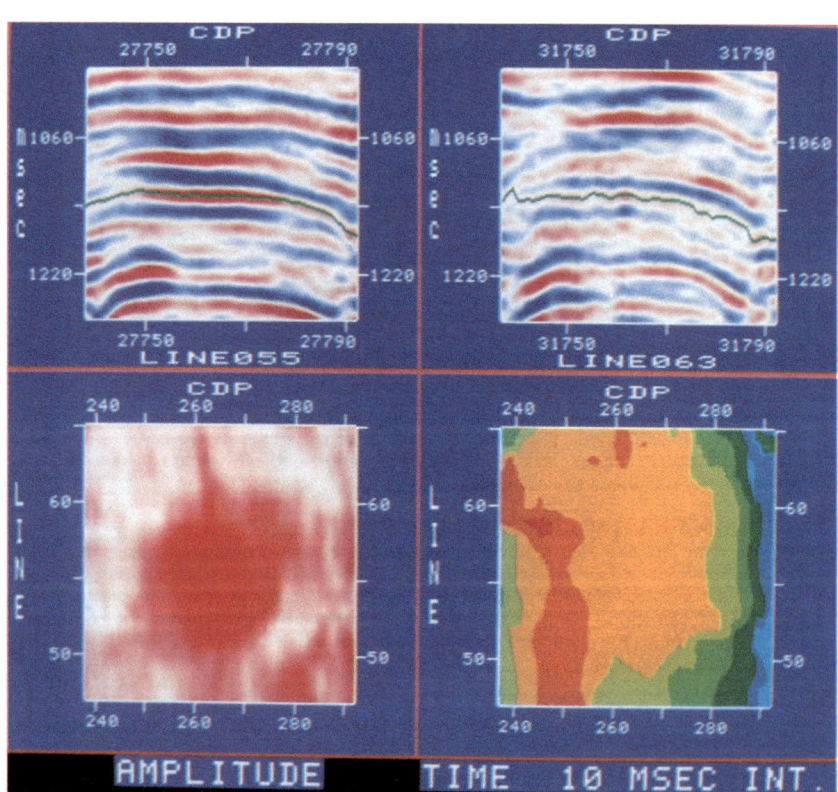

Figure 7. Composite display of two vertical sections, horizon Seiscrop section and time structure map showing extent of suspected hydrocarbon bright spot. (Courtesy Texas Pacific Oil Company, Inc.)

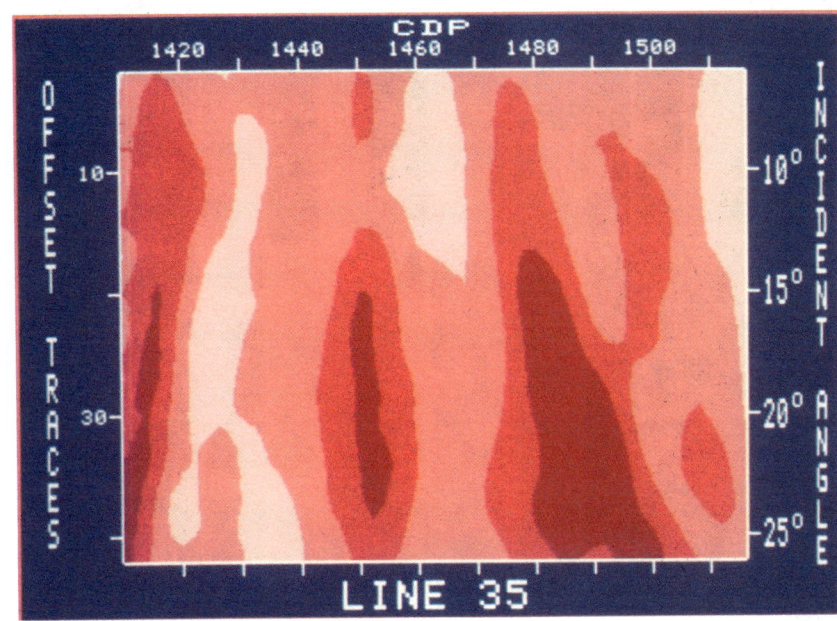

Figure 8. Horizon offset section showing variation of amplitude with offset for one horizon along a seismic line. The general increase in amplitude with offset is a strong indicator of gas.

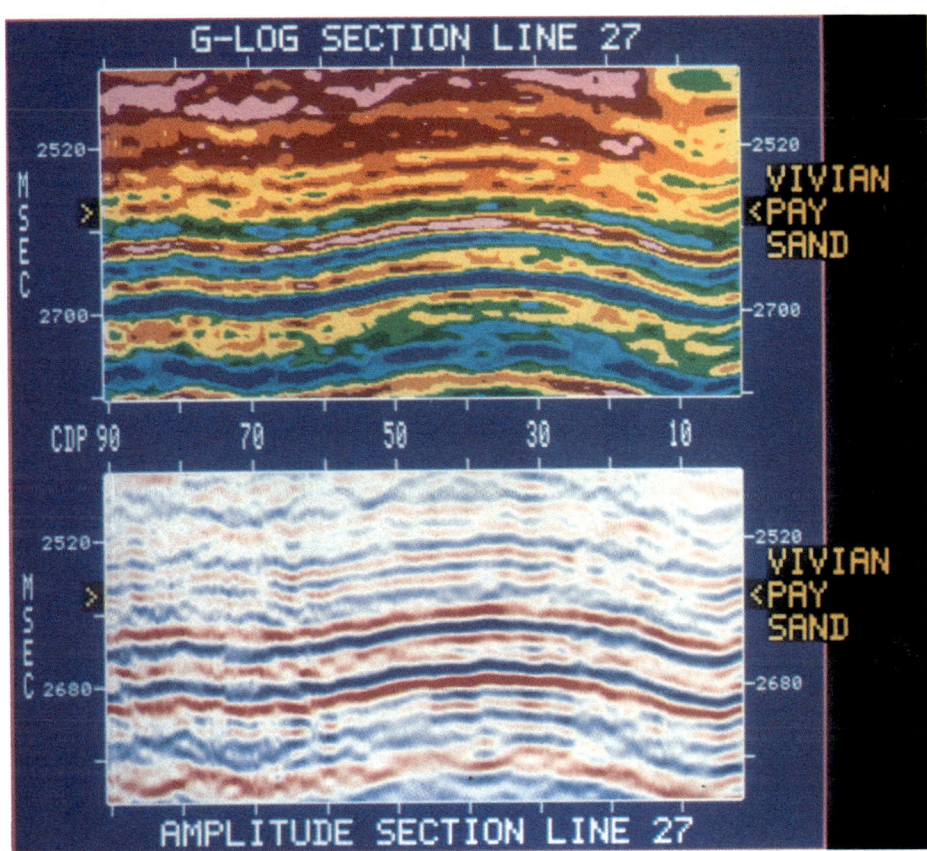

Figure 9. G-LOG velocity section and color amplitude section from 3D survey in Peru indicating Vivian Pay Sand. (Courtesy Occidental Exploration and Production Company.)

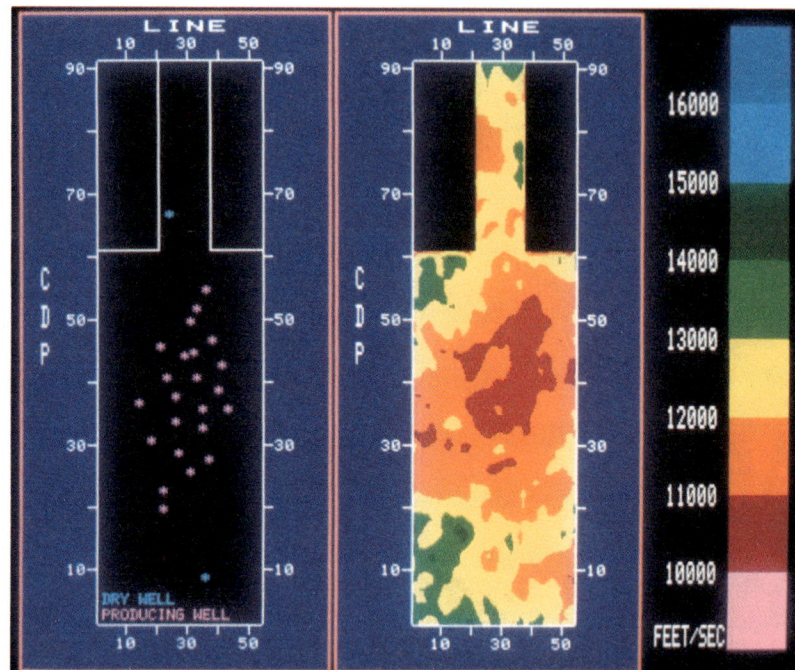

Figure 10. Horizon Seiscrop velocity section through Vivian Pay Sand showing low velocity zone enclosing the area of producing wells. (Courtesy Occidental Exploration and Production Company.)

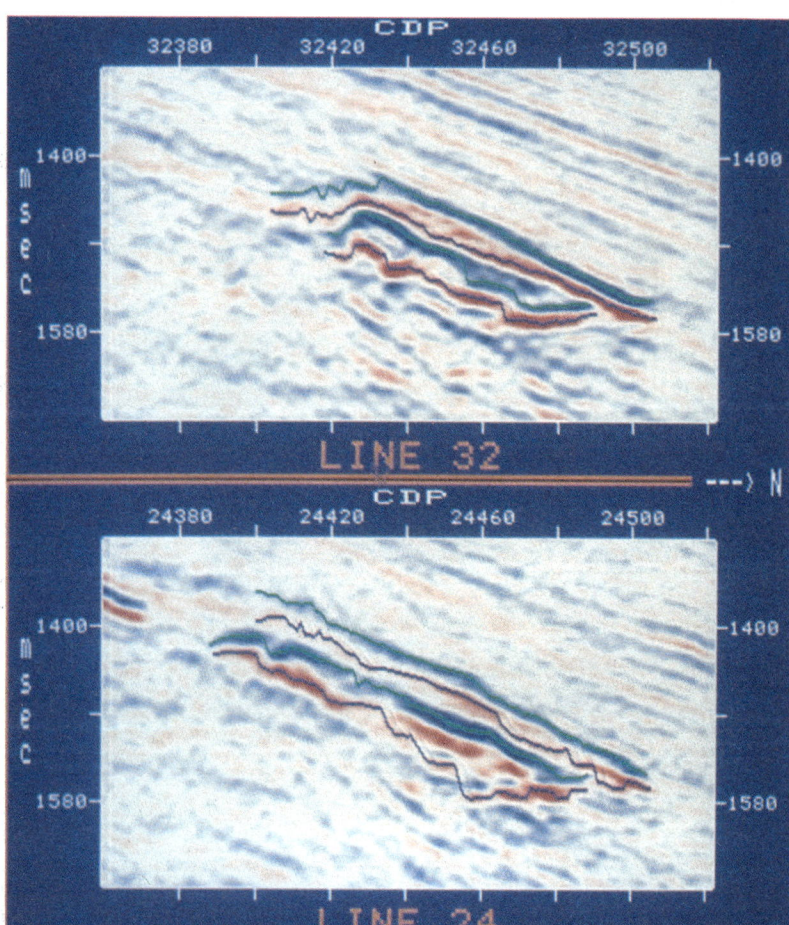

Figure 11. Two lines from 3D survey in the Garden Banks area of the Gulf of Mexico showing bright and flat reflections associated with known gas sands. (Courtesy Chevron U.S.A., Inc.)

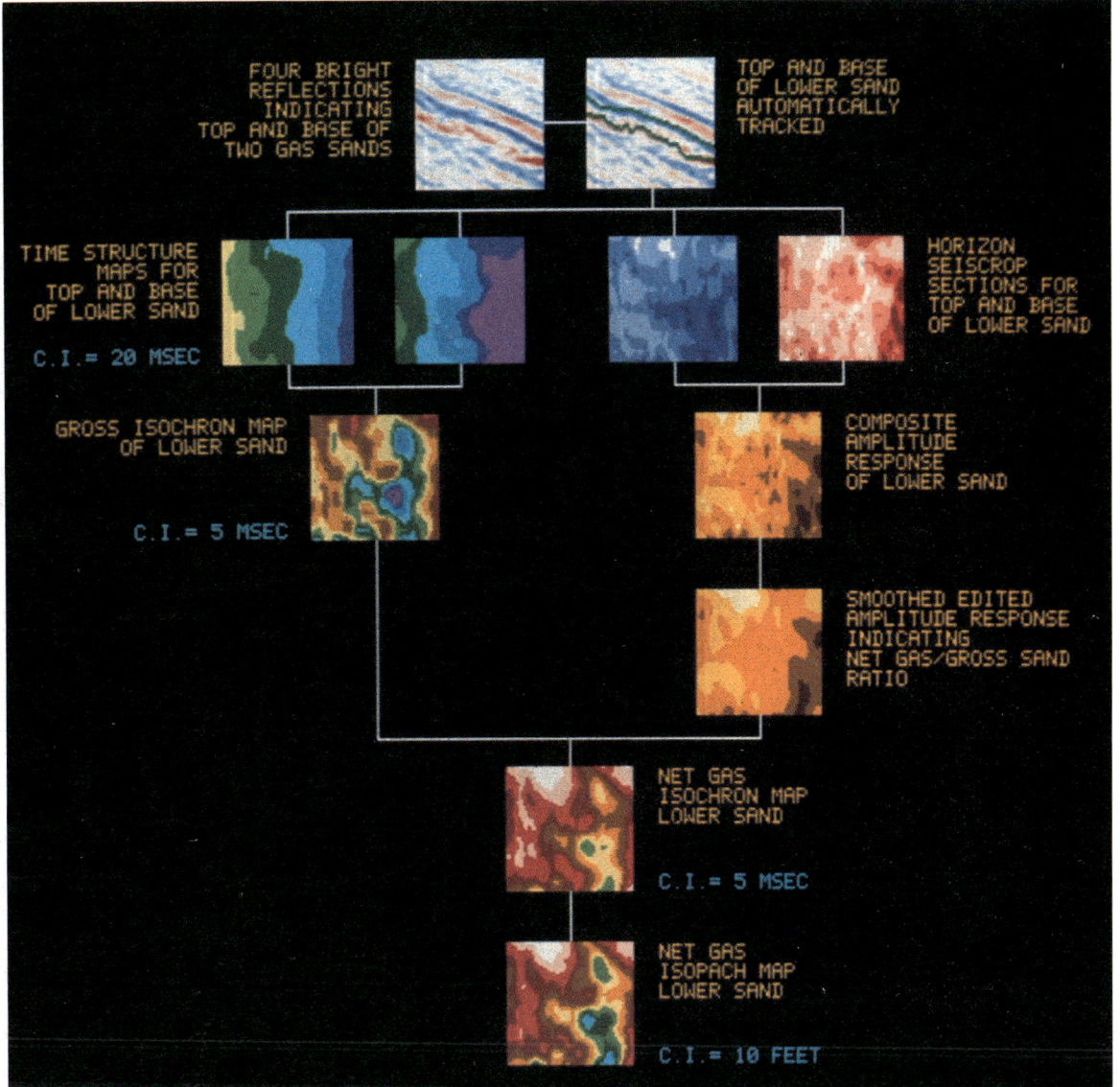

Figure 12. Interpretive sequence used and intermediate products generated in the course of deriving net producible gas sand maps of the Garden Banks reservoirs. (C. I. = Contour Interval.) Time structure maps show dip down to the right; the purple area is the flat spot at the base of the Lower Sand. The isochron and isopach maps have greens and blues indicating the thicker zones. All the four amplitude products have darker colors indicating the higher amplitudes. (Courtesy Chevron U.S.A., Inc.)

The above procedure was followed for both sand members and the net isopach maps were added together to produce a total net producible gas sand isopach map over the area under study (Figure 13). This map ties the net gas sand in the five wells in the area with a standard deviation of 8 ft (2.4 m). By integration of this map the total volume of net producible gas sand in this 4.4 sq km area was computed to be 42 million cu m (34,000 acre ft).

COLOR AND INTERACTIVE GRAPHICS

Every figure in this paper was generated on an interactive seismic interpretation system (Gerhardstein and Brown, 1984) and is displayed in color. Both of these facilities are of great importance in stratigraphic interpretation today.

A 3D data volume requires that the interpreter study a great deal of data in making an interpretation; an interactive system helps with his data management. Automatic tracking and hence the generation of horizon Seiscrop sections is of particular benefit to the stratigraphic interpreter. Flexible display capabilities are very important as they permit a zooming and focusing of attention on stratigraphic detail. The interpreter can also compose images with different but related displays, such as intersecting vertical and horizontal sections. The interactive interpreter can devote a large proportion of time to critical interpretive decision making and, because of the rapid response time of the system, can more readily maintain an idea flow.

Color display, when thoughtfully used, can increase the useable optical dynamic range of any seismic section (Brown, 1983b). With increased amplitude preservation in modern seismic processing, the interpreter needs to be able to view a greater range of amplitudes on one section; with variable area/wiggle traces,

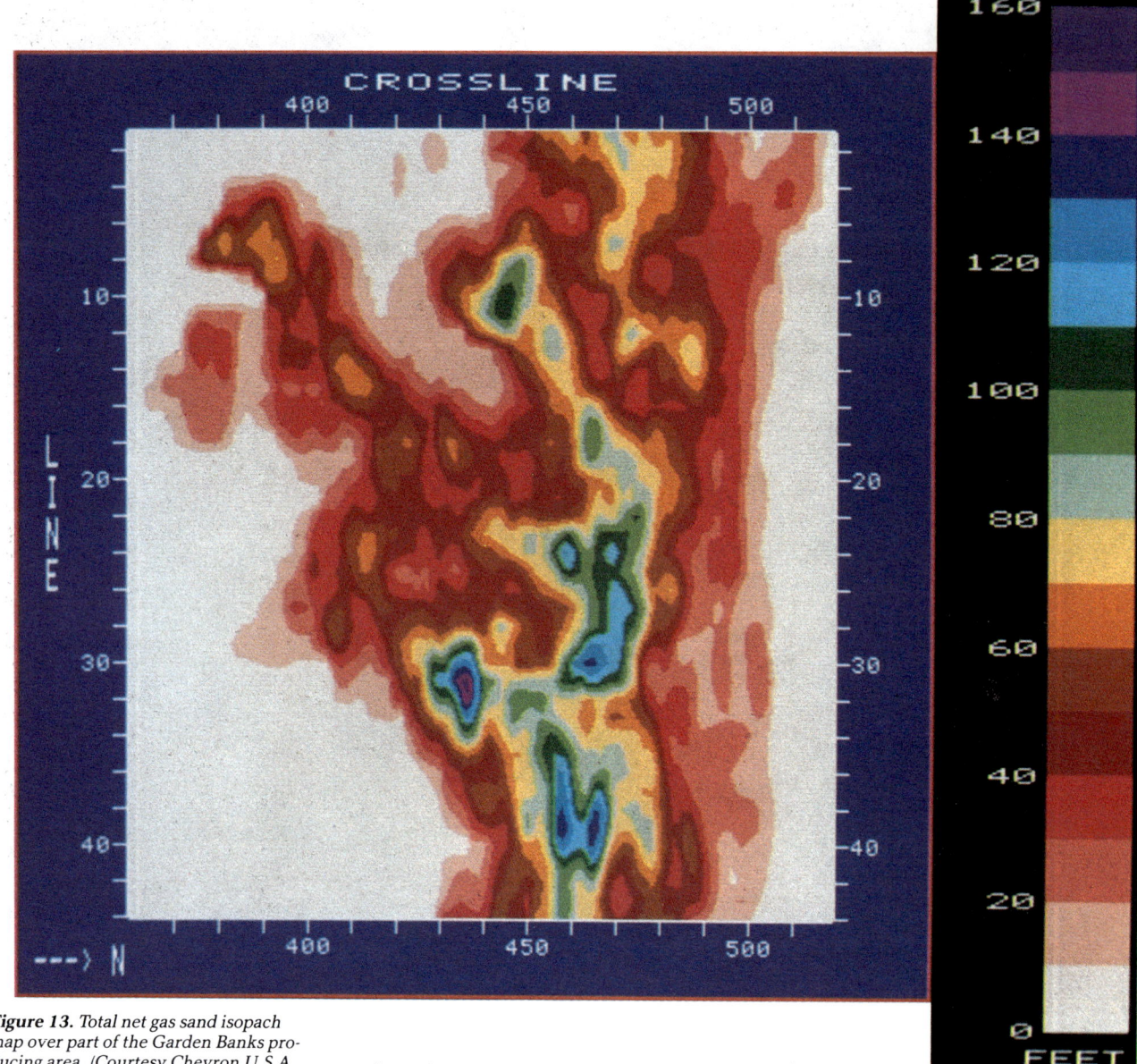

Figure 13. Total net gas sand isopach map over part of the Garden Banks producing area. (Courtesy Chevron U.S.A., Inc.)

high amplitude events are often saturated or low amplitude events are not visible. Color display helps to solve this problem and is particularly important for the study of stratigraphic detail.

In this paper gradational color schemes are used when shapes and trends are being studied (for example, Figs. 1, 2, 3, 4, 7, and 8). Contrasting schemes are used when quantitative analysis of the displays is intended (for example, Figs. 7, 10, and 13). In Figure 6 the color scheme is primarily gradational but an increased contrast is added for the highest amplitudes to accentuate the zone of probable gas. Many different color uses are illustrated in Figure 12.

CONCLUSION

The value of horizontal sections, especially those which follow seismic horizons, is well established both for the delineation of stratigraphic features and the synthesis of horizon properties. Horizon Seiscrop sections are becoming routine stratigraphic tools as interactive interpretation generates them as natural products of horizon tracking.

ACKNOWLEDGMENTS

Several oil companies have released data examples to provide the figures for this paper. They are individually recognized in the figure captions.

Mr. K. D. Burkart, Mrs. C. J. Young, and Mr. E. C. Brede, GSI, provided significant contributions to this paper.

REFERENCES CITED

Backus, M. M., and N. R. Goins, 1984, The change in reflectivity with offset; 1983 research workshop special report: Geophysics, v. 49, p. 838–839.

Bone, M. R., B. F. Giles, and E. R.

Tegland, 1983, Analysis of seismic data using horizontal cross sections: Geophysics, v. 48, p. 1172-1178.

Brown, A. R., 1978, 3-D seismic interpretation methods: San Francisco, Proceedings, Society of Exploration Geophysicists 48th Annual Meeting. (Later published as, 3-D seismic survey gives better data: Oil and Gas Journal, v. 77, n. 45, p. 57-71.)

———, 1983a, Structural interpretation from horizontal seismic sections: Geophysics, v. 48, p. 1179-1194.

———, 1983b, Discussion on "Seismic data display and reflection perceptibility": Geophysics, v. 48, p. 1291.

———, and R. G. McBeath, 1980, Three-D seismic surveying for field development comes of age: Oil and Gas Journal, v. 78, n. 46, p. 63-65.

———, C. G. Dahm, and R. J. Graebner, 1981, A stratigraphic case history using three-dimensional seismic data in the Gulf of Thailand: Geophysical Prospecting, v. 29, p. 327-349.

———, R. J. Graebner, and C. G. Dahm, 1982, Use of horizontal seismic sections to identify subtle traps, *in* M. T. Halbouty, ed., The deliberate search for the subtle trap: AAPG Memoir 32, p. 47-56.

———, et al, 1984, Interactive seismic mapping of net producible gas sand in the Gulf of Mexico: Geophysics, v. 49, p. 686-714.

French, W. S., 1974, Two-dimensional and three-dimensional migration of model-experiment reflection profiles: Geophysics, v. 39, p. 265-277.

Gerhardstein, A. C., and A. R. Brown, 1984, Interactive interpretation of seismic data: Geophysics, v. 49, p. 353-363.

Graebner, R. J., G. Steel, and C. B. Wason, 1980, Evolution of seismic technology into the 1980's: Australian Petroleum Exploration Association Journal, v. 20, p. 110-120.

Hilterman, F. J., 1982, Interpretative lessons from three-dimensional modeling: Geophysics, v. 47, p. 784-808.

Johnson, J. P., and M. R. Bone, 1980, Understanding field development history utilizing 3D seismic: Offshore Technology Conference, Paper 3849, p. 473-475.

Kurfess, J. A., B. F. Giles, and M. R. Bone, 1977, Field development with 3D seismic methods — a case history: Calgary, Proceedings, Society of Exploration Geophysicists 47th Annual Meeting. (Abstract published in Geophysics, v. 42, p. 1517.)

Meckel, L. D., Jr., and A. K. Nath, 1977, Geologic considerations for stratigraphic modeling and interpretation, *in* C. E. Payton, ed., Seismic stratigraphy - applications to hydrocarbon exploration: AAPG Memoir 26, p. 417-438.

Ostrander, W. J., 1984, Plane-wave reflection coefficients for gas sands at non-normal angles of incidence: Geophysics, v. 49, p. 1637-1648.

Sanders, J. I., and G. Steel, 1982, Improved structural resolution from 3D surveys in Australia: Australian Petroleum Exploration Association Journal, v. 22, p. 17-41.

Seismic Stratigraphy and Global Changes of Sea Level, Part 10: Seismic Recognition of Carbonate Buildups[1]

J. N. BUBB[2] and W. G. HATLELID[3]

Abstract Carbonate buildups, including reefs and banks, are ideally suited for stratigraphic interpretation from reflection seismic data because of pronounced differences in depositional or bedding characteristics between the buildups and enveloping strata. Geophysical criteria that allow recognition of buildups can be either direct—those seismic parameters that directly outline buildups such as reflections from the boundaries of the buildups, onlap of overlying cycles, or seismic facies changes between the buildups and enveloping beds; or indirect—those seismic parameters that indirectly outline or indicate the presence of buildups such as drape, velocity anomalies, and spurious events. Use of basin architecture is an additional indirect, but generally geologic, line of evidence to infer locations of buildups. All available geologic and geophysical data should be used; the techniques of seismic stratigraphic and seismic facies analysis provide the framework for this interpretation.

INTRODUCTION

This paper provides seismic interpreters with: (1) criteria for recognizing carbonate buildups on seismic sections, and (2) several examples of the seismic expression of documented carbonate buildups for comparison with seismic data from their own areas of interest.

Carbonate buildups, including reefs and banks, form important and prolific hydrocarbon reservoirs in many operating areas of the world, particularly in the United States, Canada, North Africa, Mexico, southeastern Asia and the Middle East. Their recognition and proper interpretation are important because of variations in reservoir characteristics of strata within and associated with the buildups, and because structural closure on prospects is commonly due to the topography generated during deposition of the buildups. Seismic stratigraphic interpretation of carbonate buildups is enhanced by depositional topography and contrasts in lithology, interval velocity, density, and bedding characteristics between the buildups and enveloping strata.

The seismic interpretation procedure recommended is that outlined in Part 6 (Vail et al, this volume). Emphasis is placed here on visual interpretation of reflection configuration and on other seismic parameters such as amplitude, frequency, continuity, and interval velocity. Modern computer processing also aids in presenting more sophisticated graphic displays of these parameters.

REEFS, BANKS, AND BUILDUPS

The term *carbonate buildup* as used here, is a general term for all sedimentary carbonate deposits that form positive bathymetric features. This inclusive term is used because seismic data do not easily differentiate between the deposits conventionally described as reefs and banks by many authors following Lowenstam (1950), Nelson et al (1967), and Klement (1967). The term *bank,* a descriptive term with genetic implications, denotes a bathymetrically positive sediment accumulation formed by the gregareous growth of organisms which cause and contribute to sediment deposition but do not form a rigid structure. The term *reef* is used for bathymetrically positive rigid structures formed by sedentary, intergrowing organisms. A reef is commonly a bioherm (Cumings, 1932), that is, a mound or lens-shaped feature of organic origin which is lithologically discordant with surrounding deposits. A bank can be a bioherm as above, or a biostrome, which is a layer or bed of coarse skeletal remains which grades into surrounding deposits of different lithology. Although we recommend the use of the terms reef and bank where possible, we have found that the subdivision of carbonate buildups given below is also very useful where reefs and banks can not be separated seismically.

For purposes of seismic analysis, a wide variety of carbonate buildups may be grouped into four major types illustrated in Figure 1: barrier buildups are linear, with relatively deep water on both sides during deposition; pinnacle buildups are roughly equidimensional and were surrounded by deep water during deposition; shelf-margin buildups are linear, with deep water on one side, and shallow water on the other; and patch buildups form in shallow water, either in close proximity to shelf margins, or over broad, shallow seas. Careful analysis of an appropriate grid of seismic data is required to define the shapes and depositional environments of these features.

---

[1]Manuscript received, January 6, 1977; accepted, June 13, 1977.

[2]Exxon Production Malaysia, Inc., Kuala Lumpur, Malaysia.

[3]Imperial Oil, Ltd., Calgary, Canada.

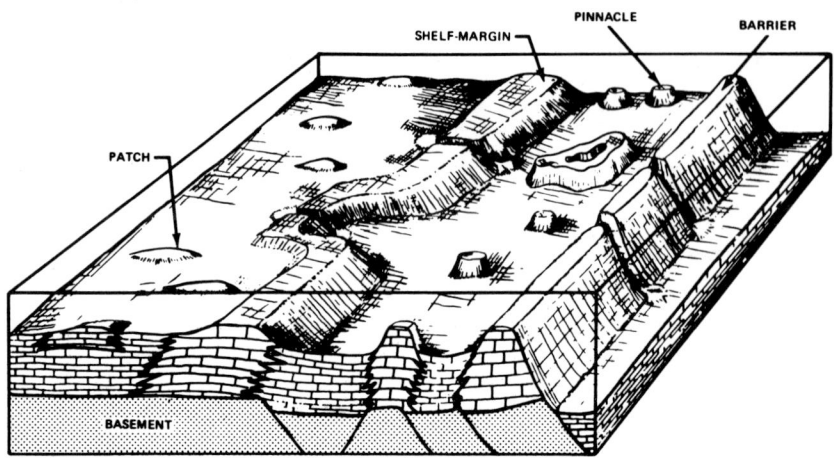

FIG. 1—Types of carbonate buildups most easily recognized from seismic interpretation. Conventional classification of reefs and banks, although preferred, is not easily applicable to seismic data.

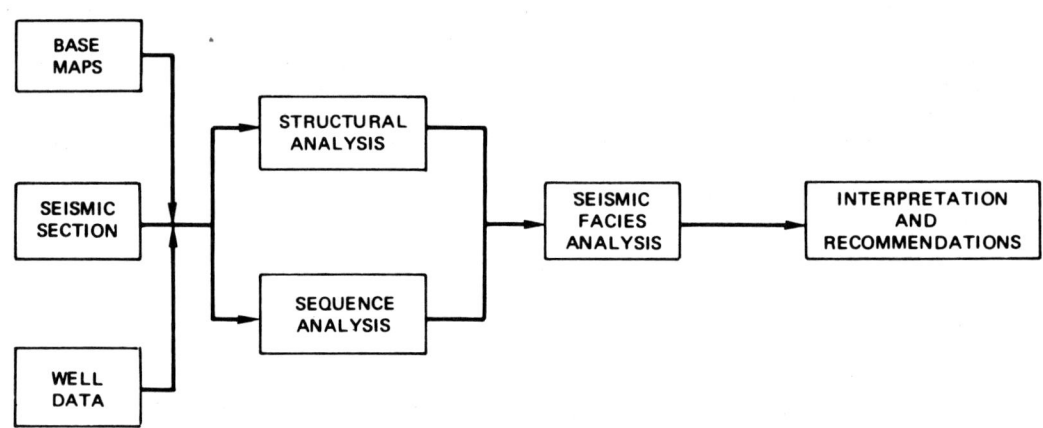

FIG. 2—Diagrammatic procedure for stratigraphic interpretation of seismic data. Three general steps include (1) conversion of all geologic data to a form compatible with seismic data, (2) seismic sequence and structural analysis, and (3) seismic facies analysis, interpretation.

Carbonate buildups as identified from seismic data commonly are composed of a variety of carbonate microfacies. For instance, a feature may have fore-reef facies changing laterally updip to synchronous reef talus and in-situ reef facies which may, in turn, grade to various back reef, lagoonal, or tidal-flat facies. Each facies has its own characteristic carbonate compositions, grain types, textures, and sedimentary structures. Resolution of most of these carbonate microfacies is beyond the limits of most conventional seismic data.

PROCEDURE FOR SEISMIC RECOGNITION OF CARBONATE BUILDUPS

Procedures recommended for the interpretation of carbonate buildups from a grid of seismic data follow the seismic stratigraphic techniques described in Part 6 (Vail et al, this volume). Basically, this interpretation procedure has three general steps (Fig. 2). As a preliminary step, all available geologic data are converted to a format compatable for interpretation with the seismic data. This usually requires construction of various time-linear well logs and overlays of paleontologic data, paleobathymetry, and other well information. Available outcrop data should be projected into the line of the seismic section.

Next, depositional sequences (Part 2, Vail et al, this volume) are interpreted from seismic data through study of the systematic patterns of cycle terminations. The sequences are correlated with well data, their ages determined, and their boundaries extended through seismic grids to complete the stratigraphic and structural framework.

Where the sequence framework is established, the third step is the recognition of seismic facies units within sequences. Seismic parameters are objectively defined, mapped, and correlated to well data where possible, and they are interpreted in terms of depositional processes, environments, and possible lithology.

GEOPHYSICAL CRITERIA FOR RECOGNIZING CARBONATE BUILDUPS

The criteria for recognizing carbonate buildups on seismic sections include seismic parameters that directly outline the buildup, and those that indirectly outline the buildup or infer its presence. Figure 3 diagrammatically illustrates these criteria; each of the diagrams was taken from an actual example.

Direct Criteria

Boundary Outline—Commonly, reflection configurations directly define the boundary of the buildup. These include reflections from the top and sides of the depositional feature, and onlap of overlying reflections onto the buildup. Depositional topography must be sufficiently great for these criteria to be evident on the reflection seismic record.

Seismic Facies Change—Changes may occur in amplitude, frequency, or continuity of reflections from within the buildup, or between the buildup and the laterally adjacent time-synchronous or younger onlapping reflections. Such changes would result where differences in characteristics of bedding continuity, density and/or velocity exist between strata within the buildup, or between the buildup and the strata enclosing the buildup.

Indirect Criteria

Drape—Drape commonly occurs in reflections overlying the buildup because of differential compaction of strata in the buildup and the enveloping strata. This phenomenon is generally most pronounced where a strong contrast exists in lithology of buildup and off-buildup sediments, such as with a limestone buildup surrounded by shale. The effects of drape generally die out stratigraphically upwards.

Velocity Anomalies—A pronounced velocity contrast commonly exists between the buildup and adjacent strata, resulting in differences in seismic travel time through these strata. For instance, reflections from strata beneath a limestone buildup with a higher velocity than laterally adjacent shales would be "pulled up" in time, compared to reflection time from the same strata beneath the shales (see Fig 13). Similarly, reflections below a buildup with slower interval velocities than those of the surrounding strata would be "pulled down" below the buildup (see Fig. 5). The amount of the velocity anomaly is directly related to the contrast of interval velocity between buildup and laterally adjacent strata, and the thickness of the buildup or strata that have the contrasting velocities. Carefully constructed isochron maps of time-stratigraphic units containing buildups are useful exploration tools where these velocity contrasts occur. Such maps were extensively used by industry in Devonian reef exploration in Canada.

Spurious Events—The edges of the buildup commonly are marked by termination of surrounding beds or abrupt changes in internal bedding geometry. These edges can be sites for development of diffractions or odd events. Mapping of such seismic events may offer a clue to the presence and distribution of otherwise hard-to-see carbonate buildups.

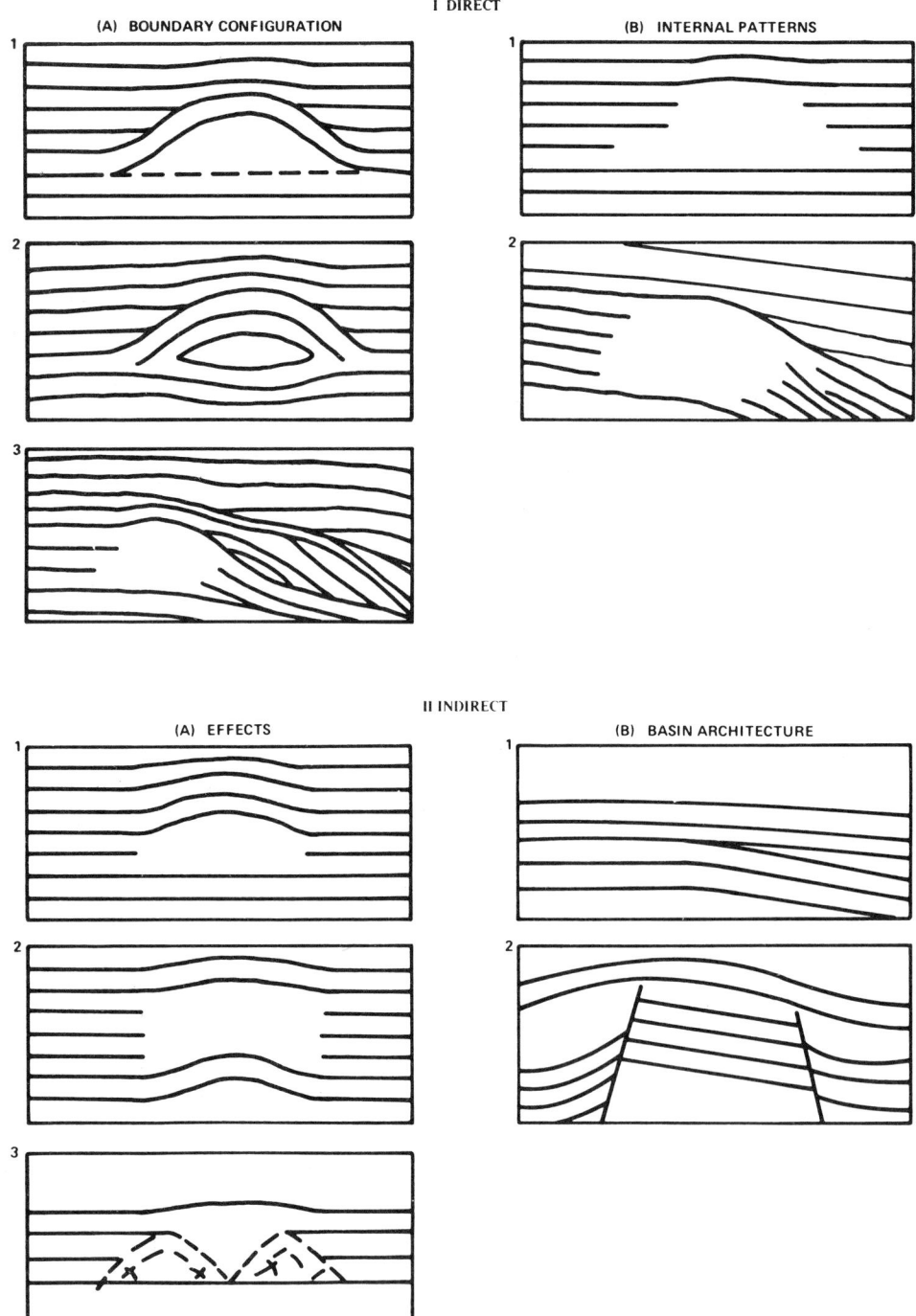

FIG. 3—Seismic criteria for recognizing carbonate buildups. Criteria for directly outlining buildups include reflections from top and sides of buildups and onlap of overlying reflections onto buildups (I-A); and patterns of seismic facies change between buildup and enclosing strata (I-B). Criteria that indirectly outline or infer presence of buildup include drape, velocity anomalies, and spurious events (II-A), and determination of optimum basin positions for buildups (II-B).

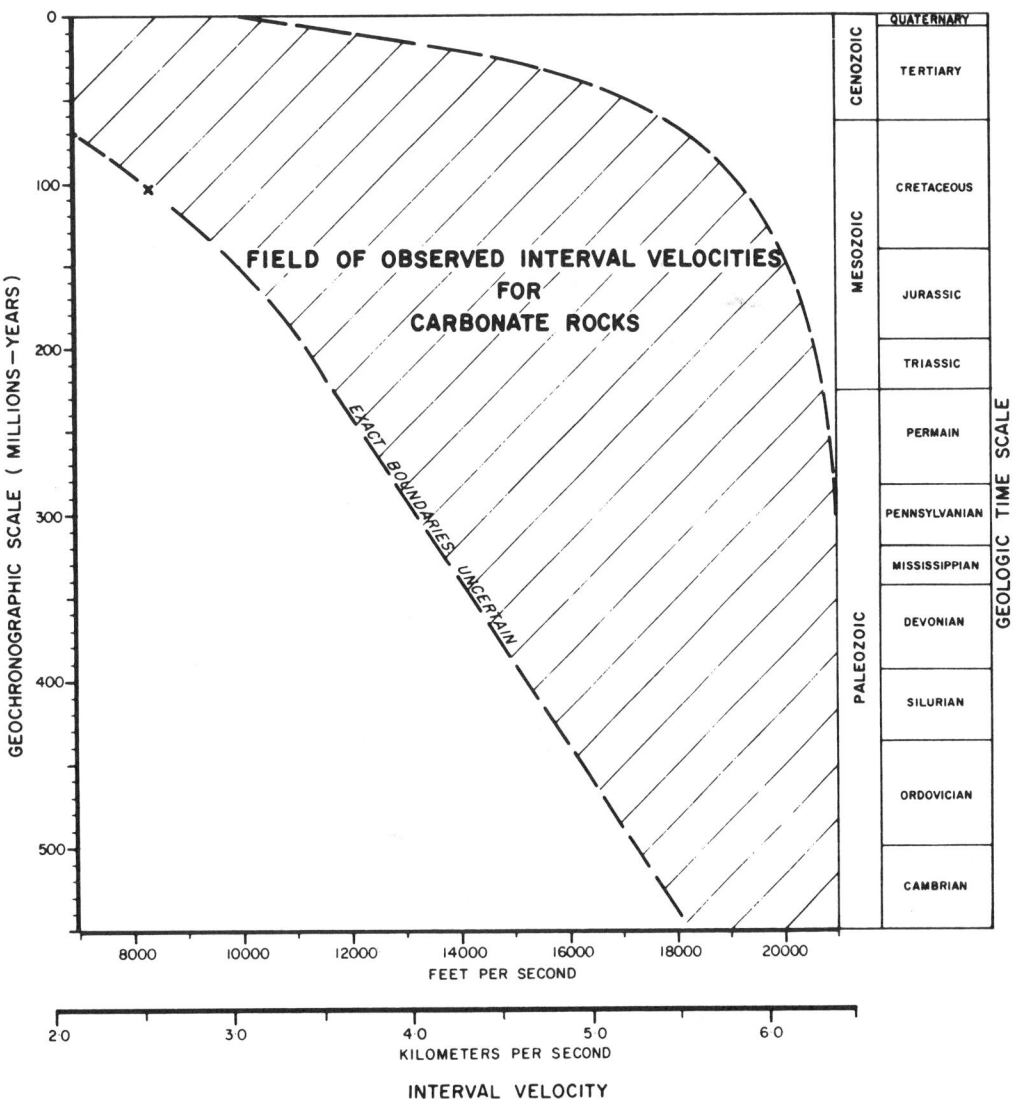

FIG. 4—Field of observed interval velocities from well data for carbonate rocks plotted against geologic age.

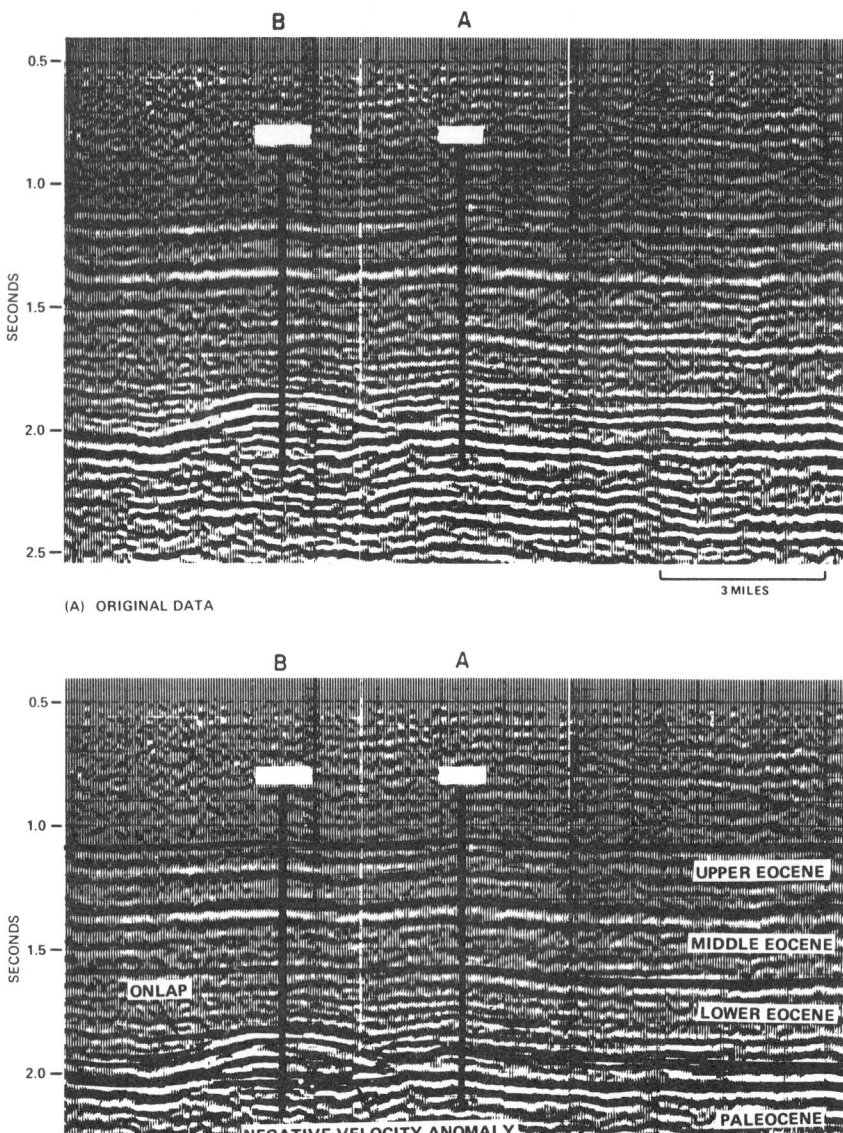

FIG. 5—North Africa (12-fold CDP thumper data). This high-quality seismic line shows anomaly interpreted as pinnacle reef by (1) a reflection outlining tops and sides of buildup, (2) three cycles of onlap, (3) drape in overlying beds, and (4) negative velocity anomaly (off-reef micritic limestones and shales have higher interval velocity than porous, lightly cemented, Tertiary reef carbonates). Overall aspect of this pattern is so-called "eye effect".

Location A was first to be drilled, based on poorer seismic data, on closure at Eocene level above Paleocene reef unit. Well encountered about 60 m of gas-filled pay in Paleocene and was abandoned; acreage in area was subsequently dropped. Another company picked up acreage, obtained high-quality seismic data that showed Paleocene reef, and drilled well B as discovery well. Well encountered about 300 m of porous algal-foraminiferal and coralline limestone. Pay section was 293 m thick. Oil flowed on test at rate of more than 40,000 bbl/day. Estimated recoverable reserves in this field are approximately 1.5 billion bbl of oil.

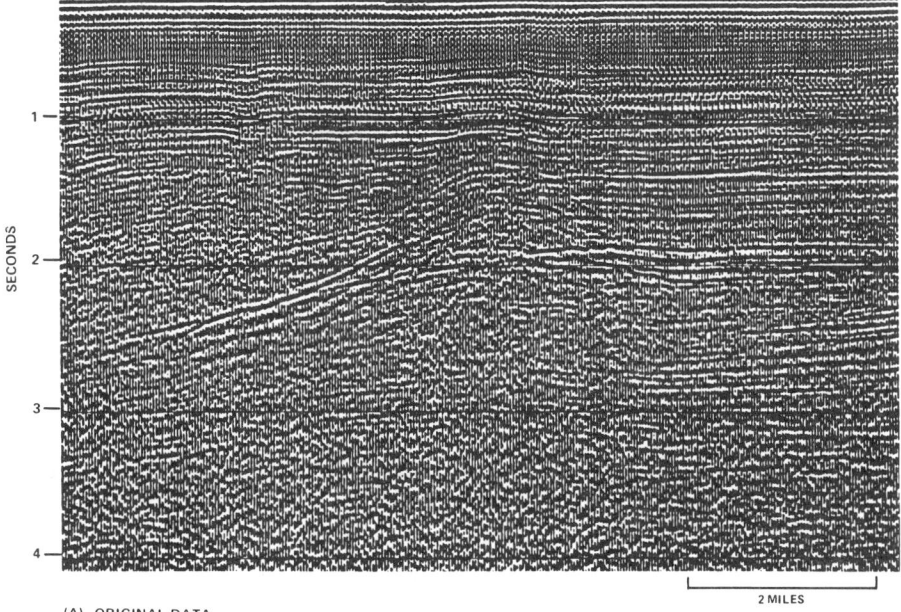

(A) ORIGINAL DATA

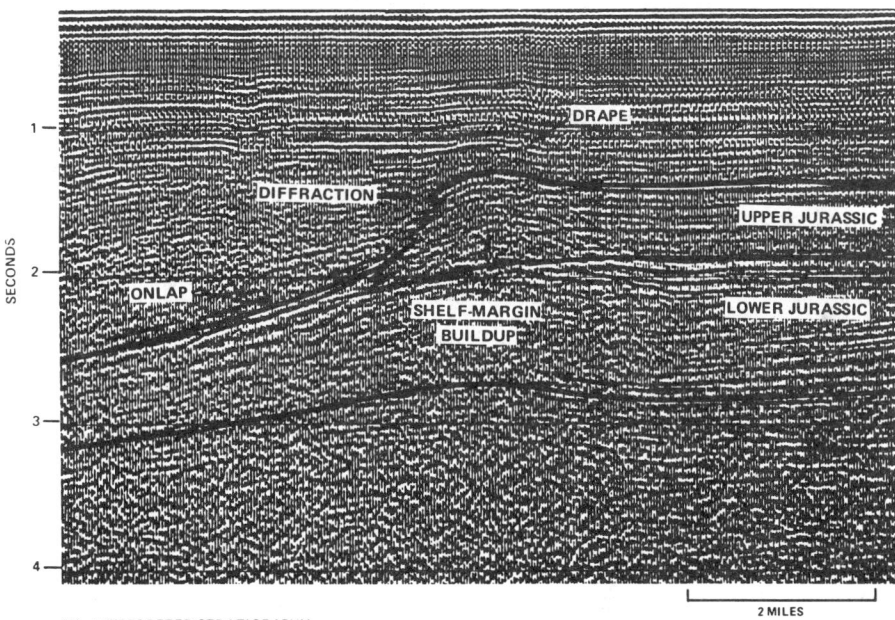

(B) INTERPRETED STRATIGRAPHY

FIG. 6—Offshore West Africa (12-fold CDP Aquapulse® data). Shelf-margin carbonate buildup can be seen by (1) reflection from top and front of buildup, (2) onlap of cycles onto buildup, (3) change from continuous, parallel reflectors into discontinuous reflectors, (4) numerous diffractions, (5) drape over buildup, and (6) abrupt change in dip of reflectors.

Wells encountered series of Mesozoic shelf-margin buildups along eastern Atlantic continental margin off Africa. Buildup displayed on this line is interpreted as Late Jurassic.

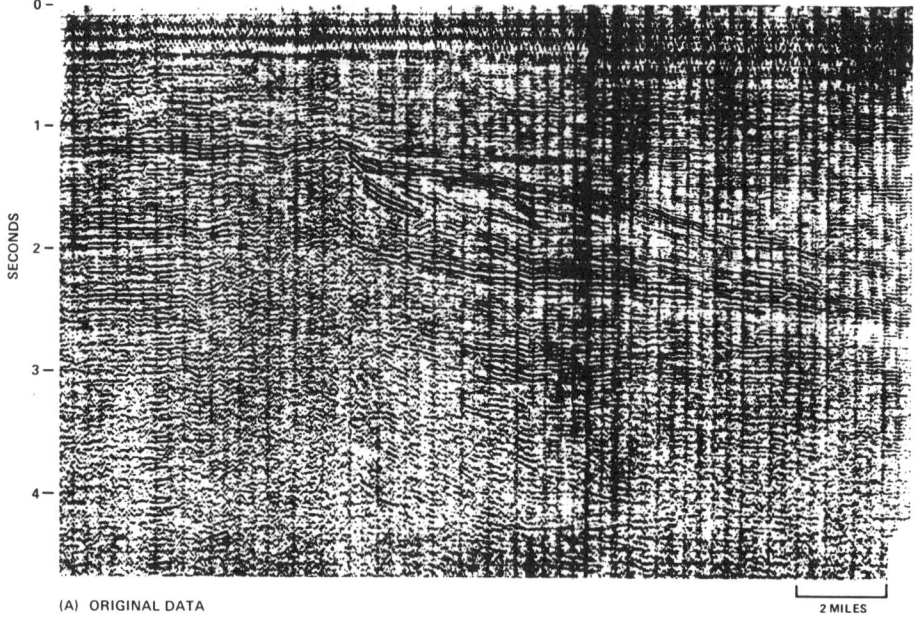

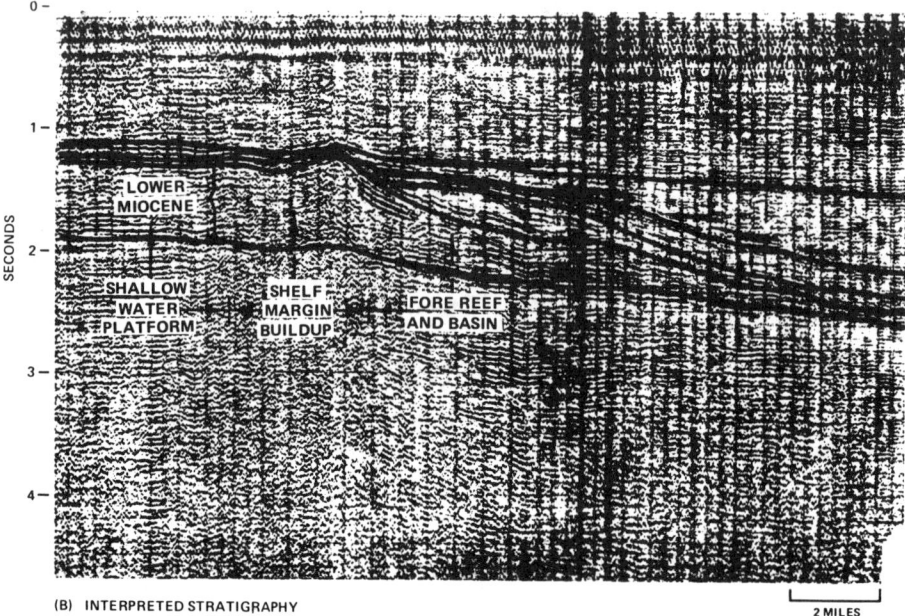

FIG. 7—Gulf of Papua (6-fold CDP dynamite data). Shelf-margin reef is interpreted on basis of (1) abrupt changes in slope of reflectors at shelf edge, (2) onlap of slope and basin units onto shelf edge, and (3) seismic facies pattern change from zone of more continuous, parallel reflectors (back reef–lagoon) to discontinuous, nearly reflection-free zone (reef facies) to thinner zone of dipping, convergent cycles (fore reef). Interesting prograding lens of convergent to sigmoid cycles, either younger or in part synchronous with reef, has partially infilled basin seaward of shelf-margin buildup. This lens, untested by wells, is interpreted to be shale and/or micritic limestone.

Early Miocene reefs, both shelf-margin and pinnacle types (some with gas) have been encountered in Papua basin. Each prospective area should be considered in terms of (1) closure, (2) seal (sometimes shelf-margin reef has porous facies over it), and (3) freshwater flushing.

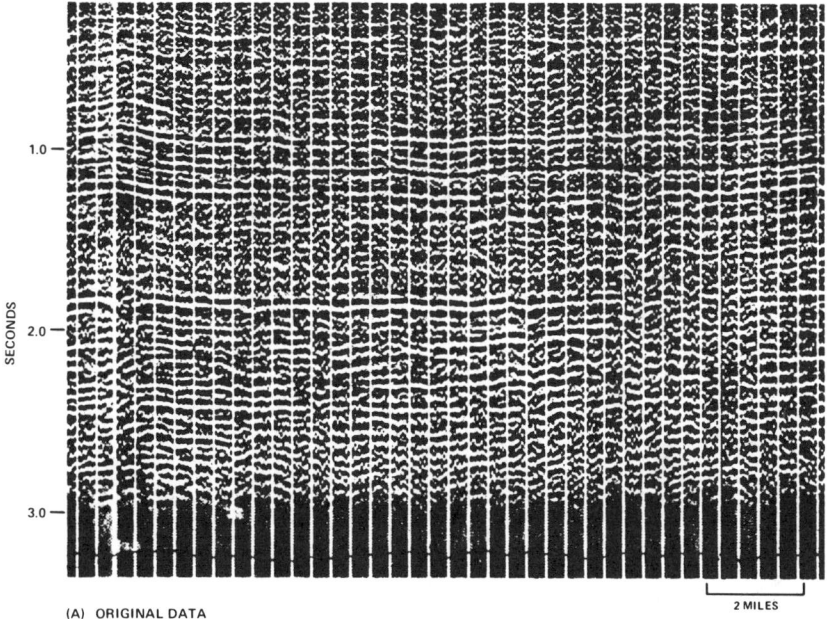

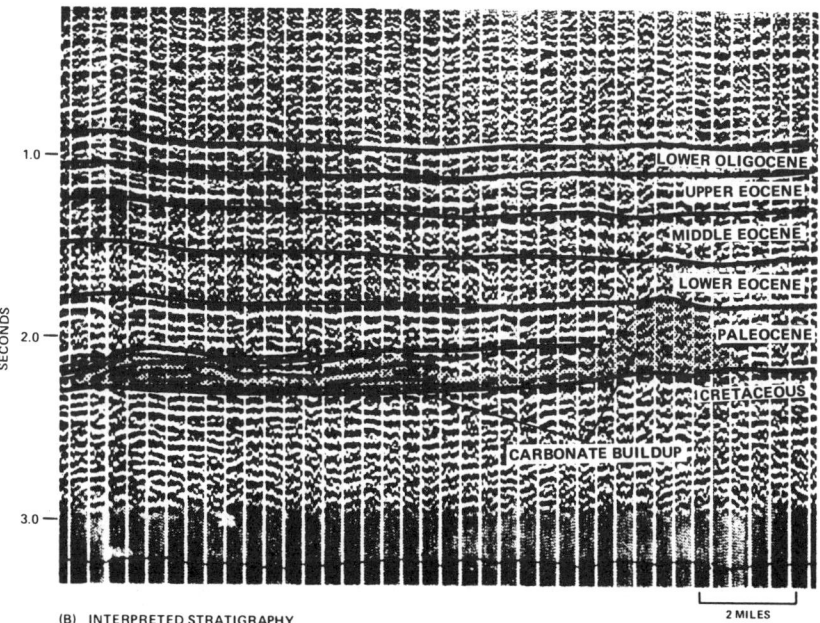

FIG. 8—North Africa (12-fold CDP thumper data). Carbonate buildups, both low platformlike banks and pinnacles, are interpreted on basis of (1) seismic facies change from continuous parallel reflectors to mainly reflection-free to very discontinuous reflection zone, and (2) one to two cycles of onlap of overlying units onto buildups (extreme left of section).

Well south of line confirmed buildup seen on right. It encountered shallow-water carbonates of Paleocene age in part of basin normally having shale and deep-water limestone. Only a thin gas pay, apparently associated with small amount of drape, was encountered. Trapping problem was lack of seal. A sheetlike porous carbonate grainstone extends over this part of basin near top of Paleocene sequence and intersects buildup to right. This contact may have allowed drainage of hydrocarbons from buildup tested in well.

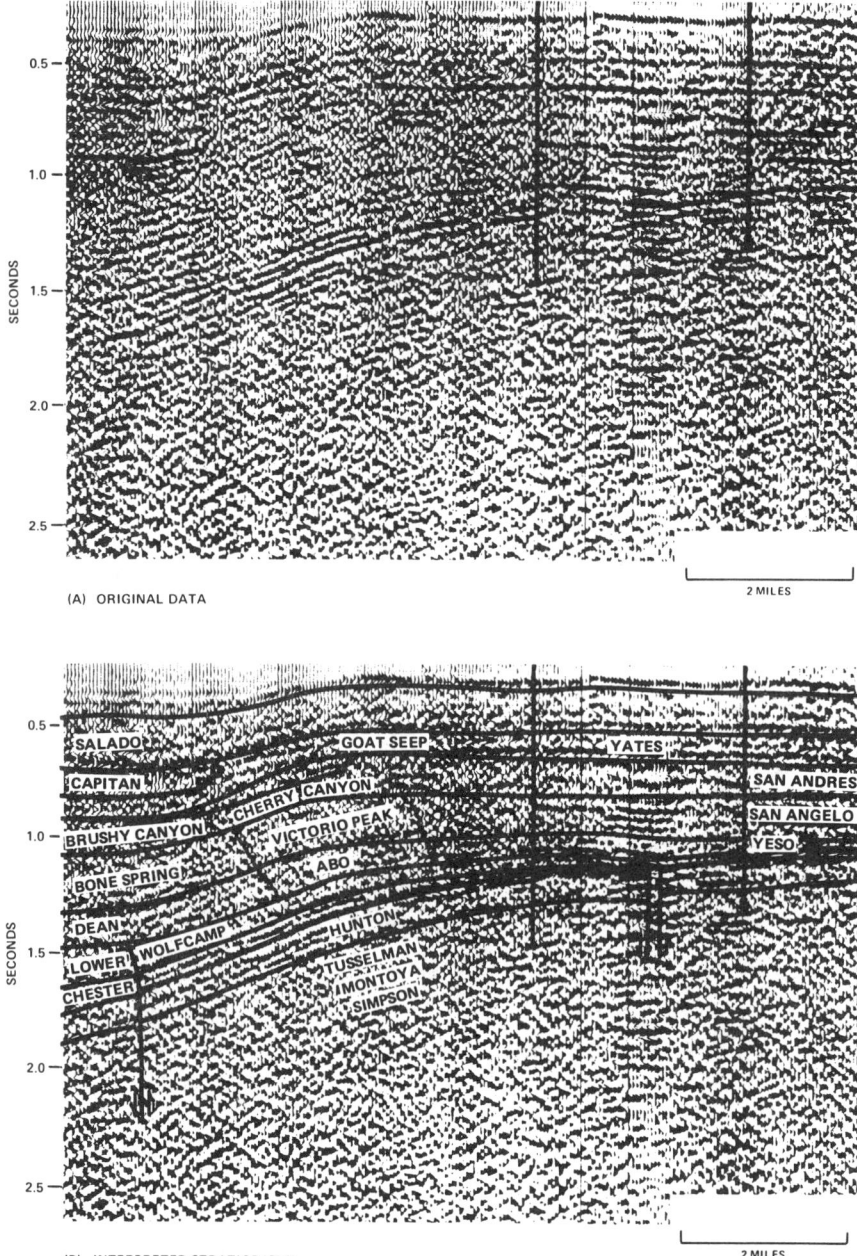

FIG. 9—Central basin platform, Lea County, New Mexico (12-fold CDP Vibroseis® data). Shelf-margin carbonate-bank buildup on this line is indicated by (1) abrupt change in dip at shelf edge, and (2) seismic facies change from high-amplitude, continuous reflections to low-amplitude to nearly reflection-free zone at shelf edge.

Leonardian and Guadalupian shelf-margin banks, composed mainly of dolomitized skeletal limestones of the Abo, Victorio Peak, Goat Seep, Getaway, and Capitan Formations, are documented by wells in this part of Permian basin. Basinward of shallow-water banks are siltstones, shales, and micritic limestones of Dean, Bone Spring, Brushy Canyon, and Cherry Canyon Formations; shelfward of banks are thin-bedded, dolomitized micritic and dolomitized algal-laminated limestones and sandstones of the Yates, Seven Rivers, Queen, Grayburg, San Andres, San Angelo, and Yeso Formations.

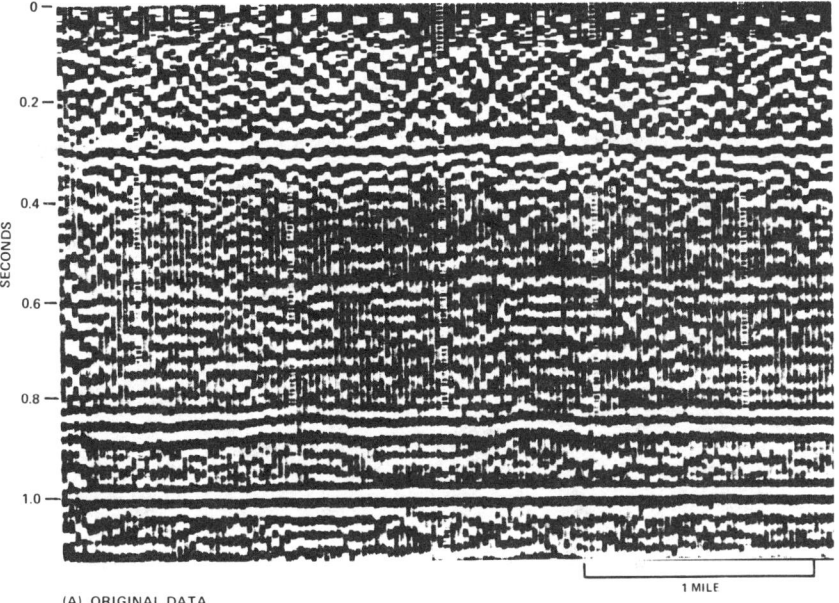

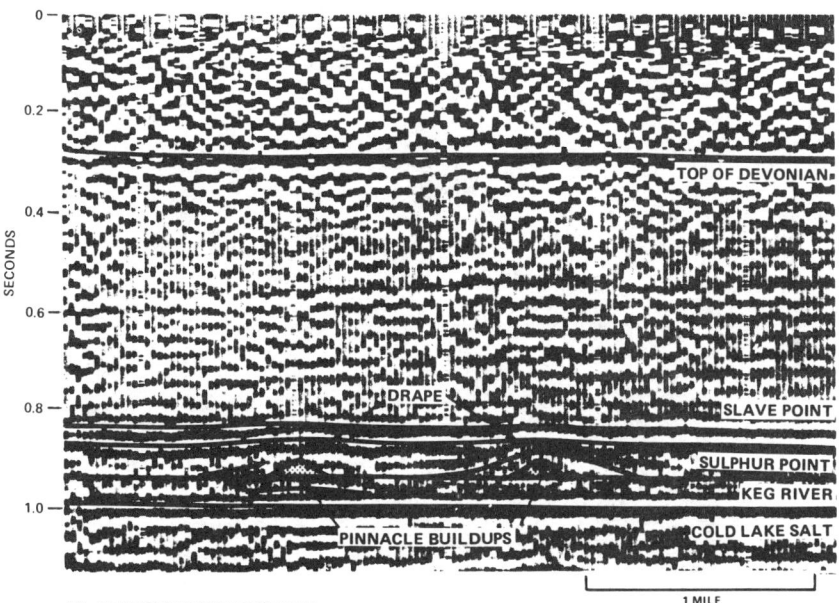

FIG. 10—Rainbow-Zama area, Alberta (6-fold CDP dynamite data, section datumed on Cold Lake salt reflection—a near-basement horizon). Presence of two small pinnacle reefs is indicated by small amount of drape in reflectors above reefs and by one cycle of onlap (?).

Buildups indicated on section are Middle Devonian Keg River Formation pinnacle reefs. Drape seen in overlying Sulphur Point and Slave Point Formations is probably due both to differential compaction and to subsequent removal of Muskeg salt deposited in interreef areas. Recoverable reserves in Rainbow-Zama area are estimated at 800 million bbl of oil. Pool size is variable; some are smaller than 8 ha., others are larger than 2 sq km, but average is 20 to 30 ha. Pay thickness is variable.

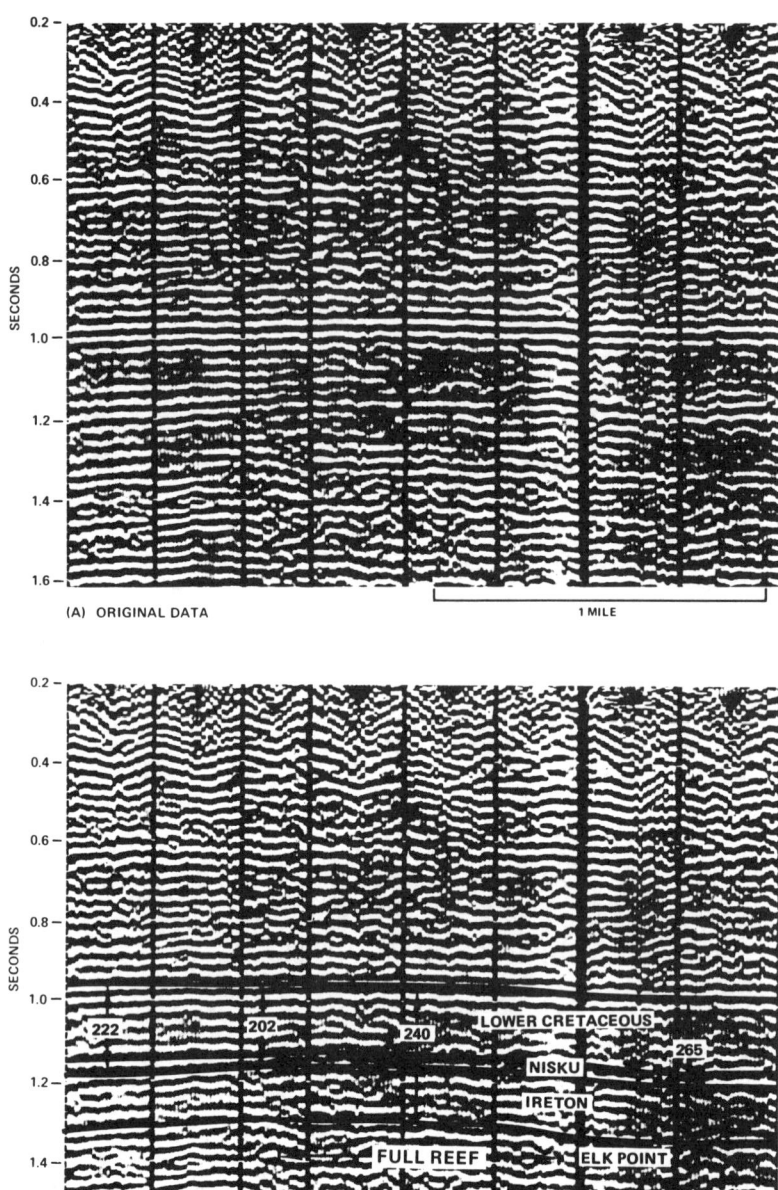

FIG. 11—Golden Spike, Alberta (singlefold conventional seismic data). Carbonate buildup indicated by (1) drape, and (2) velocity anomaly. Drape effect due to differential compaction shown by about 23 msec of thinning in Lower Cretaceous to Ireton isochron. The 30 msec of thinning in Lower Cretaceous to Elk Point isochron, which encompasses reef, is caused by faster velocity in reef limestones than in laterally adjacent shales.

Golden Spike (discovered 1949) is areally small (10 sq km) pinnacle reef, with more than 300 million bbl of oil in place. It formed on Cooking Lake bank seaward of Leduc-Rimbey barrier reef trend and is surrounded by shales of Ireton and Duvernay formations.

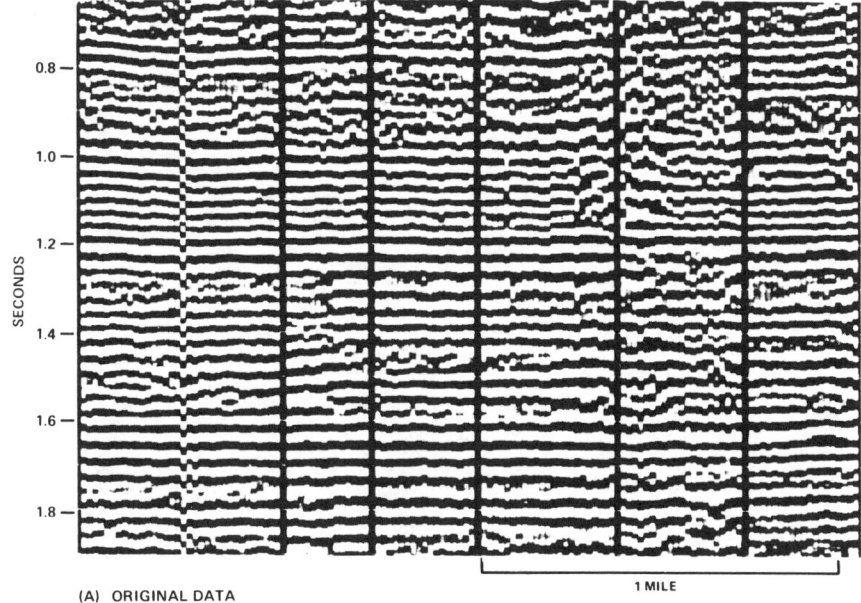

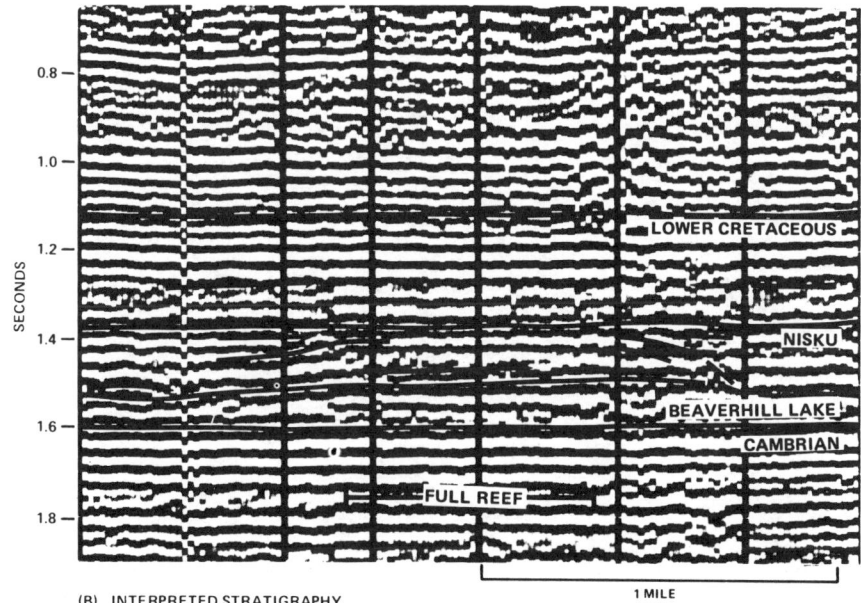

FIG. 12—Innisfail, Alberta (singlefold conventional seismic data). Only indications of carbonate buildup are vague, odd dips and diffractions. After reef was discovered, mapping of spurious events produced good outline of reef which was useful in development stages of drilling. Method was not successful in predicting reef prospects in adjacent areas.

Innisfail field, a Leduc reef on west side of Bashaw complex in southwest Alberta, is in an area where off-reef Ireton has high lime content. Hence, no measurable velocity anomaly or drape effect is found. Innisfail field (discovered 1957), produced about 20 million bbl of oil up to December 1968; total reserves near 75 million bbl.

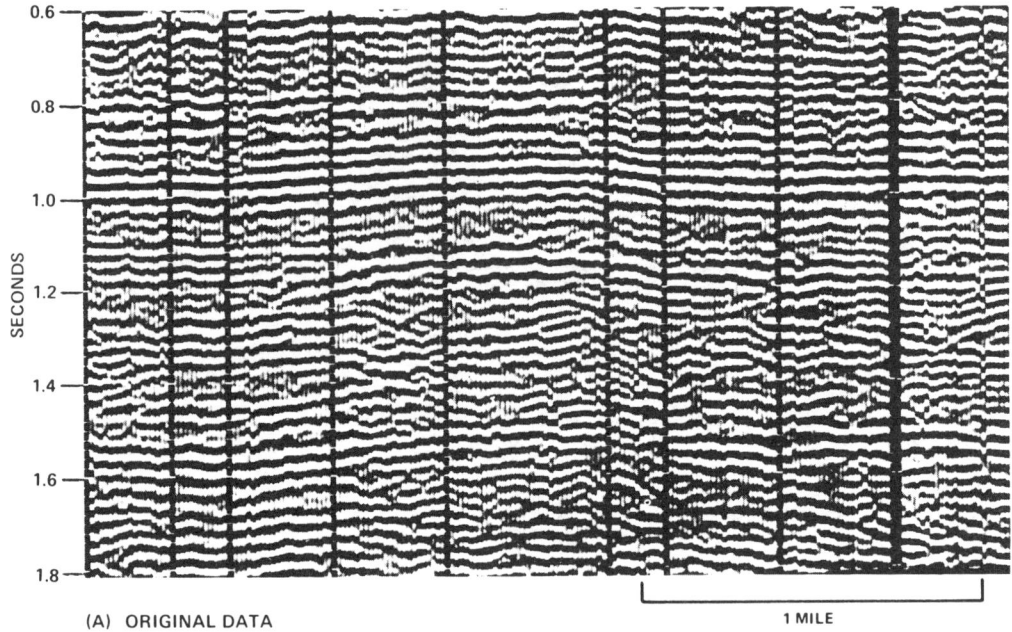

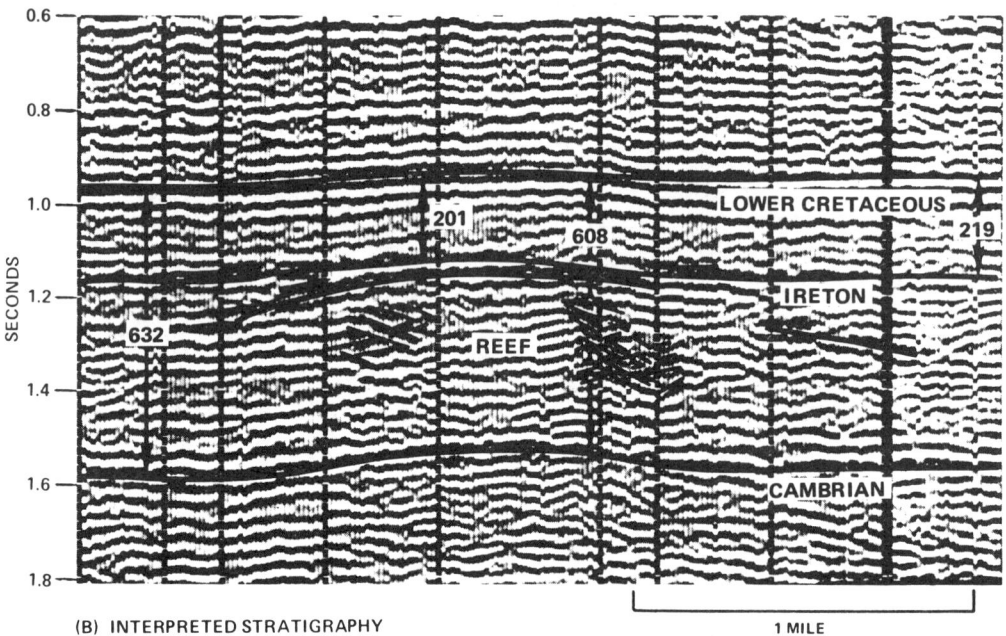

FIG. 13—Yekau Lake, Alberta (singlefold conventional seismic data). This buildup is defined by (1) about 18 msec drape measured in Lower Cretaceous to Ireton interval, (2) about 25 msec velocity anomaly measured in Lower Cretaceous to Cambrian isochron, and (3) spurious events, probably diffractions, near edges of buildup.

Yekau Lake reef is small Leduc reef within Leduc-Rimbey barrier trend. Off-reef Ireton-Duvernay in this area is mainly shale, giving rise to pronounced velocity anomaly and drape effects. Field size is slightly more than 2 sq km, pay thickness averages 6.5 m, and recoverable reserves estimated at 3.6 million bbl.

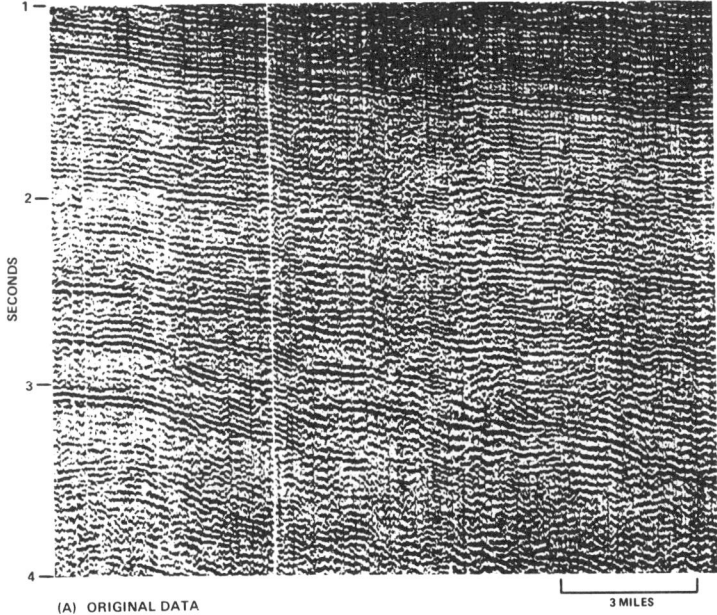

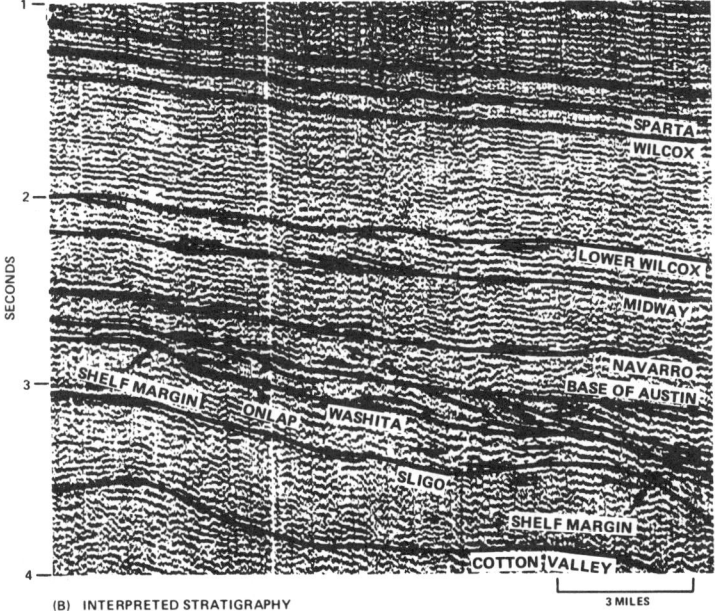

FIG. 14—Washita reef trend, south Texas (6-fold CDP seismic data). Shelf-margin carbonate buildup inferred adjacent to hinge line as indicated by (1) abrupt thinning in Washita-Sligo interval and (2) onlap of about two cycles onto Washita edge. Sligo shelf margin also suggested on right-hand (south) side of section.

Wells encountered Washita-Fredericksburg rudistid reefs and banks that pass seaward into dense deeper water carbonate mud. Exploration problem is that of updip closure. Back-reef and lagoonal sediments behind shelf-margin buildups are commonly porous, and may have permitted hydrocarbons to escape.

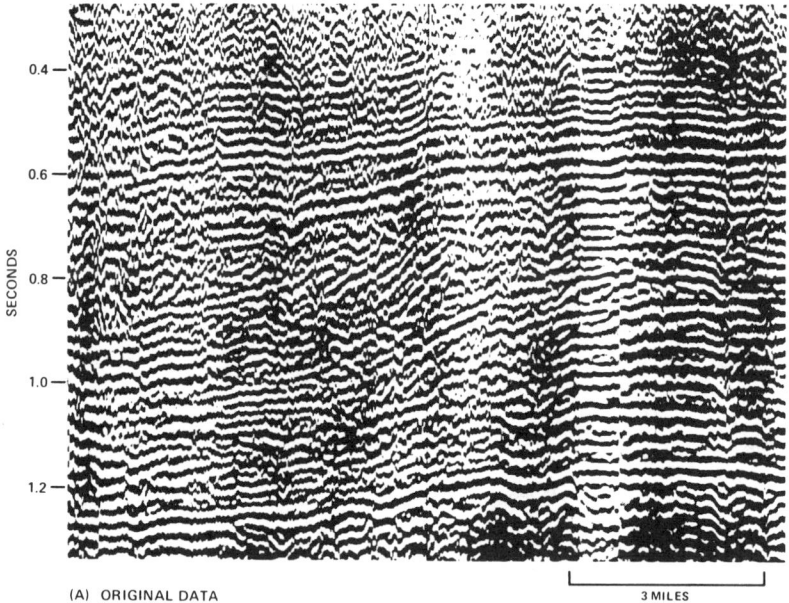

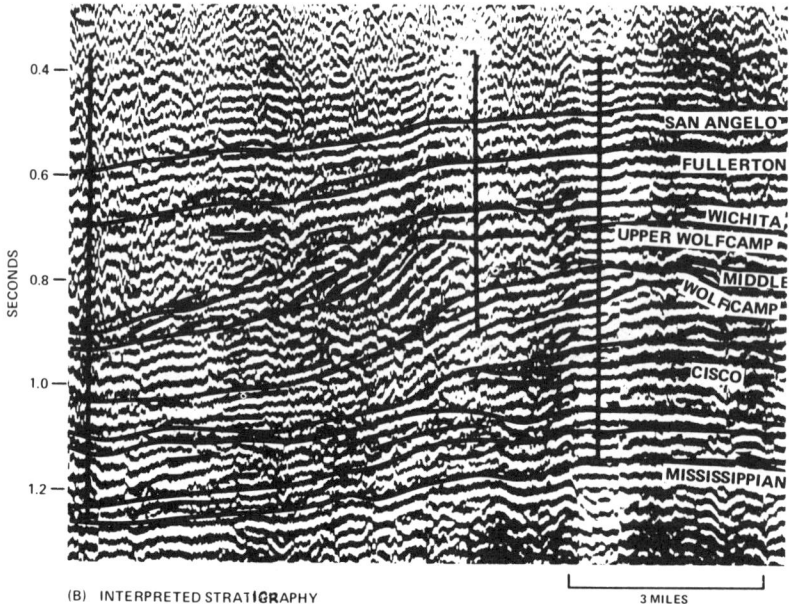

FIG. 15—Eastern shelf area, Permian basin, Crosby County, Texas (singlefold conventional seismic data). Series of stacked carbonate-shelf-edge buildups were detected here after seismic sequences were carefully defined to show (1) abrupt change in slope, (2) onlap onto shelf edges, and (3) thinning of sequences.

Upper Pennsylvanian (Cisco-Canyon) and Lower Permian bank edges are stacked in transgressive series through middle Wolfcamp, then in regressive fashion through late Wolfcamp and Wichita. Carbonate banks and bank edges are porous in this area, but lack of lateral closure prevented hydrocarbon accumulation. Banks of same age are prolific producers in Horseshoe atoll area, and locally in Central basin platform.

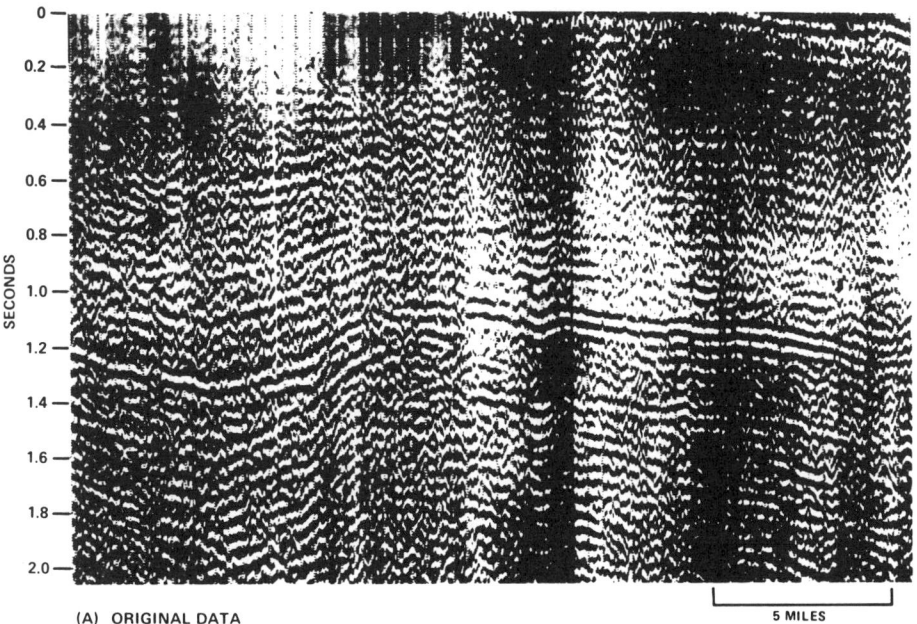

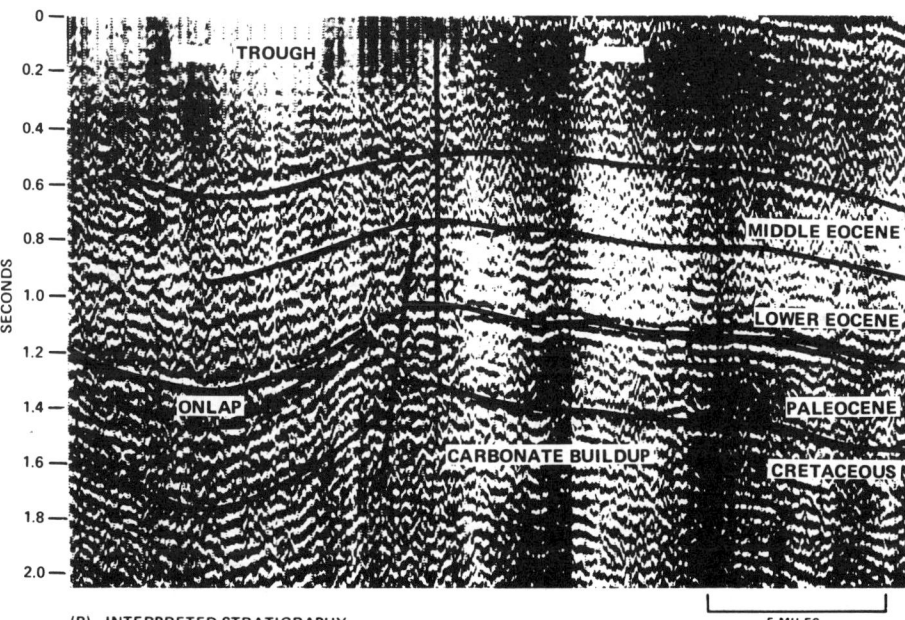

FIG. 16—North Africa (3-fold CDP seismic data). Presence of carbonate buildup can only be inferred from basin architecture. Upthrown side of normal fault block would be logical position for shallow-water carbonates to accumulate under proper conditions. In seismic data, no parameter change was noted along fault block even though there is pronounced facies change from porous shallow-water carbonates on platform to shales and micritic limestones in trough.

Field produces from shallow-water Paleocene carbonate "pipe" or buildup in formations on upthrown side of major fault that separates platform from trough. Cumulative production of over 1 billion bbl of oil.

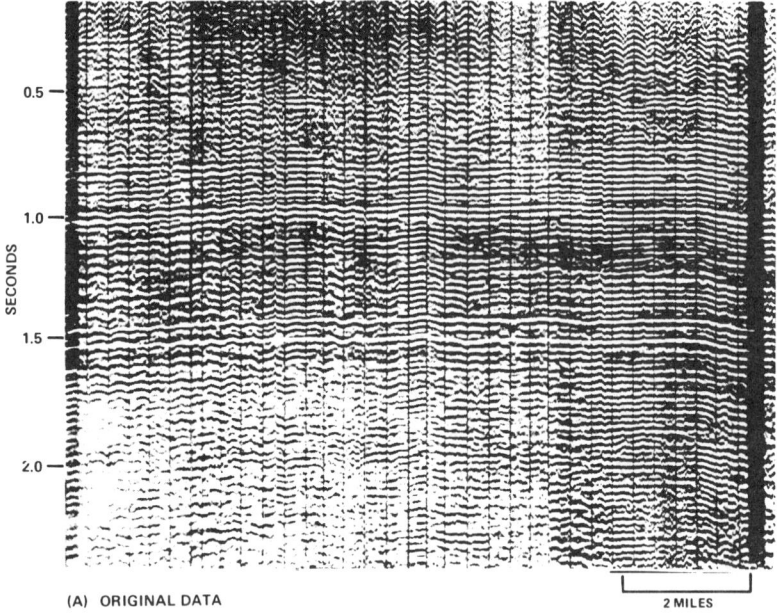

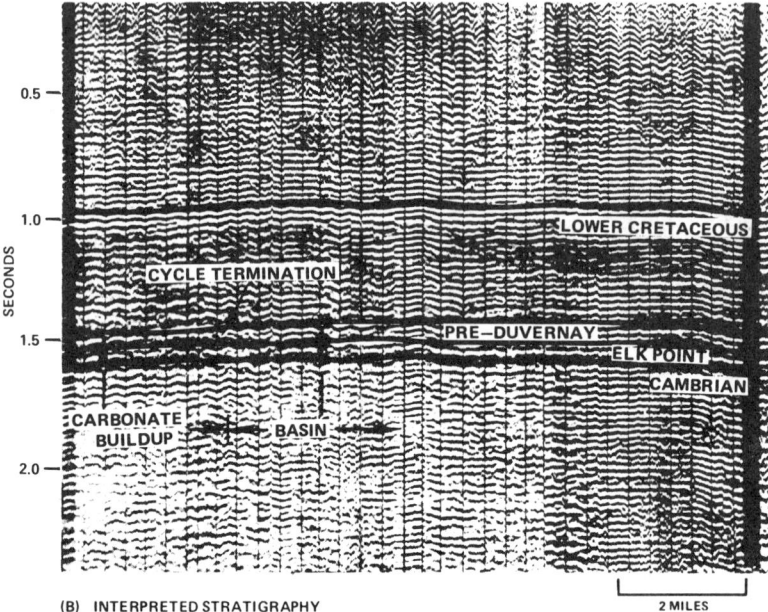

FIG. 17—Judy Creek area, Alberta (singlefold conventional seismic data). Presence of shallow-water carbonate buildup indicated only by termination of single cycle. Termination occurs near point where, to southwest in structurally high position, shallow-water carbonates formed. To northeastward, or basinward within same time span, shale was deposited. After discovery well was drilled, fair success was achieved in outlining limits of this and similar nearby buildups using cycle termination as diagnostic parameter.

Judy Creek reef and bank complex is thin (75 m thick) carbonate mound within Beaverhill Lake Group (Upper Devonian) encased by argillaceous limestones and shales. Judy Creek field (discovered 1958) had cumulative production to June 1968 of 43 million bbl of oil; estimated recoverable reserves are about 131 million bbl of oil. Field covers about 115 sq km, and pay thickness averages 21 m.

Seismic Recognition of Carbonate Buildups

Basin Architecture—In some instances, carbonate buildups can be inferred as likely to occur in a preferred location along a seismic profile, based on seismic and other geologic evidence of basin architecture, such as fault-block edges, position of hinge line, or contemporaneous structural highs. The interpretation is enhanced by a thorough knowledge of the geologic setting and history of the area, and of those time-stratigraphic units prone to the development of buildups or marked by extensive or local carbonate sedimentation.

INTERVAL VELOCITY IN CARBONATE SECTIONS

The development of high speed computers and the advent of widespread digital processing have facilitated the estimation of seismic interval velocity and its use for interpretation of lithology. Although applied mostly to estimating sandstone/shale ratios, some interval velocity work has been done with stratigraphic sections containing carbonate, sandstone, and shale. Carbonates do not have unique interval velocities, and the interval velocity of a particular buildup is dependent on a wide variety of factors including porosity and burial history. A search of the literature on carbonates in a variety of basins around the world shows a wide variation in the interval velocity of carbonate rocks (Fig. 4). Within each local area, therefore, all available well data should be analyzed carefully to aid in the interpretation of stratigraphy from interval velocity.

SOME LIMITATIONS AND QUALIFICATIONS

Many seismic reflection configurations interpreted as carbonate buildups are not unique, but may be exhibited by other geologic features similar in size or shape. Examples of such features include salt intrusions or pillows, igneous intrusions, volcanic cones, unconformity surfaces, and slump deposits. Interpretation of carbonate buildups from seismic data must be merged with all available geologic and geophysical data of the area. In particular, data on the occurrence and distribution of known buildups and carbonate units in adjacent parts of the basin are useful.

Resolution limitations must be recognized. This is particularly true in older carbonate sequences with high interval velocities. Interval velocities of 4,500 to 6,400 m/sec are common for many Paleozoic carbonate strata. Thus, a carbonate buildup 130 m thick with an interval velocity of 5,500 m/sec would be represented by only 47 msec, two-way time. There may not be sufficient response for recognition of this 130-m buildup, depending on such parameters as (1) the frequency content of the input pulse, (2) the amount of attenuation of frequency of the pulse with increasing travel time, and (3) the processing and filtering both in the field and before display of the final seismic section. A cycle breadth of 50 msec for 20-Hz data might not be uncommon and the buildup described would be represented by one cycle. Reflections from directly overlying beds might mask entirely any response of the seismic data to the buildups.

EXAMPLES OF CARBONATE BUILDUPS ON SEISMIC SECTIONS

Thirteen seismic profiles (Figs. 5-17), mostly documented by wells, illustrate the geophysical criteria for recognizing carbonate buildups shown diagrammatically on Figure 3. These criteria are summarized in Table 1 and are discussed in each figure description, along with pertinent geologic and hydrocarbon-production data. These exam-

Table 1. Summary Sheet of Seismic Examples.

Examples	Criteria for Recognizing Buildup									Type of Buildup		Production Associated with Buildup	
	Direct			Indirect									
			Amplitude Frequency, or Continuity Change	Effects			Basin Architecture						
Figure Number	Reflection Outline	Onlap		Drape	Velocity	Spurious Events	Change in Slope	Basin Position	Shelf Margin	Barrier	Pinnacle	Yes	No
5. North Africa	X	X		X	X						X	X	
6. Offshore W. Africa	X	X	X	X		X	X		X				X
7. Gulf of Papua	X	X	X				X		X				X
8. North Africa		X	X								X		X
9. Cen. Basin Plat.		X					X		X				X
10. Zama				X							X	X	
11. Golden Spike				X	X						X	X	
12. Innisfail						X					X	X	
13. Yekau Lake				X	X	X				X		X	
14. South Texas		X					X	X	X				X
15. Crosby Co., Texas		X					X	X	X				X
16. North Africa							X	X				X	
17. Judy Creek							X	X				X	

ples provide a spectrum of criteria for recognition of carbonate buildups.

REFERENCES CITED

Cumings, Edgar R., 1932, Reefs or bioherms?: Geol. Soc. America Bull., v. 43, p. 331-357.

Klement, K. W., 1967, Practical classification of reefs and banks, bioherms and biostromes (abs.): AAPG Bull., v. 51, no. 1, p. 167.

Lowenstam, H. A., 1950, Niagaran reefs of the Great Lakes area: Jour. Geology, v. 58, no. 4, p. 430-487.

Nelson, H. F., C. W. Brown, and J. H. Brineman, 1962, Skeletal limestone classification *in* W. E. Ham, ed., Classification of carbonate rocks—a symposium: AAPG Memoir 1, p. 224-252.

The American Association of Petroleum Geologists Bulletin
V. 71, No. 3 (March 1987), P. 281-297, 19 Figs., 1 Table

Seismic Interpretation of Carbonate Depositional Environments[1]

J. M. FONTAINE,[2] R. CUSSEY,[3] J. LACAZE,[2] R. LANAUD,[4] and L. YAPAUDJIAN[5]

ABSTRACT

Most seismic stratigraphic or seismic lithologic interpretations have been applied to clastic depositional systems. Because of their unique sedimentologic and mineralogic characteristics, carbonate rocks are more difficult to study using seismic data. Increased knowledge of carbonate deposits and their petrophysical parameters, and higher quality geophysical tools now permit a better understanding of carbonate rocks using seismic images.

In this paper, seismic facies of carbonate rocks are characterized, step by step, from the basin to the supratidal environment. (1) Pelagic deposits (shales and micritic limestone layers) exhibit continuous, parallel reflections with an apparently high frequency. Chalk deposits display continuous high-amplitude reflections at the top and base with an internal reflection-free zone. (2) Talus deposits are characterized by discontinuous, oblique reflections of high amplitude. Carbonate debris flow generates chaotic reflections with hummocky surfaces. Channels exhibit erosional truncations. (3) Reef barriers are mound-shaped biogenic deposits that display marginal onlapping reflections. Overlying reflections drape the reefs, and underlying reflections exhibit pull-up or pull-down effects. Hyperbolic diffractions also may occur. (4) Platform border sands are mound-shaped oblique reflections of moderately high amplitude. (5) Inner shelf strata are characterized by parallel, continuous reflections typically displaying low frequency. Patch reefs may be distinguished within the inner shelf system (mound shape, drape of overlying reflections, velocity anomalies, and spurious events). (6) Where dolomitized, the intertidal to supratidal facies exhibit a "marbled" zone—a practically reflection-free zone with a few discontinuous reflections. Diagenetic events, such as paleokarst zones, may be delineated using special analyses involving amplitude offset variations.

The study of carbonate depositional environments and petrophysical characteristics provides a more complete geologic insight, including relative changes in sea level, within a sedimentary basin. Such studies are fundamental in analyzing carbonate reservoirs and their paleogeographic settings.

©Copyright 1987. The American Association of Petroleum Geologists. All rights reserved.
[1]Manuscript received, January 3, 1986; accepted, October 28, 1986.
[2]Elf Aquitaine, Tour Elf, 92078 Paris-la-Défense, France.
[3]Elf Aquitaine, Centre Micouleau, 64000 Pau, France.
[4]Elf Aquitaine, Avenue des Lilas, 64000 Pau, France.
[5]Elf Aquitaine, Ets Boussens, 31360 Saint-Martory, France.
We thank the Société Nationale Elf Aquitaine for permission to publish this paper; S. Jardine, P. Masse, P. Louis, and L. F. Brown for critically reading the manuscript and offering many suggestions; and M. Morice, P. Arditty, and G. Drullion for their contributions. Several figures have been modified from unpublished Elf Aquitaine studies by J. Dumay, J. P. Bancelin, M. Coulon, and others.

INTRODUCTION

Current seismic interpretations use three principal approaches. The first method, structural mapping, involves picking seismic horizons and analyzing seismic velocities to determine geologic structure. Most of the hydrocarbon reserves known in the world today were discovered using this approach. The second method, seismic stratigraphy, is an analytical technique, developed during the 1970s, that permits the extraction of stratigraphic information from seismic data. In this approach, seismic sequences and seismic facies are analyzed to recognize depositional sequences and, in turn, to predict gross lithology and hydrocarbon potential throughout a basin. Methods and applications were presented in Payton (1977). In the third method, seismic lithology, individual seismic reflections are analyzed to determine acoustic and elastic parameters at the seismic interface. This method requires high-quality reflection data, thus permitting a better characterization of the lithology and secondary factors such as porosity, fluid content, and reservoir thickness (Ostrander, 1983).

Most advances in seismic stratigraphy and seismic lithology have involved the study of clastic rocks. Less attention has been given to similar analyses of carbonate rocks. This paper summarizes our knowledge of seismic interpretation of carbonate strata.

GEOLOGIC AND GEOPHYSICAL CHARACTERISTICS OF CARBONATE ROCKS

In comparison with clastic rocks, carbonate rocks exhibit particular geologic characteristics, ranging from sedimentation processes to mineralogic states. Carbonate rocks are normally generated in situ, except for turbidites and carbonate mass flows. Carbonate rocks are autochthonous, whereas clastic sediments are totally allochthonous. As a result, the depositional environment and its modifications directly affect the nature of the deposit and its postdepositional diagenesis. The nature and homogeneity of the deposits are affected by climate. In the world today, carbonate platforms are located between lat. 30°N and 30°S. Gradual relative variations in sea level generally cause vertical geometric changes in carbonate platforms. For example, a reefal structure will display considerable vertical accretion to adapt to a gradual relative sea level rise (Kendall, 1981). This vertical growth contrasts with the essentially horizontal displacement that relative eustatism imposes on clastic deposition. Subsequent to their deposition, carbonates may undergo diagenetic transformations that modify their mineralogic nature and alter their original texture by means of recrystallization, dolo-

Table 1. Elastic Moduli and Velocities of Nonporous Sedimentary Rocks*

Rock Type	Moduli (megabar)			Poisson's Ratio (σ)	Density (ρ) (g/cm^3)	Velocities (ft/sec)	
	Young's (E)	Bulk (k)	Shear (μ)			Compressional (V_p)	Shear (V_s)
Dolomite	1.129	0.801	0.446	0.265	2.84	23,000	13,000
Limestone	0.805	0.709	0.307	0.311	2.73	21,000	11,000
Sandstone	0.772	0.408	0.326	0.185	2.65	18,500	11,500
Anhydrite		0.54	0.31	0.260	2.96	20,000	11,300

*Compiled from Domenico (1983) and Gardner et al (1974).

mitization, creation of porous networks, and leaching (Choquette and Pray, 1970; Bathurst, 1971; Murray, 1979; Wanless, 1979). These processes may profoundly change the characteristics of the depositional sequence by superimposing diagenetic effects on primary characteristics. Consequently, diagenesis can either modify petrographic fabrics and mineralogy or create new ones.

Mineralogy, petrophysical characteristics, and mechanical behavior of carbonates result from their sedimentation and diagenesis. The mineralogic and lithologic parameters may be determined by analyzing physical properties. Table 1 lists obvious differences in the physical and mechanical properties of carbonate and other rocks.

Such parameters may be determined in situ using EVA (evaluation of velocity and attenuation), an acoustic full waveform logging tool developed and operated by Elf Aquitaine. EVA data are processed to give P, S, and Stoneley wave characteristics such as velocity and amplitude. These parameters enable us to compute, for example, the elastic moduli (E, K, μ, σ) and the petrophysical parameters (porosity, shale content of the formation, ρ), and to obtain a fully automatic lithologic interpretation as represented in Figure 1. This information may be obtained in open or cased holes.

Because of their physical differences, carbonate beds in a sedimentary series produce an extremely high reflectivity, five to ten times greater than the average reflectivity of clastic rocks. On the basis of this contrast, carbonate rocks can be analyzed independently within a sedimentary succession (Figure 2). Conversely, the multiples created by carbonate beds will be of an amplitude comparable to the primary amplitude generated by the clastic strata. Therefore, these multiples are difficult to remove during seismic processing without distorting the seismic information generated by the clastic strata.

Within a carbonate succession, considerable differences in velocity related to different lithologies can be distinguished. Velocities within Middle Jurassic rocks of the Paris basin (Figure 3) are shown as an example. In a regressive carbonate sequence, from shallow marine shoal deposits to restricted lagoonal facies, velocities range from 13,780 ft/sec (4,200 m/sec) in oolitic limestone at the base to 18,050 ft/sec (5,500 m/sec) in sublithographic limestone at the top (Dumay and Kenaan, 1983). Differences in petrophysical properties resulting from sedimentation and diagenetic phenomena help differentiate carbonate depositional facies from seismic data (Delaplanche and Michon, 1978; Maureau and Van Wijne, 1979; Angeleri and Carpi, 1982).

To obtain the best possible seismic representation of carbonate facies, specific precautions must be taken regarding sequences of seismic processing, essentially those of amplitude control (gain recovery, equalization window). Care also must be taken when dealing with demultiplication processing programs.

CARBONATE DEPOSITIONAL PALEOENVIRONMENTS AND THEIR SEISMIC ANALYSIS

In this study, we examine several carbonate depositional environments using the Wilson (1975) theoretical model (Figure 4), to produce plausible seismic criteria for recognizing the facies deposited in each. Seismic examples are associated to each depositional environment from basinal to supratidal deposits.

Basinal Pelagic Deposits

Deposited in a low hydrodynamic energy environment below wave base, carbonate pelagic sediments normally occur either as homogeneous micritic limestones (chalk) or as interlayered carbonate and shale beds. Two parallel, continuous, high-amplitude reflections bound the chalk and are uniform in frequency, phase, and amplitude. A virtually reflection-free zone exists between the two reflections. Two examples are the Upper Cretaceous of Louisiana (Bally, 1983), and the Upper Cretaceous chalk of the North Sea (Figure 5). Hydrocarbons generally reduce the amount of porosity lost in the chalk. Then particular seismic effects can be detected such as several amplitude anomalies (Van den Bark and Thomas, 1980, Ekofisk field; Munns, 1985, Valhall field) or a distinct reflection below the top of the chalk reflector. The reflection corresponds to the decrease in porosity between porous and nonporous chalk. No such reflection occurs where the chalk is unproductive (nonporous). An example is given by a seismic section through the Harlingen field, northern Netherlands (Figure 6). Harlingen field produces gas from the uppermost part of the Upper Cretaceous chalk (Van den Bosch, 1983). Thin interlayered beds of limestone and shale typically yield continuous parallel reflections with an apparent frequency related to the bed-thickness to signal-wavelength ratio. An apparently high frequency is commonly observed (signal derivation due to thin layers) (Figure 7). In contrast, a predominantly shaly zone will

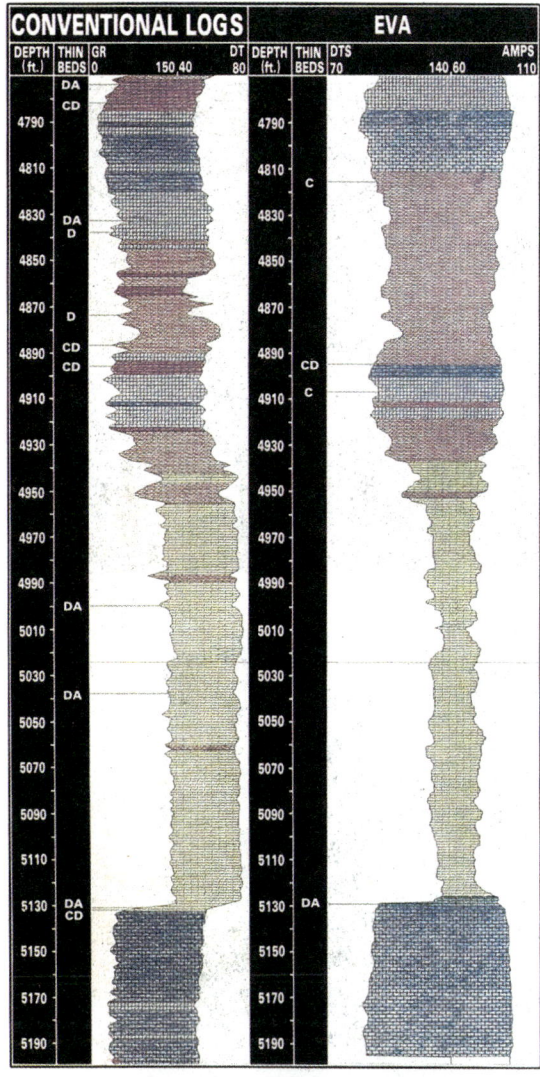

Figure 1—Lithologic columns obtained by automatic interpretation technique using either conventional logging data or EVA data. AMPS = S-wave amplitudes; DTS = S-wave interval transit time; GR = gamma ray; and DT = interval transit time. See text for more detailed discussion on EVA tool.

exhibit an apparently low frequency. Generally, the pelagic reflections terminate by onlap against the paleoshelf edge.

Talus Deposits

Talus deposits typically mark the boundary of a carbonate shelf. They define a transitional zone between generally neritic and pelagic deposition. Two types of slopes may occur at a carbonate platform margin. A gentle slope (less than 1°) in a carbonate ramp (Read, 1981) is represented on seismic sections by low-angle sigmoid or shingled reflections becoming tangential to, or downlapping at, the base of the sequence. These reflections commonly have a low frequency (Figure 8). A steep slope (from a few degrees to over 45°), as in a rimmed carbonate shelf (Read, 1981), will normally exhibit gravity phenomena (sliding, slumping). On seismic sections, such periplatform talus is represented by oblique, discontinuous, and in places, chaotic reflections with high amplitudes (Figure 9). Seismic examples of rimmed shelves were shown by Vail et al (1977) (western Africa), Bubb and Hatlelid (1977) (Gulf of Papua, offshore west Africa), and Gamboa et al (1985) (Jurassic System of Baltimore Canyon Trough).

Slumping effects may be identified by slightly discontinuous reflections of irregular geometry. Where substantial debris flows occur, seismic sections exhibit chaotic reflections interposed between the parallel continuous reflections of pelagic deposits (Ravenne et al, 1985). Where these debris deposits become more regular, structures are created (submarine cones) that may have varying morphologies, thus disturbing the more uniform pelagic deposits. On seismic sections, these turbiditic structures can be recognized by high-amplitude, mound-shaped reflections; therefore, the envelope of the facies is an important recognition criterion. Sediments may be transported from shallow to deep environments through channels, which on seismic lines are represented by erosional truncations of underlying reflections (Brown and Fisher, 1980). The channels may be filled by clastic carbonates (chaotic reflections) or pelagic sediments (onlap fill, parallel reflections).

Shelf Deposits

Typically, the shelf lies beneath shallow water, from 0 to 200 m (615 ft) deep. Consequently, the sea floor is subjected to various hydrodynamic conditions. In the low energy zones (inner shelf, lagoon), continuous, horizontal strata are deposited, such as shales and limestones or limestones only. Accordingly, seismic reflections are continuous, parallel, and horizontal with little change in amplitude, frequency (quite low), and phase (Figures 8, 9).

In the high-energy zones, sediments should be analyzed according to their particular sedimentologic criteria, such

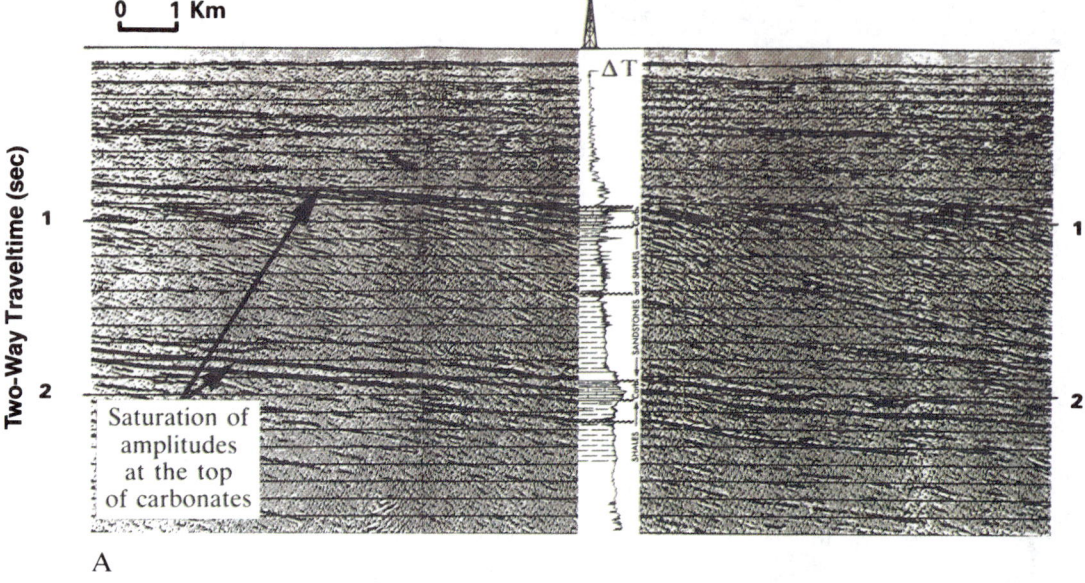

Figure 2—Presence of carbonate deposits within clastic series (east Africa). (A) Migrated section. (B) Image processing: same section (with one of every two traces) with amplitude codification. Green and yellow (highest amplitude values) show limestone layers. (Image processing developed by N. Keskes, 1982-1984; G. Sibille, 1985.)

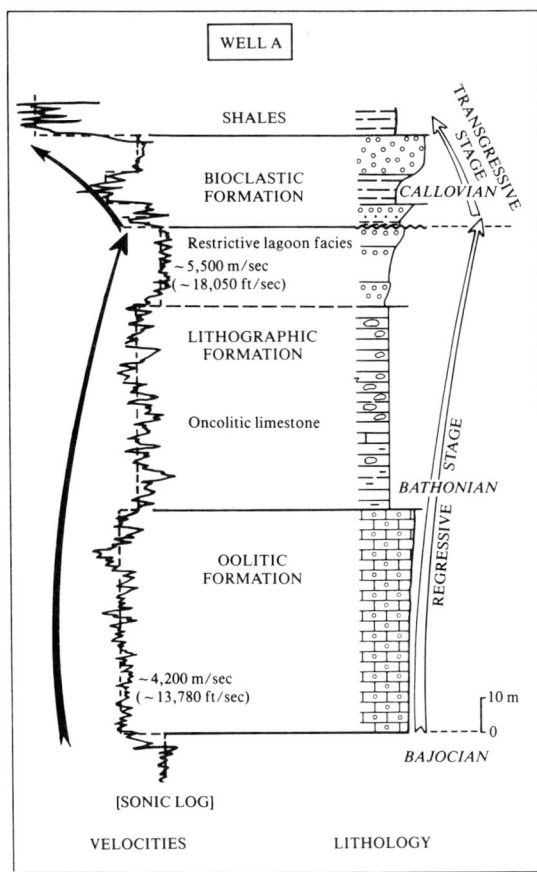

Figure 3—Variations of velocities within a carbonate succession, from Middle Jurassic, Paris basin, France (from J. Dumay and T. Keenan, 1983).

as reefal constructions, oolites, or bioclastic sands. High-energy zones may be continuous or discontinuous, and they may create a barrier that isolates paleogeographic regions (Friedman, 1969; Elf Aquitaine, 1975, 1977; Wilson, 1975; Reading, 1978; Playford, 1980; Purser, 1980; Scholle et al, 1983). The favorable location for a barrier is normally the shelf edge. With or without the presence of a barrier, however, high-energy conditions can exist on the shelf, creating patch reefs and submarine shoals. These high hydrodynamic energy zones are important for hydrocarbon prospecting because of their good reservoir properties.

Reefal Buildup

Reefs in the neritic setting may develop either as a nearly continuous barrier or as patch reefs. Because they are significant in petroleum exploration, these structures have been the subject of numerous seismic studies (McClintock, 1976; Burbury, 1977; Kenyon, 1977; Chevron Standard Limited, 1979; Hriskevich et al, 1980; Baria et al, 1982; Mundt, 1982; Bally, 1983; Frost et al, 1983). Bubb and Hatlelid (1977) proposed the following criteria for recognizing such deposits:

1. Direct criteria—boundary outline (reflection configurations and onlap of overlying reflections) and seismic facies changes defining the buildup.
2. Indirect criteria—drape, velocity anomalies (pull-up and pull-down), spurious events, and basin architecture (fault-block edges, structural highs).

Two types of reef complexes are indicated on Figure 10. In the lower part of the seismic section, dolomitized reefal units are developed on a fault-controlled uplifted block. The reef extension is limited to the high zone, and the surrounding sediments are of pelagic origin. This reefal construction is seismically characterized by a mound shape and onlap of surrounding reflections. Higher in the series, a shelf-margin buildup separates shelf deposits (shallow marine limestone) from basin sediments (shaly limestone). This shelf edge can be located seismically and controlled by wells along a large continuous area. Shelf-margin buildup is interpreted on the basis of: (1) seismic facies pattern change from a zone of continuous, parallel reflectors (back-reef lagoon) to a nearly reflection-free zone (reef facies) to continuous, parallel reflectors (basin deposits); and (2) onlap of basin units over the shelf edge.

Reef complexes could develop on ridges in a pelagic environment if the ridges reach favorable bathymetric conditions (Figure 11). Growth of this type of reefal buildup is controlled by relative movements of sea level (essentially the results of diapirism, subsidence, and possibly eustatism). These movements can be analyzed in detail using color display of the "instantaneous phase." The instantaneous phase is obtained by analyzing a seismic trace as a component of an analytic signal (Taner and Sheriff, 1977). Such processing and color displays help locate unconformities, pinch-outs, and seismic sequence boundaries (Figure 12).

Carbonate Sand Bank

Because of their thinness in relation to the signal wavelength, carbonate sand facies are usually difficult to detect on conventional seismic sections. Consequently, tools especially suited to this problem must be used. The inversion of seismic traces may be a feasible approach, if good quality seismic sections are obtained and rigorous methodology is ensured. The method involves the following steps:

1. Model a "pseudo-sonic log" synthetic trace in the seismic frequency domain using well logs (sonic, density), to determine whether the sedimentary phenomenon is visible at a seismic level and what type of pattern it produces on the trace.
2. Control the signal's lateral stability among various wells.
3. Reprocess sections with operators extracted from well data (zero-phase section followed by pseudo-sonic log).
4. Interpret pseudo-sonic log sections, controlling interpretation by modeling.

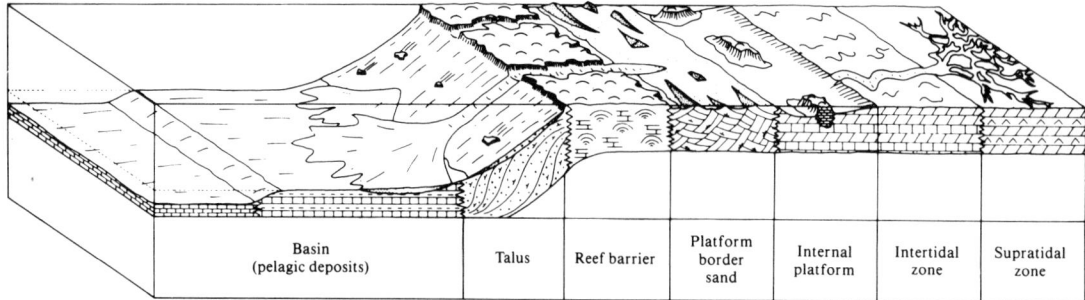

Figure 4—Synthetic and theoretical geologic model of carbonate depositional environments (from J. L. Wilson, 1975).

Such methodology has been used in the Paris basin to delimit an oolitic shoal in Bathonian-lower Callovian strata. A geologic (sedimentologic) model is proposed between wells A and B (Figure 13A). The sonic log and lithostratigraphic data of well A are presented on Figure 3. On well A, the oolitic limestones are overlain by oncolitic limestones (back-shoal facies), which are absent in well B. The restrictive lagoonal facies terminates the Bathonian regressive sequence. Porous oolitic limestones are well developed in well B, and restricted lagoonal facies lie above.

The problem is to obtain a seismic representation of this geologic model, recognize the different sedimentologic events on the model and on real seismic lines, and map the extension of the thick porous oolitic limestone (oolitic shoal). The synthetic pseudo-sonic log section, based on well data and on the geologic model, shows the seismic response of these sedimentologic features (Figure 13B). Although they are difficult to recognize on the seismic stack section (Figure 13C), these seismic events can be followed on the pseudo-sonic log sections resulting from trace integration[6] (by using Velog[7] processing) (Figure 13D). Thus, the thick porous oolitic limestone can be mapped.

As the depositional environments become increasingly restricted, sometimes undergoing temporary emergence (intertidal, supratidal zone), new sedimentologic and diagenetic factors are introduced (such as dolomitization and karstification, possible presence of evaporites).

---

[6]Use and methodology of trace integration are explained in detail in Sheriff (1980).
[7]Velog is a trademark of Compagnie Générale de Géophysique.

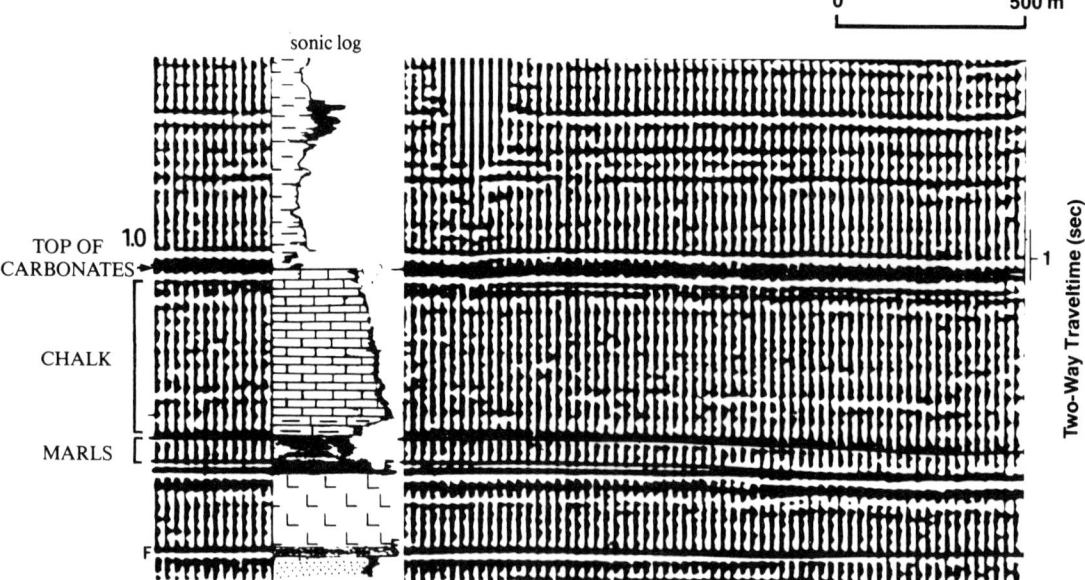

Figure 5—Seismic facies of pelagic carbonate deposits from homogeneous limestones of Upper Cretaceous chalk, western Europe. Stacked section.

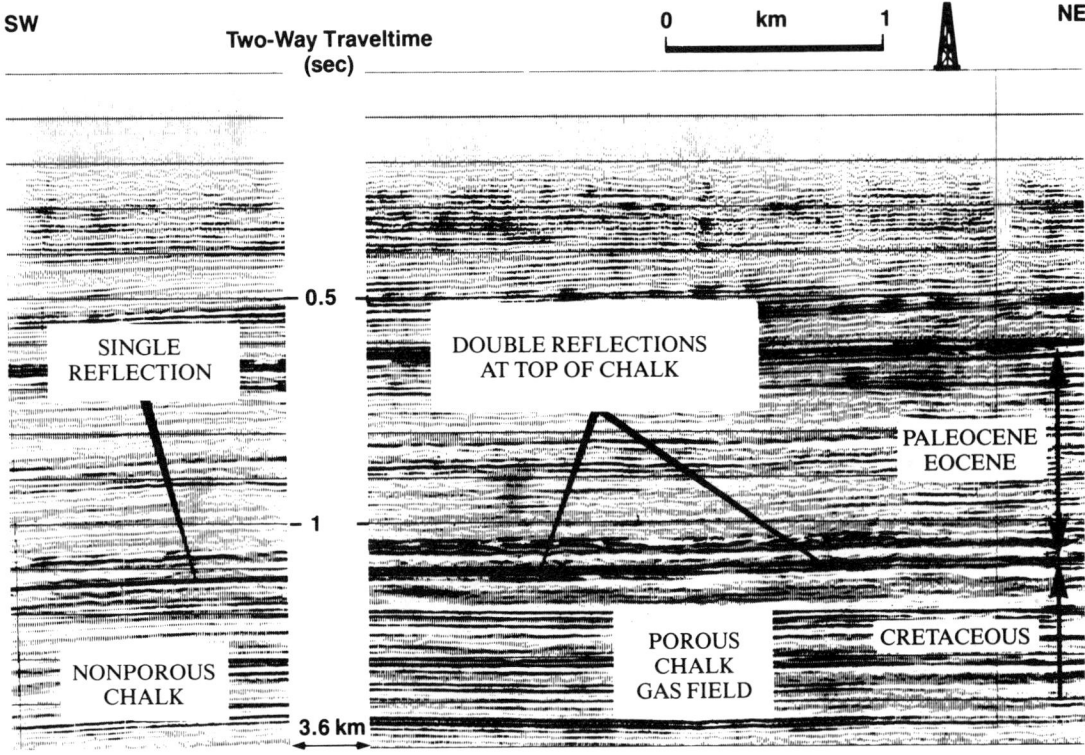

Figure 6—Seismic stacked (zero-phase) section through Harlingen field (Netherlands). Double reflection developed at top of Upper Cretaceous chalk is caused by presence of porous chalk.

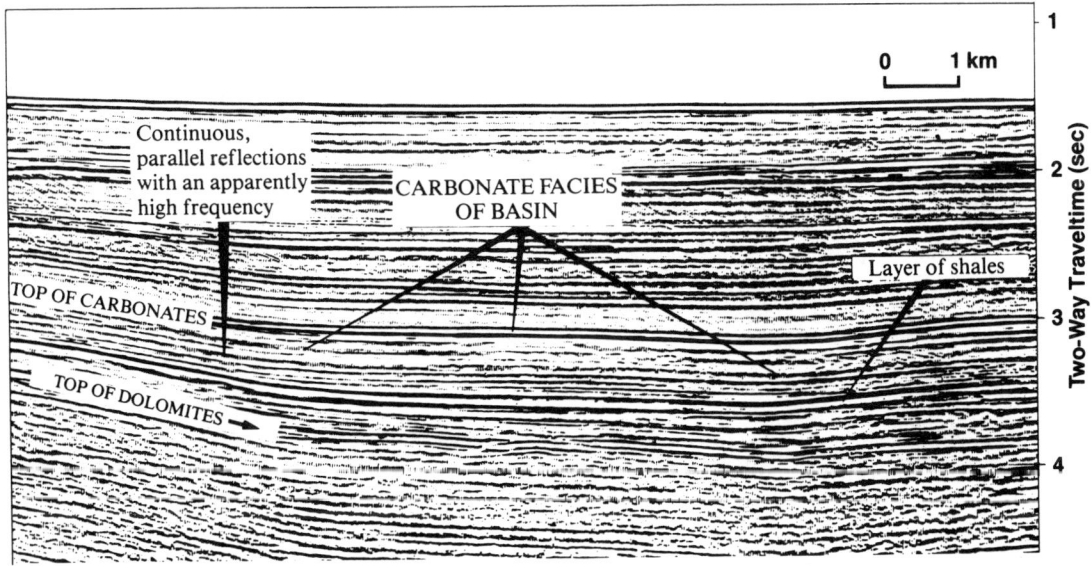

Figure 7—Seismic facies of pelagic carbonate deposits from alternating beds of shales and limestones of Lower Cretaceous, southern Europe. Stacked section.

288   Carbonate Depositional Environments

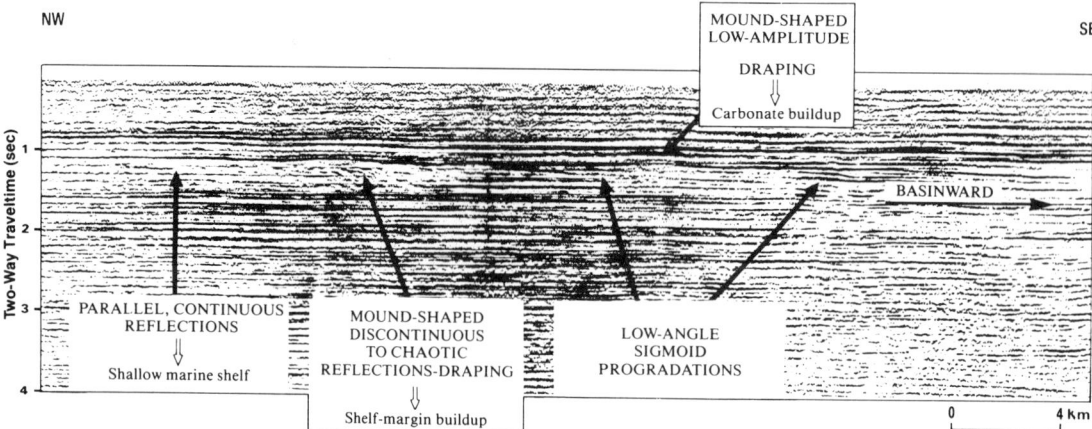

Figure 8—Seismic stacked section from east Africa, showing low-angle carbonate progradations. Evolution, essentially made by progradation with no aggradation, occurs during period of constant relative sea level or slow fall.

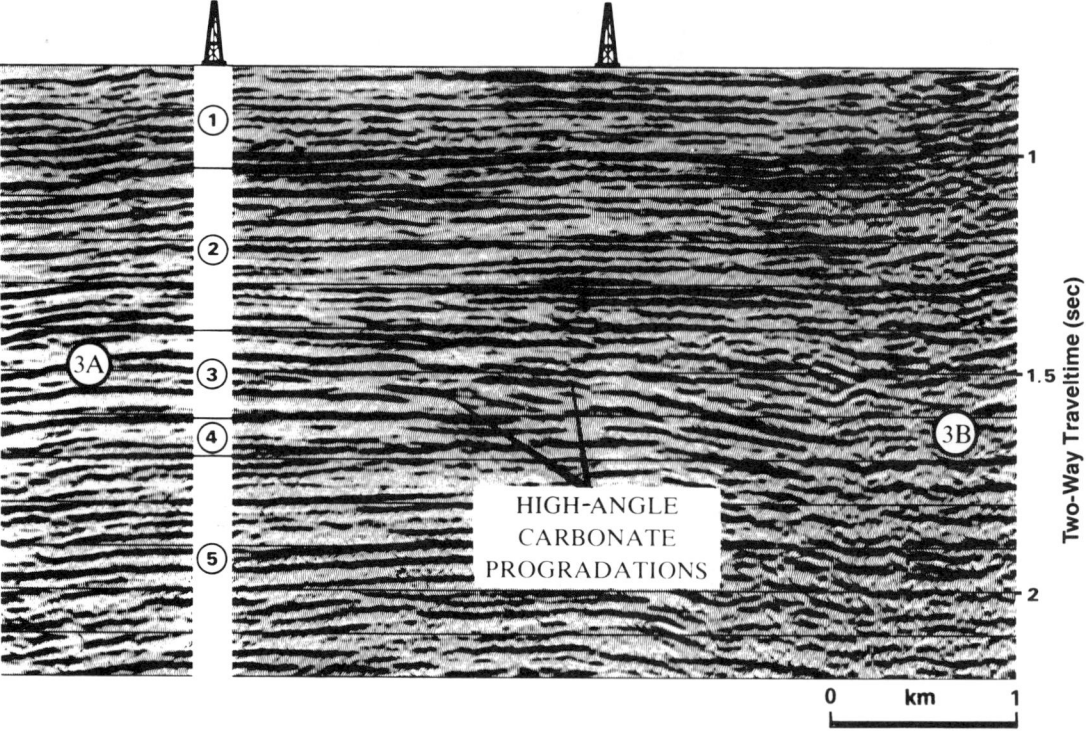

Figure 9—Seismic stacked section showing high-angle oblique carbonate progradations. Evolution, made by progradation and aggradation, occurred during slow rise of relative sea level (Miocene, offshore north Africa). (1) Sandstone; (2) shales with limestone layers; (3) limestone; (3A) shelf deposits; (3B) pelagic shales and limestones; (4) shales; and (5) alternating beds of shales and limestone.

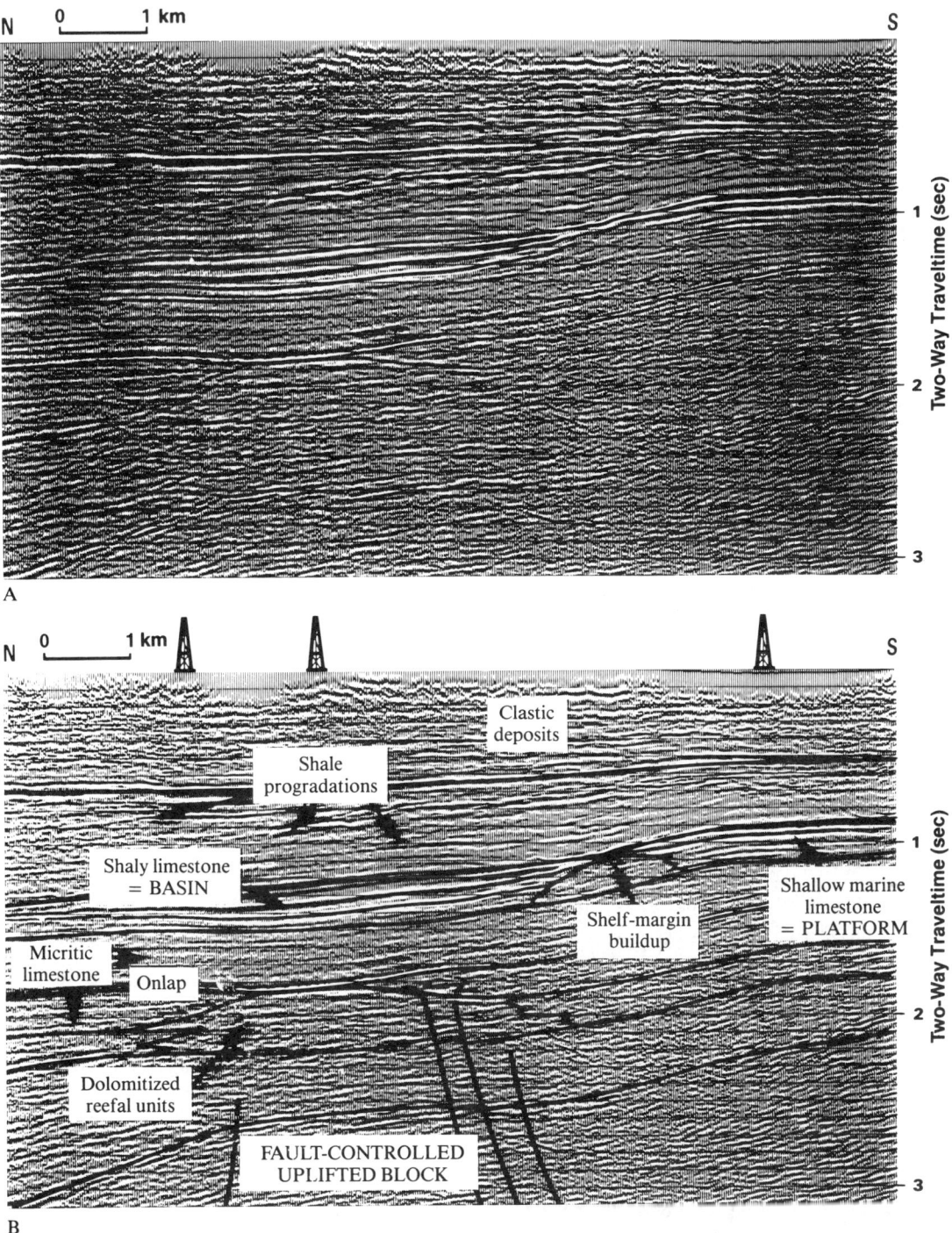

Figure 10—Seismic section (migrated) from Aquitaine basin (southwestern France). In lower part of section (Cretaceous sequences), dolomitized reefal units developed on fault-controlled uplifted blocks. Units are seismically characterized by mound shape and onlaps of surrounding reflections. Higher shelf-margin buildup (Paleocene) is shown, interpreted on basis of: (1) seismic facies pattern change from zone of continuous, parallel reflectors (platform deposits) to nearly reflection-free zone (reef facies) to continuous, parallel reflectors (basin deposits); and (2) onlap of basin units over shelf edge.

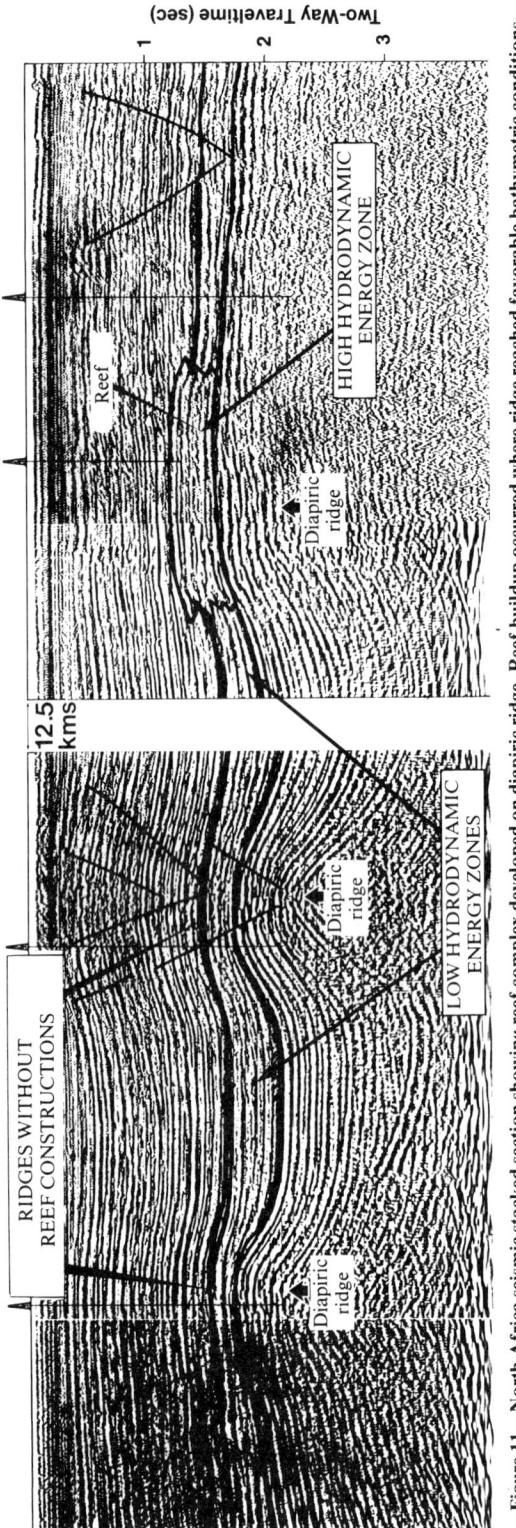

Figure 11—North Africa seismic stacked section showing reef complex developed on diapiric ridge. Reef buildup occurred where ridge reached favorable bathymetric conditions.

DIAGENESIS OF CARBONATE DEPOSITS

After sedimentation, carbonate sediments may undergo diagenetic alterations closely connected to the physical and chemical factors affecting the environment (Pray and Murray, 1965). This physico-chemical action changes the texture or mineralogy of the sediment. We discuss two types of diagenetic evolution of carbonates: dolomitization and karstification.

Dolomitization

Measurements made in the laboratory and from well logs show that petrophysical qualities of dolomites differ from those of calcitic carbonates (Gregory, 1977; Rafavich et al, 1984); for example, dolomites have higher density and higher acoustic velocities. Therefore, we would assume that a seismic section would reveal these differences as acoustic impedance contrasts at the limestone-dolomite interface. Unfortunately, these differences are not always visible, but occasionally dolomite and limestone can be differentiated on seismic facies.

An example from the Mesozoic carbonate shelf of southern Europe shows distinctive facies calibrated by the well data (Figure 14). Shallow marine limestone with continuous parallel reflections appears in the upper part. Dolomitized beds in the lower part are represented by a "marbled" zone—a practically reflection-free zone with a few discontinuous reflections that imply original stratification. This marbled seismic facies may be associated with the well-developed dolomitic layers throughout the basin (see also Figure 7 below the reflector "top of dolomites") and now seems to be recognized in other regions in large-scale postdepositional dolomitization.

Karstification

Large-scale karstification results in caves within carbonate rocks. Thus, the karst can form a particular type of reservoir, characterized by a high anisotropy and high permeability. On two-dimensional seismic sections, karst areas are commonly difficult to distinguish from compact limestone. They generally do not have a distinct seismic facies pattern. The first step in detecting karst zones is to recognize paleotopographic highs or possible subaerial exposure on seismic lines by determining (1) erosional truncation (del Olmo and Esteban, 1983, their Figure 1 from offshore northeastern Spain), (2) structural highs onlapped by overlying sediments (Figure 15), and (3) abundant irregularities affecting the seismic marker (suggesting local collapse features) (Jenyon, 1984). Horizontal seismic sections, which are the normal product of three-dimensional seismic surveys, help reconstruct karst paleotopography. The utility of horizontal sections in karst areas was first demonstrated by Brown (1985), who presented examples from Mackerel field in offshore Gippsland basin, southeastern Australia.

Amplitude analyses can be used to determine the lateral extension of a karst area. Vandenberghe et al (1983) noted that the magnitude of the reflection as a function of offset is affected by Poisson's ratio, and they theoretically proved that a relationship exists between values of Poisson's ratio and changes in limestone diagenesis. Also, the relationship between the amplitudes of near and far traces, on various offset stacks, typically highlights these differences (Figure 16). This process, involving analyses of near and far traces, has been applied to a Barremian hydrocarbon field "A" in the Mediterranean basin (Figure 17), where the karst strata pass laterally into compact limestones (well C).

These carbonate strata are covered by a prograding clastic succession ranging in age from late Miocene to Pliocene-Quaternary. Along a seismic line joining the hydrocarbon field to well C, we compared the amplitudes corresponding to the acoustic impedance contrasts in a 200-msec window, with 6 partial coverings. C1 covers the carbonate strata with near offsets, and C6 covers the strata with far offsets. The C6/C1 and C2/C1 ratios clearly differentiate between the karst and nonkarst zones (Figure 18). A similar study was conducted in northern Belgium to recognize karst in Dinantian limestones (Vandenberghe et al, 1983).

CONCLUSION

The relationships between seismic facies and carbonate depositional environments, based on the analysis of various examples, are summarized in Figure 19. Having modeled the horizontal and vertical changes of the seismic facies, we can attempt to identify the carbonate depositional sequences within sedimentary basins and evaluate the relative changes in sea level, a fundamental parameter in carbonate sedimentation and postsedimentary diagenetic transformations (Kendall, 1981). An example from the African coast (Figure 12) shows the data that can be obtained by analyzing a seismic section in terms of relative variations in sea level. For carbonates, the depositional sequence concept must be linked to the diagenetic sequence concept because of diagenetic discontinuities created in carbonate sediments. Several lithofacies characterized by specific diagenetic conditions may develop between these discontinuities.

Seismic sections provide an overall view of the vertical evolution of these diagenetic sequences, and seismic facies in a carbonate domain can be recognized on the basis of two often inseparable factors: sedimentation and diagenesis.

Seismic facies analysis facilitates the interpretation of carbonate depositional environments. This seismic stratigraphic approach uses several geophysical tools in addition to conventional P-wave reflection data (two- or three-dimensional acquisition): e.g., velocity analysis; analysis of the seismic trace parameters such as instantaneous phase (see Figure 12B), and instantaneous amplitude and frequency (Robertson and Nogami, 1984); zero-phase seismic section, true-amplitude seismic sections (Figure 13C), and seismogram inversion (Figure 13D); and computer-assisted interpretation such as modeling (forward or inverse) (Sheriff, 1980) and image processing (Figure 2).

Seismic stratigraphic analysis should not overshadow the precise study of the elastic and acoustic parameters of rocks. Seismic lithology appears to be particularly well-suited to the seismic analysis of carbonates because of their postsedimentary diagenetic modifications. Diagenesis alters the petrophysical nature of rocks and, hence, their seismic response. Many types of analyses may be conducted to define these phenomena better: e.g., amplitude variations with offsets (Vandenberghe et al, 1983); S-wave reflection (Domenico, 1983; Arditty et al, 1985) (Figure 1); laboratory measurements (Grady et al, 1979; Larson, 1980; Rafavich et al, 1984); and downhole measurements such as density, sonic, and EVA logs (Arditty et al, 1985) (Figures 1, 3).

By studying both depositional environments (seismic stratigraphic approach) and petrophysical characteristics (seismic lithologic approach), we can more completely define the geologic history and the variations in sea level by which a basin has been affected, which will help us investigate and comprehend carbonate reservoirs and their paleogeography.

SELECTED REFERENCES

Angeleri, G. P., and R. Carpi, 1982, Porosity prediction from seismic data: Geophysical Prospecting, v. 30, p. 580-607.

Arditty, P. C., G. Arens, and P. Staron, 1985, Improvement of formation properties evaluation through the processing and interpretation results of the EVA tool recordings (abs.): Geophysics, v. 50, p. 268.

Bally, A. W., ed., 1983, Seismic expression of structural styles—a picture and work atlas: AAPG Studies in Geology Series 15, 3 volumes.

Baria, L. R., D. L. Stoudt, P. M. Harris, and P. E. Crevello, 1982, Upper Jurassic reefs of Smackover Formation, United States Gulf Coast: AAPG Bulletin, v. 66, p. 1449-1482.

Bathurst, R. G. C., 1971, Carbonate sediments and their diagenesis: Amsterdam, Elsevier, Developments in Sedimentology 12, 677 p.

Berg, O. R., and D. G. Woolverton, eds., 1985, Seismic stratigraphy II: an integrated approach to hydrocarbon exploration: AAPG Memoir 39, 276 p.

Bosselini, A., 1984, Progradation geometries of carbonate platforms: examples from the Triassic of the Dolomites, northern Italy: Sedimentology, v. 31, p. 1-24.

Brown, A. R., 1985, The role of horizontal seismic sections in stratigraphic interpretation, in O. R. Berg and D. G. Woolverton, eds., Seismic stratigraphy II: an integrated approach to hydrocarbon exploration: AAPG Memoir 39, p. 37-47.

Brown, L. F., Jr., and W. L. Fisher, 1980, Seismic stratigraphic interpretation and petroleum exploration: AAPG Continuing Education Course Note Series 16, p. 1-125.

Bubb, J. N., and W. G. Hatlelid, 1977, Seismic stratigraphy and global changes of sea level, part 10: seismic recognition of carbonate buildups, in C. E. Payton, ed., Seismic stratigraphy—applications to hydrocarbon exploration: AAPG Memoir 26, p. 185-204.

Burbury, J. E., 1977, Expression of carbonate buildups, north-west Java basin: Indonesian Petroleum Association 6th Annual Congress, p. 239-268.

Chevron Standard Limited, Exploration Staff, 1979, The geology, geophysics, and significance of the Nisku reef discoveries, West Pembina area, Alberta, Canada: Bulletin of Canadian Petroleum Geology, v. 27, p. 326-359.

Choquette, P. W., and L. C. Pray, 1970, Geological nomenclature and classification of porosity in sedimentary carbonates: AAPG Bulletin, v. 54, p. 207-250.

Delaplanche, J., and D. Michon, 1978, Les carbonates en géophysique: Pétrole et Techniques, v. 254, p. 15-32.

Del Olmo, W. M., and M. Esteban, 1983, Paleokarst development, in P. A. Scholle, D. G. Bebout, and C. H. Moore, eds., Carbonate depositional environments: AAPG Memoir 33, p. 93-95.

Domenico, S. N., 1983, Sandstone and limestone porosity determination from shear and compressional wave velocity: Australian Society of Exploration Geophysics Bulletin, v. 14, no. 3-4, p. 81-90.

(References continued on p. 296.)

292   **Carbonate Depositional Environments**

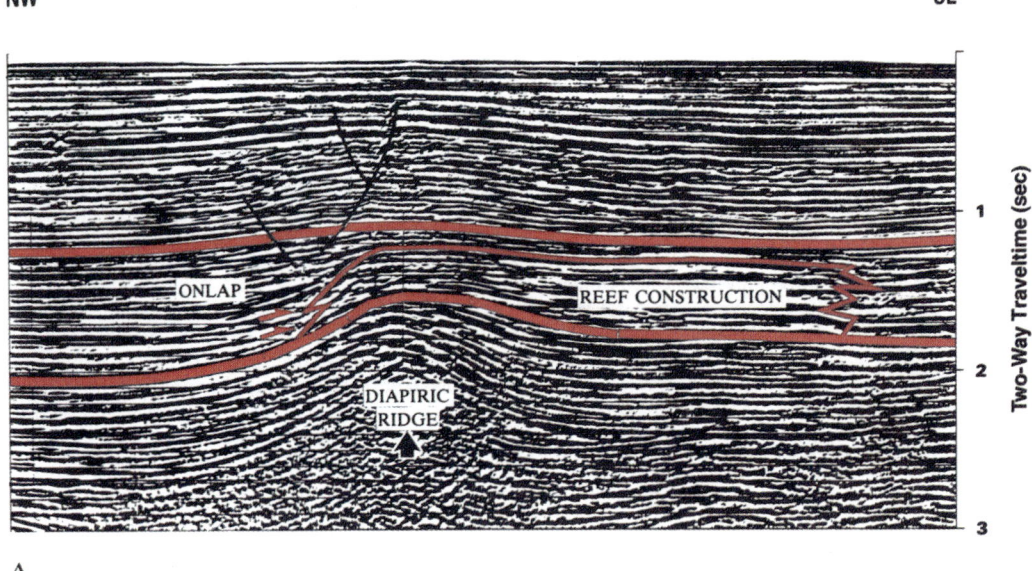

A

B

Figure 12—North Africa reefal buildup developed on diapiric ridge: (A) seismic stacked section; (B) instantaneous phase section; (C) chronostratigraphic interpretation based on instantaneous phase section analysis; and (D) interpretation of successive phases of building. (1) Ridge reaches favorable bathymetric conditions, and growth of reef buildup occurs. (2) Diapirism is dominant; no deposits appear on top of diapir, but buildup extends laterally. (3) Reef buildup extends. (4) Diapirism is dominant (last diapiric event occurs, and some erosion appears). (5) Deposits accumulate because of continued subsidence, and sea level may rise. (6) Sea level probably drops, and erosion occurs. Downslope deposits accumulate. (7) Deposits accumulate because of continued subsidence, and sea level may rise. (8) Sea level drops, and erosion occurs. (9) Final diagram of section.

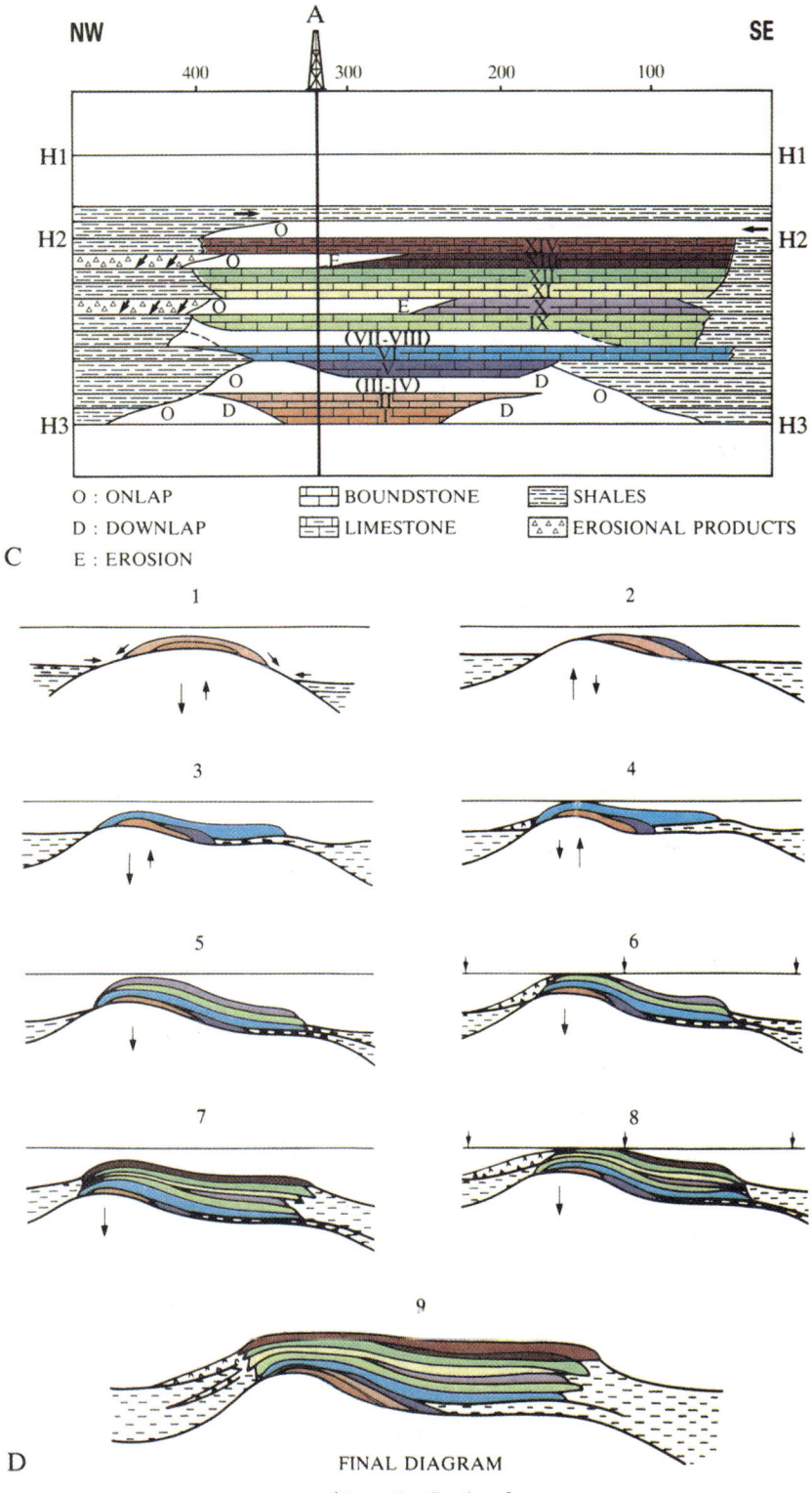

Figure 12 — Continued.

Carbonate Depositional Environments

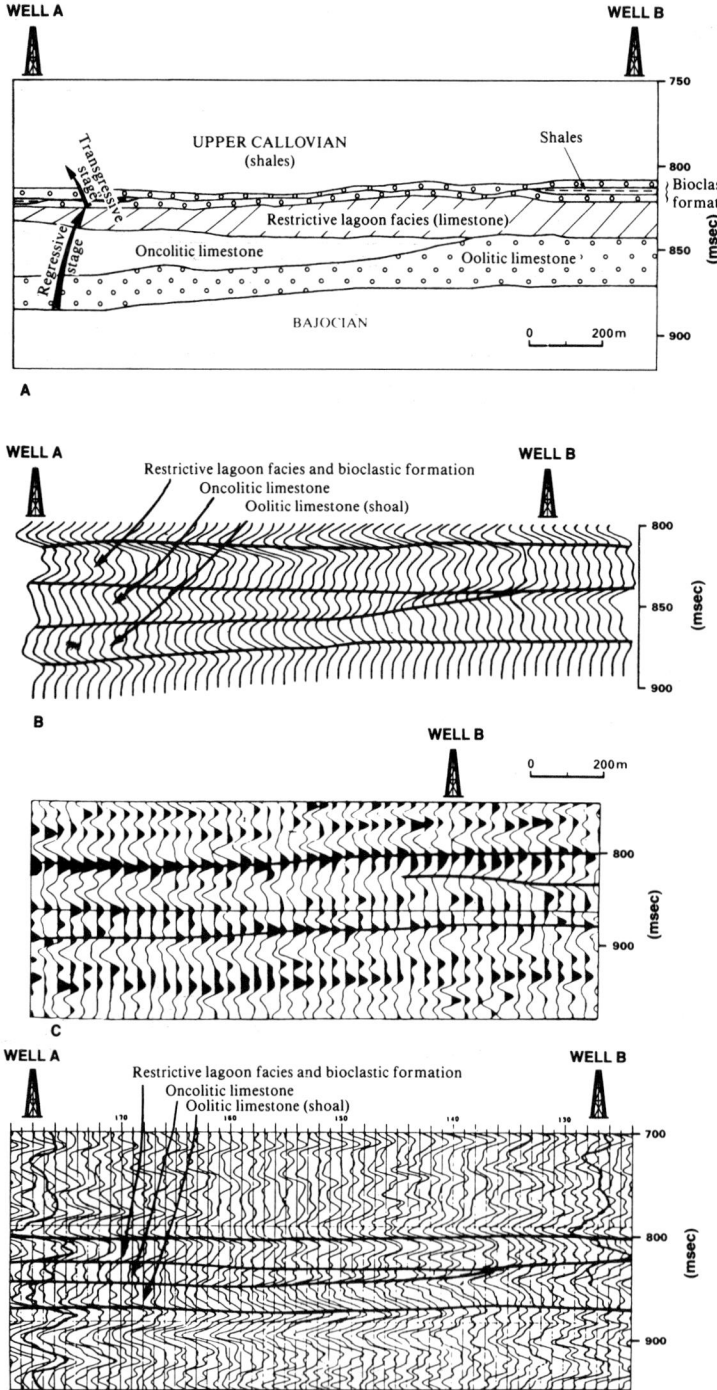

Figure 13—Seismic interpretation of submarine oolitic shoal using trace integration processing and modeling. (A) Geologic (sedimentologic) model between wells A and B. (B) Synthetic pseudo-sonic log sections. (C) Migrated section (true amplitude). (D) Pseudo-sonic log sections obtained from trace integration, Paris basin (France), Bathonian to lower Callovian strata. See text for more detailed discussion.

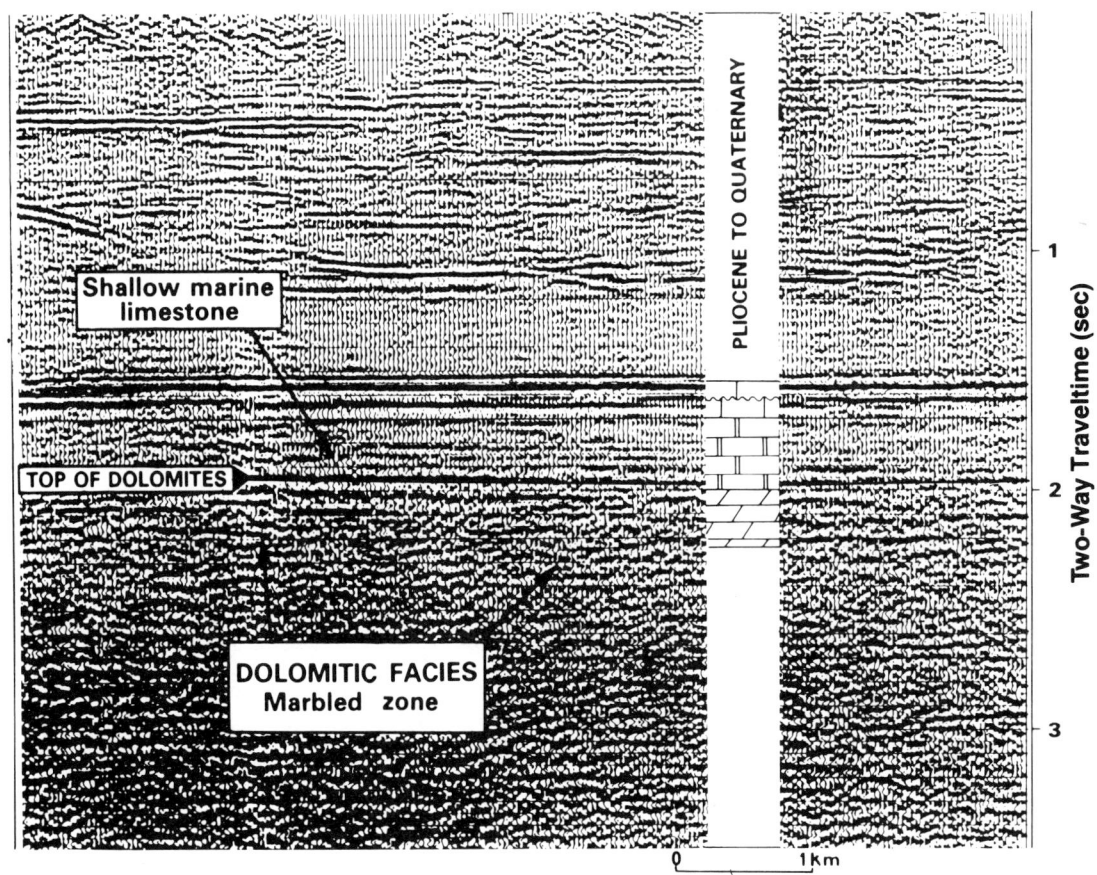

Figure 14—Seismic facies of large-scale postdepositional dolomites (stacked section), Jurassic to Cretaceous strata, southern Europe.

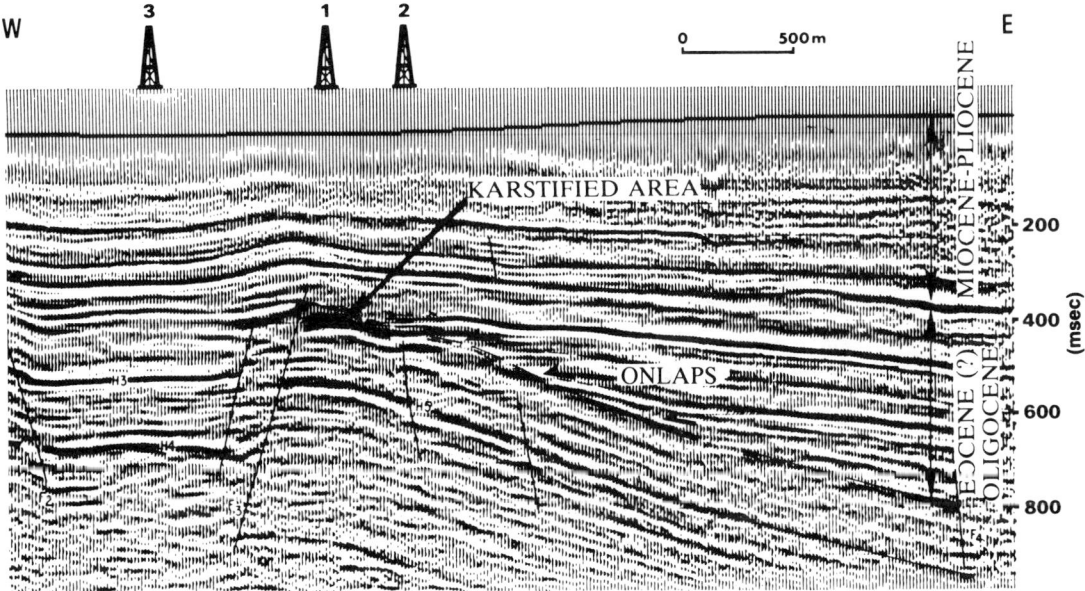

Figure 15—Seismic stacked section throughout Bresse basin (France) showing an uplifted block onlapped by Eocene(?)–Oligocene sediments. Wells 1 and 2 encountered Jurassic karst limestone below Oligocene sediments. Well 3 encountered unkarstified Mesozoic limestone. Karst area seems to correspond to upper part of tectonic high, the last zone onlapped by Tertiary sediments. H3 = Intra-Oligocene reflector; H4 = top of Mesozoic carbonates; H5 = intra-Jurassic reflector; F1, F2, F3 = faults. (Reprinted with permission of Gaz de France.)

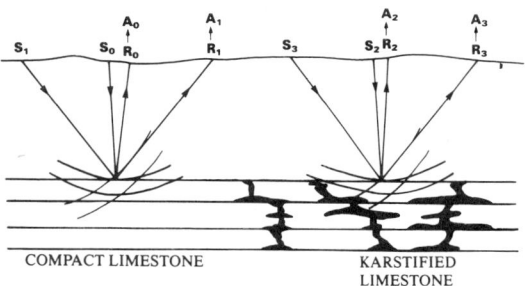

Figure 16—Amplitude analysis as function of offset for compact and karst limestone. A_0, A_1, A_2, A_3 = magnitudes of reflection for S_0-R_0 (near offset), S_1-R_1 (far offset), S_2-R_2 (near offset), S_3-R_3 (far offset), respectively.

Dumay, J., and T. A. Kenaan, 1983, Interprétation stratigraphique des données sismiques de l'ensemble Bathonien-Callovien inférieur dans le bassin de Paris à l'aide de l'inversion des traces sismiques: Pétrole et Techniques, v. 297 (Avril), p. 7-12.

Elf Aquitaine, 1975-1977, Essai de caractérisation sédimentologique des dépôts carbonatés: Elf Aquitaine Centres de Recherches de Boussens et de Pau, tome I, 172 p., tome II, 215 p.

Friedman, G. M., ed., 1969, Depositional environments in carbonate rocks: SEPM Special Publication 14, 209 p.

Frost, S. H., D. M. Bliefnick, and P. H. Harris, 1983, Definition and porosity evolution of a Lower Cretaceous rudist buildup: Shuaiba Formation of eastern Arabian Peninsula, *in* P. M. Harris, ed., Carbonate buildups—a core workshop: SEPM Core Workshop 4, p. 381-410.

Gamboa, L. A., M. Truchan, and P. L. Stoffa, 1985, Middle and Upper Jurassic depositional environments at outer shelf and slope of Baltimore Canyon Trough: AAPG Bulletin, v. 69, p. 610-621.

Gardner, G. M. G., L. W. Gardner, and A. R. Gregory, 1974, Formation velocity and density: the diagnostic basis for stratigraphic traps: Geophysics, v. 39, p. 770-780.

Grady, D. E., R. E. Hollenbach, and K. W. Schuler, 1978, Compression wave studies on calcite rock: Journal of Geophysical Research, v. 83, p. 2839-2849.

Gregory, A. R., 1977, Aspects of rock physics from laboratory and log data that are important to seismic interpretation, *in* C. E. Payton, ed., Seismic stratigraphy—applications to hydrocarbon exploration: AAPG Memoir 26, p. 15-46.

Hriskevich, M. E., J. M. Faber, and J. R. Langton, 1980, Strachan and Ricinus West gas fields, Alberta, Canada, *in* M. T. Halbouty, ed., Giant oil and gas fields of the decade 1968-1978: AAPG Memoir 30, p. 315-327.

Jenyon, M. K., 1984, Seismic response to collapse structures in the southern North Sea: Marine and Petroleum Geology, v. 1, p. 27-36.

Kenyon, C. S., 1977, Distribution and morphology of early Miocene reefs: East Java Sea Indonesian Petroleum Association 6th Annual Congress, p. 215-238.

Kendall, C. G. St. C., 1981, Carbonates and relative changes in sea level: Marine Geology, v. 44, p. 181-212.

Keskes, N., O. Faugeras, and A. Boulanouar, 1982, Application of image analysis techniques to seismic data: Paris, Proceedings of the International Conference of Acoustic Speech and Signal Processing, p. 855-858.

Larson, D. B., 1980, Shock wave studies in Blair dolomite: Journal of Geophysical Research, v. 85, no. B1, p. 293-297.

Maureau, G. T. F. R., and D. H. Van Wijne, 1979, The prediction of porosity in the Permian (Zechstein 2) carbonate of the eastern Netherlands using seismic data: Geophysics, v. 44, p. 1502-1517.

McClintock, P. L., 1977, Seismic data-processing techniques in exploration for reefs, northern Michigan, *in* J. H. Fisher, ed., Reefs and evaporites—concepts and depositional models: AAPG Studies in Geology 5, p. 111-124.

Mundt, P. A., 1982, Miocene reefs, offshore north Sumatra: exploration III, geology session: Offshore South East Asia 1982 Conference, February 9-12, Singapore, p. 1-11.

Munns, J. W., 1985, The Valhall field: a geological overview: Marine and Petroleum Geology, v. 2 (February), p. 23-43.

Murray, R. C., 1960, Origin of porosity in carbonate rocks: Journal of Sedimentary Petrology, v. 30, p. 59-84.

Ostrander, B., 1983, Seismic lithology, section 1: historical development: Society of Exploration Geophysicists Continuing Education, p. 1-12.

Payton, C. E., ed., 1977, Seismic stratigraphy—applications to hydrocarbon exploration: AAPG Memoir 26, 516 p.

Playford, P. E., 1980, Devonian "Great Barrier Reef" of Canning basin Western Australia: AAPG Bulletin, v. 64, p. 814-840.

Pray, L. C., and R. C. Murray, eds., 1965, Dolomitization and limestone diagenesis, a symposium: SEPM Special Publication 13, 180 p.

Purser, B. H., 1980, Sédimentation et diagenèse des carbonates néritiques récents: Editions Technip, tome I, 366 p.

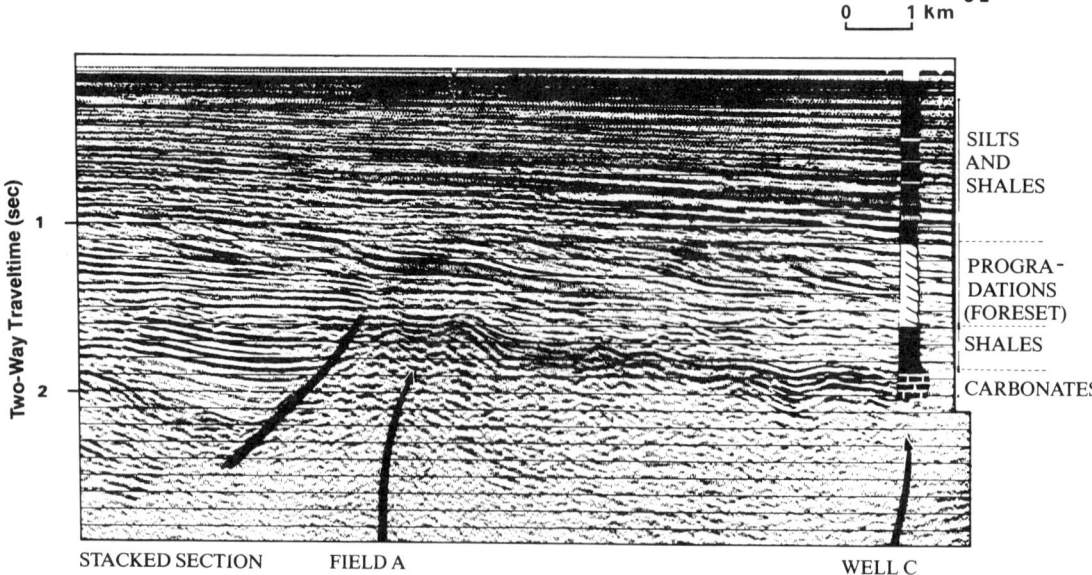

Figure 17—Seismic section through hydrocarbon field producing from Barremian karst limestone. Well C encountered Barremian compact limestone, offshore southern Europe.

Rafavich, F., C. G. St. C. Kendall, and T. P. Todd, 1984, The relationship between acoustic properties and petrographic character of carbonate rocks: Geophysics, v. 49, p. 1622-1636.

Ravenne, C., P. le Quellec, and P. Valery, 1985, Dépôts carbonatés profonds des Bahamas: Géodynamique des Caraïbes, Symposium Paris, 5-8 Févrie, Editions Technip, p. 255-270.

Read, J. F., 1981, Carbonate platforms of passive (extensional) continental margins: types, characteristics and evolution: Tectonophysics, v. 81, p. 195-212.

———— 1985, Carbonate platform facies models: AAPG Bulletin, v. 69, p. 1-21.

Reading, H. G., 1978, Sedimentary environments and facies: Oxford, Blackwell Scientific Publications, 557 p.

Robertson, J. D., and H. H. Nogami, 1984, Complex seismic trace analysis of thin beds: Geophysics, v. 49, p. 344-352.

Schlee, J. S., ed., 1984, Interregional unconformities and hydrocarbon accumulation: AAPG Memoir 36, 184 p.

Scholle, P. A., D. G. Bebout, and C. H. Moore, eds., 1983, Carbonate depositional environments: AAPG Memoir 33, 708 p.

Sheriff, R. E., 1980, Seismic stratigraphy: Boston, IHRDC, 227 p.

Sibille, G., N. Keskes, J. M. Fontaine, R. Lanaud, and J. L. Lequeux, 1985, Enhancement of the perception of seismic facies and sequences by image analysis techniques (abs.): Geophysics, v. 50, no. 2, p. 323.

Taner, M. T., and R. E. Sheriff, 1977, Application of amplitude, frequency, and other attributes to stratigraphic and hydrocarbon determination, in C. E. Payton, ed., Seismic stratigraphy—applications to hydrocarbon exploration: AAPG Memoir 26, p. 301-327.

Vail, P. R., R. M. Mitchum, Jr., R. G. Todd, J. M. Widmier, S. Thompson, III, J. B. Sangree, J. N. Bubb, and W. G. Hatlelid, 1977, Seismic stratigraphy and global changes of sea level, in C. E. Payton, ed., Seismic stratigraphy—applications to hydrocarbon exploration: AAPG Memoir 26, p. 49-212.

Van den Bark, E., and O. D. Thomas, 1980, Ekofisk: first of the giant oil fields in western Europe, in M. T. Halbouty, ed., Giant oil and gas fields of the decade 1968-1978: AAPG Memoir 30, p. 195-224.

Vandenberghe, N., E. Poggiagliolmi, and E. Schwarz, 1983, Analysis of reflection seismic data for the detection of karstified limestones at depth: Third International Seminar, European Geothermal Update, Munich, November 29-December 1, 1983, p. 122-133.

Van den Bosch, W. J., 1983, The Harlingen field, the only gas field in the Upper Cretaceous chalk of the Netherlands: Géologie en Mijnbouw, v. 62, p. 145-156.

Wanless, H. R., 1979, Limestone response to stress: pressure solution and dolomitization: Journal of Sedimentology Petrology, v. 49, p. 437-462.

Wilson, J. L., 1975, Carbonate facies in geological history: New York, Springer-Verlag, 471 p.

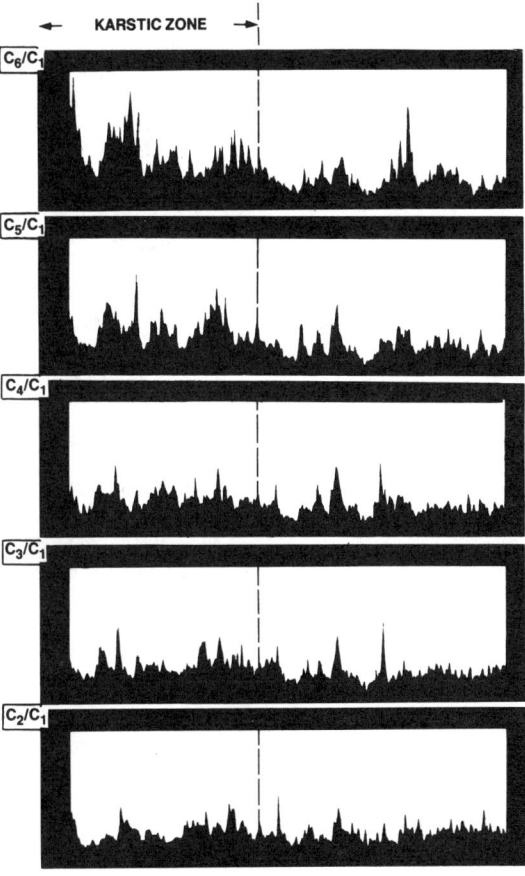

Figure 18—Relative amplitude analysis along seismic line of Figure 17 from near offset (C1) to far offsets (C6). C6/C1 and C2/C1 ratios clearly differentiate between karst and compact limestone, offshore southern Europe.

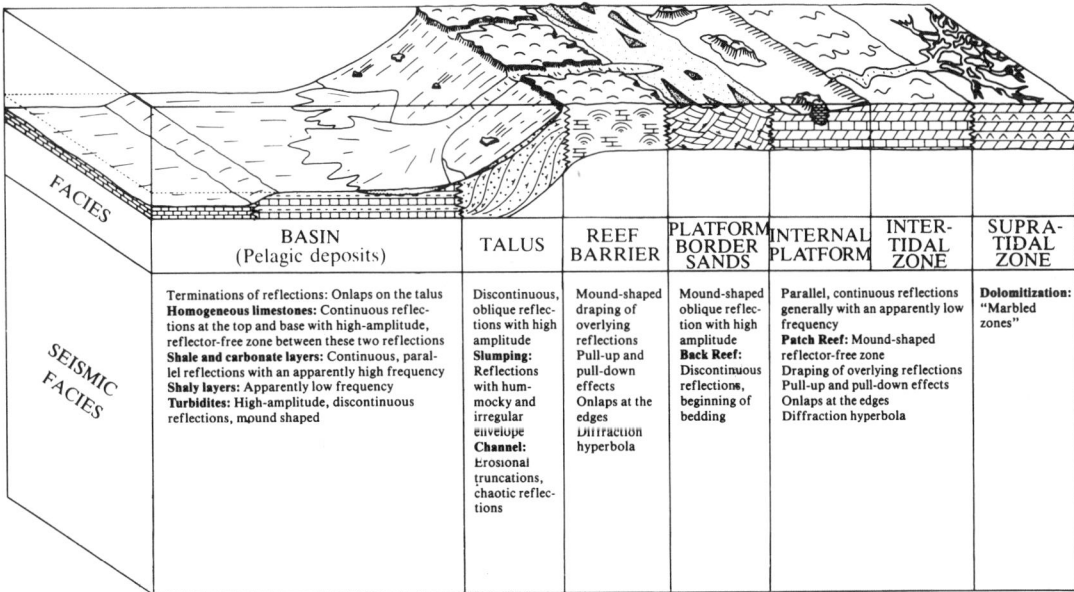

Figure 19—Seismic facies of carbonate depositional environments (geologic model modified from J. L. Wilson, 1975).

SEISMIC EXPRESSION OF CARBONATE BUILD-UPS NORTHWEST JAVA BASIN

J.E. BURBURY *)

ABSTRACT. Deposition of carbonates in the Northwest Java Basin occurred at four different stratigraphic levels during the Tertiary.

Widespread carbonate deposition occurred during Oligocene-lower Miocene and late middle Miocene time intervals, while more localized deposition occurred during two intervals within the lower to middle Miocene.

During each of these times associated carbonate build-ups were developed which can be recognized and mapped in detail from the excellent seismic data in the area.

The size, shape and disposition of all build-ups, except those developed during late middle Miocene, is shown to be related to the tectonic framework, depositional history and local structural features of the basin. The late middle Miocene build-ups appear to be unrelated to palaeo-structural features, indicating deposition on a base-levelled surface at a time of structural maturity and quiescence.

The ability of the seismic data to indicate variations in porosity and to detect both direct and indirect indications of hydrocarbons within the build-ups is illustrated. An excellent example is shown in which the presence of a large volume of gas within a late middle Miocene carbonate build-up can be interpreted directly from the seismic data.

INTRODUCTION

In most of the foreland basins of Indonesia a period of depositional quiescene developed within the Oligocene-early Miocene time interval. Clastic supply deminished to virtually nil and a period of carbonate sedimentation with associated reef growth ensued and continued intermittently into Miocene time, Fig. 1. Re-elevation of clastic source areas, during mid Miocene, resulted in a flood of mostly fine clastics being deposited over the Miocene carbonates and reefs, thus setting up potential hydrocarbon traps in the early Miocene reefs of these basins.

So far, this situation has proved to be productive in three widely spaced basins of Indonesia, with very significant hydrocarbon reserves found in the Arun gas field of N. Sumatra, the Rama oil field of N.W. Java and oil fields of the Salawati Basin. The gas fields of the Pauan basin also fit into this category.

Subsequent to this regionally widespread early Tertiary carbonate deposition, clastic sedimentation prevailed in most areas.

In the offshore Northwest Java Basin, deposition of the initial widespread carbonates and associated build-ups of Batu Raja limestone commenced during Oligocene and continued into lower Miocene, Fig. 2.

During the remaining period of the lower Miocene and most of middle Miocene, clastic deposition of sands, shales and siltstones of the Upper Cibulakan formation predominated. Laterally restricted carbonates and associated carbonate build-ups developed at two stratigraphic levels in structurally controlled areas. These carbonates are referred to by the informal names of "Mid–Main Carbonate" and "Pre-Parigi Limestone"

In late middle Miocene regional quiescence returned and widespread deposition of carbonates of the Parigi limestone occurred with prolific development of build-ups in local areas.

This second period of widespread carbonate deposition was followed by deposition of mostly fine clastics of the Cisubuh claystone with further carbonates developed only in the Thousand Islands area during recent time.

From the high density of good quality seismic data available in the offshore north-

*) Atlantic Richfield Indonesia, Inc.

west Java area it can be shown that the build-ups associated with the deposition of the Batu Raja limestone, "Mid-Main Carbonate" and the "Pre Parigi Limestone" are developed and localized according to the regional tectonic setting within the basin and the presence of palaeo-structural features. Build-ups associated with the Parigi limestone appear to be unrelated to palaeo-structural features, indicating deposition on a base levelled surface at a time of structural maturity and quiescence.

The good quality of seismic data available in the offshore northwest Java area allows for more than just the physical mapping of size and shape of the build-ups. The data can be used to obtain indications of porosity and to detect both direct and indirect indications of hydrocarbons within the build-ups. The following measurements and observations can be used to this end:

1. Interval velocity within the build-up derived from seismic stacking velocities, as an indicator of porosity.
2. The presence and amount of either positive or negative velocity pull-up, indicating interval velocity relative to surrounding sediments and hence porosity estimation.
3. Decrease in amplitude of the top reflection indicating increase in porosity and/or the presence of high gas saturations, "dull spots".
4. The presence of "bright spots" in sands above the build-up indicating gas and possible diffusion of this gas from a hydrocarbon accumulation within the build-up.
5. The presence of a "flat spot" within the build-up indicating a gas/liquid interface — liquid could be either oil or water since the acoustic impedance of a reservoir containing either is approximately the same.

BATU RAJA LIMESTONE

The Batu Raja limestone was deposited regionally during the initial marine transgression in the offshore northwest Java area, over a surface base-leveled by the prior deposition of marine and non-marine clastics of the Talang Akar formation.

At this time the major structural elements of the basin were established and continuing to develop, influencing the deposition and facies variations of the Batu Raja limestone. These major structural features are listed below and illustrated on Figures 3 and 4.

1. Fault bounded Sunda Sub-basin
2. Seribu platform
3. Ardjuna Sub-basin
4. Central and Eastern platforms

Figure 5 is the isopach of Batu Raja limestone which, when viewed in relation to the isopach of Talang Akar formation, illustrates the following significant features.

1. The Batu Raja limestone is thin over the stable Seribu platform with some areas of non-deposition on the crest of this feature.
2. Slight thickening to the east over the gently shelving east flank of the Seribu Platform.
   Buildups are small both aerially and vertically and all are developed over paleo structures.
3. On the west flank of the Ardjuna trough and over the eastern part of the Ardjuna Sub-basin, a greater rate of subsidence promoted more rapid build-up growth resulting in large build-ups in these areas, all of which overlie old basement and Talang Akar highs.
4. Limestones thin towards the axis of the Ardjuna trough.

Over the high, flat-topped part of the Seribu platform, thin Batu Raja limestone was deposited over thin to absent Talang Akar formation. Figure 6a is typical of this part of the area where the strongest seismic events recorded originate from the top and base of the Batu Raja limestone. On Figure 6a a very slight thickening or build-up of the Batu Raja, to the extent of approximately 10 milliseconds or 30 ft., can be noted south of the basement hinge line.

In itself this very small scale build-up appears rather insignificant. However, the relative amplitude playout of the same section shows a distinct amplitude loss on the top Batu Raja reflection and this could be significant.

This loss of amplitude, which does not appear to be due to absorbtion or interference effects within the section above, means the

reflection coefficient at this interface is lower due to a drop in velocity or density or both and indicates one or a combination of the following :
1. Facies change to a limestone of lower velocity and/or density.
2. Increase in porosity.
3. Presence of gas.

This slight build-up and loss of amplitude effect can be mapped aerially and is seen to be consistantly related to the structural hinge on basement. It could be reasonably interpreted as a zone of slight build-up and porosity in the Batu Raja limestone fringing the crestal area of the Seribu platform.

East and north east from the crest of the Seribu platform the Batu Raja limestone was deposited on a gently shelving part of the platform. A scattering of isolated build-ups occurs in this area, all of which show north westerly elongation and maximum vertical build-up of 50 millisecond or approximately 150 feet, as illustrated on Figure 7.

Significant amplitude anomalies are not observed over these build-ups and the small separation between top and bottom reflections (100 ms or less) is not great enough for seismic velocity analyses to give reliable estimates of interval velocity and hence indications of porosity.

Either regional structural elevation or the presence of local structural highs can be seen to be responsible for the initiation of build-up growth over this area.

Over the western plunging flank and the eastern area of the Ardjuna sub-basin a greater rate of subsidence promoted more rapid build-up growth. All of these build-ups occur over pre-existing basement and Talang Akar structural highs as illustrated by Figures 3, 4 and 5.

A concentration of build-ups developed in the eastern area of the Ardjuna Sub-basin and that shown on Figure 8 is typical for the area. It covers an area of approximately 6 x 2 miles and shows growth of 110 milliseconds, approximately 800 feet. The presence of an excellent reflection from the base of the build-up allows for precise mapping.

Indications of porosity and/or hydrocarbons associated with Batu Raja build-ups in this area are not apparent on the seismic sections. However, with build-up of this size, in excess of 200 milliseconds thick, it is possible to derive from the seismic reliable interval velocities as an indicator of porosity.

Figure 9 shows the results of velocity analyses from the seismic data between SP's 56 and 61 on Figure 8. Very distinct breaks in the time-velocity curve occur at the top and base of the Batu Raja limestone reflecting the high interval velocity of this formation compared with surrounding sediments. Using the method given by Dix, 1955, "Seismic Velocities from Surface Measurements", Geophisics, we obtain an interval velocity of approximately 15,000 feet/sec., which is indicative of a low porosity limestone.

Most Batu Raja build-ups analyzed seismically show interval velocities in the range 14,000 to 16,000 feet/sec. Well velocity surveys and sonic logs confirm that velocities lie within this range.

Some of the Batu Raja build-ups of the type here discussed have been found to possess a basic coral framework indicating true reef environment, however, the rocks have suffered moderate to severe diagenetic alteration resulting in limestones with interval velocities of 14,000 to 16,000 ft/sec and generally low porosity and permeability.

"MID–MAIN CARBONATE" AND "PRE–PARIGI LIMESTONE"

During the clastic phase of deposition that followed the Batu Raja limestones in upper Cibulakan time some carbonate build-ups were developed in structurally selective areas at two stratigraphic levels. See Figures 2 and 4.

These carbonates are not widespread but occur as isolated build-ups that grade laterally into deeper marine silts and muds with limestone stringers as illustrated on Figure 10.

Figure 11 shows the restricted distribution of the carbonates developed during these times. The "Mid Main Carbonates" are restricted entirely to the south eastern shelf and shelf edge of the Seribu platform while the "Pre Parigi limestones" occur in this same area and also over some old highs on the west flank of the Ardjuna sub-basin.

Figure 12a illustrates typical carbonate build-ups developed at these stratigraphic levels near the shelf edge of the Seribu platform. A direct coincidence between both the "Mid-Main" and Pre-Parigi" build-ups and an old structural high can be seen on this section.

Seismically determined interval velocities are low for carbonates indicating high porosity. This has been confirmed by well surveys. Figure 12b shows the same "Mid-Main" build-up as that on Figure 12a, but near the structural crest of the build-up. A definite character change and loss of amplitude of the top reflection is apparent on the crestal line which could be due to the presence of gas, at the crest of the feature. No gas/liquid "flat spot" reflection can be seen but other seismic lines show the character and amplitude change to occur only over a small crestal area.

PARIGI LIMESTONE

Towards the end of middle Miocene time widepsread carbonate deposition occured again in the area with the deposition of the Parigi limestone.

The isopach of this formation Figure 13, shows regional thickening to the south with a depositional thin extending west-southwesterly across the area. There is no relationship between the isopach of Parigi limestone and older structural trends, indicating that the Parigi limestone was depositing during a period of maturity and quiescence of the old structural trends.

A chain of Parigi build-ups developed along the northern flank of the isopach thin, all of which show a similar amount of vertical growth and no relationship to underlying structure. Build-ups to the south of the isopach thin are more sporadic in distribution and quite large.

Figure 14 shows some of the Parigi build-ups that occur on the northern flank of the isopach thin, and clearly illustrates the lack of any relationship between the location of build-ups and old structural features.

With such a number of build-ups as shown on Figure 13 presenting themselves as potential hydrocarbon reservoirs, we must turn to the seismic data in detail to attempt to isolate those which show signs of porosity and indications, both direct and indirect, of the presence of hydrocarbons.

On Figure 14 the seismic character of Parigi build-up "A" differs from the other two. It shows a low amplitude discontinuous reflection at the top and an unusual, approxiamtely flat reflection at 740 ms, that appears to be disconformable with both top and base — a possible gas/liquid interface.

Following up this lead, another section over the same Parigi build-up "A" shows on Figure 15a:

1. Weak discontinuous reflection from top
2. "Bright spot" in sand above the build-up.
3. Strong disconformable flat reflection, again at about 740 ms., that extends across the full width of the build-up.

Figure 15b is another section over the same Parigi build-up "A" and the same anomalous features are apparent:

1. Weak arched reflection from the top.
2. Fault terminated bright spots in sands above the build-up.
3. "Flat spot" at 740 ms., complicated by faults and interference at the western edge of the build-up.

Figure 15c shows the "Flat spot" at 740 milliseconds extending the full length of the long axis of the build-up. As a gas and possible oil prospect this particular build-up is looking good. On Figure 16 there appears to be 70 milliseconds of gas, which at an interval velocity lower than normal because we do interpret gas in the reservoir, say 7000 ft/sec., we could prognose 250 ft of gas overlying a liquid which could be either oil or water.

The results of well No.3 drilled at SP 130 on this section, are shown by the sonic log on Figure 17. A gas column of 277 feet overlying water was encountered in the top of the Parigi. The reasons for the absence or low amplitude of the top reflection and the presence of a strong gas/liquid reflection are clearly shown by the sonic log.

Now let us compare in detail the seismic expression and hydrocarbon indications in some other Parigi build-ups in the same general area.

On Figures 17 and 18:

Well No. 1 : Penetrated Parigi between build-ups.

Well No. 2 : Located on a Parigi build-up which seismically shows no indications of hydrocarbons and encountered thick porous Parigi reef with no hydrocarbons. The build-up to the east shows some signs of gas and was found to have a few feet of gas in the top Parigi only.

On Figure 17 and 16 :

Well No. 3 : Penetrated the Parigi build-up "A" previously discussed. The build-up to the east, Parigi build-up "B" lies between wells 2 and 3 and has no hydrocarbon indications. We will return to this one later.

On Figure 17 and 19 :

Well Nos. 4 and 5 :

Encountered thin gas columns in the top Parigi and the presence of this gas is indicated by bright spots above. However, we can note that the top reflection is strong and perfectly flat, which is unusual. The sonic logs show that the main velocity break is at the gas/liquid interface and that the seismic time interval through the thin gas would be between 0 and 12 milliseconds only. Hence it could be reasonably interpreted that the flat 'top' reflection is in fact the gas/liquid interface with some interference from the weaker event originating from the top of the carbonate only a few milliseconds above.

On Figure 19 there is a good example of a fault which cuts all horizons except the interpreted gas/liquid interface, which is what would be expected in this thick partly gas filled reservoir. The sonic two way time to the gas/liquid interface in both wells is 670 milliseconds which fits very well with the seismic time of 680–690 milliseconds on Figure 19 allowing for reflection lag. Both Well Nos. 4 and 5 penetrated the Parigi build-up on the downthrown side of the fault. Faulting at the base of the Parigi and at the "bright spot" above the Parigi is to the extent of approximately 20 milliseconds or 60 feet. Hence a well located on the high side of the fault could be expected to encounter a gas column thicker by 60 feet than that which occurs on the low side of the fault.

Returning to Parigi build-up "B", between Wells 2 and 3 on Figure 16, the top is seen to be slightly arched and hydrocarbon indicators are not apparent. Note that the top of the build-up is at 720 milliseconds. Approximately 4 miles to the south the same Parigi build-up "B" on Figure 20 shows a strong "bright spot" above and perfectly flat top reflection at a time of 690 milliseconds. This could be interpreted similarly to the previous one, as a build-up with a thin crestal gas cap, expressed by the flat "top" reflection, the lateral extent of which can be mapped, overlying a liquid that could be either oil or water. It is interesting to note here the very limited lateral migration of the gas in the sands above the build-ups.

CONCLUSIONS

Carbonate build-ups that occur at four different stratigraphic levels in the offshore northwest Java area are all clearly expressed by the good quality seismic data available.

In the foregoing discussion I have attempted to demonstrate that more than just the size, shape and distribution of these buildups can be determined from the seismic data. In some cases it is possible to answer the question "Does it, or does it not contain hydrocarbons?", or at least "Could it, or could it not contain hydrocarbons?"

When pushed to the limit the seismic method can be a very powerful exploration tool. With the present day quality of data available to the industry, it is up to the interpreter to extract all possible information from the data. More than ever before geophysicists are "Wiggle Pickers" who must recognise that every wiggle bears some relation to the subsurface we are exploring.

ACKNOWLEDGEMENTS

I wish to thank the management of Atlantic Richfield Co. for time made available and permission to publish this paper. My special thanks go to Mr. Bill Edwards, who first recognized the direct indication of gas in the Parigi "A" build-up and initiated a study of same, and to Mr. Dan Taiclet for his critical editing and reading of the paper on my behalf.

REFERENCES

DIX, 1955 : "Seismic Velocities from Surface Measurements", Geophysics.

FLETCHER, G.L. and SOEPARJADI, R.A., 1977 : "Indonesia's Tertiary Basins — The Land of Plenty" : Proceedings Indonesian Petroleum Association, Fifth Annual Convention.

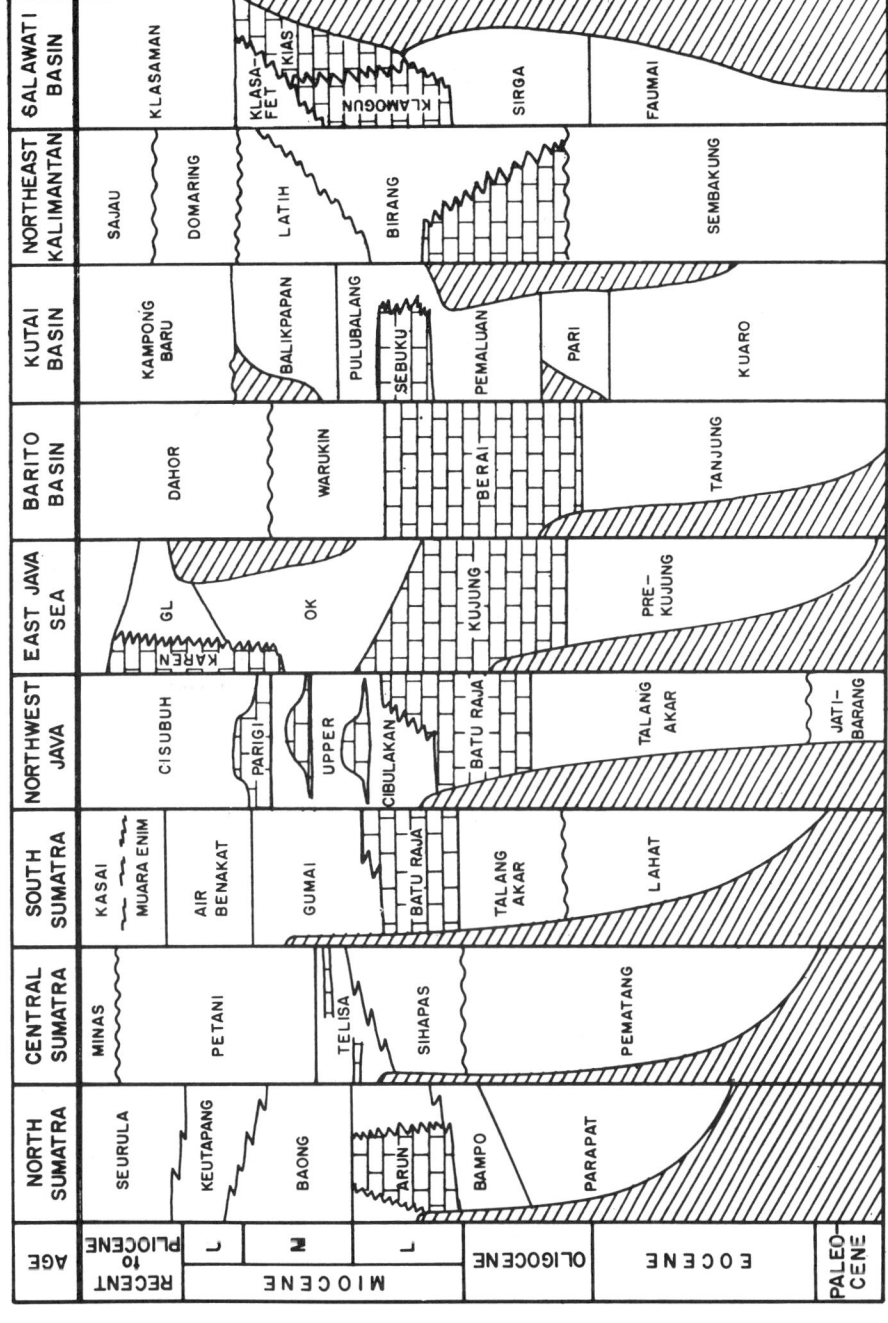

Figure 1 – Generalized Stratigraphic Correlation

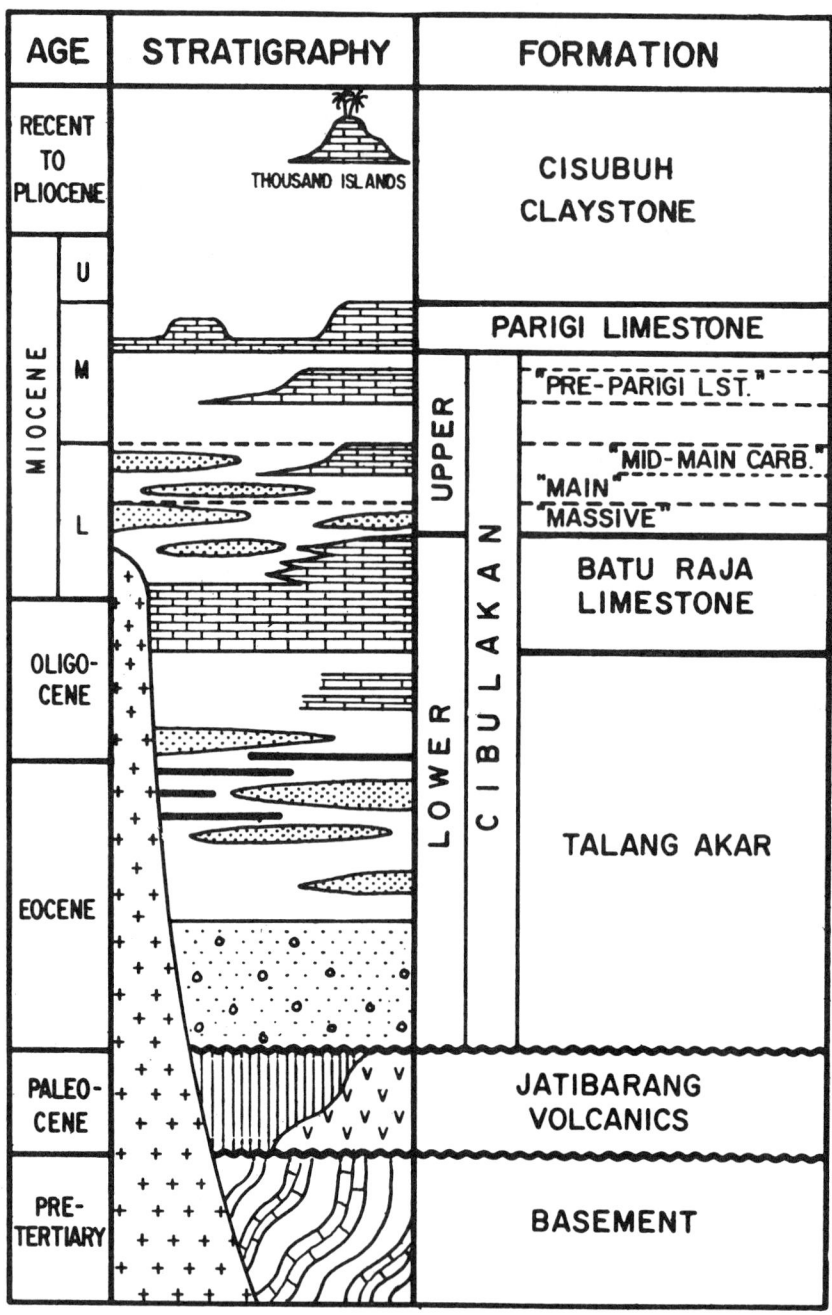

Figure 2 — Stratigraphic Column, Offshore N.W. Java

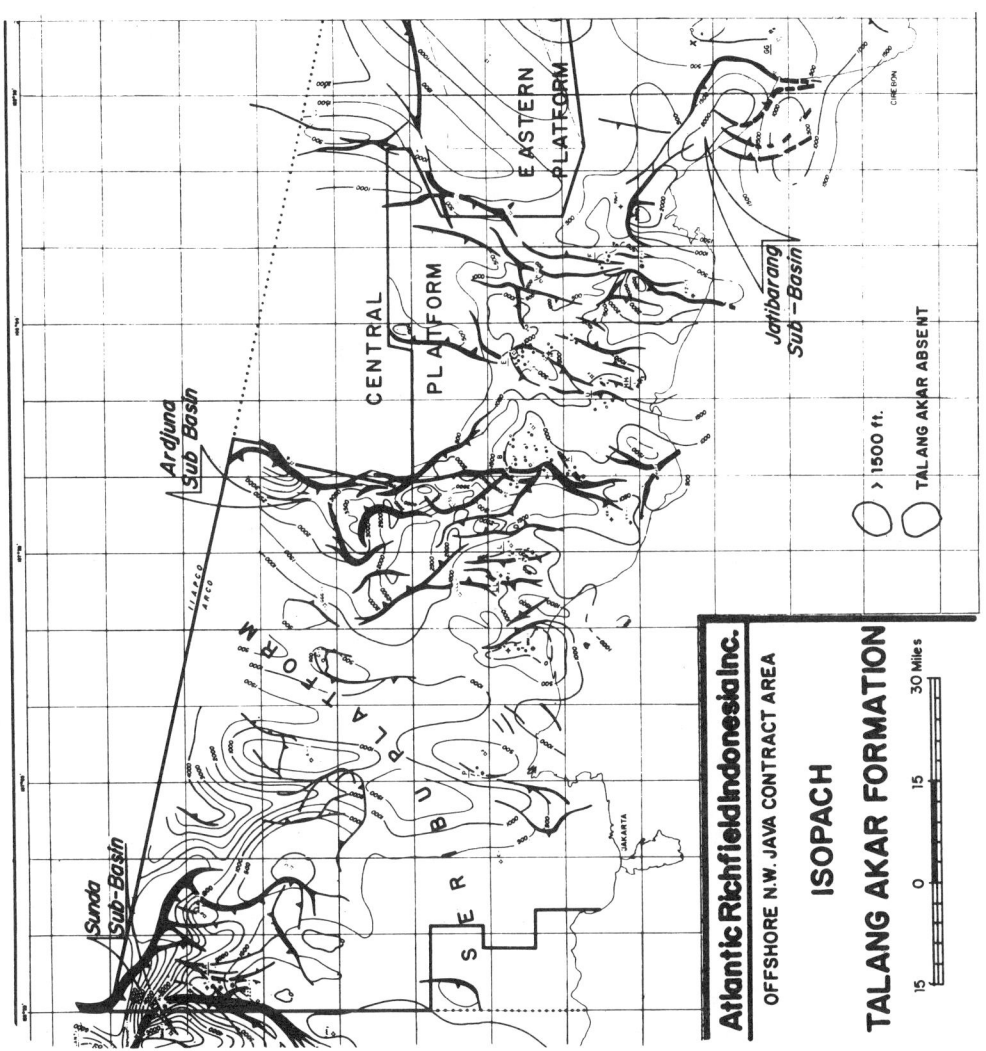

Figure 3 – Isopach Talang Akar Formation

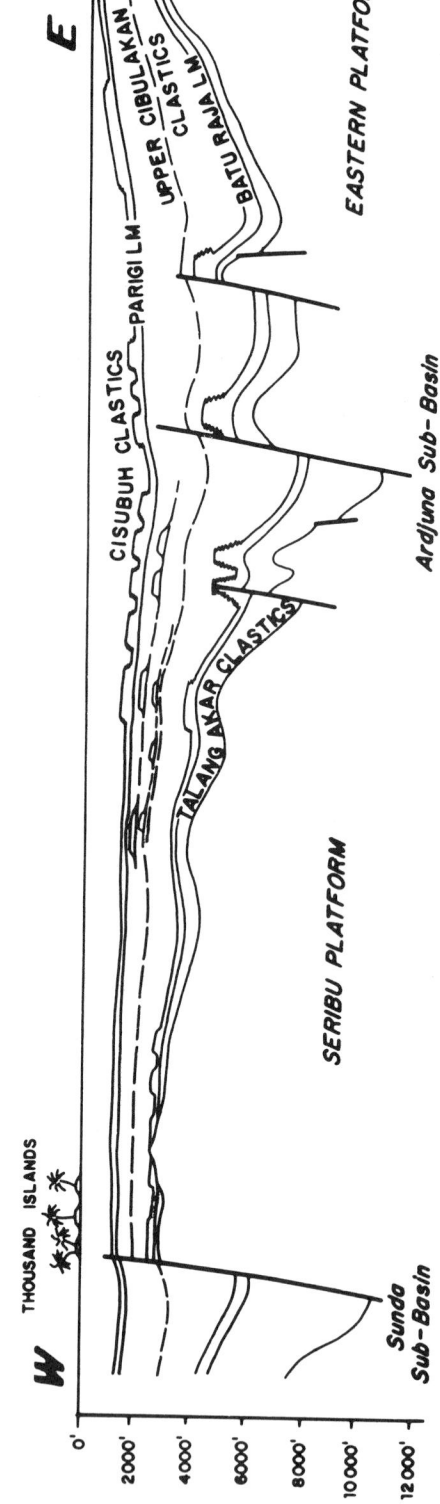

Figure 4 — Diagrammatic Cross Section

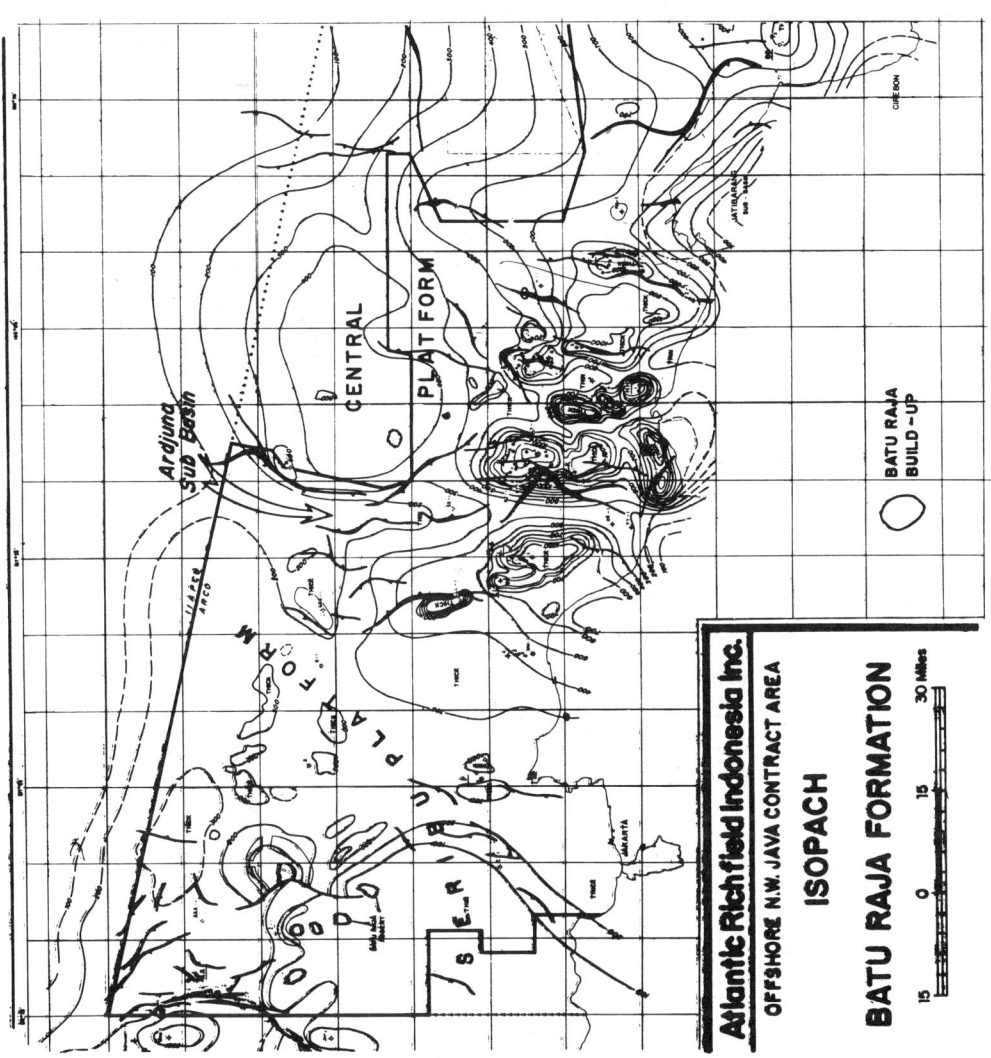

Figure 5 — Isopach Batu Raja Limestone

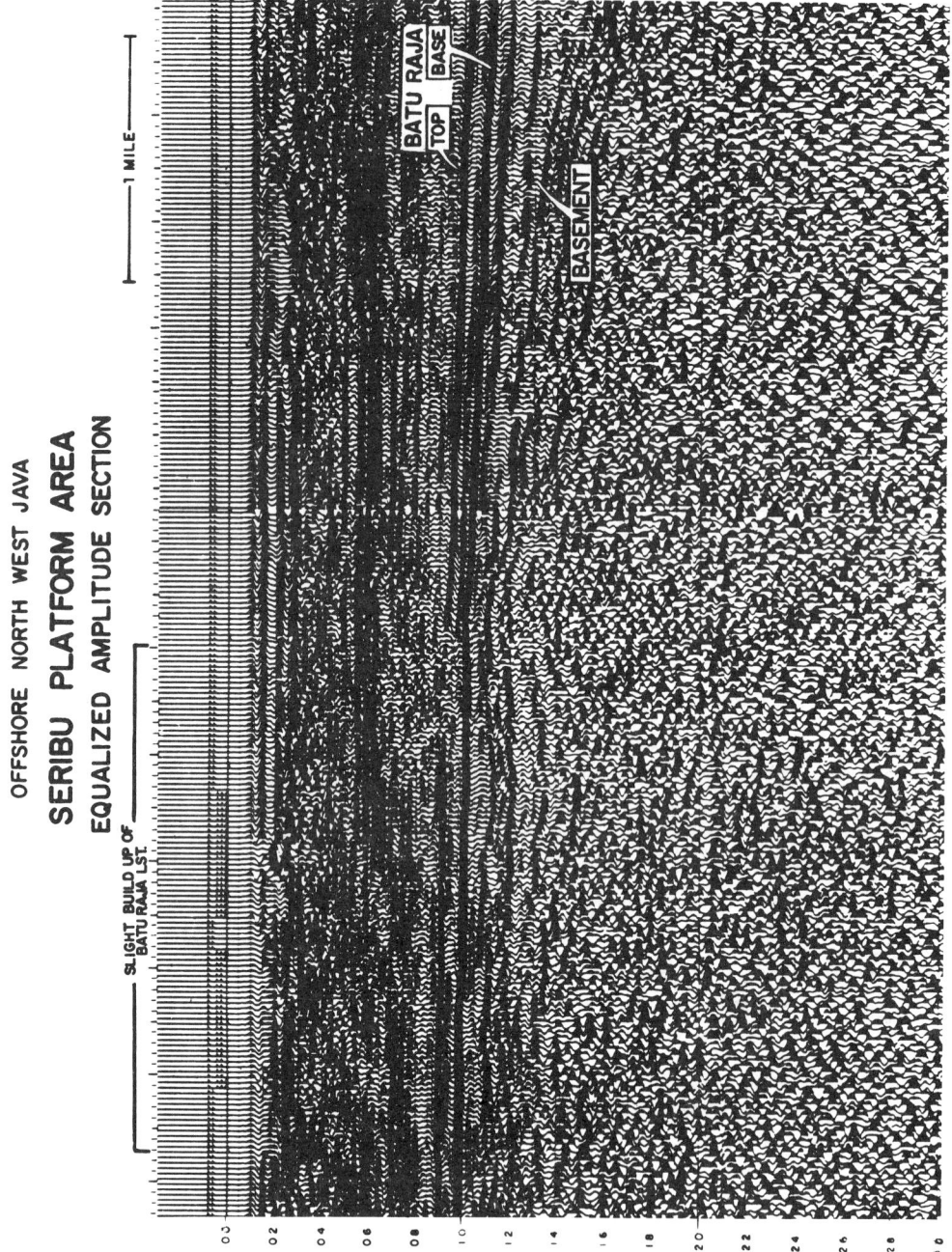

Figure 6a — Batu Raja Build-up — Seribu Platform

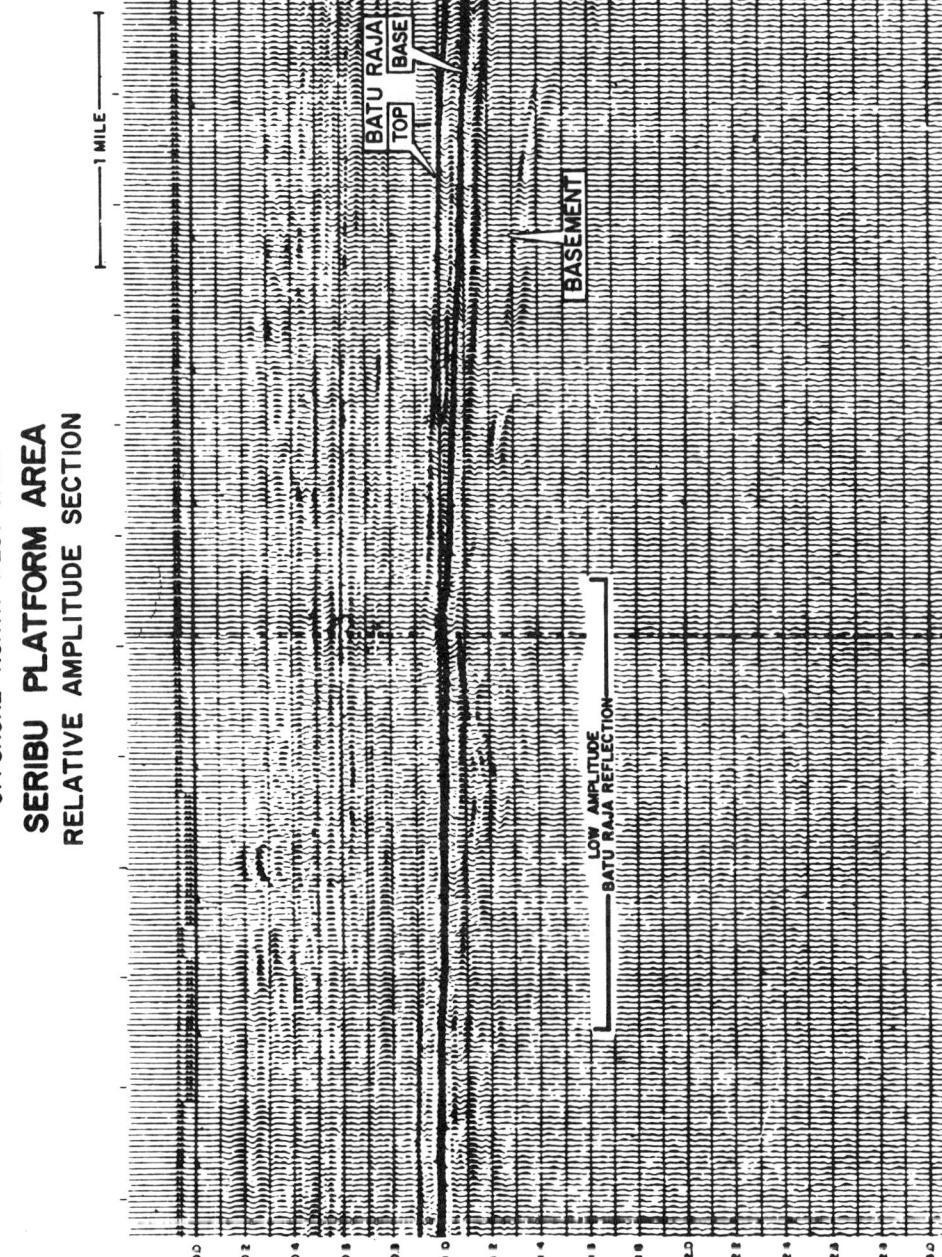

Figure 6b — Batu Raja Build-up — Seribu Platform

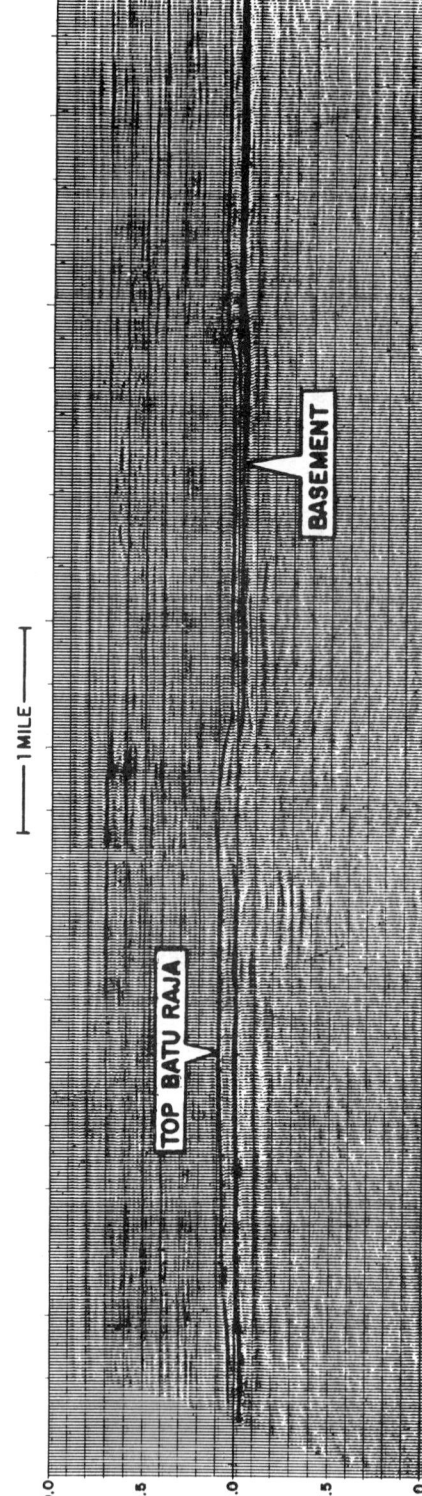

Figure 7 — Batu Raja Build-up — Seribu Platform

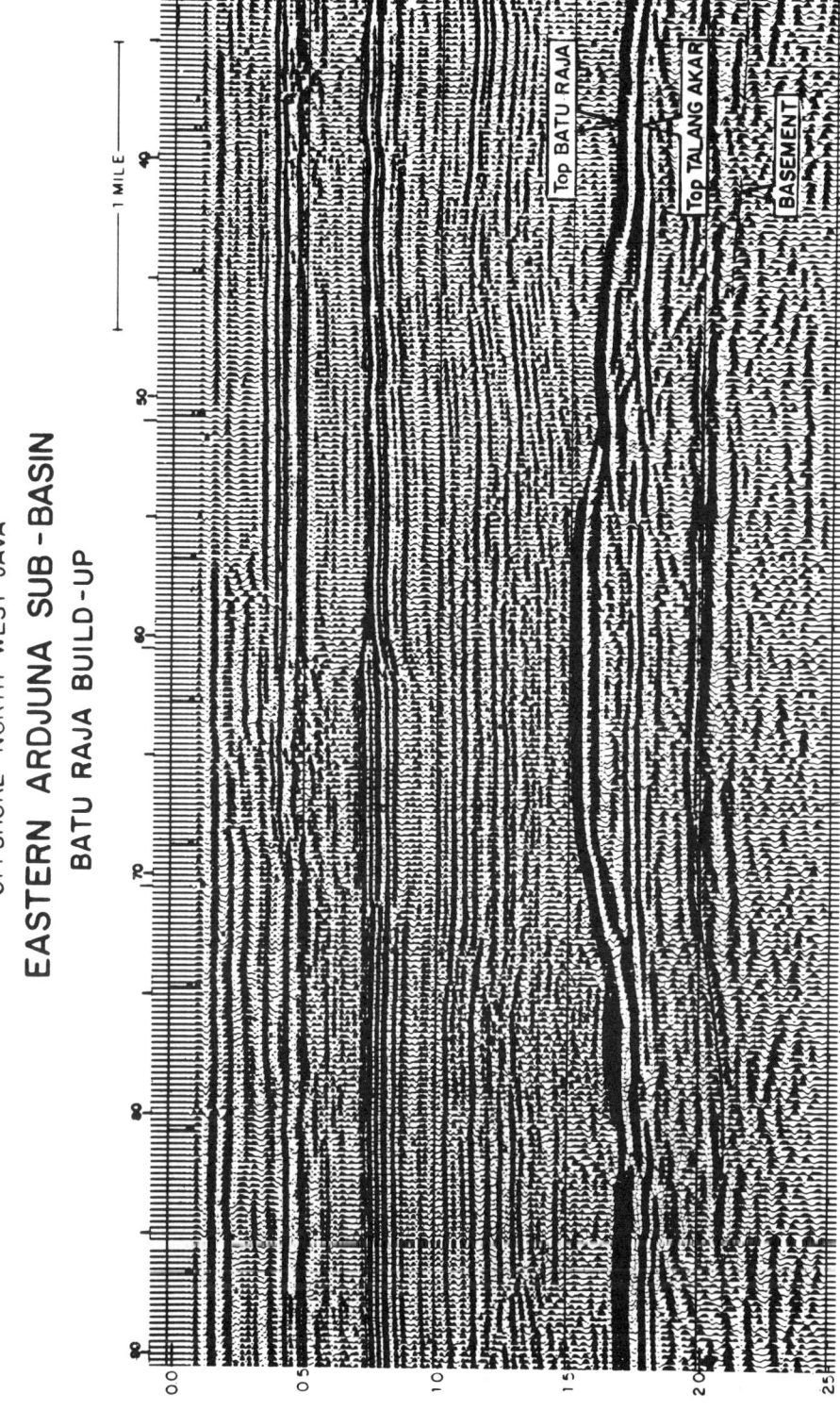

Figure 8 — Batu Raja Build-up — Eastern Ardjuna Sub-basin

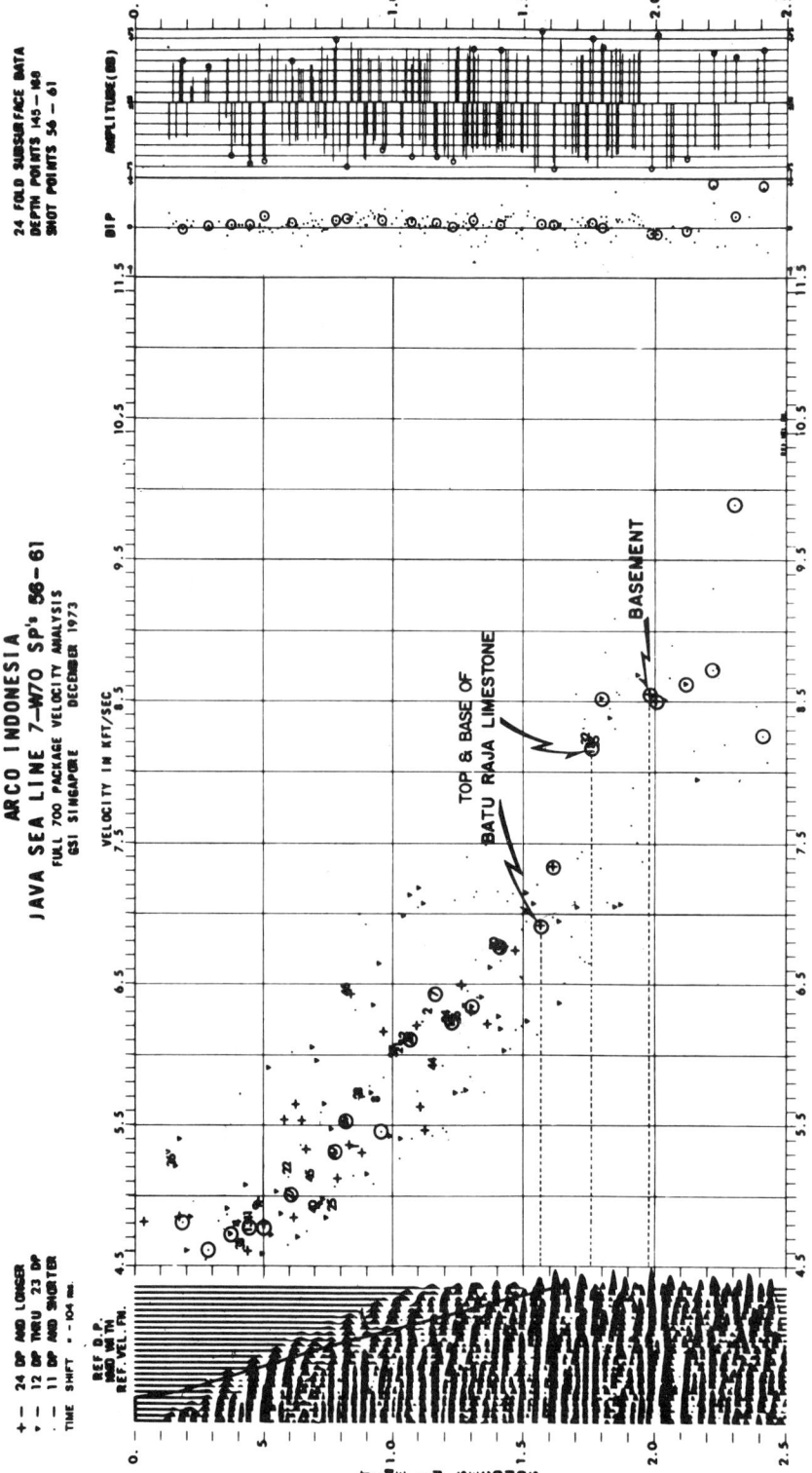

Figure 9 — Velocity Analysis — Line 7

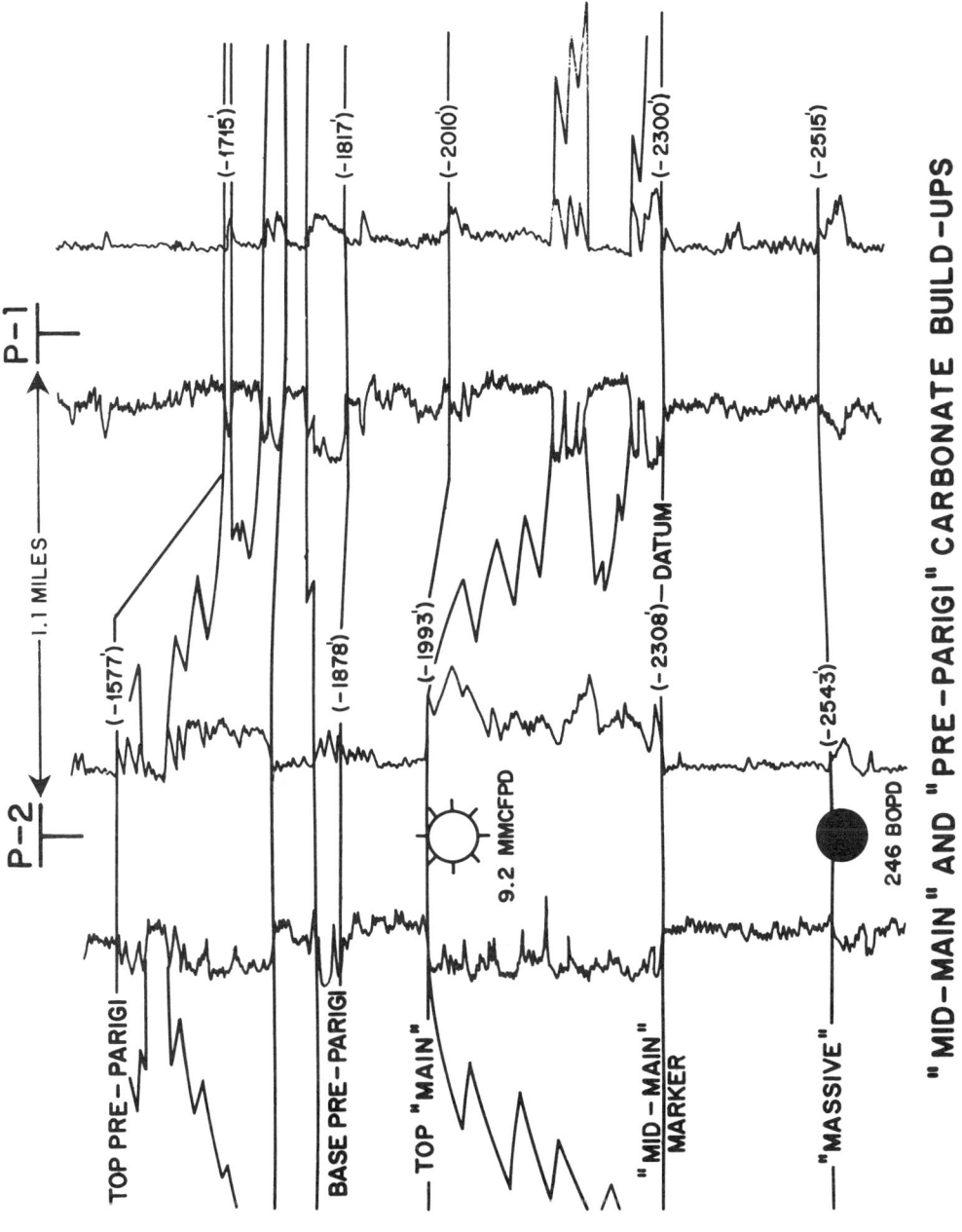

Figure 10 — "Mid-Main" and "Pre-Parigi" Build-ups – section

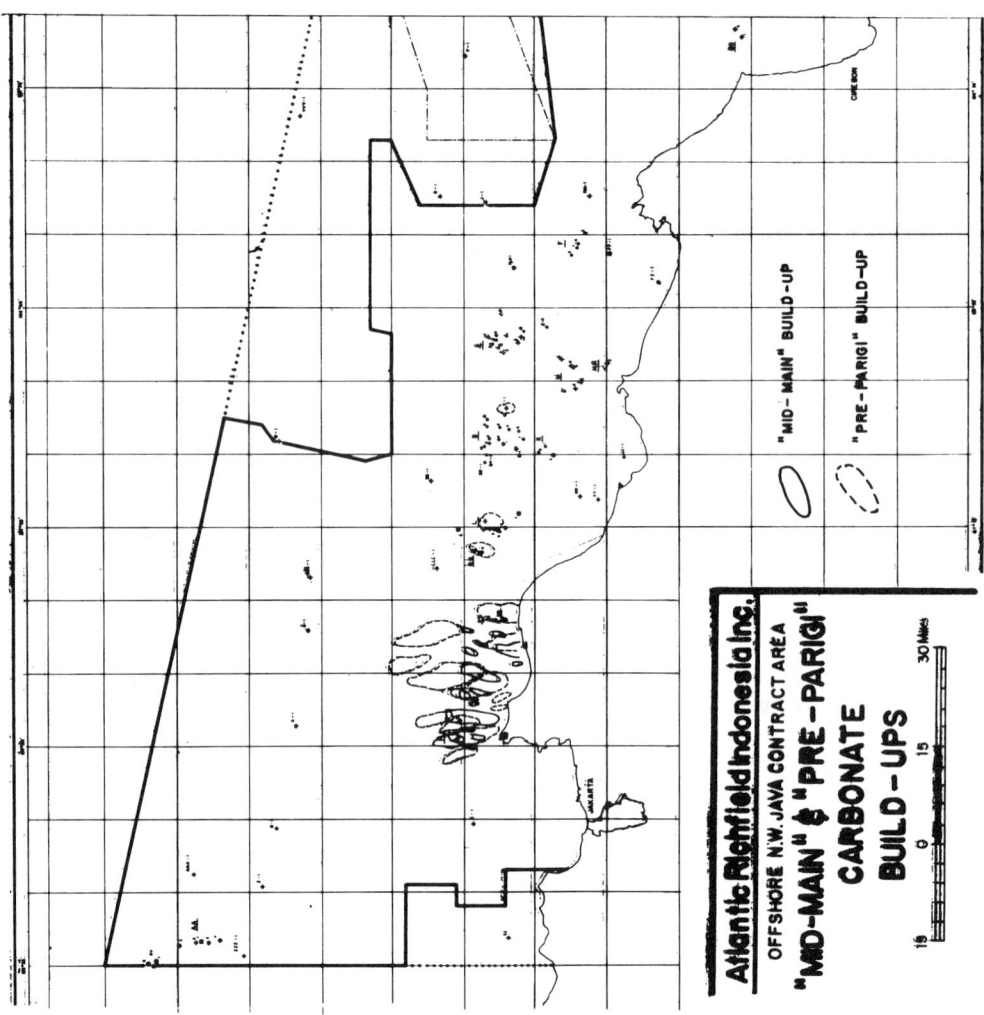

Figure 11 — "Mid-Main" and "Pre-Parigi" Build-ups — map

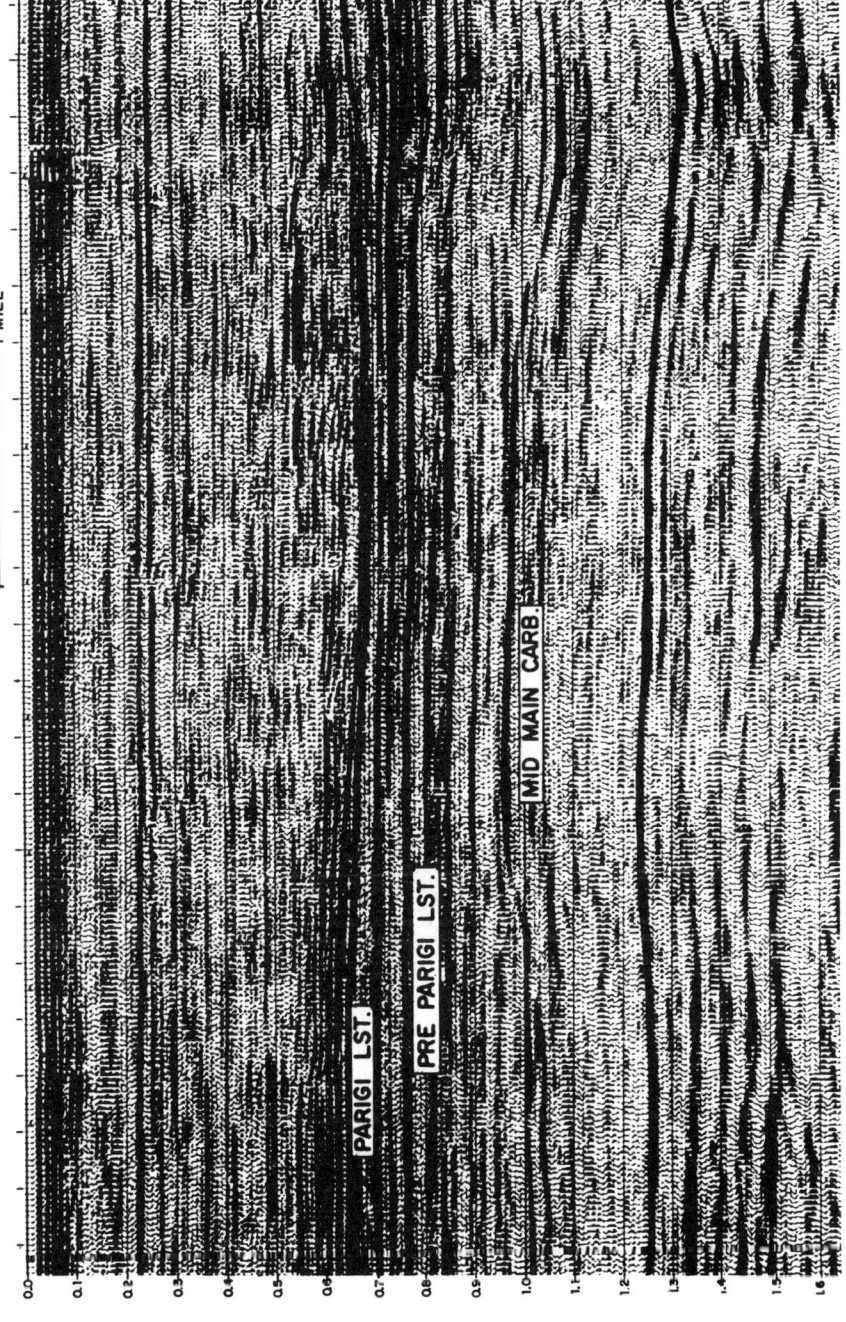

Figure 12a — "Mid-Main" Build-up, East Seribu Platform

Figure 12b – "Mid-Main" Build-up, East Seribu Platform

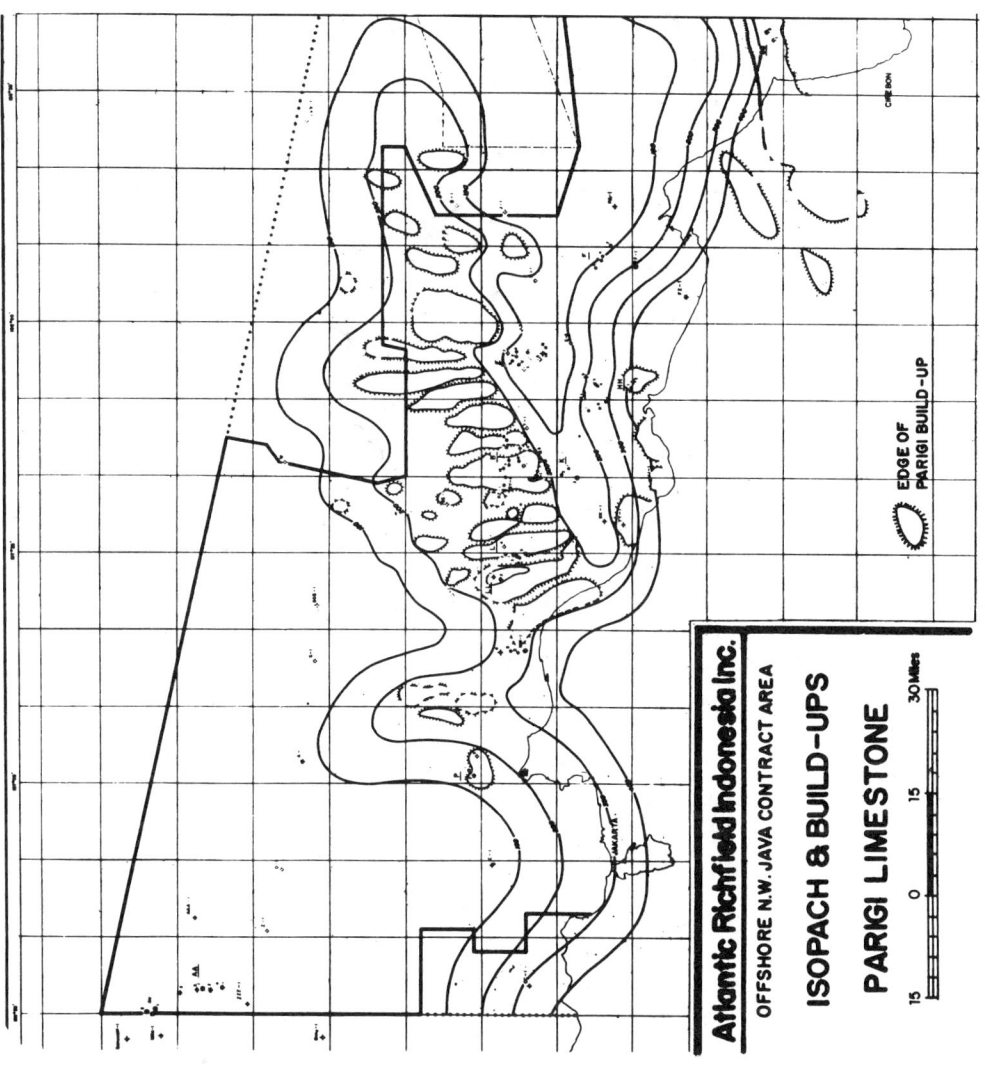

Figure 13 — Isopach and Build-ups — Parigi Limestone

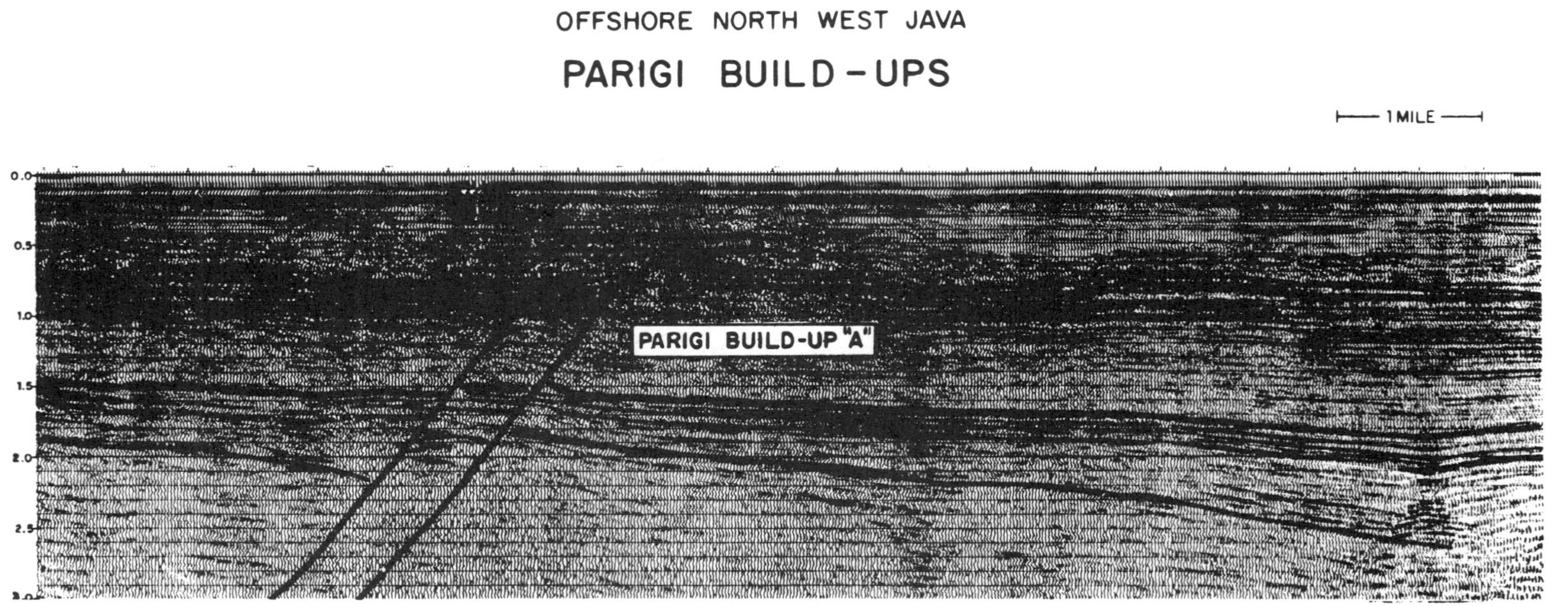

OFFSHORE NORTH WEST JAVA
PARIGI BUILD-UPS

PARIGI BUILD-UP "A"

Figure 14 — Seismic Section — Parigi Build-ups

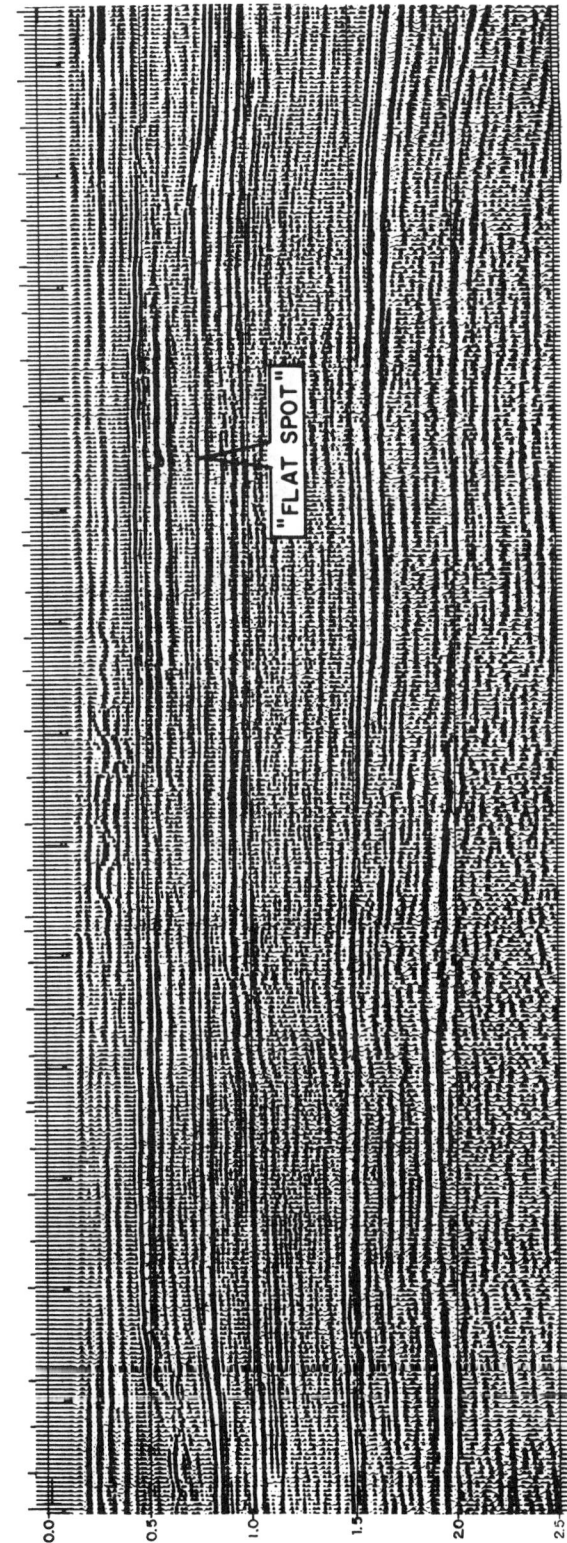

Figure 15a — Seismic Section — Parigi Build-up "A"

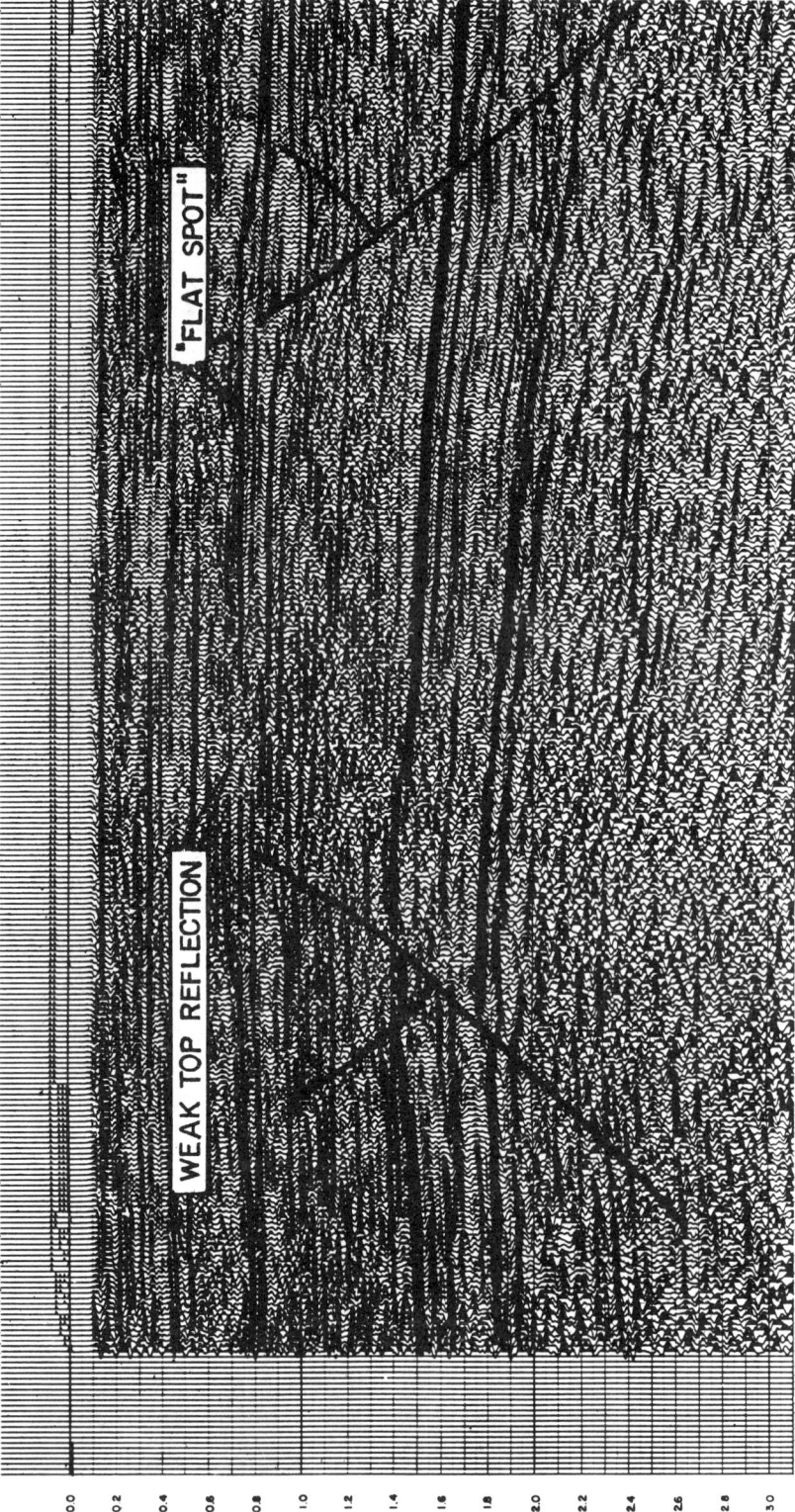

Figure 15b — Seismic Section — Parigi Build-up "A"

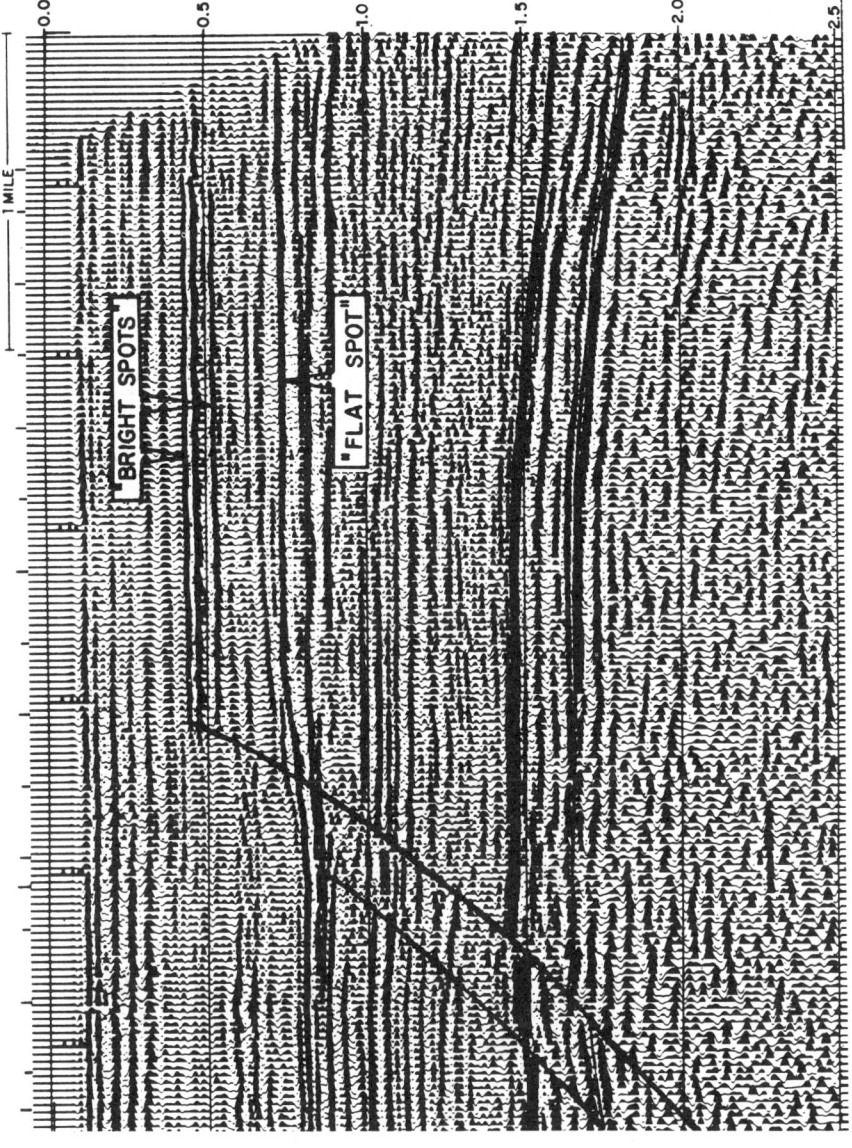

Figure 15c — Seismic Section — Parigi Build-up "A"

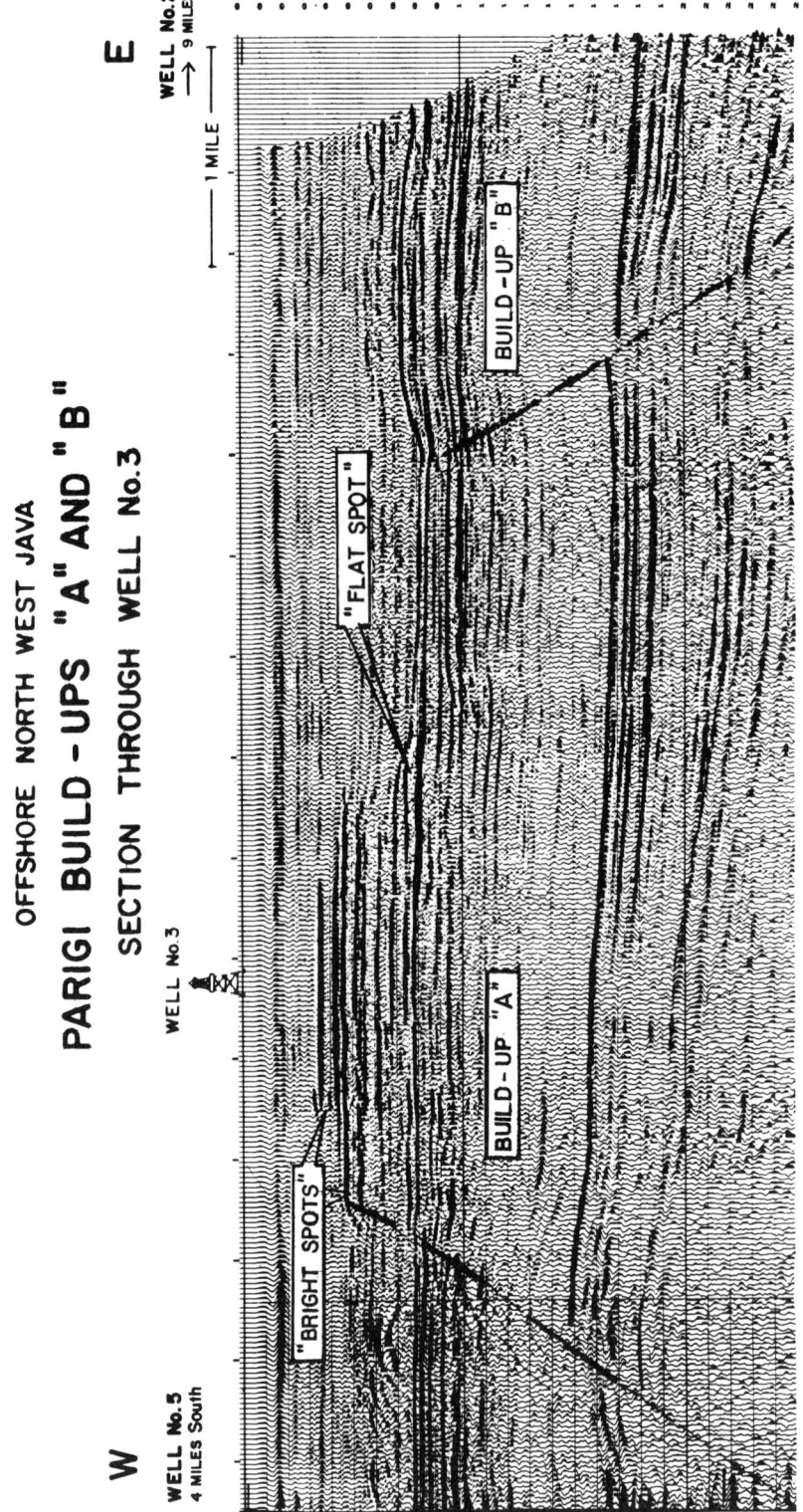

Figure 16 - Seismic Section - through Well No. 3

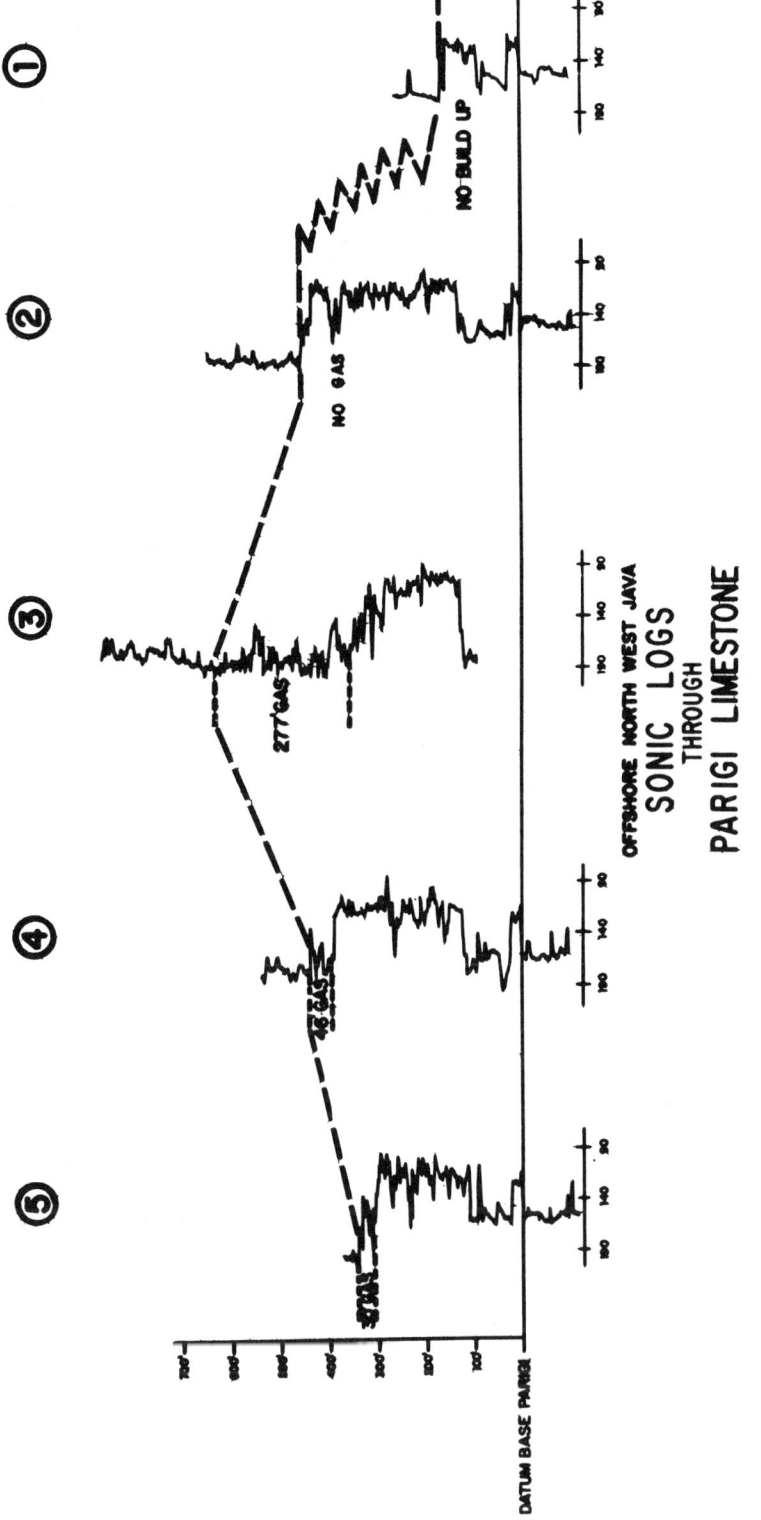

Figure 17 — Sonic Logs through Parigi limestone

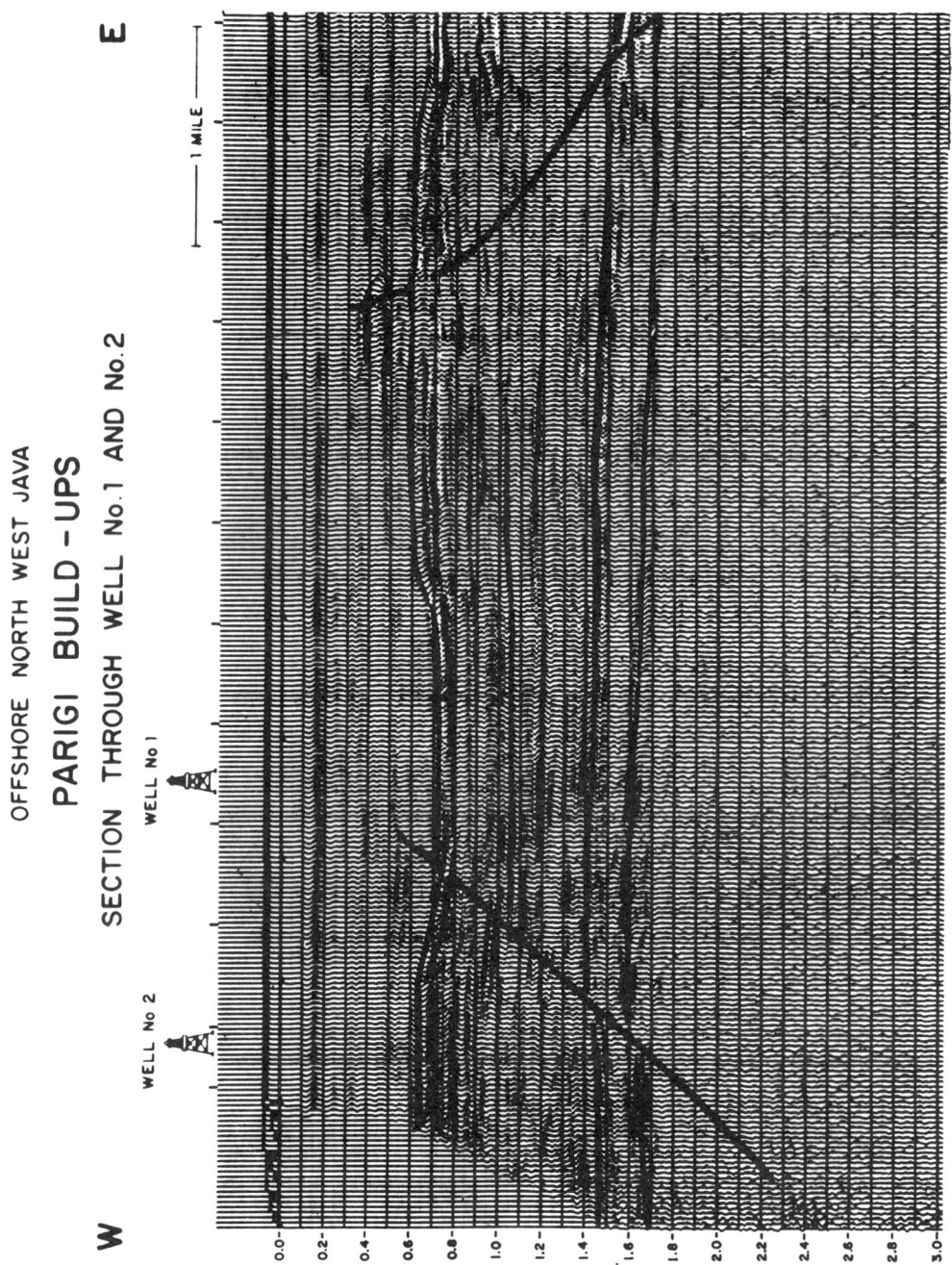

Figure 18 — Seismic Section through Well Nos 1 and 2

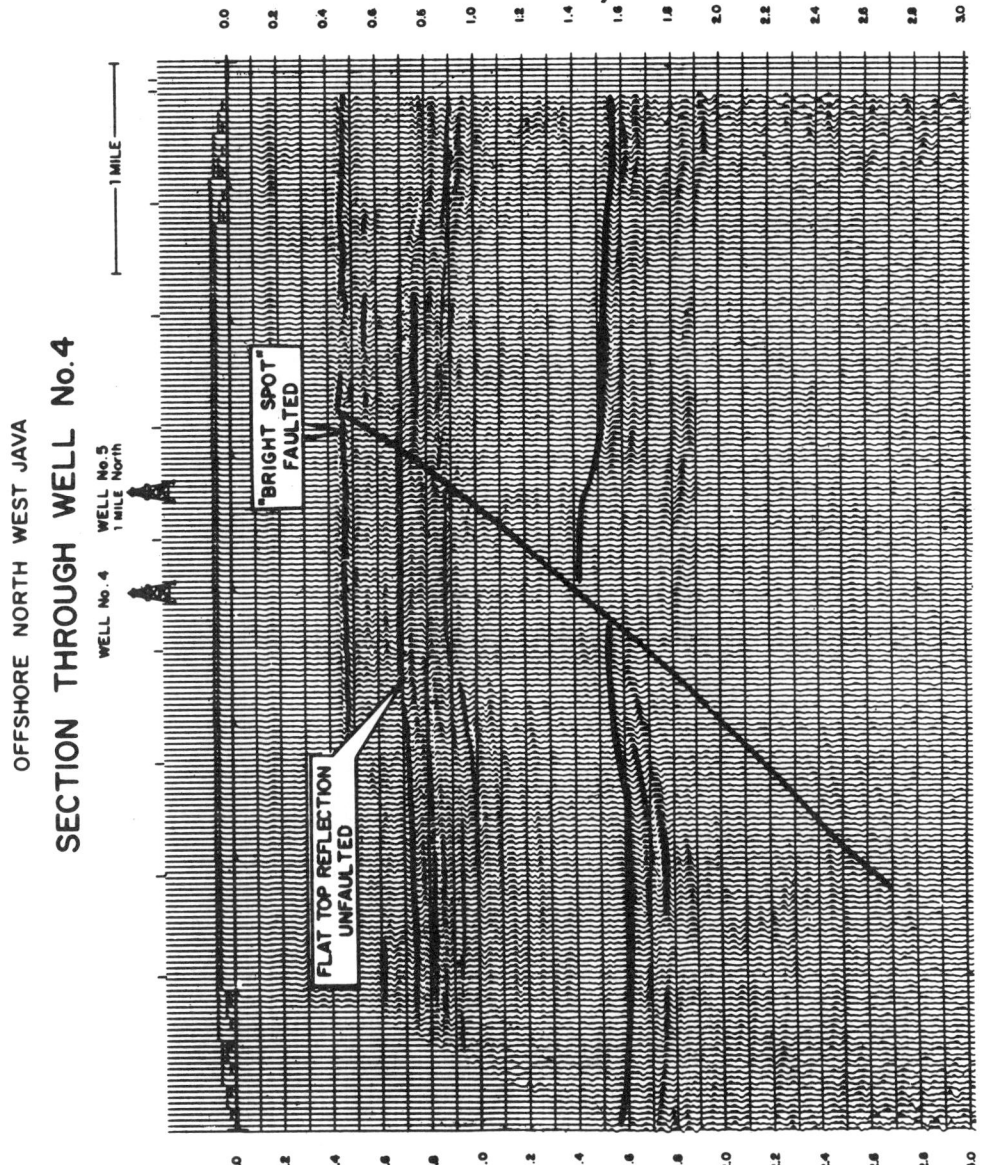

Figure 19 — Seismic Section through Well No. 4

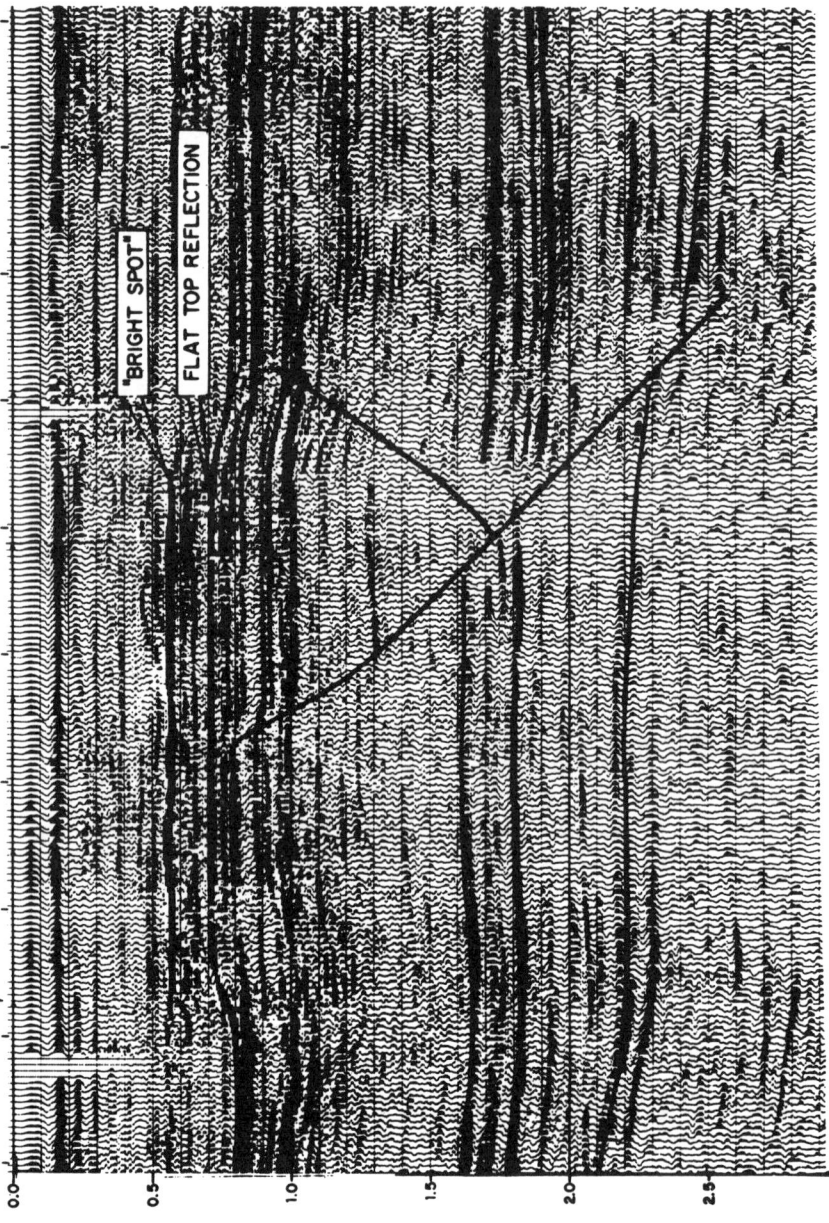

Figure 20 — Seismic Section — Parigi Build-up "B"

Volume 47 February 1982 Number 2

GEOPHYSICS

Field development with three-dimensional seismic methods in the Gulf of Thailand—A case history

C. G. Dahm[*] and R. J. Graebner[‡]

ABSTRACT

A three-dimensional (3-D) marine seismic survey was conducted in the Gulf of Thailand to aid in the development of a gas field indicated by three wildcat wells which had been located by seismic reconnaissance programs shot over a period of several years. The key to successful exploration in the area, basically a hinge line play, was a detailed understanding of the complex faulting controlling the hydrocarbon traps. Since the prospect lies 160–220 km offshore, some specialized surveying techniques were employed to achieve the required positioning accuracy. About 1280 km of seismic data were recorded at 100-m line spacing over a roughly rectangular block covering about 120 km^2. The 48-fold data were processed using a 3-D wave equation migration algorithm yielding a set of seismic traces representing the data vertically below a grid of depth points spaced at 33⅓ m by 100 m.

The results of the 3-D program showed greater fault resolution and structural delineation. The interpretation developed from a series of horizontal slices provided by the 3-D processing further improved fault resolution. Five wells, drilled on the basis of the 3-D survey, are productive and closely tie the seismic data. Initial studies of amplitude patterns of key reflectors, combined with interval velocities from seismic derived logs, appear to offer the potential of direct detection of productive gas zones thicker than 25 to 30 ft. The 3-D seismic data are being utilized for planning additional development wells and potential platform locations.

INTRODUCTION

The role of the three-dimensional (3-D) seismic systems is a fundamental one since, finally, all representations of the subsurface must be in a 3-D form, even if only in the form of a contour map in the final stage of prospect evaluation. However, when the geology is complex or resolution requirements are great, there is often no substitute for the treatment of the 3-D nature of the problem in the earliest stages of a survey, beginning with data collection. Relative to conventional practices in conducting seismic surveys, 3-D systems imply a major increase in the density of seismic data points, a density adequate to satisfy the sampling theorem in space as well as in time.

Advantages of the 3-D approach are positioning of the dipping and curvilinear subsurface reflectors in their proper vertical and horizontal locations, collapsing of diffractions from faults and other sharp discontinuities to their point of origin, and restoration of the signal strengths lost to the defocusing action of subsurface scatterers.

These advantages, coupled with the increased density of control which allows more detailed mapping of the subsurface 3-D systems, are beginning to have their greatest appeal in field development applications.

We report a 3-D field development project in the Gulf of Thailand where Texas Pacific engaged GSI to carry out a 3-D survey in August, 1977. The paper includes an examination of the results in terms of the wells drilled in the area before, during, and after the 3-D survey.

PROSPECT BACKGROUND

In late 1975, Texas Pacific Oil Company took a farmout on four blocks in the Gulf of Thailand (Figure 1), blocks 14 and 15 from Tenneco and blocks 16 and 17 from British Petroleum. Along with these farmouts came about 8500 miles of seismic data covering a number of years of data collection. Altogether there were seven different sets of seismic data collection, all of it digital, covering the years from 1967 through 1973.

Dr. C. Achalabhuti (1978), Director of the Natural Gas Organization of Thailand, at the Circum-Pacific Energy Conference in August 1978, pointed out that exploration in the Texas Pacific concession in blocks 15 and 16 had passed into a field development stage with reserves estimated at 2 to 3.4 trillion cubic feet of gas. Since that paper was given, three more successful gas wells have been drilled on the concession. In 1979, a consulting engineering firm estimated the proved and probable reserves at 5.8 trillion cubic feet.

The Gulf of Thailand is a Tertiary basin and is of tectonic origin. The thickness of the sedimentary section overlying the basement, shown in Figure 2, is expressed in terms of the two way reflection time. The area being developed is a gas area located in the extreme northwest part of the Central basin. The sediments in this part of the basin were derived at least partially from the erosion of the Narathi Wat high, a granite basement

---

Manuscript received by the Editor December 23, 1980; revised manuscript received August 3, 1981.
*Texas Pacific Oil Company, Inc., 800 Glen Lakes Tower, Lockbox 101, 9400 N. Central Expressway, Dallas, TX 75321.
‡Geophysical Service Inc., P.O. Box 225621, MS 3970, Dallas, TX 75265.
0016-8033/82/0201—149$03.00. © 1982 Society of Exploration Geophysicists. All rights reserved.

ridge, and from other similar granite ridges. The section of interest is Miocene and possibly Oligocene in age. While the principal interest is limited to the upper 8000 ft of section (about 2 sec in reflection time), there is perhaps as much as 15,000 ft (3 sec) of Tertiary and Mesozoic section overlying a strong reflector which in places tops the basement and in other places may represent the top of the Paleozoic.

A diagrammatic section between two wells in the prospect, separated by only 3 miles (Figure 3), illustrates one aspect of the exploration problem. The Miocene section of interest in these wells consists primarily of a shale section interbedded with relatively thin sand beds. The 15-B-4X well at the left encountered a total of 227 ft of sands of reservoir quality in 15 separate sands, the thickest being 35 ft. This illustration also suggests that many of the sand bodies are lenticular. It was thought that the development of the sand buildups was probably controlled by faulting and local structural highs.

When Texas Pacific became involved in the Gulf of Thailand prospect, Tenneco had already drilled three wells on block 15 at the northern end of the area of interest (Figure 4). Two of these wells had good gas shows and some oil. Then British Petroleum drilled the 16-B-1 well at the southern end of the gas area in the northern part of block 16. It also had good shows and some condensate. At this time it was concluded that the basic prospect was a hinge line play, the 15-B-1X and 2X wells being on the hinge line and the 16-B-1 some 16 miles southeast on the hinge

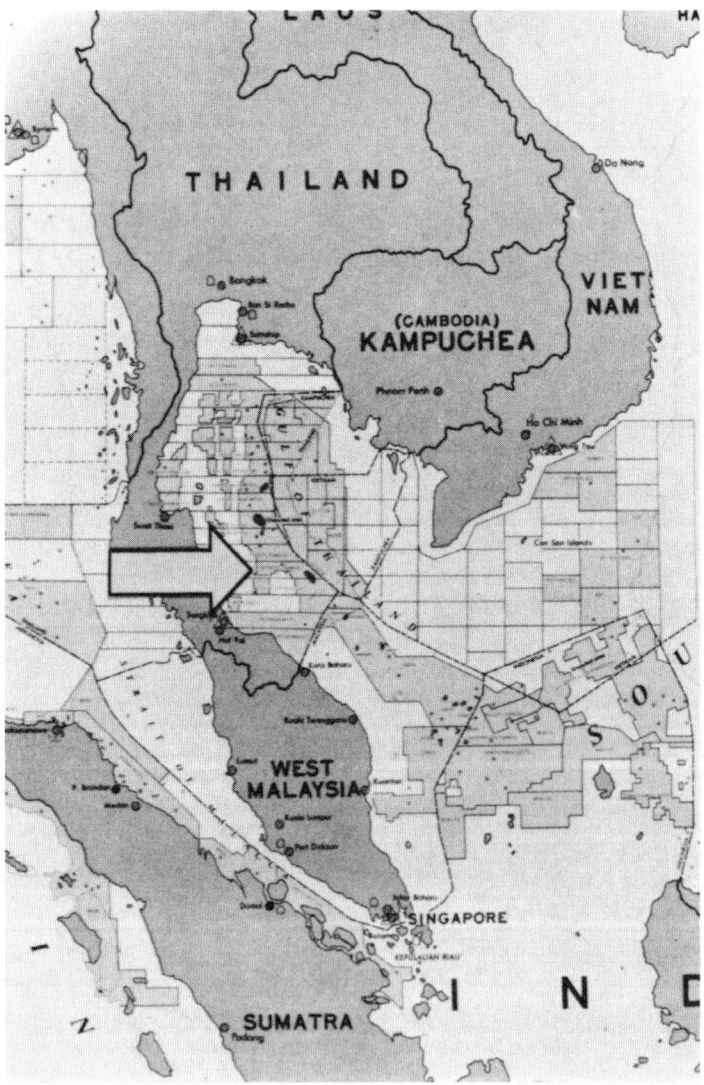

FIG. 1. Map of survey area.

line. Therefore, we concentrated our efforts along the hinge line in the area between the 15-B-2X and 16-B-1 wells. It is within this area that the 3-D survey was conducted.

Prior to drilling of the 16D-1 well, the last one drilled on the old two-dimensional (2-D) data, the limitations of the existing seismic information were realized, due both to the coarseness of the grid and to the fact that the information represented a composite of several seismic surveys distributed over several years.

New information was needed—first, to establish with confidence the relationship between blocks 15 and 16; second, to obtain better data to make new locations which would be productive; and finally, to assist in determining optimum locations for platforms.

To meet all these criteria, a sizable 3-D program was shot over the gas area. At the time the survey was conducted (summer of 1977), it was one of the largest 3-D programs in the world and the first one undertaken in the Far East.

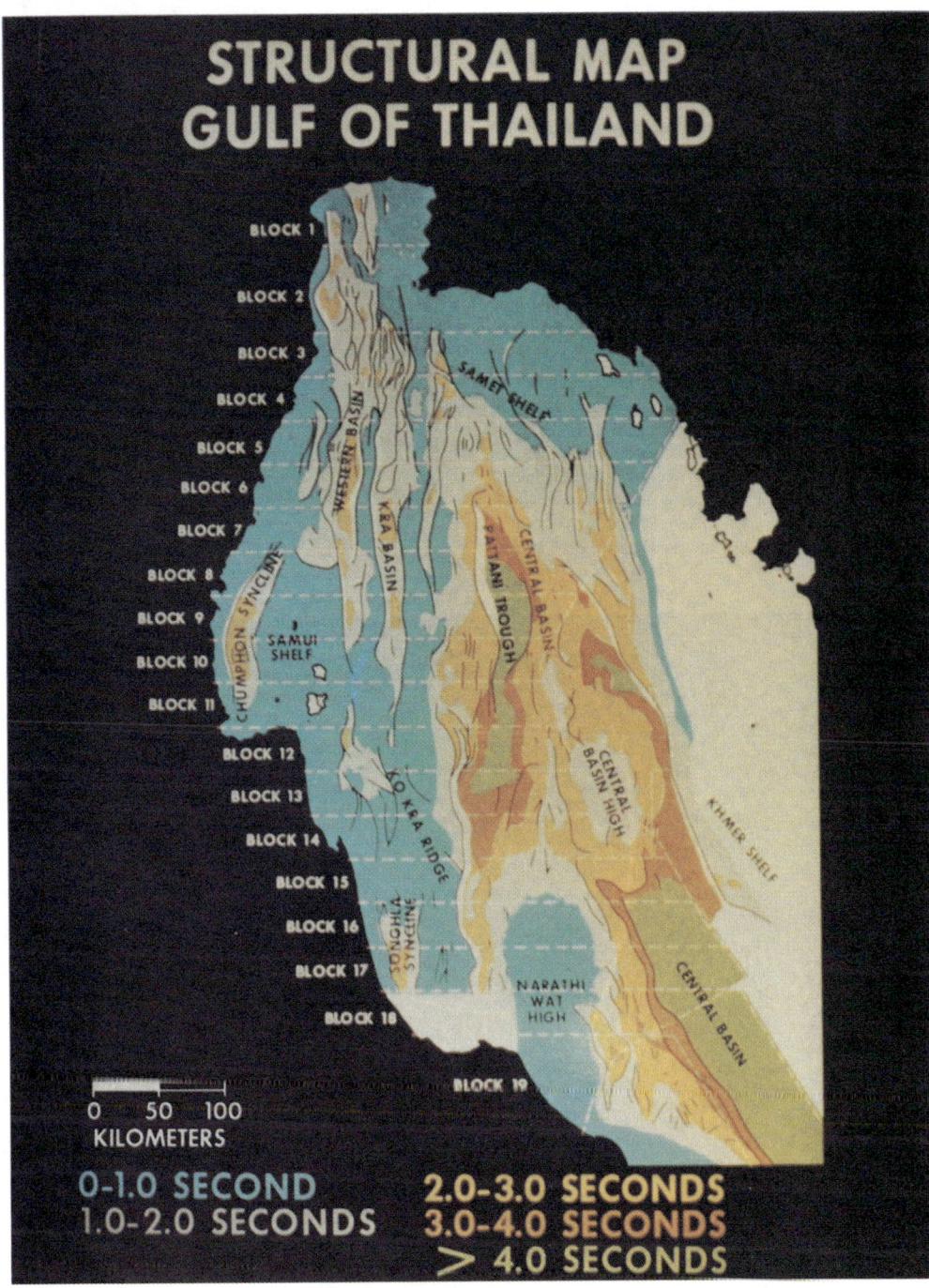

FIG. 2. Regional tectonic map of the Gulf of Thailand, with sedimentary section thickness overlying basement.

SEISMIC DATA COLLECTION AND INTERPRETATION TECHNIQUES

The 3-D program consisted of 1280 km, with the seismic lines shot at 100-m intervals, enclosing an area of 120 km^2. A total of 128 lines were shot in an east-west direction over the prospect, roughly rectangular in shape (Figure 5), and an additional ten diagonal lines running northwest were shot at the north end. The length at the top was approximately 11 km.

The shore stations used for positioning were located at distances which ranged between 160 and 200 km from the prospect. Two survey systems were employed because of these large distance ranges coupled with the unusually accurate positioning required to obtain the close line spacing required in a 3-D survey. Primary control was obtained with the Argo system, and Shoran was used simultaneously to keep track of and to verify lane count.

Data from a streamer tracking system were combined with the boat navigation data to obtain an X- and Y-coordinate for each depth point associated with every shot and seismometer group. Along each line, a 48-fold multiplicity with common-depth-point (CDP) stack was obtained at 33.3-m depth point intervals (Figure 6). This set of depth points formed the input to the 3-D migration process (Schneider, 1978). The output of this process generated a 3-D data volume, each point of which represents the amplitude of a seismic trace at an X, Y, and time coordinate. A typical east-west section (Figure 7) from this 3-D data volume and a north-south section (Figure 8), both crossing the 15-B-4X discovery well, illustrate several features characteristic of the area —a closely spaced complex mosaic of faults and the presence of several good reflectors exhibiting strongly contrasting amplitude properties.

As an aid in understanding the interpretation of the 3-D data

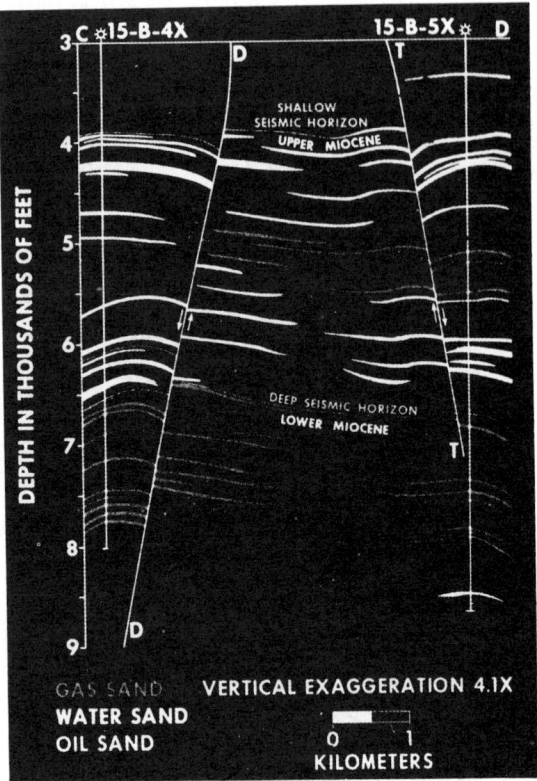

FIG. 3. Geologic cross-section (conceptual) between two wells.

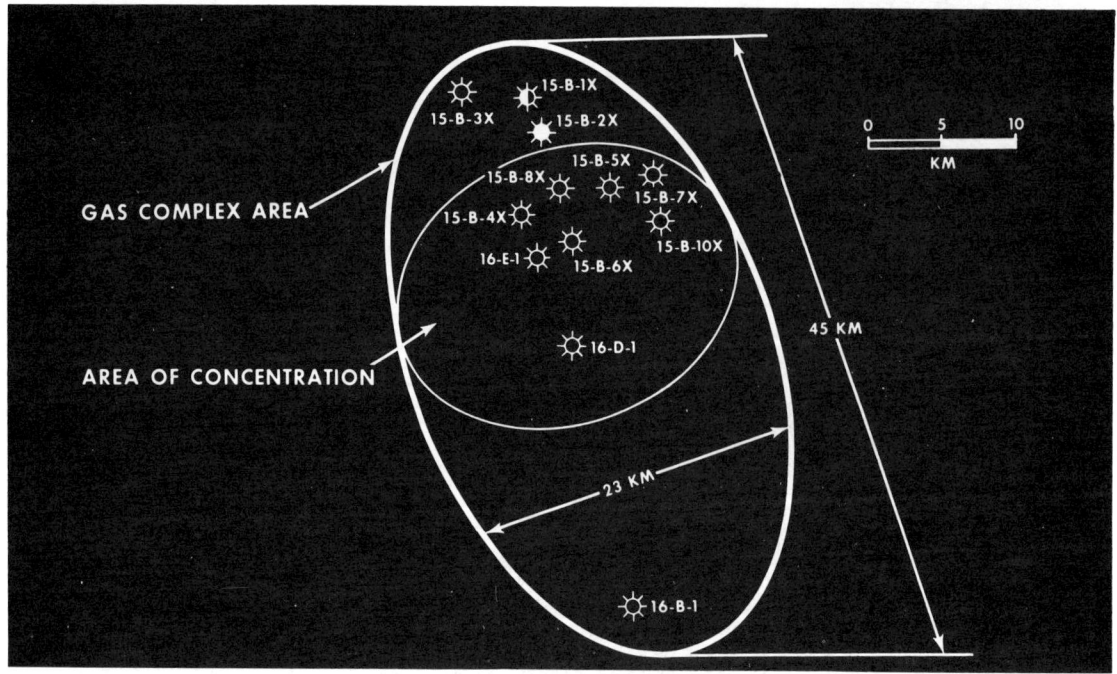

FIG. 4. Gas complex area in the Gulf of Thailand.

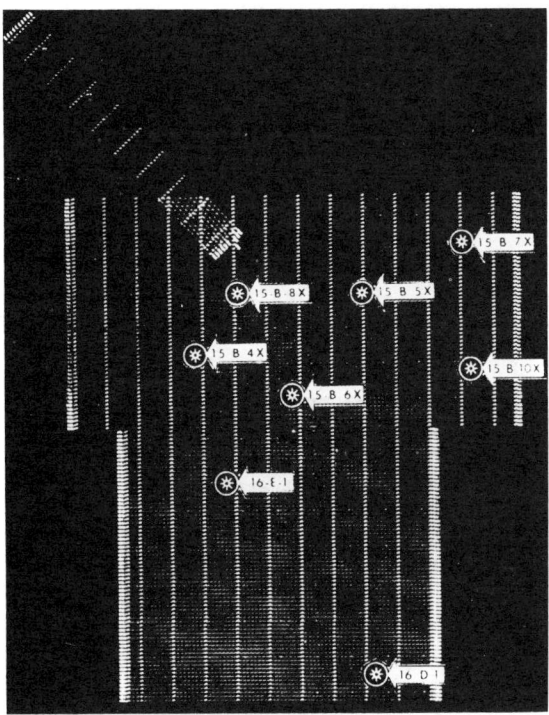

FIG. 5. 3-D depth point map.

volume, we compared results at several stages in the processing sequence. A comparison of the conventional 2-D stack stage with the 3-D migrated vertical section, whose aperture is centered at the position of the conventional stacked line (Figures 9 and 10), shows a number of differences. Observe that the fault definitions are sharpened and the reflections are somewhat better focused on the 3-D section. The conventionally stacked section is degraded by the interference of both reflections and diffractions out of the plane of the section. These "signals" appear as noise on the conventionally stacked section. This comparison is generally typical of the prospect. When one remembers that the purpose of the survey was to obtain detailed resolution on fault locations and possible sand buildups, the improved definition through better signal focusing becomes very important.

Probably a more valid way of comparing 2-D stacked data with 3-D migrated data is with a horizontal slice through the subsurface, a display which gives a view of changes in both X and Y directions (Figures 11 and 12). Both of these horizontal slices, called Seiscrop® sections (Brown, 1978), were made at a time of 1.408 sec. The black areas represent peaks and the red areas troughs. Although one cannot interpret and assess the significance of the differences at a glance, there are some obvious improvements on the Seiscrop section of the 3-D data (Figure 12).

The first noticeable characteristic is that the 3-D Seiscrop section is much more sharply focused. On the Seiscrop section made from the unmigrated data volume (Figure 11), the diffractions from the complicated faulting patterns at this level interfere with the desired signals. Further, since some of these signals

®Trademark of Geophysical Service Inc.

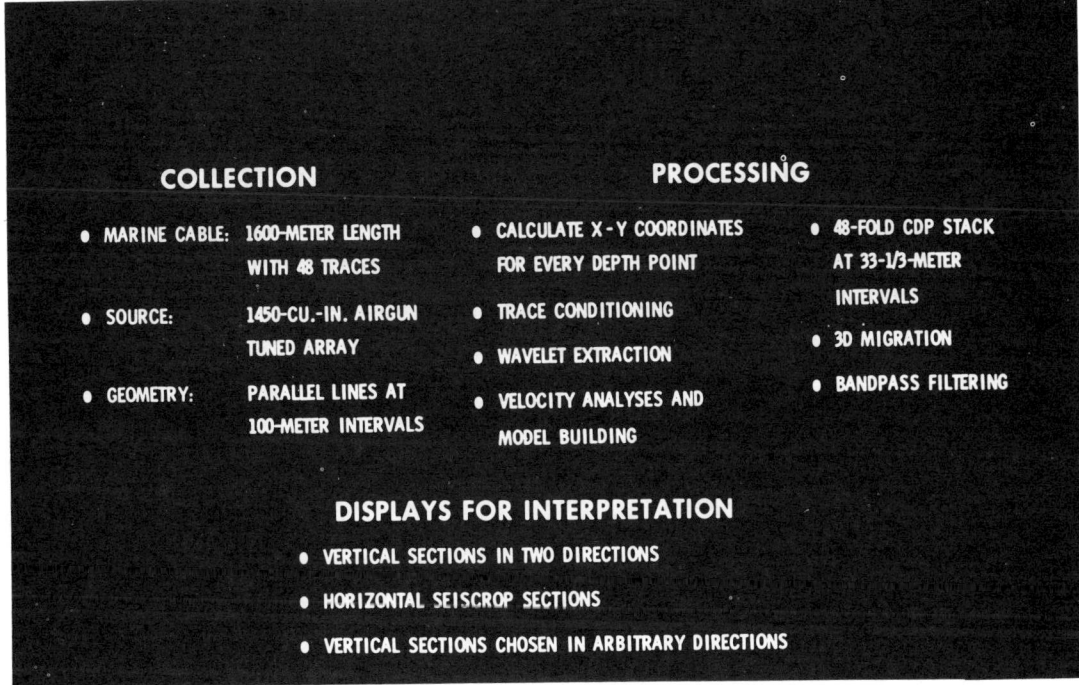

FIG. 6. 3-D system parameters.

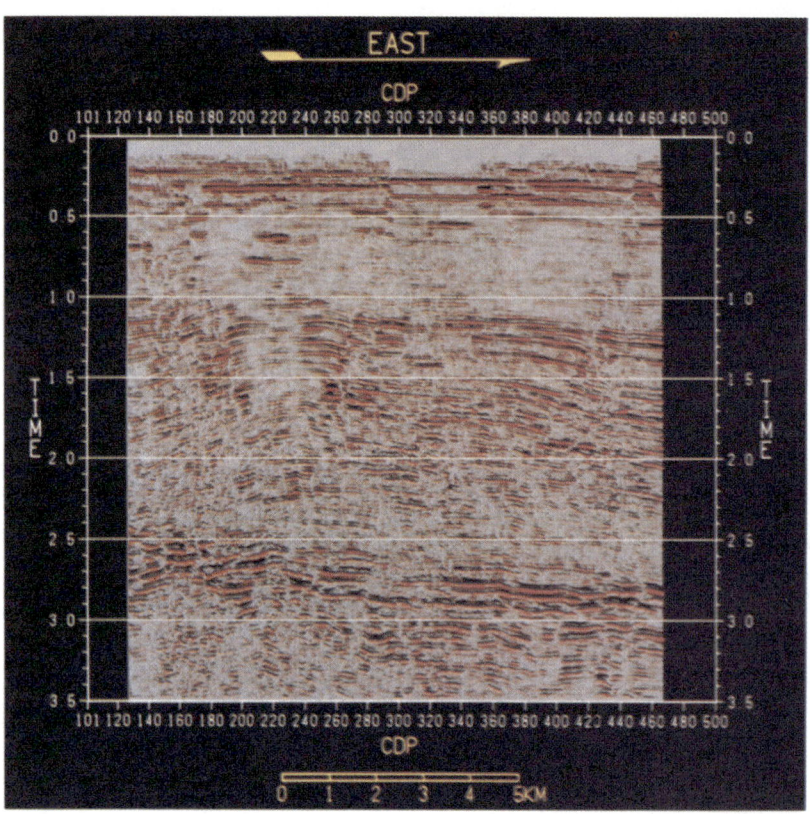

FIG. 7. 3-D migrated east-west seismic section, line 085.

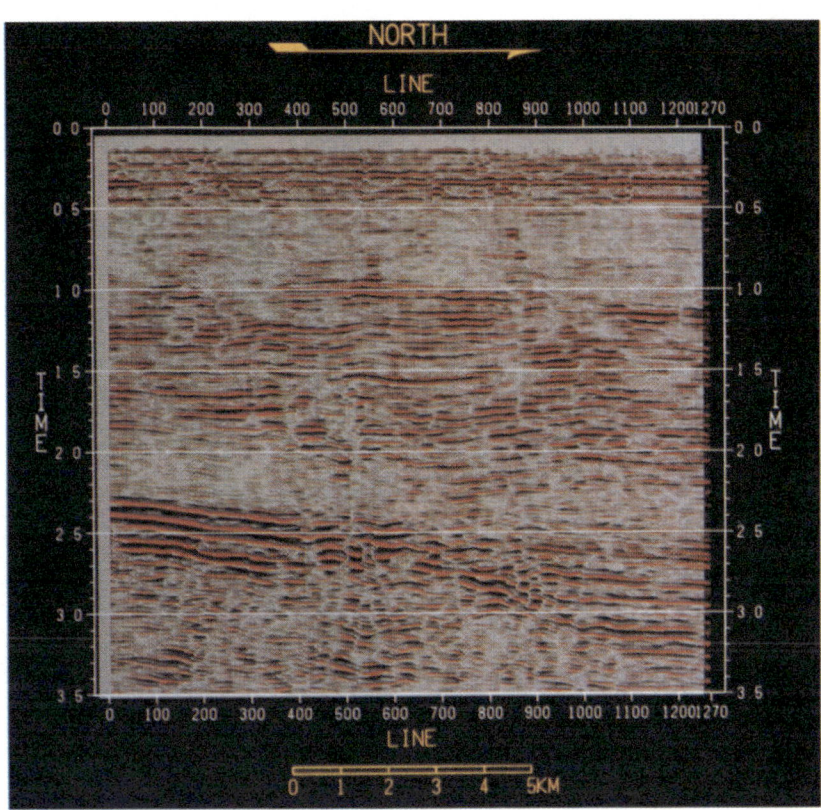

FIG. 8. 3-D migrated north-south seismic section, line 222.

appear in the wrong place, the 2-D Seiscrop section gives a somewhat blurred and distorted representation of the subsurface. In practice, these Seiscrop sections are not usually viewed statically. Rather, on a work table (Brown, 1979) from a film an interpreter can view many of them in a short time, enhancing the capability of the interpreter both to perceive small changes in reflection character and to use a large volume of seismic data efficiently.

An examination of a sequence of two consecutive Seiscrop sections (Figures 13 and 14) provides an understanding of how these sections were used in an interpretation. The coherent events, faults marked with small ticks and reflections as smooth contour lines, represent the intersection of fault planes and a number of reflecting surfaces with a horizontal plane at the indicated time. At first glance, these Seiscrop sections appear difficult to interpret because of the very complicated fault system. However, they are unusually valuable because of the fault detail they reveal when studied carefully. Following a particular reflection or fault trace on a series of Seiscrop sections made at different times, an interpreter can rapidly construct a contour map. Note how the reflection events at the upper right move outward to the right as one proceeds down 24 msec in time from 1.384 sec (Figure 13) to 1.408 sec (Figure 14).

The changes in coherent events with increasing time help distinguish reflections from fault edges. Fault planes are identified on a sequence of Seiscrop sections at the position where the pattern of reflections appears to terminate abruptly at approximately the same location on each section.

STRUCTURAL MAPS

In general, one uses the vertical 3-D sections to identify the key horizons, dips, and fault locations. Seiscrop sections yield strike, knowledge of how to connect the faults, and the spatial distribution of the geologic formations. Using one Seiscrop section, the interpreter can view the entire prospect, at one time level, on one piece of paper.

Using both the vertical and the horizontal sections from the 3-D migrated data volume, two maps were constructed (Figures 15 and 16), a shallow one near a reflection time of 1.2 sec and a deeper one near 1.8 sec. Both maps display the general shape of a more or less elongated central area striking approximately north-south with dip to the east and west from the central graben.

The shallow map represents a depth range of 3700 to 4400 ft and is based on a reflection in the upper Miocene. The deeper map is in the depth range of 6000 to 7600 ft; it is in the lower Miocene.

Maps of these same horizons were made from the 2-D data obtained from seismic surveys conducted over a period of several years prior to the 3-D survey (Figures 17 and 18). These data were collected on an average grid spacing of 2 km. Although it was suspected that the area was highly faulted, only the more prominent faults were recognized on the 2-D maps because of the coarse grid. Therefore, the 2-D map should be considered more as a form map.

The maps made from the 2-D and 3-D data both display the general shape, but in most detail aspects, the maps are quite different. Comparing the 2-D and 3-D maps (Figure 15 with 17 and Figure 16 with 18) shows that both 2-D maps exhibit excessive overall structural relief, in part due to the coarseness of the spatial sampling relative to the frequency of occurrence of faulting and to confusion in resolving the interpretation of unmigrated diffraction curves. The faults are both more numerous and differently placed on the 3-D maps. Since one of the criteria for selecting favorable drilling sites is fault location, the value

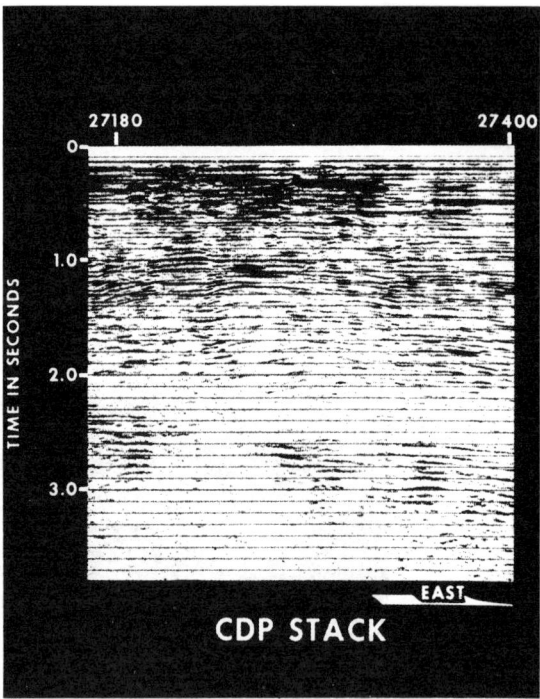

FIG. 9. Line 054, seismic section processed through common depth point stack.

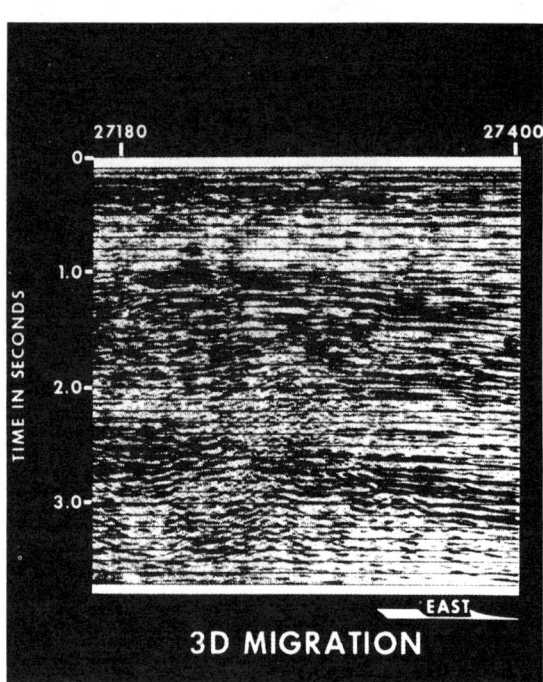

FIG 10. Line 054, 3-D migrated seismic section.

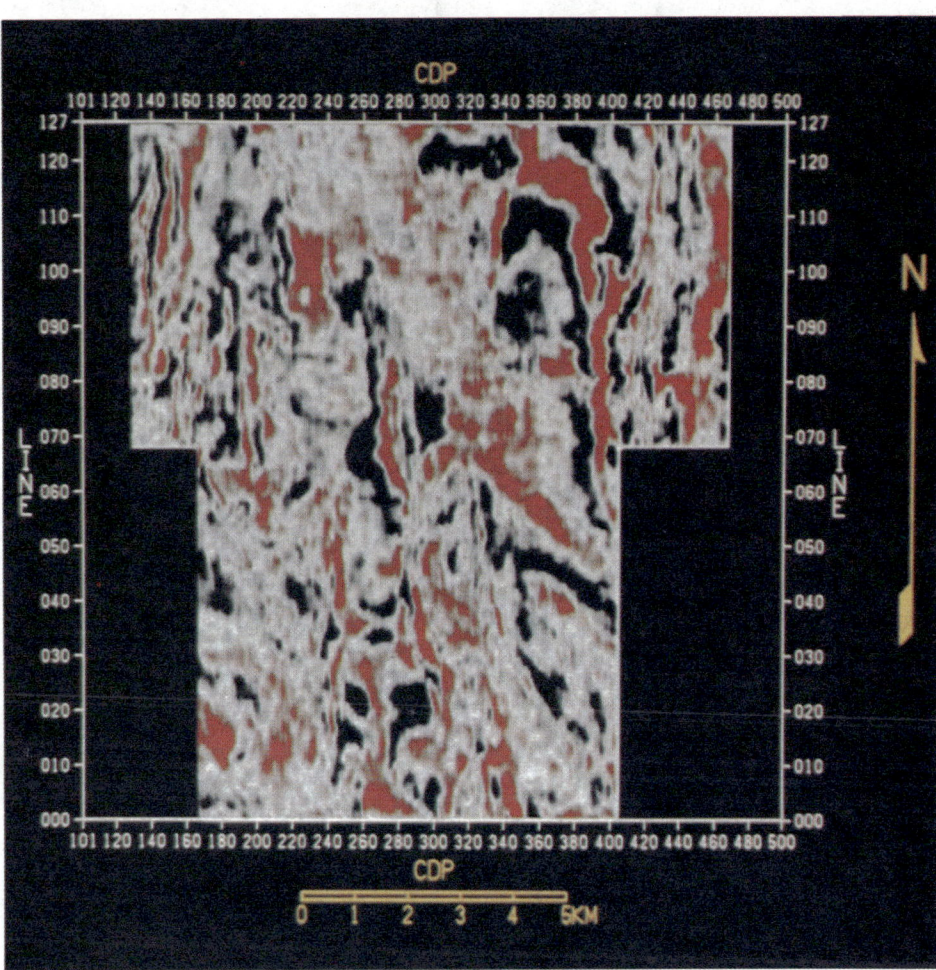

Fig. 11. Seiscrop section at 1.408 sec on the unmigrated data set.

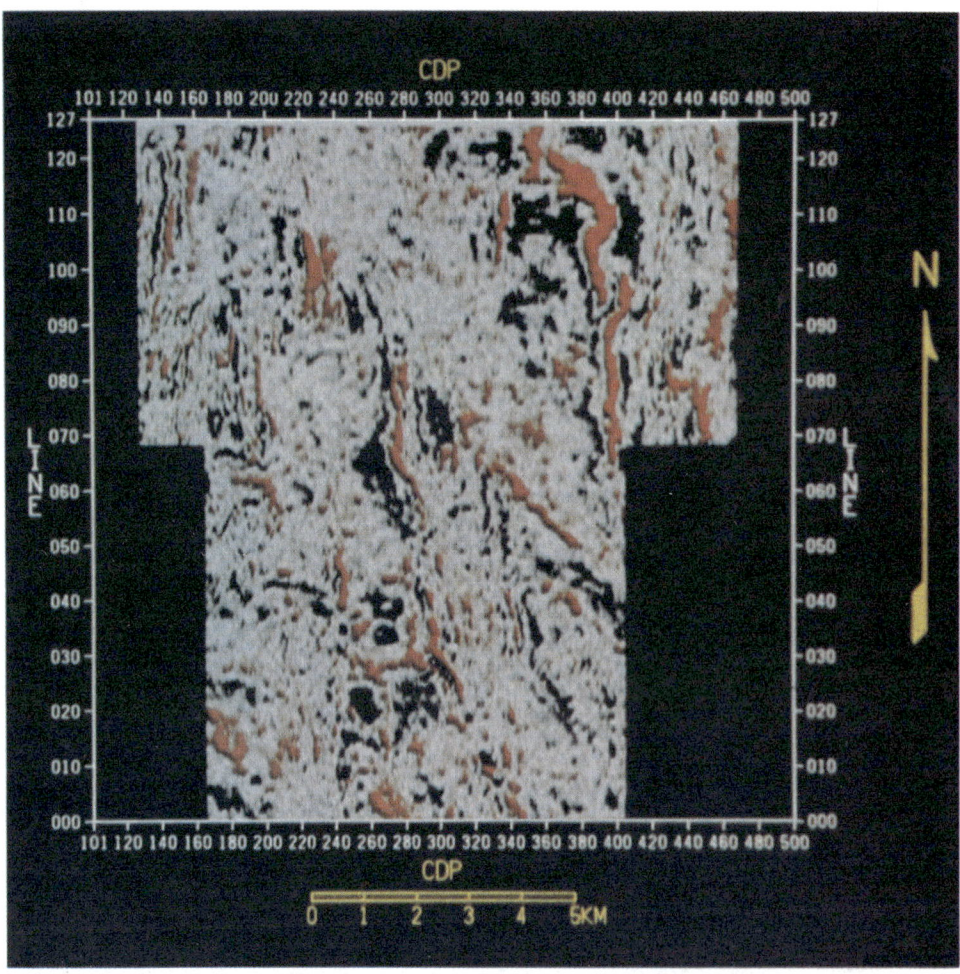

FIG. 12. Seiscrop section at 1.408 sec with 3-D migration.

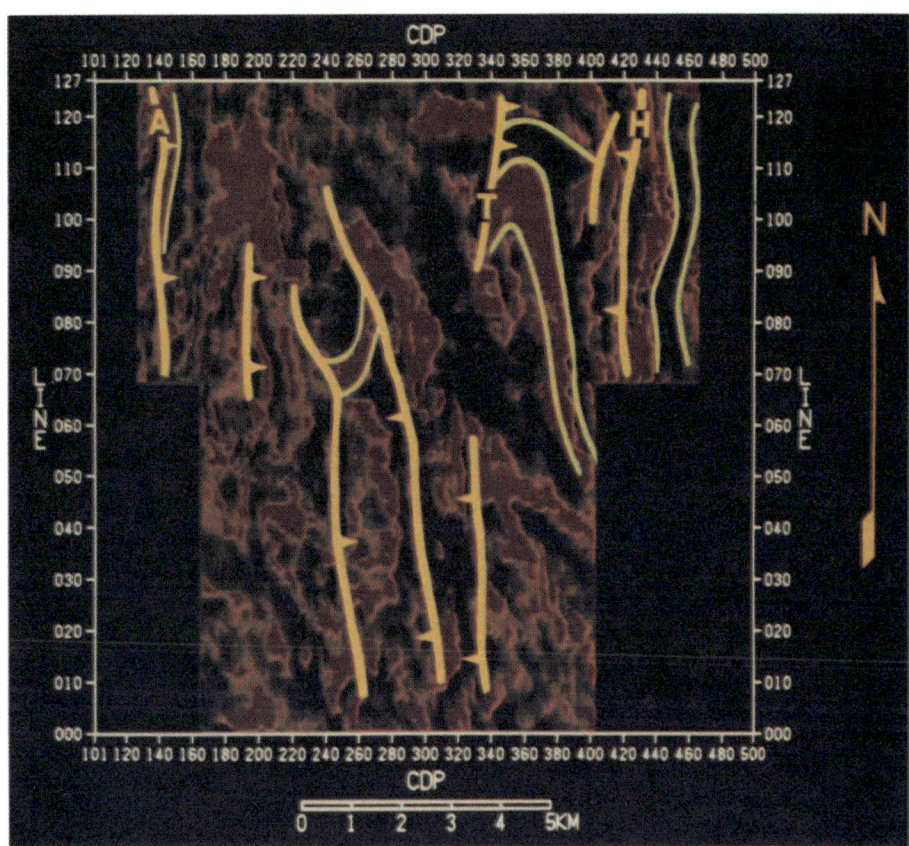

FIG. 13. Interpreted Seiscrop section at 1.384 sec after 3-D migration.

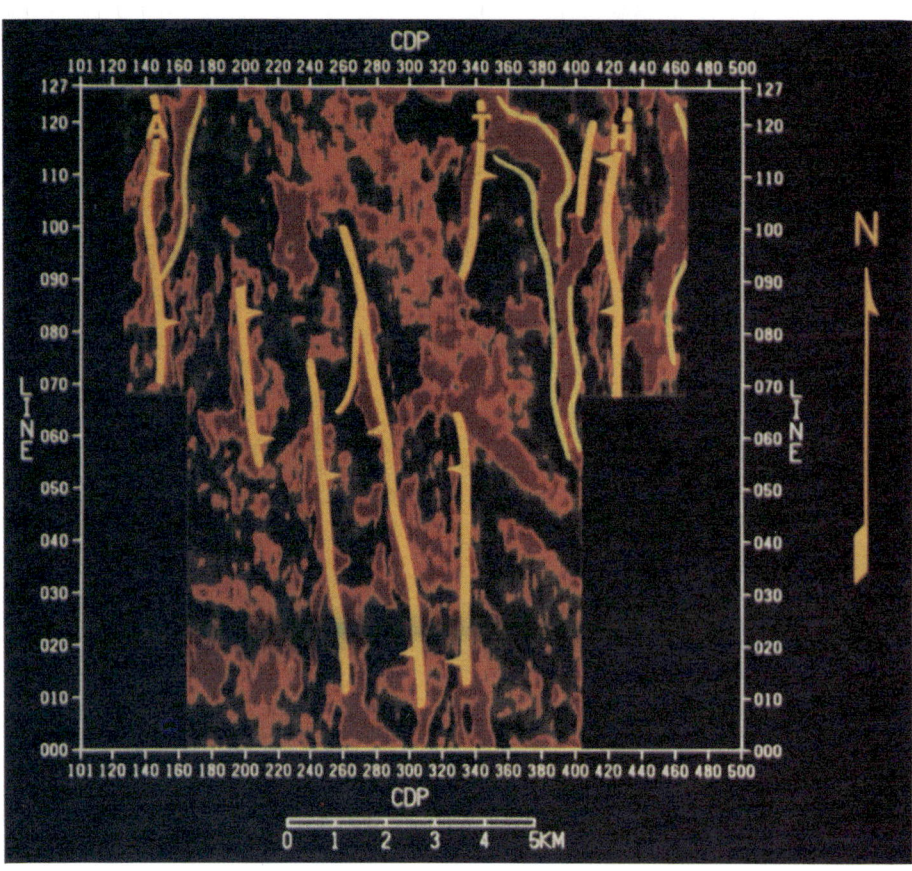

FIG. 14. Interpreted Seiscrop section at 1.408 sec after 3-D migration.

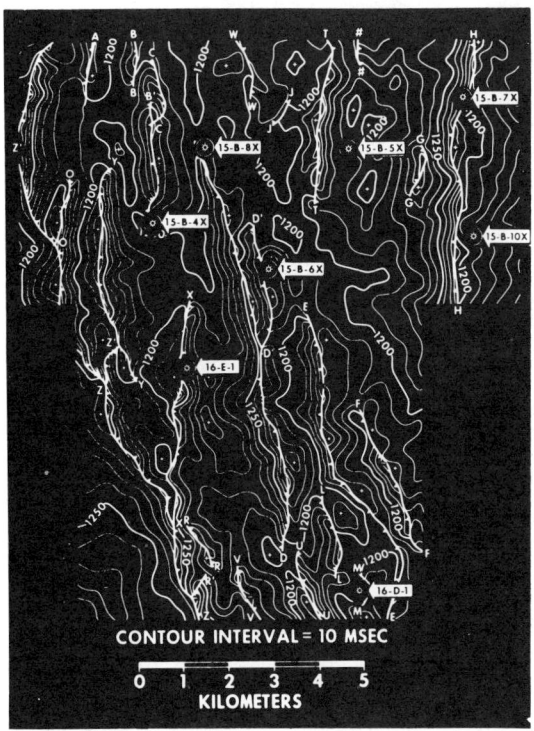

FIG. 15. Map of shallow horizon constructed from the 3-D migrated data volume.

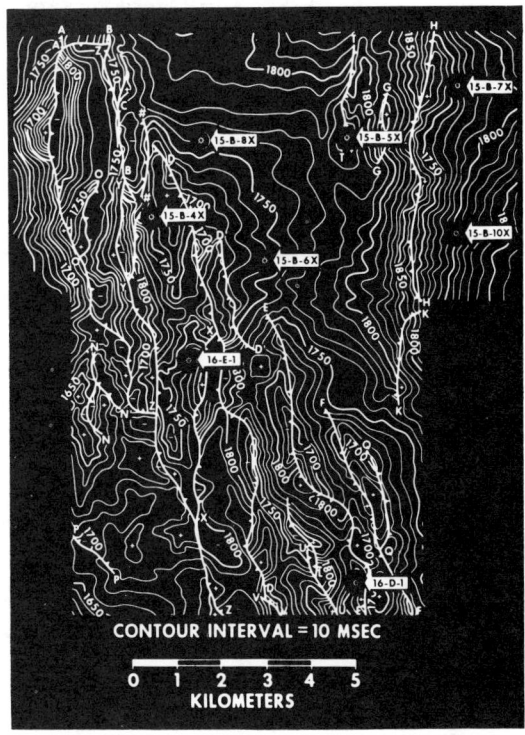

FIG. 16. Map of deep horizon constructed from the 3-D migrated data volume.

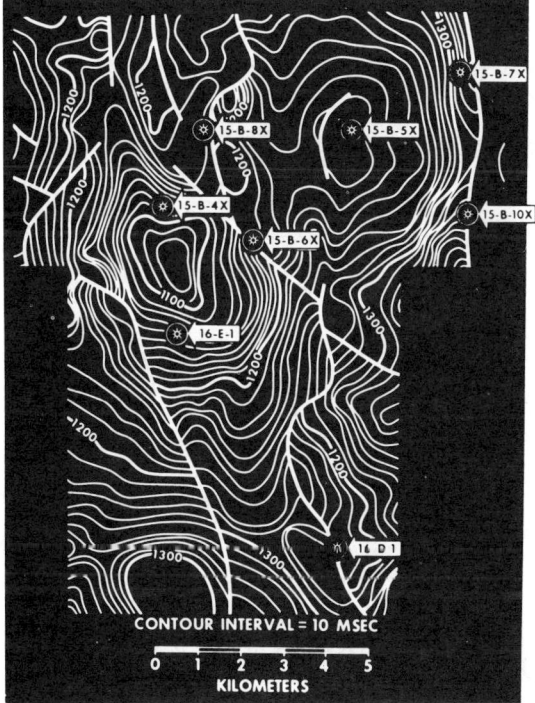

FIG. 17. Map of shallow horizon from 2-D seismic survey.

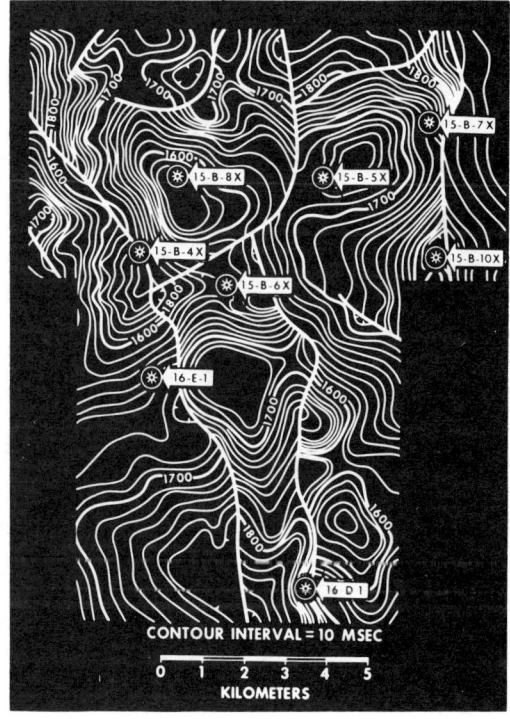

FIG. 18. Map of deep horizon from 2-D seismic survey.

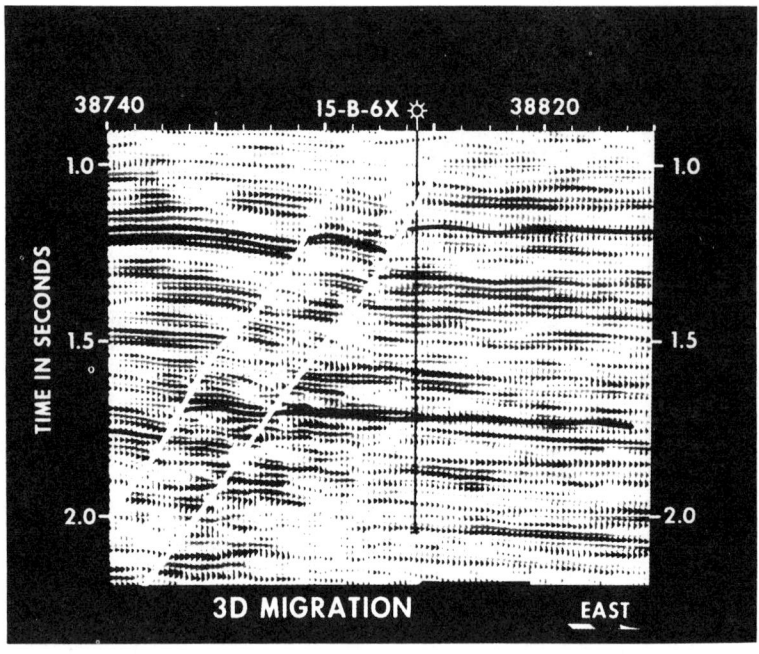

FIG. 19. East-west seismic section, line 077, at 15-B-6X well.

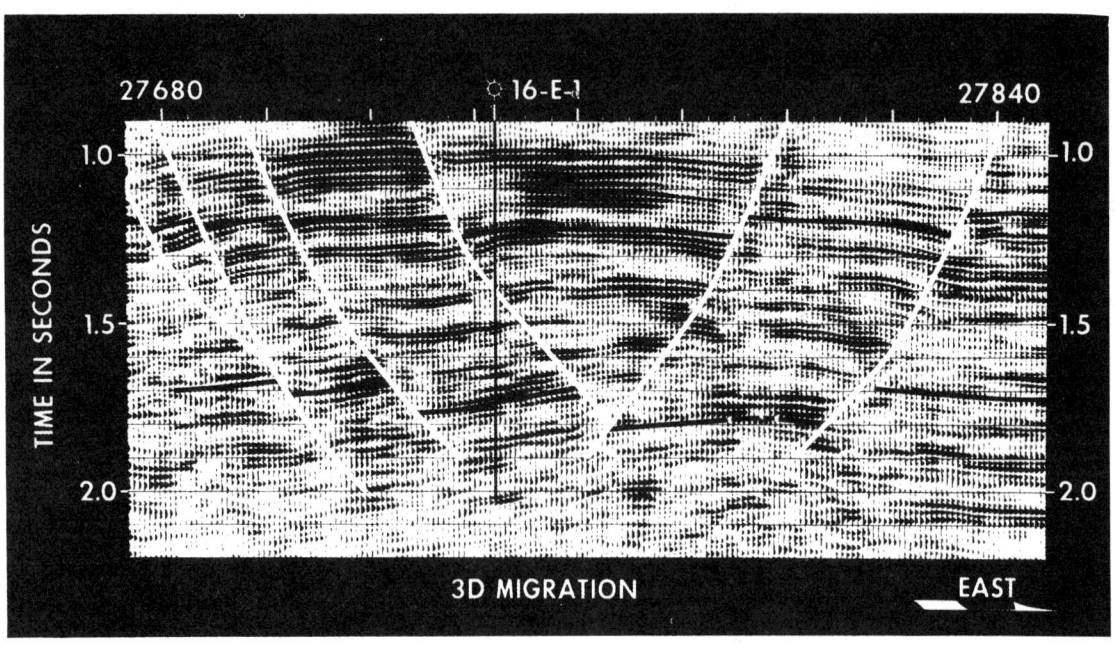

FIG. 20. East-west seismic section, line 055, at 16-E-1 well.

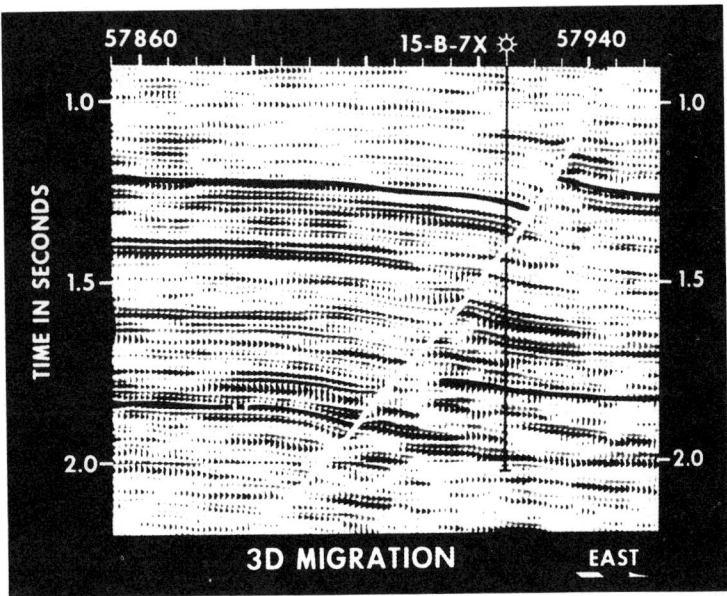

FIG. 21. East-west seismic section, line 115, at 15-B-7X well.

of the increased density coupled with 3-D migration becomes apparent.

RESULTS FROM WELLS

Five wells have been drilled on the basis of the 3-D maps. The 15-B-6X well, the first drilled (Figure 19), tested at a cumulative rate of 62 million cubic feet of gas per day and 1129 barrels of condensate from 5 separate sands, the maximum sand thickness being 57 ft.

The second well, the 16-E-1 (Figure 20), was located in the central graben where the shallow horizon is around 1.2 sec. On the deep horizon, around 1.7 sec, it was located on a closure against the fault. The well intersected the fault at 1.430 sec, at a depth within 20 ft of that predicted from the 3-D seismic data. The well tests showed 197 ft of productive gas sands and yielded a cumulative production of 55 million ft^3/day and 630 barrels of condensate per day from 6 sands. In this case, the location was chosen with the objective of drilling through the fault plane at such a position to have a full suite of the sands under the fault plane on the upthrown block, yet not too far downdip on the deeper beds.

The third well, the 15-B-7X, tested for 35 million ft^3 of gas and 800 barrels of condensate from five zones. This well was drilled with the objective of going through the fault plane at about 4750 ft (Figure 21). It actually penetrated the fault plane within 20 ft of this depth. One of the objectives of the well was to test what looked like a possible "bright spot" in the vicinity of 1.6 sec at the well. An 82 ft sand was encountered which tested for a daily rate of 12.4 million ft^3 of gas and 304 barrels of condensate per day at depths ranging between 5710 and 5792 ft.

A fourth well located on the basis of the 3-D survey, the 15-B-8X, was finished in late November of 1978. This well was drilled on line TN-135 at a point 2 miles west of the 15-B-5X well (Figure 22). It is not fault related. It is slightly updip from the 15-B-5X well and slightly downdip from the 15-B-4X well. The 15-B-8X well encountered 39 sands 10 ft, or more, thickness. The well was not tested, but of these 39 sands, there were 7 sands which indicated hydrocarbon production based on the resistivity log calculations. Excluding several sands less than 10 ft thick that may have been productive, the 7 hydrocarbon-bearing sands totaled 217 ft.

The 15-B-10X well, completed in March 1980, was drilled in the same fault block as the 15-B-7X but much farther to the south. The well encountered two productive sands. The basic objective was to see how far downdip the productive sands in that part of the section between the shallow and deep horizons could be traced. The well was also located to penetrate a shallow feature, 0.900 sec in time and 2580 ft in depth, which appeared from the seismic section (Figure 23) to be an elongated sand buildup or sandbar on the upthrown side of the H fault. The areal extent of this feature can be delineated from just a few Seiscrop sections, one of which is shown in Figure 24. Here the feature shows up as a discrete but small red event (trough). Using data from both the horizontal and vertical seismic sections, this feature was mapped (Figure 25) as an elongated sand body of three segments stretching 3.8 km along the upside of the H fault. This sand body, which may be a sand-filled stream bed, is on an average only 300 m wide and 18 ft thick. The sand tested at 5.7 million ft^3 of gas per day. The second productive zone is a 27 ft gas sand encountered at 6254 ft. This sand tested at 9.9 million ft^3 of gas per day and 500 barrels of condensate per day. This particular sand was not present in the 7X well.

SHALLOW SEISCROP SECTIONS

Somewhat peripheral to the main issue of gas exploration, but nontheless of considerable interest, are the Seiscrop sections

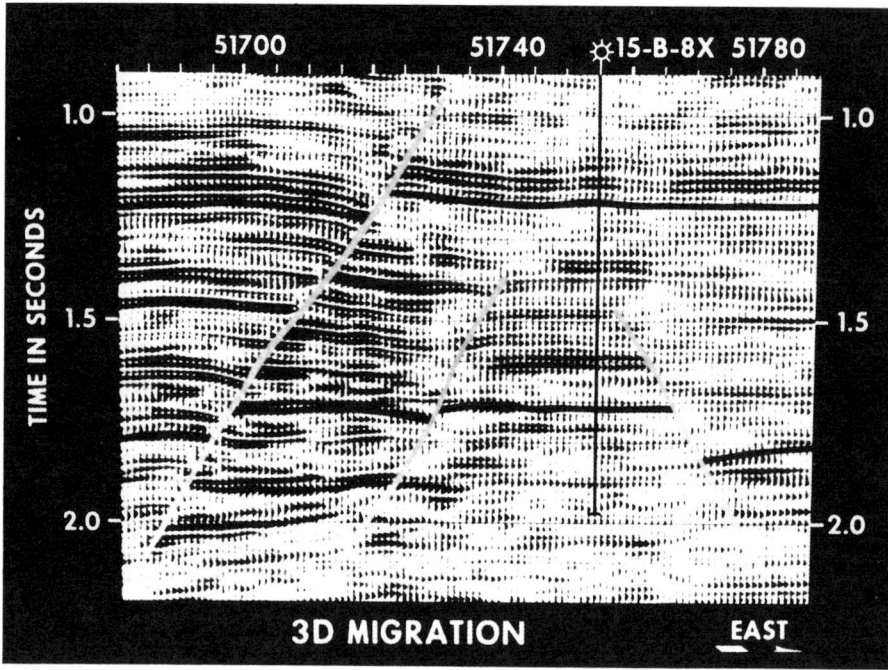

FIG. 22. East-west seismic section, line 135, at 15-B-8X well.

through the very shallow part of the section (Figures 26 and 27). The Seiscrop section at 0.200 sec (Figure 26), representing a depth of around 500 ft, shows both a meandering channel and, in the northeast part, the trace of the H fault as it comes to the surface. These features can also be seen on sparker profile (Figure 28) along a 1.9-km line crossing the well. At 500 ft, the H fault has a displacement of 20 ft. The Seiscrop section at .240 sec, corresponding to a depth of 600 ft, shows a marked change in the configuration of the channel.

These channel effects, after they were first noticed on the Seiscrops sections, can be seen on the vertical sections crossing it. Looking at the vertical sections only, these phenomena were first thought to be caused by errors in processing.

Shallow channeling as shown here could be very useful in searching for shallow tin or other mineral deposits associated with buried stream channels.

AMPLITUDE STUDIES

In the shallow part of the section from which these two examples were taken, the formations are essentially flat. The H fault, which comes to the surface, is the exception rather than the rule. The Seiscrop sections therefore represent amplitudes along the reflecting horizon or bedding plane, and the channels stand out as prominent amplitude anomalies.

The observation that key reflections exhibiting a wide range of amplitudes were present throughout the section in the time range of the productive zones, between 0.9 and 2.0 sec (Figure 20), brought up the question as to whether these amplitudes showed any systematic patterns similar, for example, to these shallow channels.

In continuing to search for additional ways to relate potential gas zones to the seismic data, amplitude maps were constructed for several reflectors. Utilizing a computer to pick and track specified events, the time amplitude pairs were contoured, with some smoothing, to produce a map displaying the amplitudes over a 10 to 1 range of values. The high-amplitude patterns observed on the shallow horizon (Figure 29) do not correlate with the structurally high locations (Figure 15), although the northwest-southeast high-amplitude trends through the data set do appear to relate to the same general structural trend.

A more precise study of the pattern of amplitudes was undertaken by constructing an amplitude map (Figure 30) on the very strong reflector penetrated by the 15-B-X well near 1.6 sec. This map, located in the northeast part of the survey, is bounded by the H fault on the west and the limit of the survey on the east. The map was made by slicing the 3-D data volume along the reflecting horizon instead of along a constant time level as shown in Figures 13 and 14. This type of section, used in several ways as a tool in stratigraphic interpretation, has been called a horizon Seiscrop section (Brown et al, 1980). The horizon chosen for the map was the trough (red waveform) following the onset of the reflection from the top of the 82-ft gas sand (Figure 31). This onset is represented by a black peak corresponding to the negative pulse generated at 1.6 sec at the well location shown in Figure 32. Amplitude levels are portrayed by the intensity of the color, high amplitudes being represented by the range of dark pink colors and the lower amplitude by the light pink and white areas.

Although the acoustic impedance contrast is large both at the top and bottom of the sand, the resolution is not adequate to obtain the base of the sand as a discrete reflector. Thus, the dis-

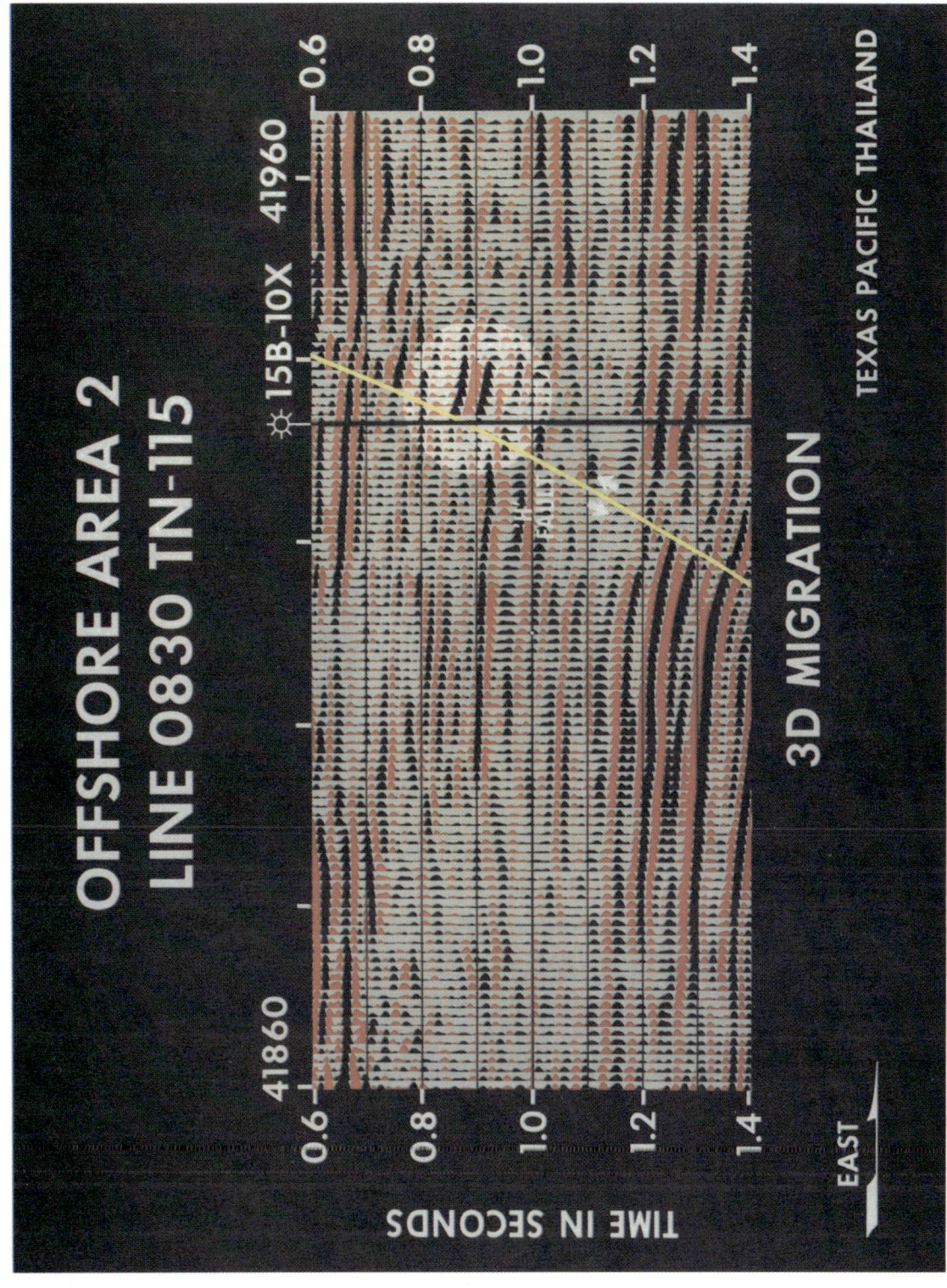

FIG. 23. East-west seismic section, line 083, at the 15-B-10X well.

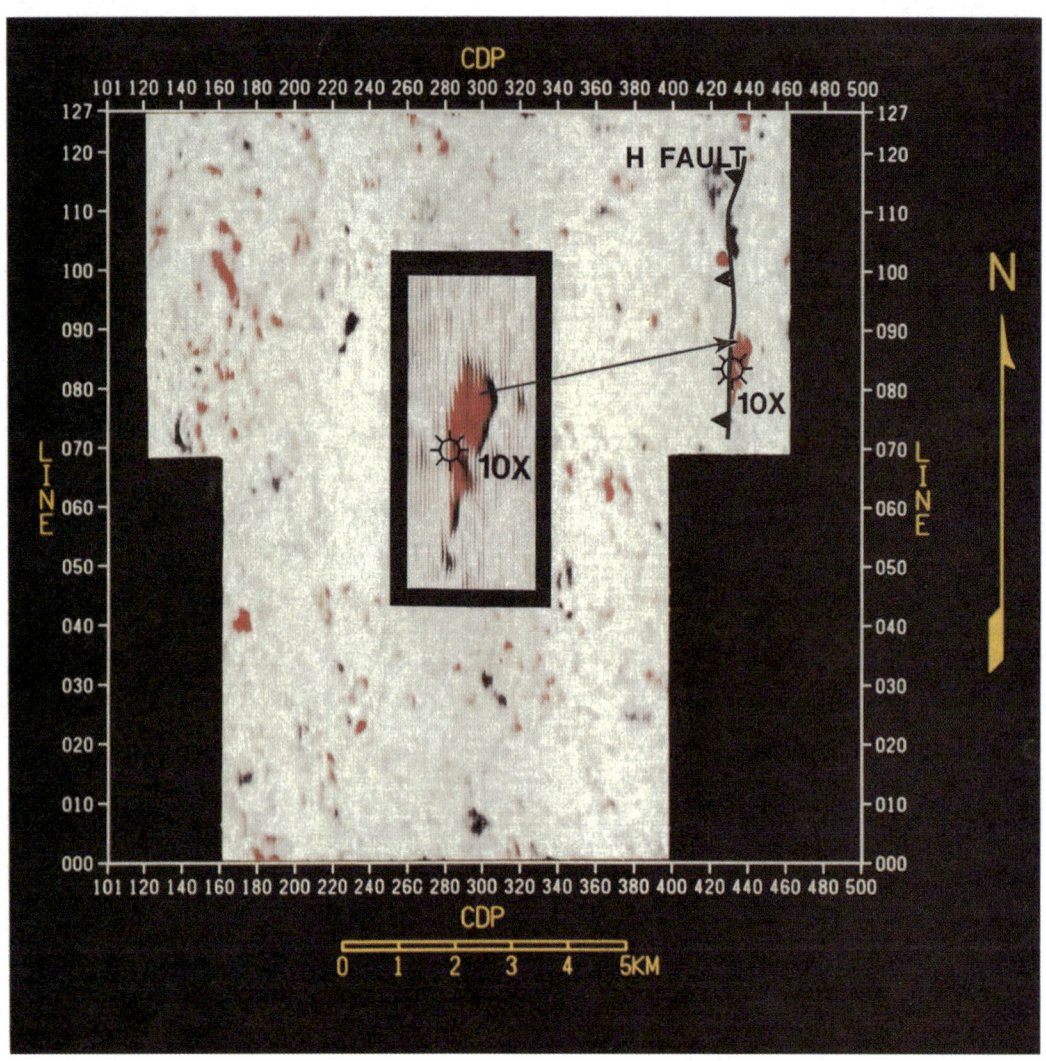

FIG. 24. Seiscrop section at 0.892 sec showing 18 ft sand east of "H" fault.

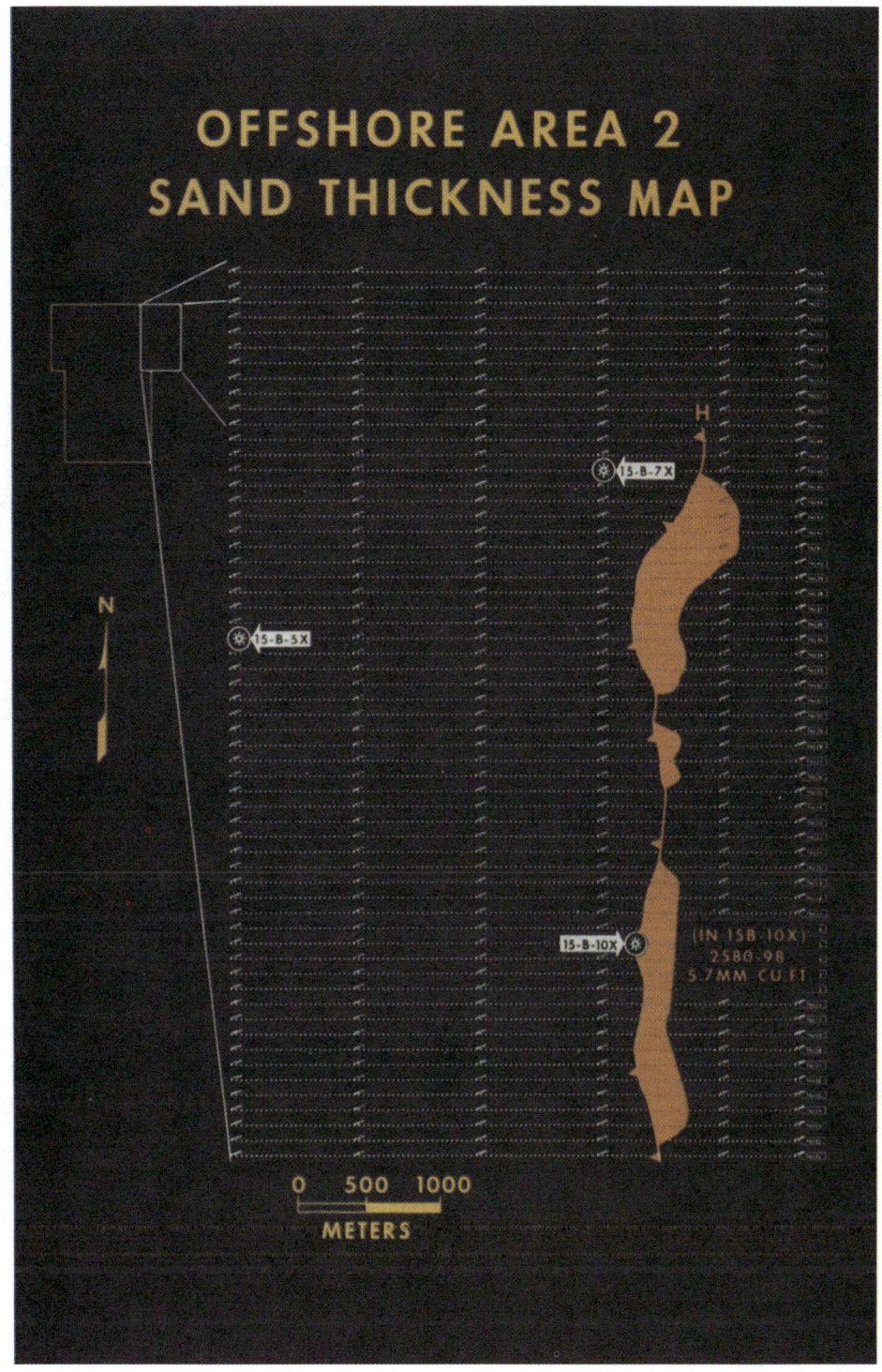

FIG. 25. Map showing areal extent of 18 ft sand encountered in the 15-B-10X well.

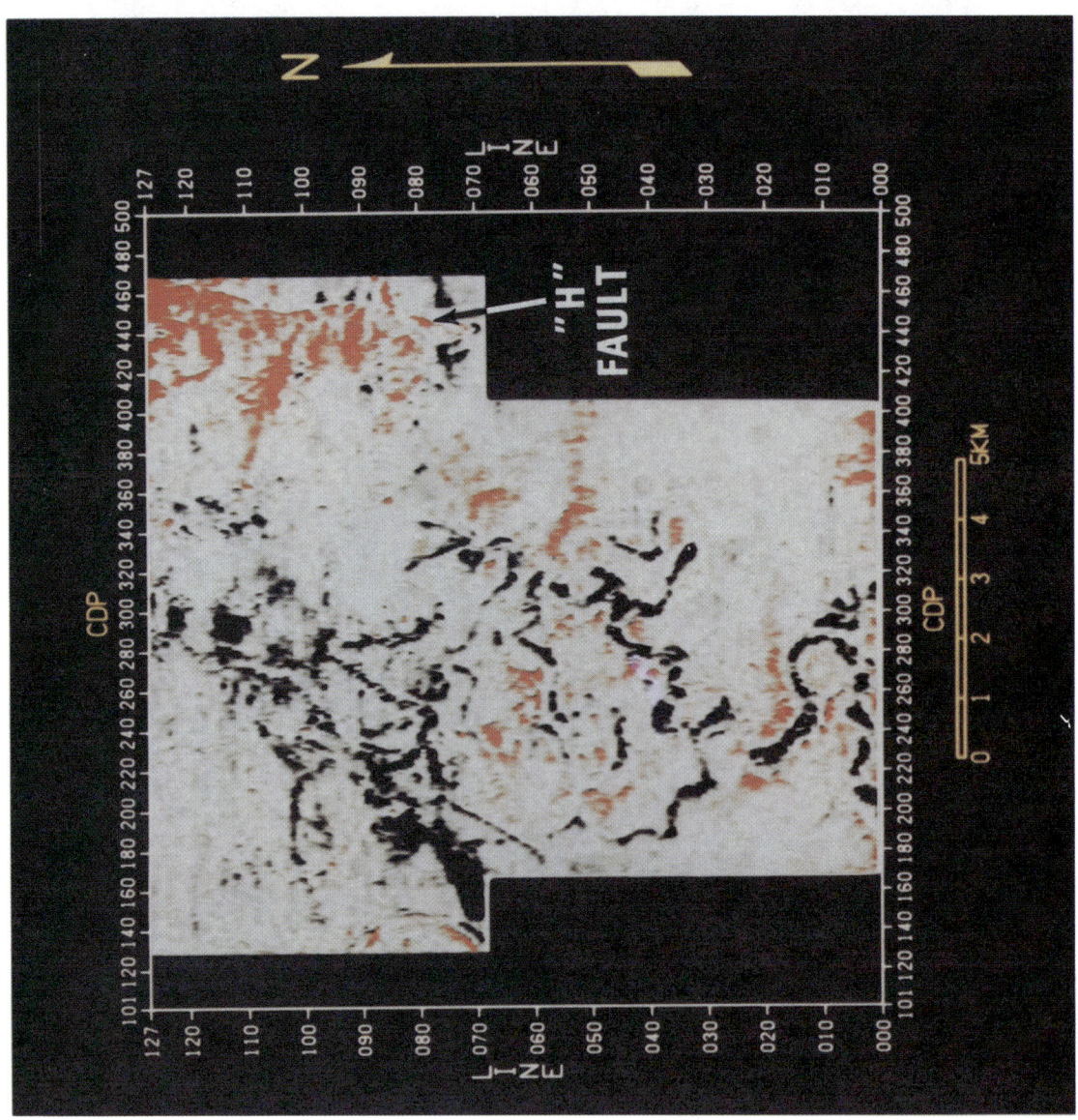

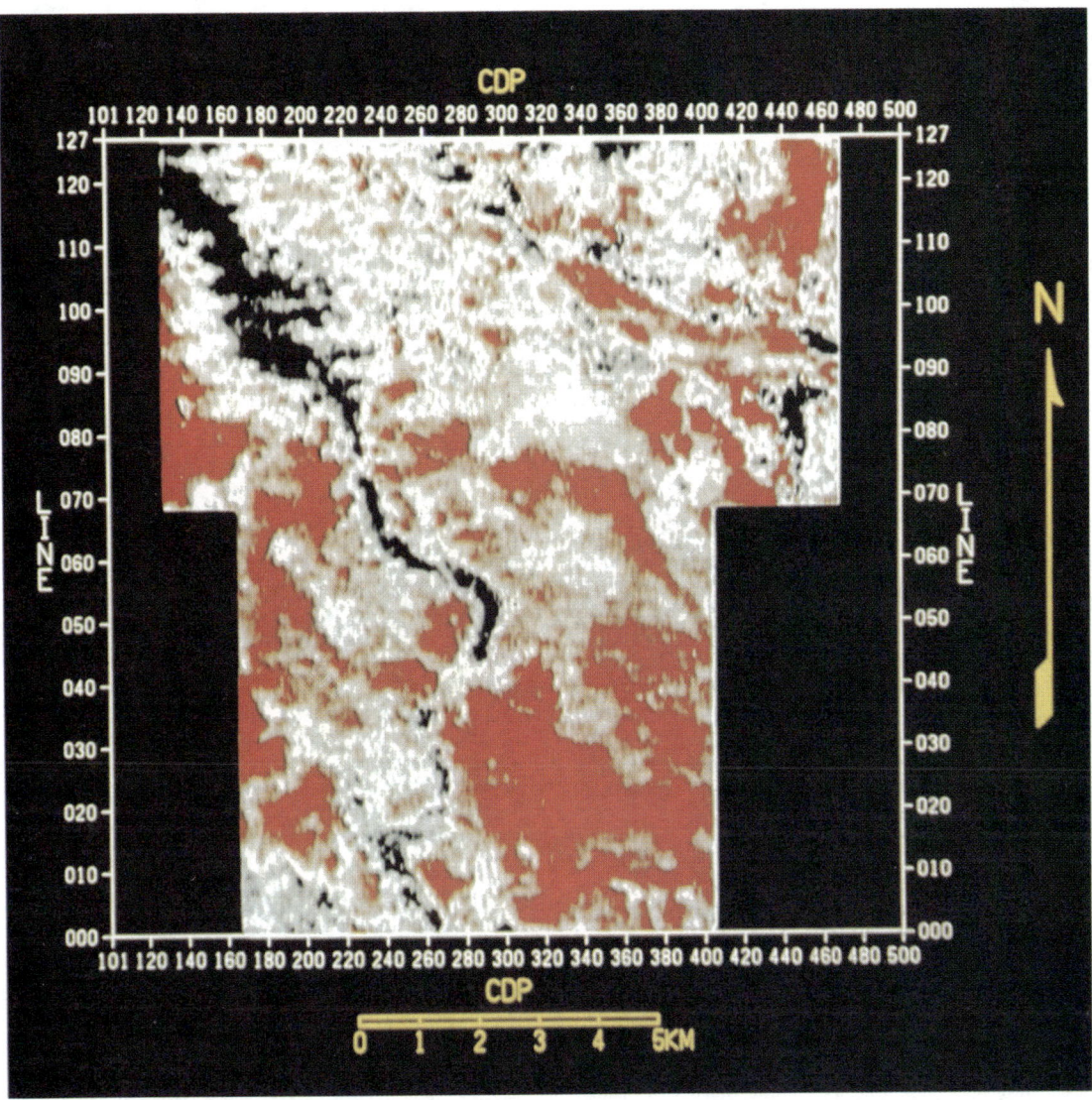

FIG. 27. Seiscrop section at 0.240 sec.

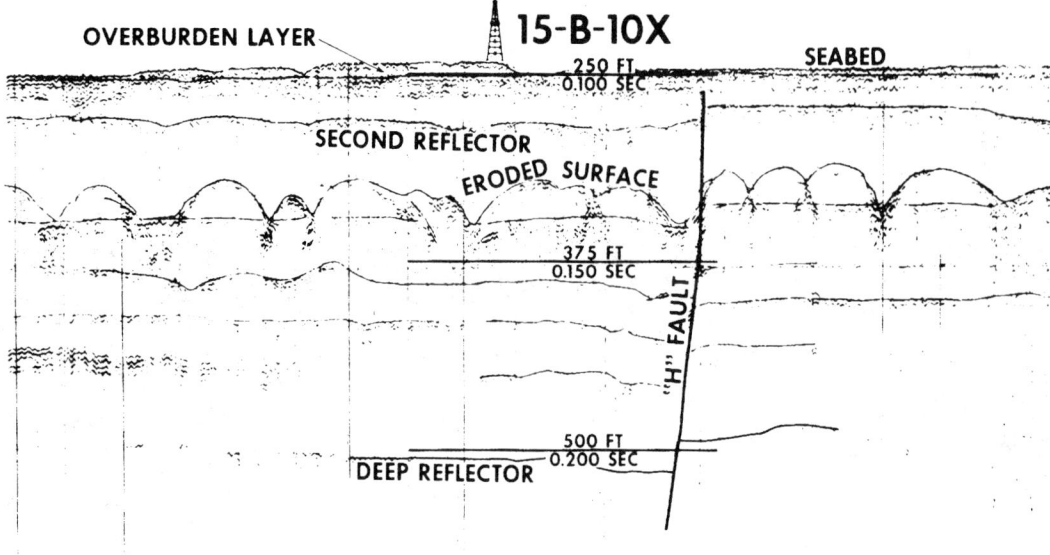

FIG. 28. East-west sparker profile through the 15-B-10X well in shallow part of section.

tinguishing feature of this sand appears to be an initial peak followed by two strong additional cycles produced by the tuning effect of the sand interfaces.

The amplitude map shows that the 15-B-7X well was drilled in the middle of an amplitude high. This well tested at 12.4 million ft^3 of gas in the time zone in the neighborhood of the reflector. This sand (Figure 31) was one of the thickest and most productive zones discovered in any of the wells. The map also shows that the 15-B-10X well, 3.1 km to the south of the 7X, was located in an area of generally low amplitude. A contour map (Figure 33) constructed on a reflector at this time interval shows no apparent correlation between the regions of high amplitudes and the structural configuration.

The observation that the gas sand discovered just above 1.6 sec in the 7X well was not encountered in the 10X well suggests the amplitude pattern displayed on the horizon Seiscrop section may delineate the outlines of this particular sand.

To relate the seismic information more directly to the well data and to understand the significance of the amplitude contrasts, a seismic log profile was constructed from the seismic data along a north-south line through the 7X well (Figure 34). These seismic derived logs represent band-limited acoustic impedance profiles and are based on a rigorous inversion of the scalar wave equation utilizing a least-mean-square technique for noise protection and a nonrecursive algorithm to avoid downward error propagation (Hays et al, 1979). On this display, the high-frequency interval velocity variations are plotted as vertical traces. The colors portray the interval velocities grouped in 500 ft/sec color-coded bins.

The seismic log profile shows a large velocity inversion in the neighborhood of the 7X well just above 1.6 sec. The inversion is displayed in the zone highlighted around 1.6 sec in Figure 34 by the area enclosing the velocities ranging between 8250 and 8750 ft/sec. This inversion correlates both with the high-amplitude reflecting sequence near 1.6 sec on the seismic section (Figure 35) and the 82 ft sand measured by the well logs (Figure 31).

This set of observations indicates that there may be a good possibility of extending the interpretation to direct gas detection from a study of amplitude patterns and seismic derived interval velocity logs.

Our present view is that the quantitative analysis of amplitude patterns promises to be a valuable technique in understanding the stratigraphic aspects of the development of the gas reserves. Continuing effort is being directed along these lines, particularly in the use of Seiscrop sections constructed along the bedding planes coupled with seismic velocity logs.

CONCLUSIONS

All five wells based on the 3-D survey have been successful and have upheld the seismic interpretation. The 16E-1, the 15B-7X, and the 15-B-10X wells intersected the fault planes within 25 ft of the depth predicted from the 3-D survey.

Several new drillable prospects have been identified. It is believed that the wells can now be located to optimize the development of the reservoir sands. Probably more important, the 3-D data set generated by the survey appears to be accurate enough (1) to locate the fault blocks defining intrafield producing boundaries, (2) to select the most favorable sites for platform locations, and (3) to assist in developing quantitative reserve estimates.

ACKNOWLEDGMENTS

The authors acknowledge with thanks permission of the managements of Texas Pacific Oil Company, Inc., Canadian Superior

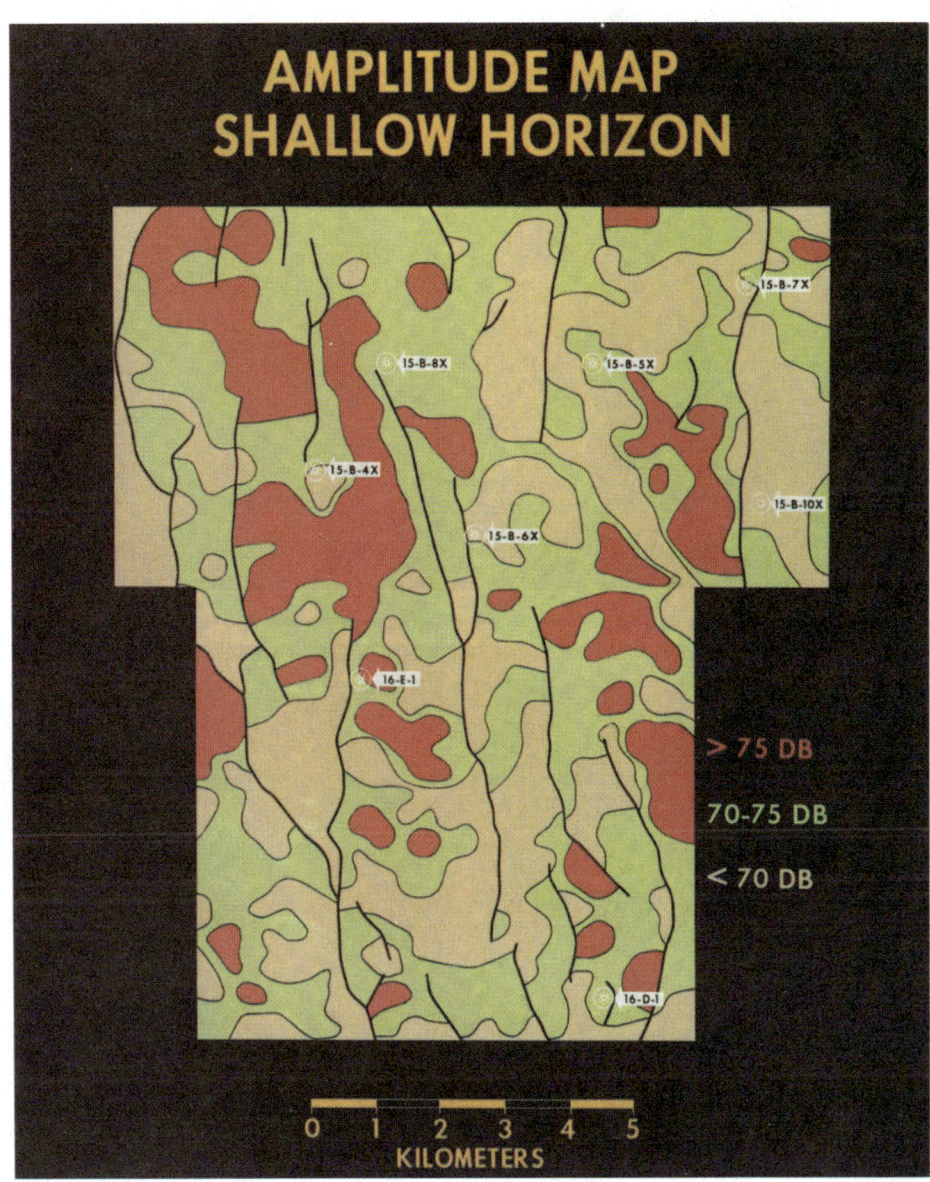

FIG. 29. Amplitude map of shallow horizon.

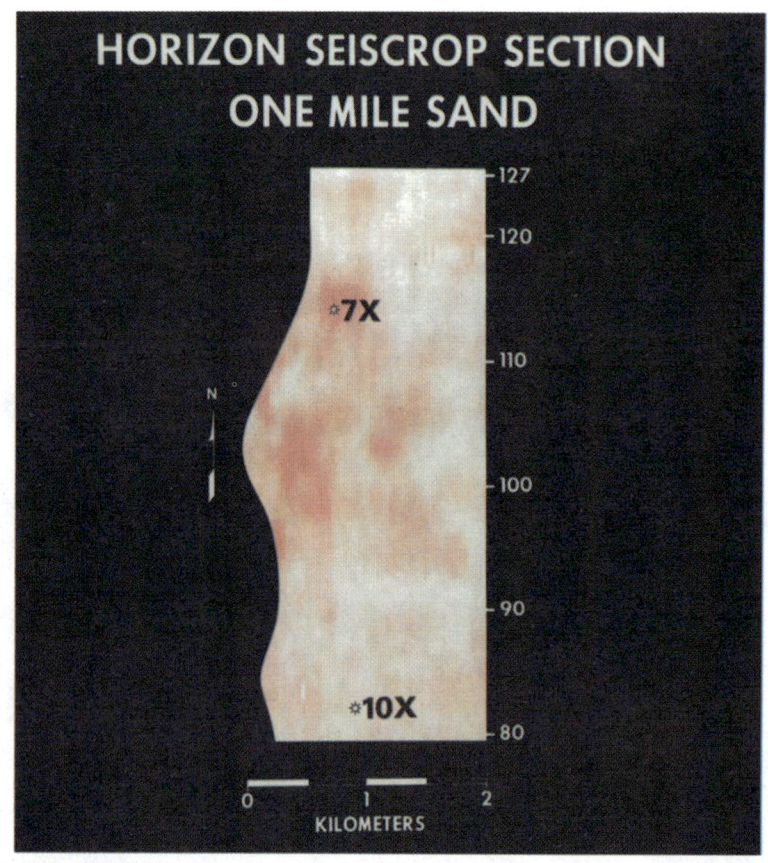

Fig. 30. Horizon Seiscrop section associated with the strong reflecting sequence identified near 1.6 sec in the 15-B-7X well.

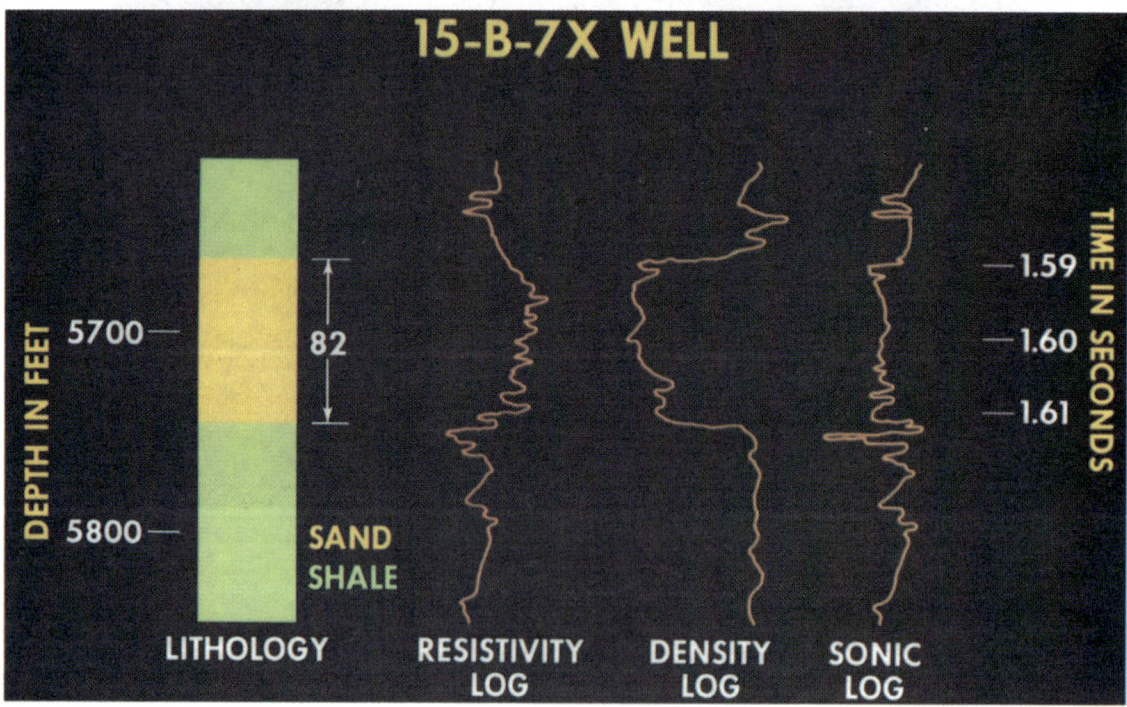

Fig. 31. Partial well logs at 15-B-7X well over the interval encompassing the strong reflector sequence in vicinity of 1.6 sec.

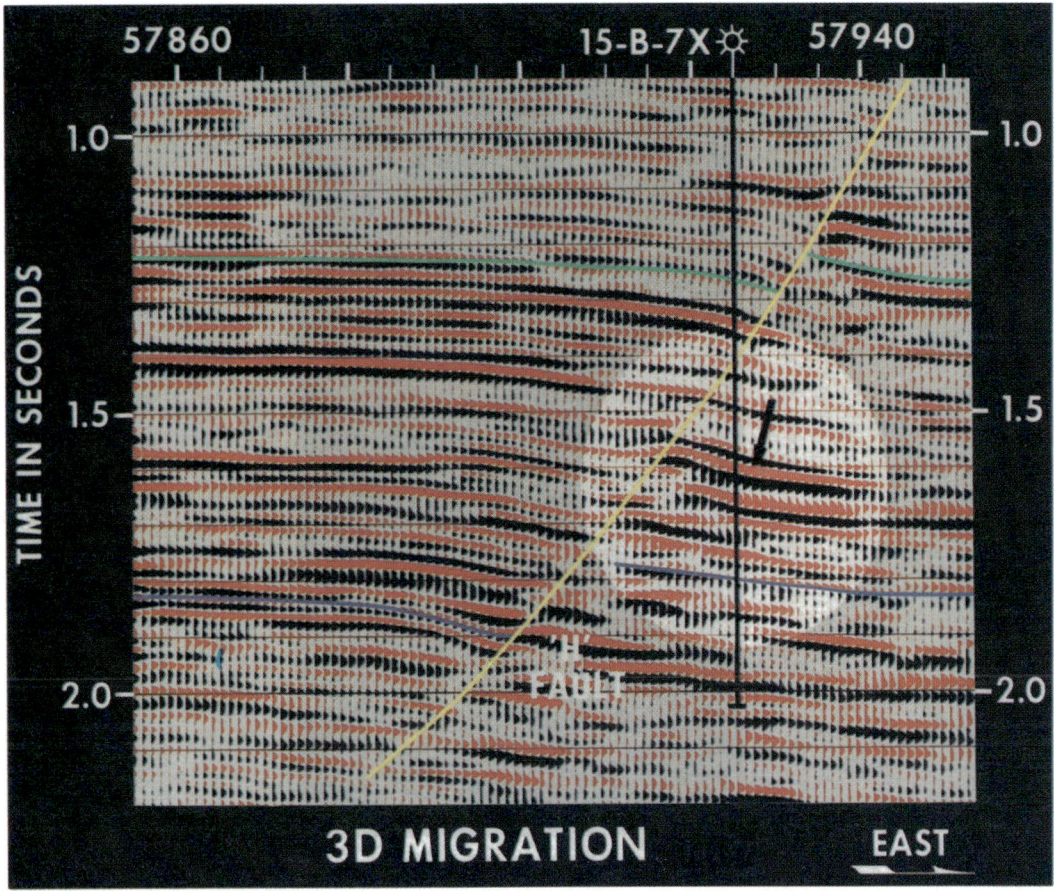

FIG. 32. East-west seismic cross-section, line 115, showing strong reflector near 1.6 sec at the 15-B-7X well on upthrown side of the "H" fault.

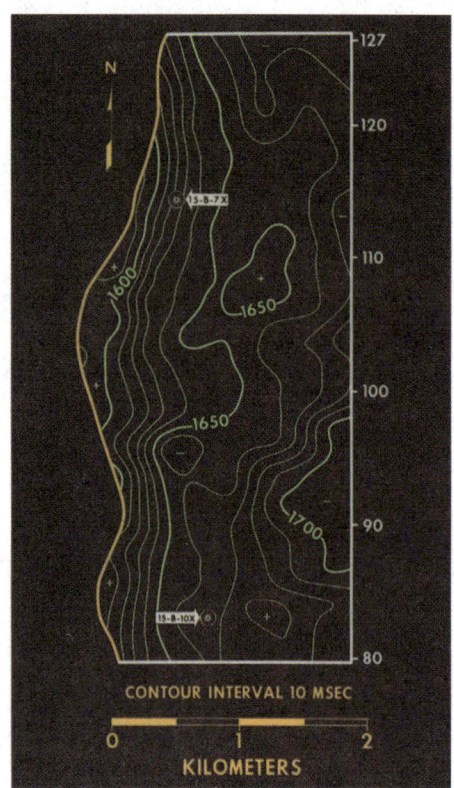

FIG. 33. Structural map on the key reflector near 1.6 sec in the 15-B-7X well.

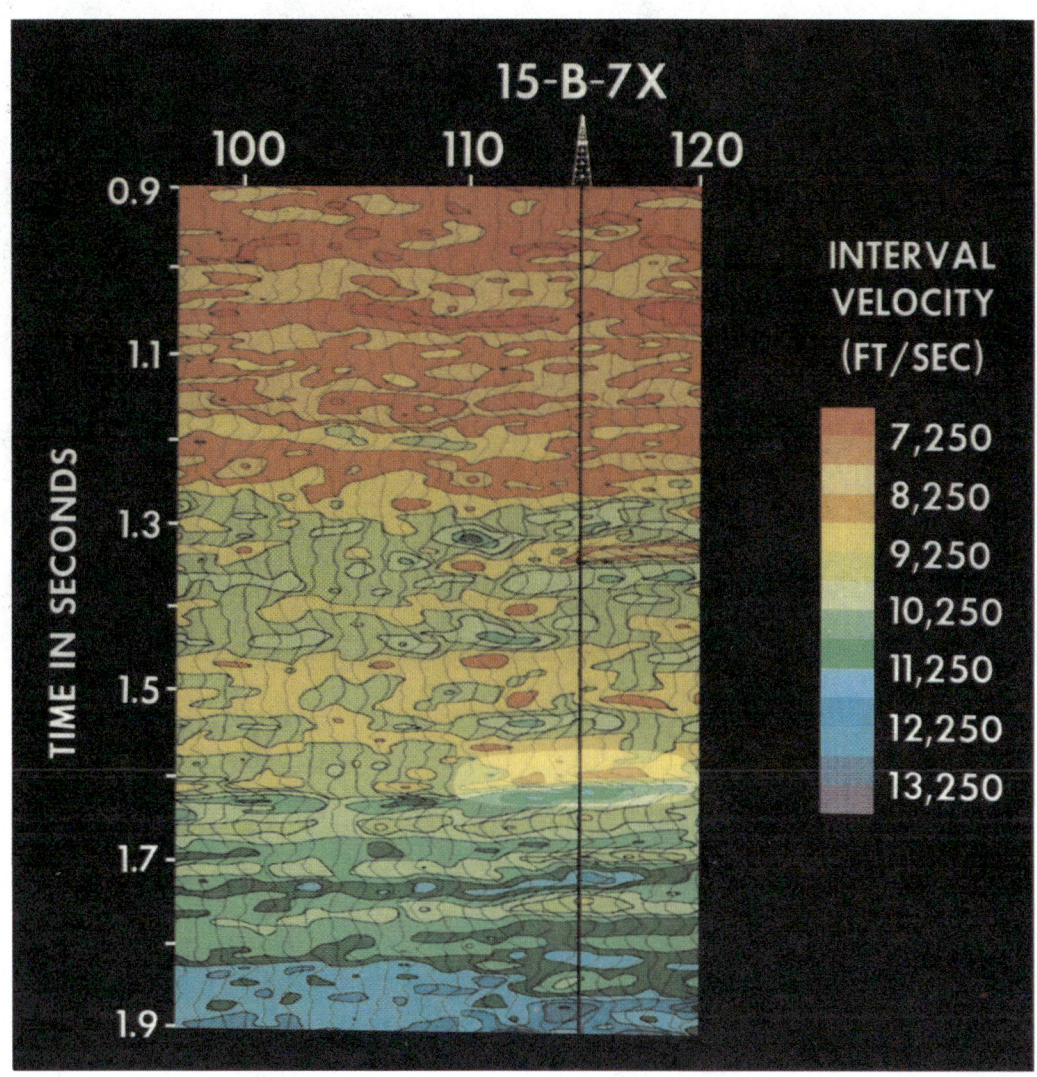

FIG. 34. Interval velocity logs obtained from inversion of seismic data on north-south line 425 through the 15-B-4X well.

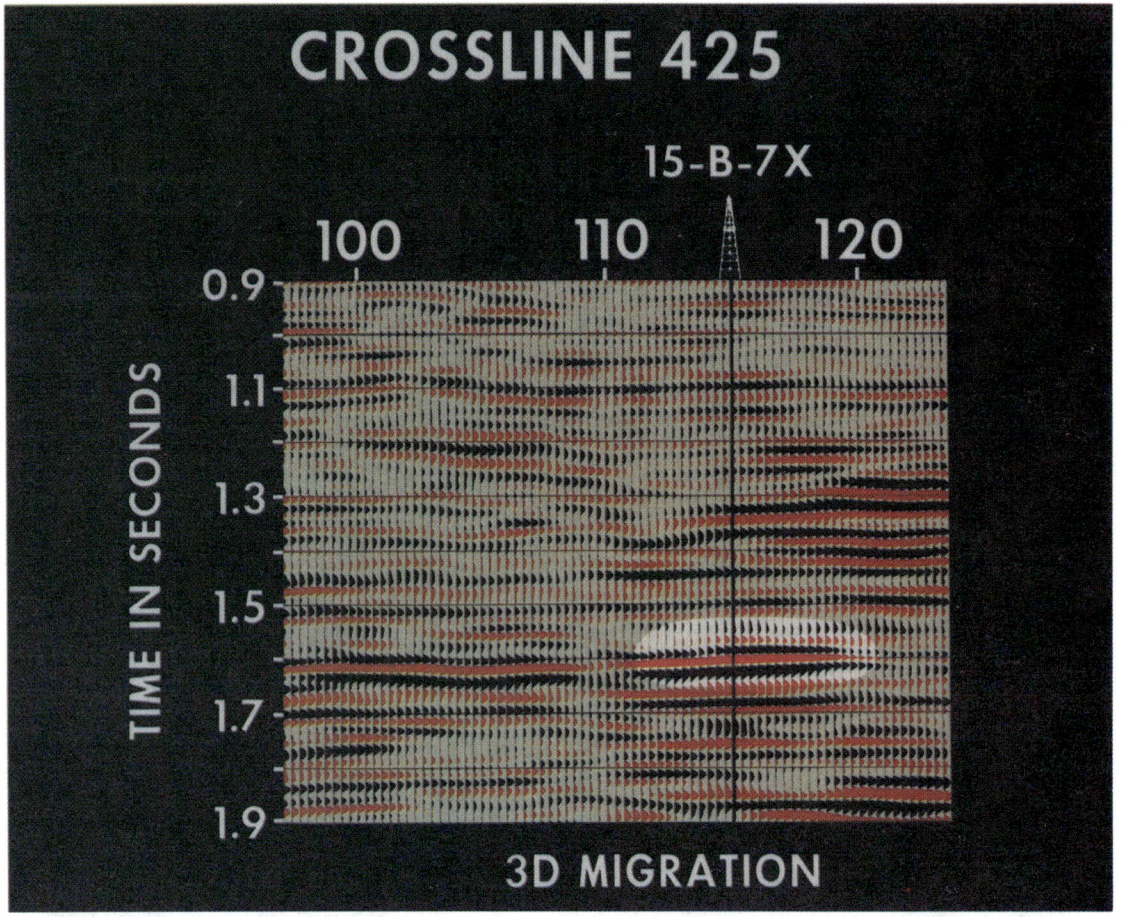

Fig. 35. Seismic section on north-south line 425 through 15-B-7X well.

Oil Company, Highland Resources, and Geophysical Service Inc. to publish this paper.

In the development of this study numerous discussions were held with several coworkers, especially with Clay Miller, Jr. of Texas Pacific Oil Company, Inc. and Alistair Brown and Marion Bone of Geophysical Service Inc. Their support and assistance are gratefully acknowledged.

REFERENCES

Achalabhuti, C., 1978, Natural gas deposits of the Gulf of Thailand: presented at the 2nd Circum-Pacific Energy and Mineral Resources Conf., August 4, in Honolulu.

Brown, A. R., 1978, 3-D seismic interpretation methods: Presented at 48th Annual International SEG Meeting November 1, in San Francisco.
——— 1979, Geophysical report: 3-D seismic survey gives better data: Oil and Gas J., November 5, p. 57–71.

Brown, A. R., Dahm, C. G., and Graebner, R. J., 1980, A stratigraphic case history using three-dimensional seismic data in the Gulf of Thailand: Presented at 42nd EAEG Meeting, June 4, in Istanbul.

Hays, D. B., Shurtleff, R. N., and Wason, C. B., 1979, Applications of rigorous seismic inversion: Presented at 49th Annual International SEG meeting November 8, in New Orleans.

Schneider, W. A., 1978, Integral formulation for migration in two and three dimensions: Geophysics, v. 43, p. 49–76.

Geophys. J. R. astr. Soc. (1983) **74**, 97–127

Aspects of seismic reflection prospecting for oil and gas

P. N. S. O'Brien *Manager, Geophysical Technical Services, BP Exploration, Britannic House, Moor Lane, London EC2Y 9BU*

Received 1982 January 4

Summary. Seismic reflection survey is a technology which has made, and is making, rapid advances by means of continuous marginal improvement over each of its subdivisions of data acquisition, signal enhancement and geological interpretation. It is wedded to the digital computer and as long as the real cost of digital computers and their peripherals continues to fall so long, at least, will reflection seismology continue to advance — for there are many algorithms waiting only for more (at the right price) computer power before they are implemented. The essence of the technique is simple echo-sounding combined with large data redundancy and (fairly) complex signal enhancement and imaging procedures. On land the source is normally a few kilograms of high explosive and at sea it is usually an array of airguns, which is a device for releasing into the water a few litres of air at high pressure. Particle velocity detectors are used on land and pressure detectors at sea, their output is digitally recorded on magnetic tape with a total dynamic range of some 180 dB, though resolution is limited to 14 bits. Arrays of sources and detectors are used and the first 10–12 stages in the signal processing chain are devoted to producing a record as close as possible to the hypothetical record which would have been obtained if the source and detector had been coincident on a horizontal datum plane and if there had been no noise and no multiply reflected echoes. Once the best such 'zero-offset' record has been obtained an imaging algorithm, based on the acoustic (not elastic) wave equation, is used in order to bring into focus as sharply as possible the seismic image of the subsurface. This is normally done for vertical slices through the Earth but increasingly attempts are being made to produce proper three-dimensional images. The models of the Earth which underlie signal enhancement procedures are grossly simplified versions of reality. A major development effort in iterative and interactive model fitting is just beginning with the aim of allowing more plausible models to be used. Interpretable echoes are commonly obtained from depths in sedimentary rocks of 5 km and more. Absorption limits penetration of the higher frequencies so that it is rare for echoes from the greater depths to have appreciable energy above 25 Hz. Some information on the nature of the rocks and their depositional environment may be obtained from the reflections but essentially nothing may be deduced about whether their pores are filled with water, oil or gas. Colour graphics work stations are just being introduced to aid in the geological interpretation of the computer enhanced signals but it will be some time before they can call up fast enough and display adequately the quantity of data involved in an average survey (10^{10} bits).

1 Introduction

As illustrated in Fig. 1 seismic reflection prospecting is simple echo-sounding. This is normally carried out at intervals of about 25 m along straight lines and a better than average result is shown in Fig. 2. By obtaining a set of such records (we call them seismic sections) over a grid of lines covering the area of interest, contour maps of the subsurface reflectors may be built up and predictions made of the subsurface geology. Measurements of the amplitudes and waveforms of the echoes may help in geological predictions but only to a minor extent. The predictions are then tested by the drill and are usually proved correct. This confident and encouraging statement should not be taken to imply that all is well because, by-and-large, the predictions made fall very far short of what is needed for the next stage in the investigation. In civil engineering, they do not extend to estimates of shear strength, plasticity, or other useful mechanical properties; in hydrocarbon exploration they do not (except under very restricted conditions) indicate whether or not hydrocarbons are present; in coal production planning they do not define minor faults; in hydrogeology they do not tell you whether the water is fresh or saline. Further, in many areas of the world, particularly on land, the reflection sections obtained are more like the one shown in Fig. 3. With such sections, where clear echoes are not detected, the uncertainty estimates attached to predictions are so large that, even if the prediction (which, of course, includes the uncertainty estimate) is correct, its use is lessened.

This paper is mainly about technical principles and therefore may have some relevance to all uses of the seismic reflection method. However, the discussion and examples all relate to exploration for oil and gas, which is the field where the method exhibits its highest degree of sophistication. It has established an indispensable place in hydrocarbon exploration due to the combination of three factors. One, there is no present method for detecting oil or gas at depth from measurements made at the ground surface. Two, oil and gas occur in sedimentary rocks which by-and-large are sub-parallel and therefore well suited to being mapped

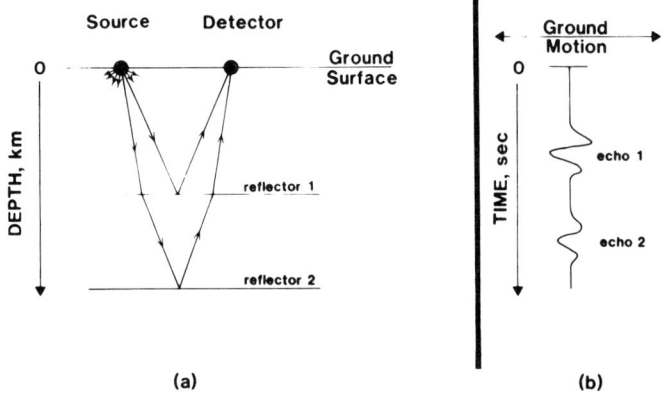

Figure 1. Seismic echo sounding. (a) Idealized Earth section. (b) Idealized reflection record.

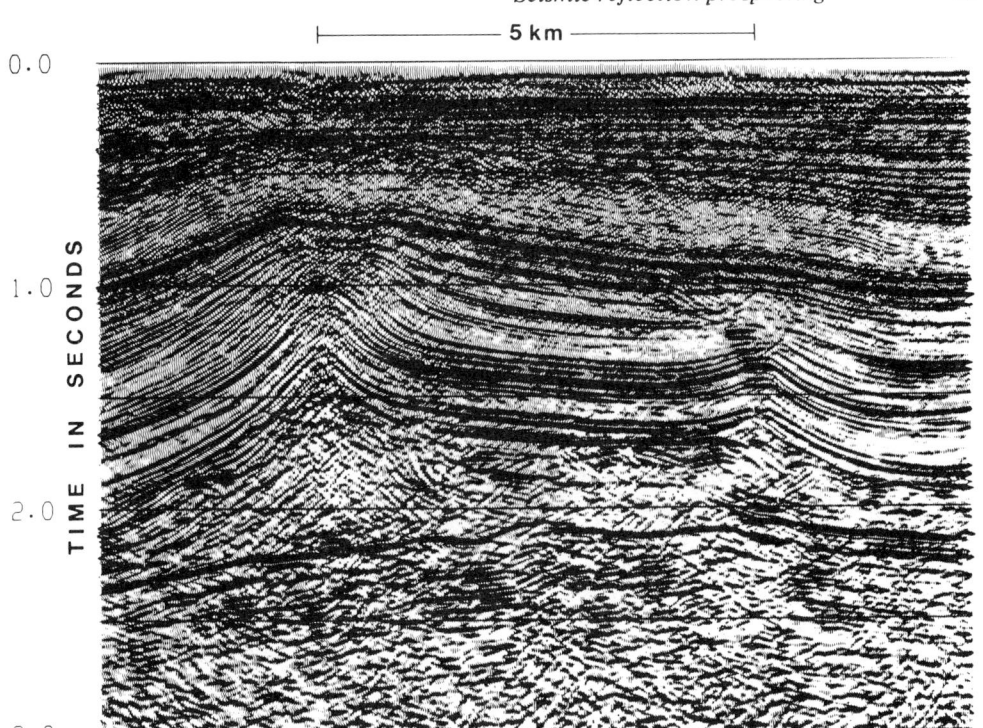

Figure 2. A better than average seismic reflection section.

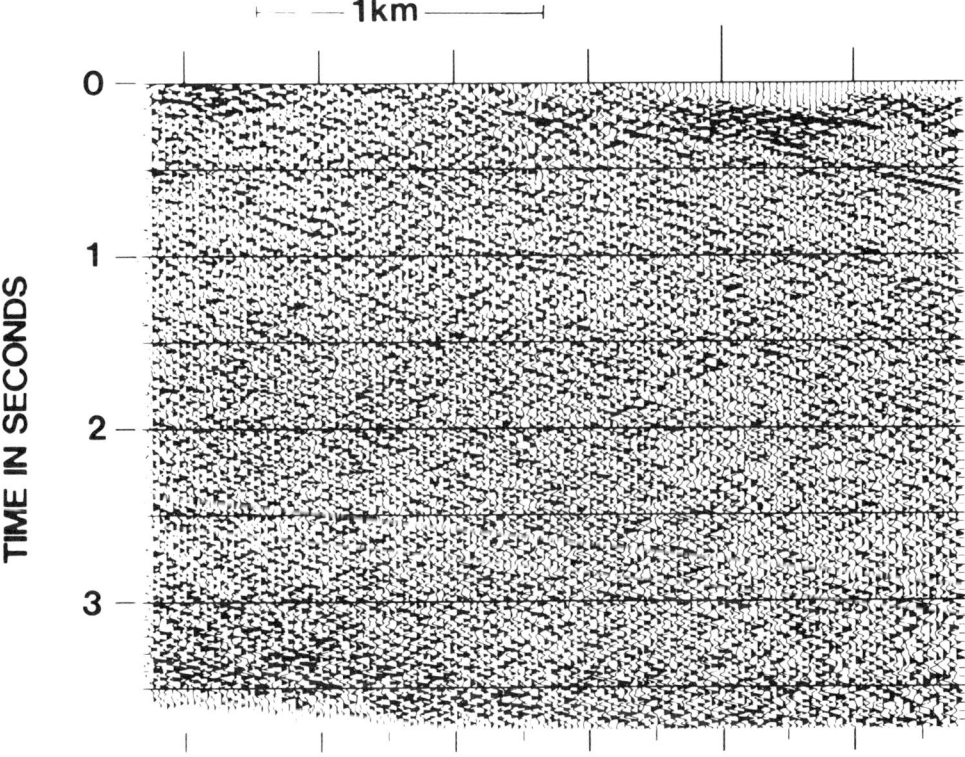

Figure 3. A worse than average seismic reflection section.

by echo-sounding. Three, oil and gas occur at depths of a few kilometres and drilling holes to those depths — which is the preferred method for most aspects of sub-surface exploration — costs several millions of pounds per hole. If cost and time were no object there would be a great many more holes drilled and a lot less seismic reflection surveying. When the depth of investigation is shallow and the information provided is further removed from what is required — as in engineering geophysics — then the need for a seismic reflection survey is much reduced.

In the non-communist world there are about 1000 seismic land crews plus about 100 seismic ships engaged in hydrocarbon exploration. They each gather data at an effective rate of about 10^8 bits km^{-1} at a cost varying widely according to local conditions but generally between £2000 and £10 000 line-km^{-1} on land and between £200 and £1000 at sea. The subsequent signal processing costs are around £500 line-km^{-1} on land and £250 km^{-1} at sea. The cost of geological interpretation, which is more manpower intensive but relies increasingly on interactive computer technique, is small in comparison, amounting to perhaps another £20 line-km^{-1}. The communist world has more land-crews, but fewer ships.

Like any technology, seismic reflection prospecting is intensely specialised, as was illustrated by a recent recruiting advertisement which listed seven different types of exploration seismologist, each of whom would be expected to stay within his own speciality throughout his career. Consequently, this review makes no claim to cover the whole of the subject, nor even that the aspects treated are those of most importance. Mainly, they are constrained to those aspects for which illustrative material is least difficult to obtain.

In spite of the listing of seven specialities as mentioned above, this paper retains the time-honoured division into Data Acquisition, Data Processing and Interpretation — it being understood, of course, that each speciality interacts significantly with the others.

2 Data acquisition

Data acquisition divides naturally into two parts, the equipment and its deployment. We start with equipment which we treat in three sections covering the source, the detector, and the recorder.

2.1 THE SOURCE

On land the archetypal source consists of a few kilograms of high explosive fired some 10–20 m below ground surface. At least one-half of all land surveys use such a source. Fig. 4(a) shows a typical example of the signal radiated from a buried explosion, the negative reflection which occurs about 20 ms after the signal onset comes from a reflector lying between the explosion and the ground surface. The next most popular source is a truck or tractor mounted vibrator, hydraulically driven and electronically controlled to radiate a signal lasting 10 s or so, with an instantaneous frequency varying from about 10 Hz to about 60 Hz. The overlapping complex of recorded echoes is then cross-correlated with the radiated signal (as in CHIRP radar) to compress the original 10 s signal to a pulse with a duration of about 0.1 s, thus enabling resolution of echoes to the same extent as if the radiated signal had been of that duration. Fig. 4(b) shows a measurement of such a compresssed pulse. These recordings were made with a detector buried 300 m below the surface (Sixta 1982). Under survey conditions the received echoes will have travelled a few thousands of metres through the absorptive Earth with the result that the explosive and vibrator pulse shapes become broadly similar, as indicated in Fig. 4(c and d), and are virtually indistinguishable when displayed on a section such as those shown in Figures 2 and 3.

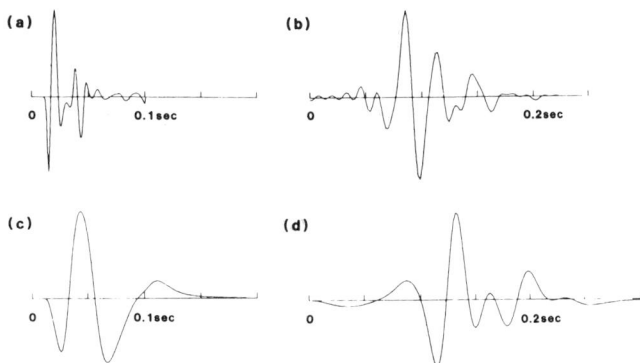

Figure 4. Radiated and received signals. (a) and (c) Radiated and received signals from 2 kg of explosive detonated at 30 m. (b) and (d) Radiated and received signals, after correlation, from a surface vibrator.

Vibrators currently provide a peak force of some 10^5 N which is insufficient to obtain detectable echoes from the greater depths and so three or four of them are operated in synchronism, each unit being activated several times. The resulting signals are then added together to overcome ambient noise. Unfortunately a very large amount of the energy radiated by a surface source is constrained to remain close to the ground surface (mainly Rayleigh waves). Consequently, the individual vibrators have to be suitably spaced to form a linear array with which to reduce the horizontally travelling waves without adversely affecting the near-vertically travelling echoes. A similar array is required at the detection location, not only for surface vibrators but for all types of source.

The geophysical literature contains a reasonably extensive treatment of the seismic signals radiated from buried explosions both theoretical (e.g. Jeffreys 1931; Blake 1952) and experimental (e.g. Sharpe 1942a, b; O'Brien 1969; White & O'Brien 1974). All those who have read that literature can be in little doubt as to the nature of the explosion radiated signal in so far as it relates to seismic prospecting. This is not the case with vibrator signals. A few papers provide analytical treatments of an ideal vibrator exciting an elastic half-space (e.g. Miller & Pursey 1954; Pursey 1956) while Lerwill (1979, 1981) uses physical insight to develop equivalent electrical circuits as analogous of the vibrator–ground interaction. However, there is not yet any consensus on what constitutes a realistic model of a practical vibrator, nor any detailed treatment of the relevant mechanical properties of those most imperfectly elastic materials — near surface soils and rocks.

There exist not a few latent PhD topics in this general area, including ones which cover the detailed specification of where sensors should be placed in order to estimate from *surface* measurements the equivalent (i.e. compressed) radiated signals (several types of wave are radiated) and their directivity patterns. But, of course, none of the results will be worth a row of beans unless the analysis is supported by appropriate measurement. In addition to Lerwill (1981) an introduction to practical vibrators is given by Waters (1978). There are a number of other land sources, most of them designed to impact the ground surface. They mainly radiate relatively low energies and therefore are useful only in areas of low ambient noise or for small depths of investigation.

Condensed explosive is now rarely used in marine reflection survey, its place having been taken by the air-gun. This is a device which, under command of the ship's position fixing equipment, releases into the water, at intervals of about 10 s, a given volume of air, typically a few litres, at a given pressure, typically 14 MPa. The immediate consequence of releasing the high pressure air is the radiation of a pressure pulse. This is followed by rapid expansion and contraction of the air bubble with the consequent radiation of a train of secondary

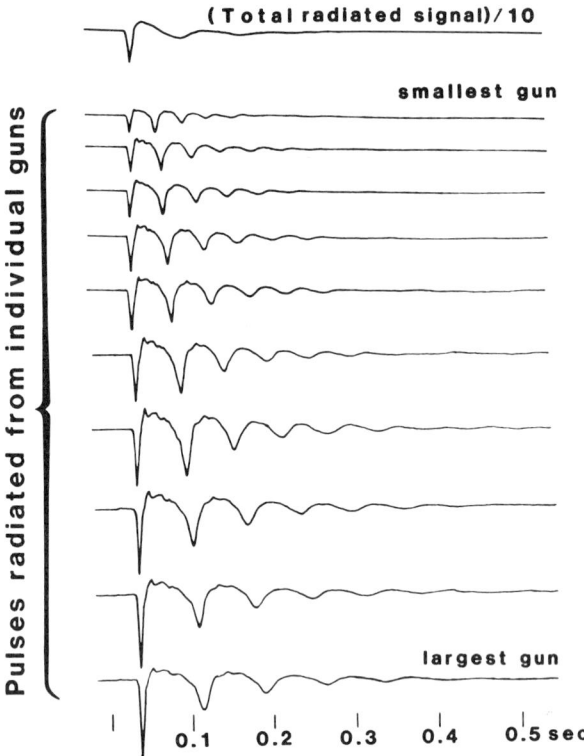

Figure 5. Design principle of an air-gun array (after Edelman 1975).

pressure pulses whose time intervals depend upon the energy in the bubble. The total radiated signal lasts 0.5 s or so and needs to be compressed in order to obtain optimum resolution of the returning echoes. This could be achieved by standard numerical techniques (signature deconvolution). However, since a single gun does not generate sufficient energy, a number of them — perhaps 20 or more — have to be used in synchronism and virtue is made of this necessity in order to reduce the duration of the radiated signal. The technique is illustrated in Fig. 5, where the radiated signals from each of 10 differently sized air-guns are shown. The upper trace is constructed by the superposition of the 10 individual traces and so corresponds to the overall vertically radiated signal, provided, of course, everything is linear. By choosing the gun volumes correctly and ensuring synchronism of firing, the initial pressure pulses add constructively while the secondary pulses add destructively, so generating a suitably short pulse. Of course, the waveforms radiated at large angles to the vertical are not so simple but this is though to be a relatively minor problem. There has been quite a lot written on air-guns, much of it rather removed from practicality. The papers of (Giles & Johnston 1973; Nooteboom 1978; Safar 1976; Ziolkowski *et al.* 1982) give a good idea of the state of the art.

There are other marine sources for which water guns, steam guns, propane-oxygen explosions in 'elastic' bags, and mechanically induced implosions are all used to some small degree. These, and some others, are described by Lugg (1979).

It is perhaps of interest to indicate the amount of energy in the seismic bandwidth which needs to be released in order to record readable echoes. Later echoes are obviously smaller than earlier ones. Their reduction is due largely to wave-front divergence (roughly inversely proportional to distance travelled) and absorption and scattering (very roughly 0.25 dB/ wavelength travelled in porous sedimentary rocks), though reduction with depth of typical reflection coefficients plays a small part. In practice, for reflections from 1 km, and after allowing for current techniques in signal-to-noise enhancement, we need maybe 50 kJ, whereas for 10 km penetration we need maybe 50 MJ for a similar bandwidth, but only one-tenth as much for an acceptable one (say 15 Hz). These figures which, since noise characteristics are so variable are not to be taken too seriously, apply to useful energy radiated by the source into the interior of the Earth. Since surface sources radiate most of their energy as surface waves, they need to generate more energy than downhole sources (explosives). The usable seismic bandwidth is controlled entirely by earth absorption and noise, and varies from about 100 Hz at 1 km to about 25 Hz at 5 km – there is no possibility of recovering significant spectral amplitudes at 100 Hz from depths of 5 km in porous rocks.

2.2 THE DETECTOR

On land, detectors are electro-magnetic, moving coil, geophones with sensitivies of about $10\,V\,m^{-1}\,s^{-1}$; at sea they are piezoelectric, ceramic, hydrophones with sensitivities of about $0.1\,mV\,Pa^{-1}$. Both types are cheap and robust – two vital characteristics when crews, far from any supplier, will be deploying a few thousand detectors. The large numbers occur because at each recording station, of which there are commonly 96 but maybe more than 1000, an array of 10–50 detectors is laid out in order to reduce horizontally travelling coherent noise. This coherent noise is mainly source generated on land and mainly ship generated cable snatch at sea; it is often 10 times the signal amplitude and may reach 100 times, proving to be a major limitation to the seismic method.

From time-to-time new detectors become available which claim to produce less distortion, fewer spurious resonances, or to have improved technical performance of one sort or another. The claims are mostly true but, equally, are mostly of little importance since the governing conditions are the quality of ground coupling and the stability of response after some tired, heavy booted, 16 stone geophysicist has walked all over them. What has just arrived (Klaassen & van Peppen 1982) is the 'electronic' geophone which by combining a higher sensitivity with lower output impedance should appreciably reduce the ever-present scourge (even in the desert!) of electromagnetic pick-up. It may also appreciably reduce harmonic distortion which, with conventional detectors, produces noise in the signal bandwidth by distorting the low frequency surface waves. On land accelerometers have the appealing property that they discriminate against the low frequency surface waves. At sea velocity sensitive hydrophones have the apparently equally appealing property that they give larger output when placed at shallower depths (the output of pressure detectors decrease as they become shallower). Neither type is much used!

It is still the norm for the detector arrays to be connected to the recording instruments via conductor pairs, even when there are 240 of them! However, more and more the signals are coded at the detector location and sent multiplexed down a cable containing a very few wires, or very occasionally sent by radio. At sea – where all the hydrophones, electronics and conductor wires are contained within a neutrally buoyant tube some 3–4 km long and of 10 cm diameter – the first fibre optic link in seismic prospecting came into operation in mid-1982. Fibre optic links on land followed close behind. Telemetry strikes another blow against cross-feed and pick-up and may remove any practical advantage of the electronic geophone mentioned above.

2.3 THE RECORDER

Of course, the signals (but mainly noise!) are recorded digitally on magnetic tape. Since I am not an electronic engineer I shall do no more than list some of the major specifications. The bandwidth is 2–500 Hz, the overall dynamic range is 180 dB and the magnetic tape packing density is usually 1600, but increasingly 6250 bits per inch. The incoming signals are band-pass filtered to reduce noise and prevent aliasing of the high frequencies, and then input to an amplifier whose gain varies sample-by-sample, commonly at 4 ms intervals, in steps of two or four. The amplified signal thus keeps within the dynamic range of the analogue-to-digital converter and, since the gain steps are also recorded, overall dynamic range is considerably increased, although the resolution is fixed as that of the converter, which is typically 14 bits. The signals are multiplexed before recording and this multiplexing normally takes place after some fixed gain pre-amplification. There are several auxiliary channels for recording other relevant information and the daily, weekly and monthly instrument checks are often under microprocessor control. There are also multichannel cameras of one sort and another for immediate visual inspection of the records.

On land the total cost of geophysical equipment for a 96 channel crew might be about £1M whereas it would be about £2M for a marine crew.

2.4 EQUIPMENT LITERATURE

Detector design is covered well, e.g. for geophones (Dennison 1953) and for hydrophones (Bruel & Kjaer; Luehrmann 1972). The rest of the equipment is covered only in manufacturers' brochures though the second edition of volume 2 of *Seismic Instruments* (Evenden & Stone 1971; under revision) should to a large extent rectify this.

2.5 MARINE POSITION FIXING

We need to know the relative positions of the ship to within a metre or two over periods of minutes and to within 5–10 m over the long term. We need the latter accuracy in absolute measure so that uncertainty in position fixing may be ignored when comparing the results of one survey with those from another or when using them to locate a drilling barge. Unless there is line-of-sight radio positioning these accuracies are usually missed by a factor of 10 or more. The ship will carry receivers to make use of the US Navy navigation satellite system, at least one, often two, and sometimes three ground based radio navigation systems, and possibly an inertial system of some sort. All these are tied together in a computer whose output steers the ship and activates the source and recording system – say one 5 s record every 25 m.

2.6 EQUIPMENT DEPLOYMENT

Choice of equipment and how to deploy it are obviously a vital part of pre-survey planning. Type of source, size and shape of detector arrays, sampling interval, source-detector spacing, sign-bit recording, etc., all require consideration. Since sign-bit recording is probably the least familiar phrase in that list I will say a little about it.

In conventional equipment each sample of the signal is recorded as a 19 bit word – one bit for direction of earth movement (up or down) and 18 bits to specify the amplitude of movement. In sign-bit recording the 18 bits are ignored and only the sign bit is recorded. That is, the record only indicates whether the ground moves up or down, not by how much it does so. Obviously, if we can get away with that we can use much simpler instrumentation

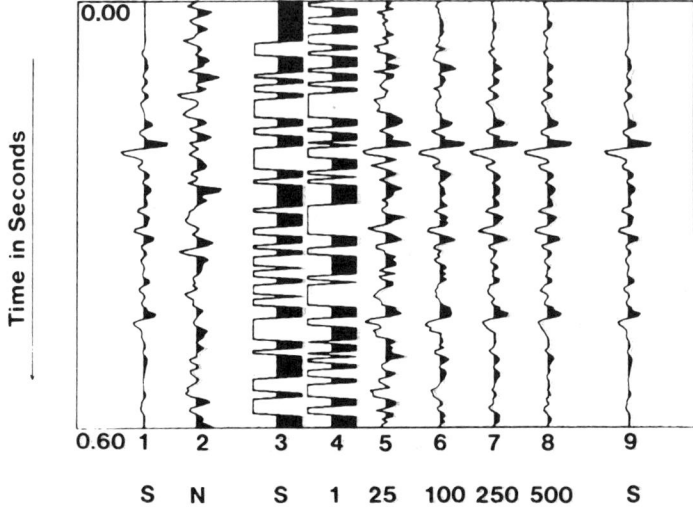

Figure 6. Sign-bit recording with a signal-to-noise ratio of 1:2. Traces 1, 3 and 9 are of the noise-free signal. Trace 2 is noise alone. Trace 4 is a single record of signals and noise superposed. Traces 5–8 are summations of 25–500 individual sign-bit records with constant signal but varying noise.

and record many more detector locations for each source location, so enabling more elaborate signal processing in the computer. When noise is a severe problem these extra detector locations may be a boon. Current sign-bit crews record from 1024 detector locations rather than the more normal 60 or 96.

Imagine that, prior to activating the source, the Earth's surface is perfectly motionless. Suppose an impulsive source is activated with the consequence that traces 1 and 9 in Fig. 6 represent the sequence of band-limited echoes that would be recorded with conventional equipment. Trace 3 would be the equivalent trace as recorded by sign-bit only. Note that all the amplitudes are now equal and no one reflection stands out from another. Suppose a whole sequence of impulsive sources are activated one after the other at the same location on an otherwise motionless earth, individually sign-bit recorded, added together after allowing for differences in source activation times, and then plotted out. Apart from a scaling factor the resulting trace will be identical to trace 3 — no additional information has been obtained. Now imagine the experiment repeated with a very low energy source activated on the real Earth whose ground surface is in continual motion. The resulting sign-bit record (trace 4) will be a similar looking train of constant amplitude, variable polarity, spikes which, if the source generated echoes are small enough, will essentially represent the ambient ground motion of the Earth — that is, noise dominates signal. Suppose the experiment to be repeated many times. Each record will contain the same amplitude very small echoes at the same times after time-zero but the ambient ground motion will differ. Addition of all the individual records, all referenced to their source activation time, will now emphasise the constant time reflections at the expense of the noise. Note that this is a non-linear process (not a linear $\sqrt{n}$ process) and that as more and more traces are added together the output will tend toward the true reflection sequence, giving relative amplitudes as well as polarity, as typified by trace 1. A few minutes' calculation adding constant amplitude bias to sets of random numbers will soon convince you that that is so. Traces 5 to 8 in Fig. 6 show how well trace 1 may be recovered when the signal-to-noise ratio for a single record is 1:2. Figures 7 and 8 allow a comparison between full-bit and sign-bit records using a vibrator

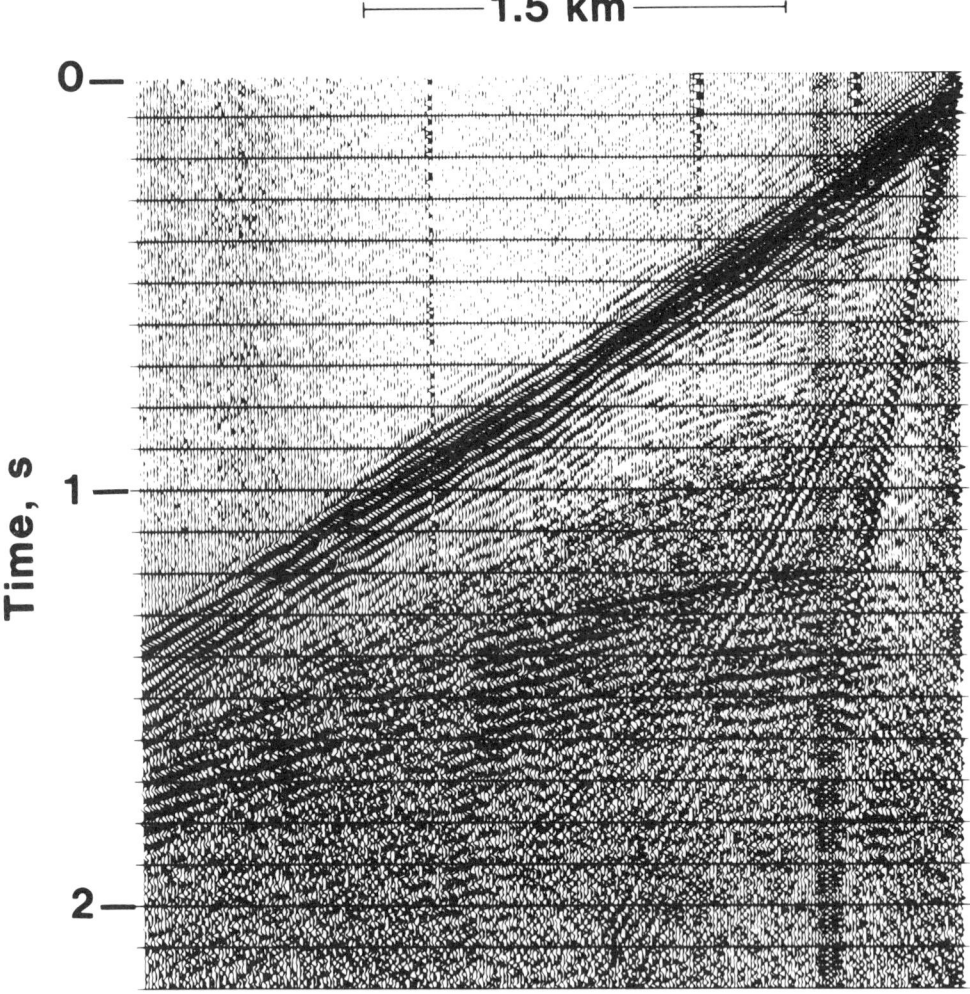

Figure 7. Vibrator source, correlated output after sign-bit recording. (Courtesy of Sohio Petroleum Company.)

source. Note that because the radiated vibration signal is long compared with its autocorrelation a single sign-bit record after cross-correlation with the VIB sweep, already looks like a full-bit record. Although sign-bit recording has been used with an impulsive source there seems little merit in so doing.

Figs 9 and 10 compare fully processed seismic sections with the full-bit section being clearly inferior. Sign-bit recording is a temporary phenomenon due to the fact that full-bit instrumentation is not yet suffcently miniaturized to handle 1000+ traces. O'Brien *et al.* (1982) give a fuller discussion.

3 Seismic data processing

There are some 12–15 separate operations in data processing, each of which requires human intervention for parameter choice and quality control. Essentially, their main purpose is

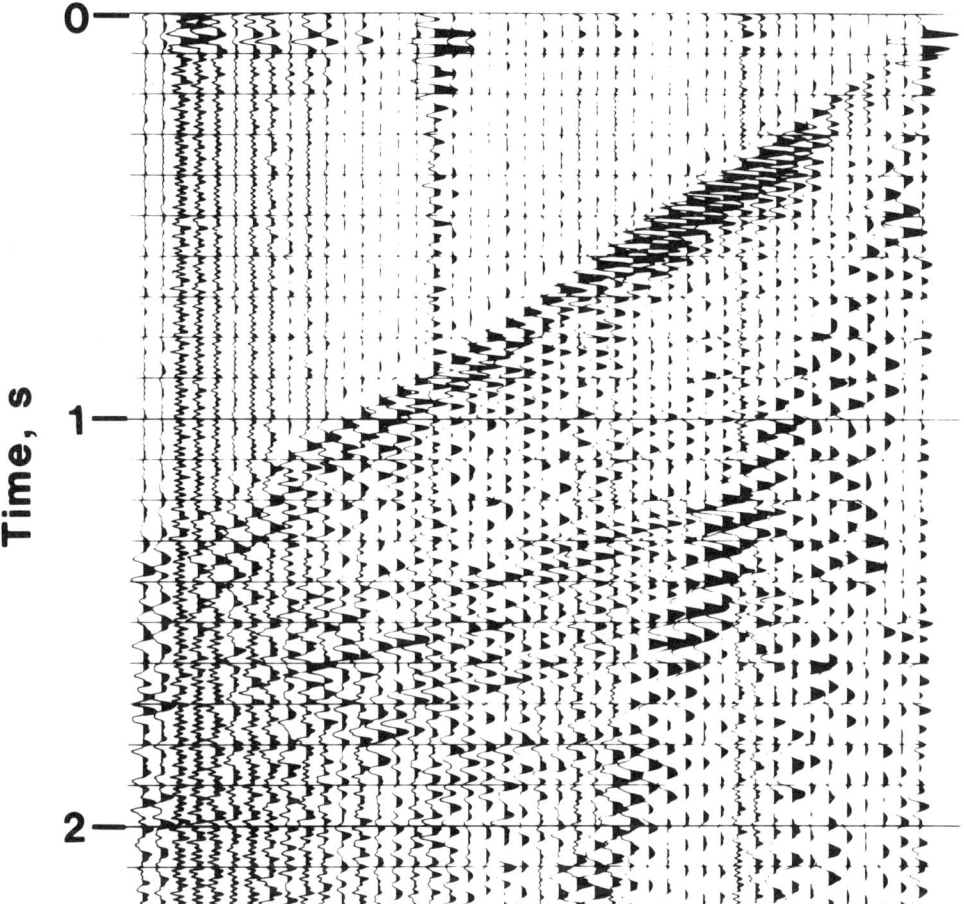

Figure 8. Vibrator source, correlated output after full-bit recording. This record and the one in Fig. 7 were both taken at the same location. (Courtesy of Sohio Petroleum Company.)

signal-to-noise enhancement, it being understood that the noise is mainly source generated and cannot be adequately reduced by field techniques. This signal-to-noise enhancement is achieved mainly by spectral analysis with subsequent 1- and 2-dimensional filtering and by exploiting measurement redundancy via the use of simple ray theory (geometrical optics). Redundancy has increased from nil in the mid-1950s, when common mid-point stacking was introduced (Mayne 1962) up to 100 and more today, though of course a factor of 100 is neither economically possible nor technically necessary under all conditions. A redundancy factor of 100 means that each reflection from an elemental portion of the sub-surface is recorded 100 times, each time with a different shot-detector spacing. This enables coherent noises of one sort and another to be reduced by multichannel filters based on ray-tracing principles. The archetypal ray-trace filter is the 'common-mid-point stack' based on 'move-out' analysis.

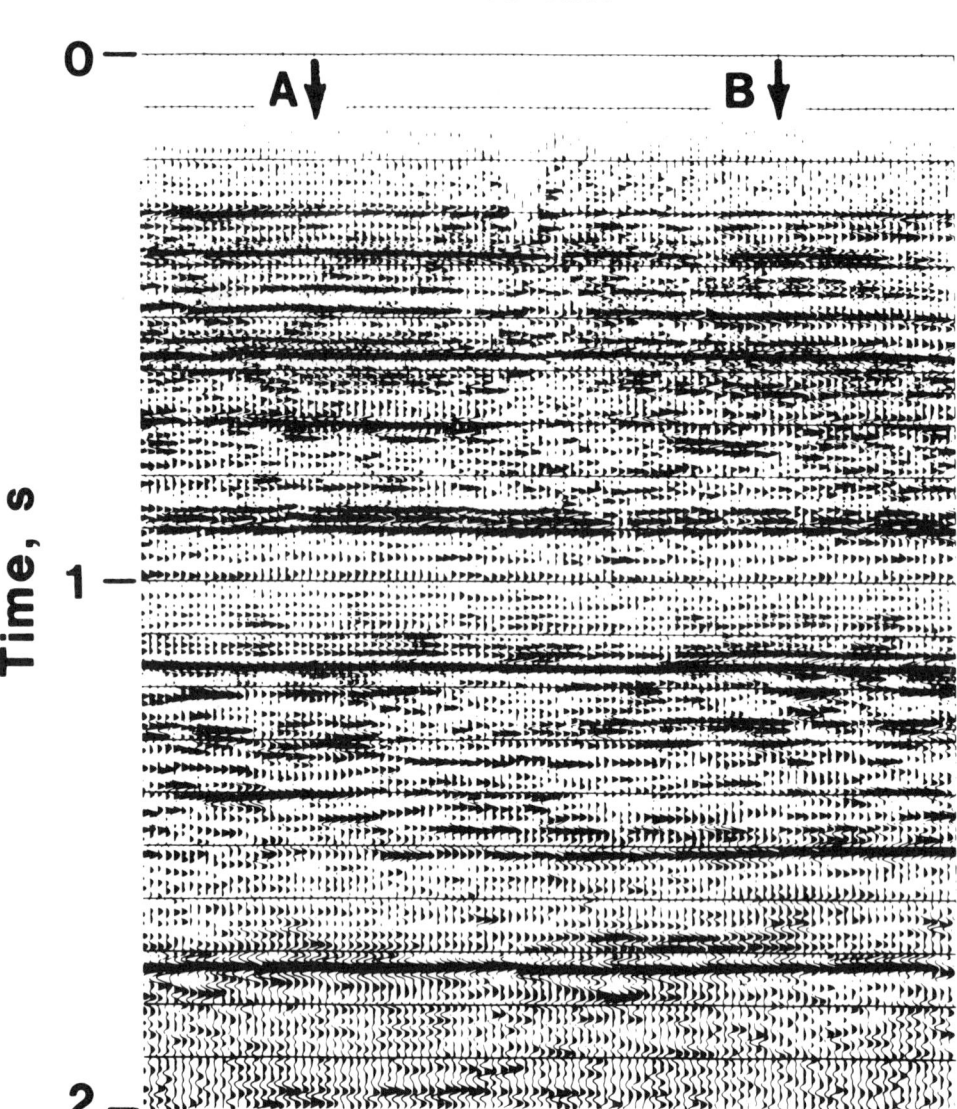

Figure 9. Sign-bit section. Eleven-fold summation after 60-fold CMP stack (Courtesy of Sohio Petroleum Company).

3.1 CMP STACK

'Common mid-point' and 'moveout' may be understood by reference to Fig. 11. A record is made with a source and detector each positioned a distance x_1 either side of the mid-point, M. Additional records are then taken with the source and detector spaced x_2, $x_3 \ldots x_m$ either side of M, which explains why it is called the *common mid-point*. For each record the reflection comes from the common *depth*-point, D. As the spacing increases so will the reflection times obviously increase, as indicated in Fig. 11. The increase in time with horizontal distance is called the moveout. Removal of the moveout time in the computer enables a simple superposition (stacking) of the traces to increase the amplitude

Seismic reflection prospecting 109

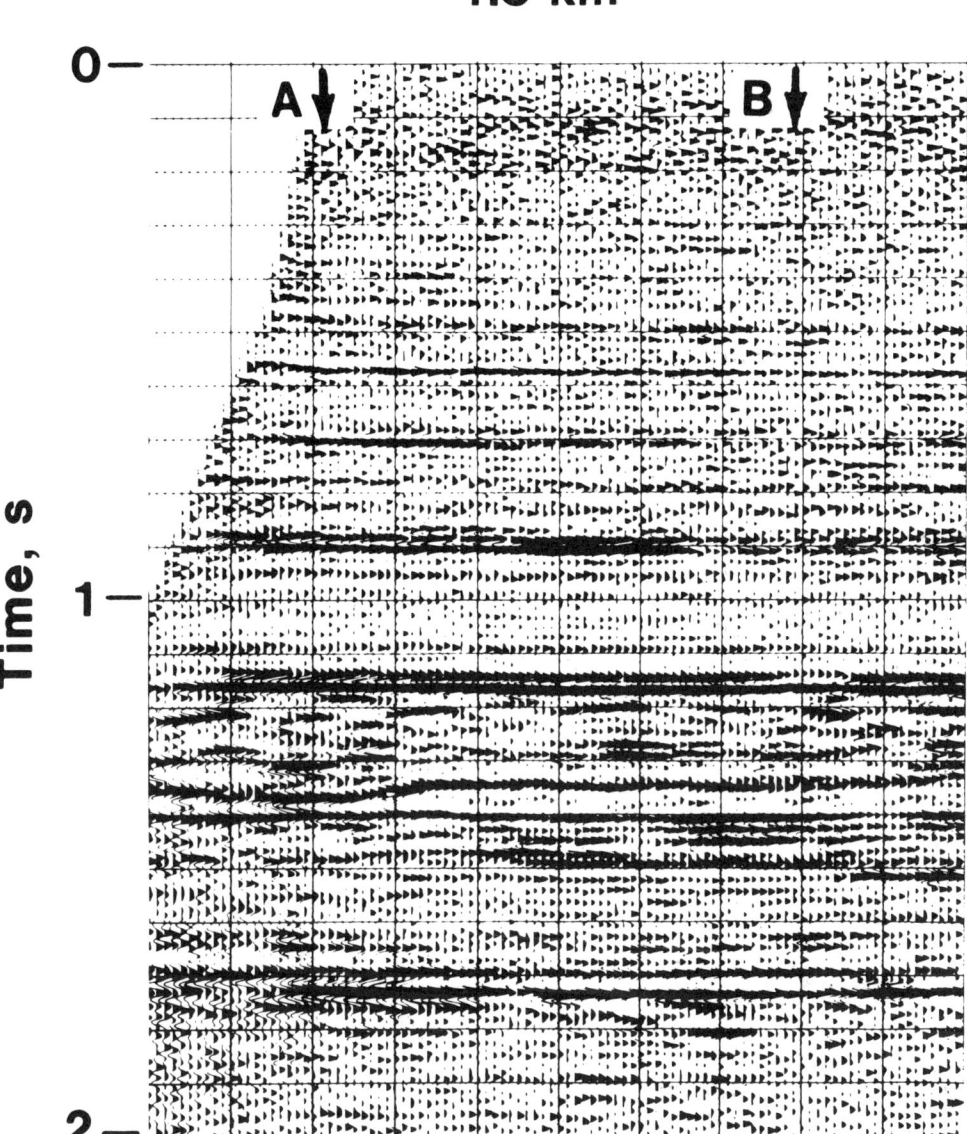

Figure 10. Full-bit section. Twenty-four-fold CMP stack for the line shown in Fig. 9 (Courtesy of the Sohio Petroleum Company).

of the reflection at the expense of any coherent noise following a different moveout curve. The moveout is measured with respect to the time which would be recorded with a coincident source-receiver pair, i.e. $x = 0$, and it is this 'zero-offset' time which is plotted on the seismic section. The mechanics of carrying out this elegantly simple and amazingly powerful technique are described in several texts (e.g. Waters 1978; Telford *et al.* 1976) as are the details of analysing the data to determine the required moveout functions (see also Schneider & Backus 1968). Fig. 12 illustrates the power of the technique. I will not attempt to run through all the other processing procedures, which are mostly described in the texts mentioned above but will select just one, record section migration.

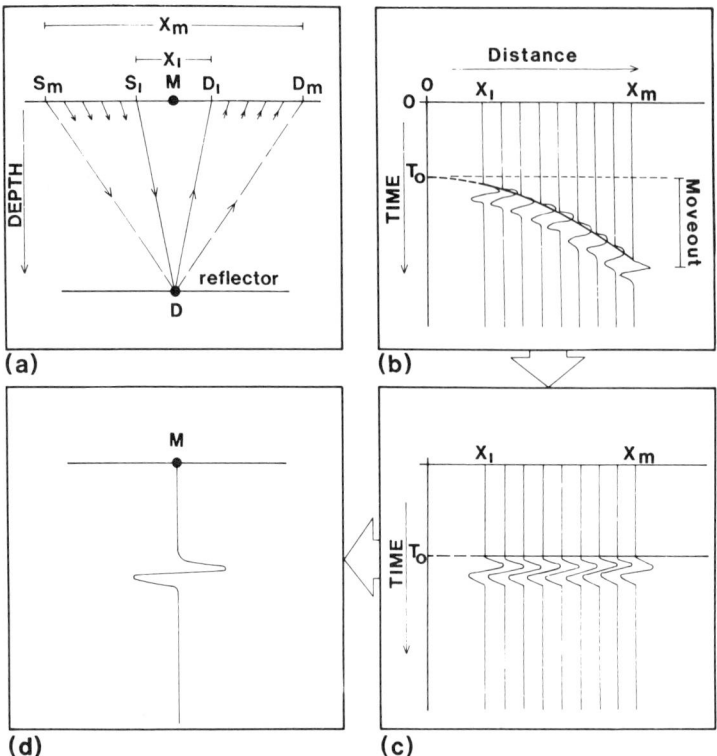

Figure 11. Common mid-point (CMP) stack. (a) Ray diagram; S = Source, D = Detector, (b) CMP record, note that each trace comes from a different field record. (c) CMP record with moveout removed. (d) Summation (stack) of all *m* traces.

3.2 RECORD SECTION MIGRATION

Seismic wave propagation is governed by the elastic wave equation. However, unlike earthquake seismologists, few exploration geophysicists would recognise the equation if they saw it and fewer still could make any use of it. Except in the trivial sense that it underlies geometrical ray theory it, as yet, plays little part in seismic prospecting. A major exception to this generalisation is record section migration, of which theoretical reviews are given by Hood (1981) and Berkhout (1980) and a philosophy of usage is given by Hosken & Deregowski (1982).

A minor application of the equation is to use it to calculate the ground motion which would be recorded at a given detector location when a given source excites a prescribed earth model. The result is called a synthetic seismogram and the procedure is illustrated in Fig. 13. It is equally possible to solve the wave equation with time running backwards and so estimate the unknown earth section from the recorded reflection time section. This latter procedure is known as record section time migration — 'record section' because that is the starting point of the process, 'migration' because the reflecting 'points' are said to 'migrate' from their recorded ground position (x) and echo time (t) to the horizontal location (x_m) of their point of origin on the reflector and the vertical travel time (t_m) to that point. And 'time' migration because the velocity function for the procedure is expressed in terms of two-way vertical travel time (t_m) and not depth.

Note first, that before the observations can be migrated it is necessary to stipulate the velocity and density functions. Of course, if we knew these precisely, we would not be carry-

Seismic reflection prospecting

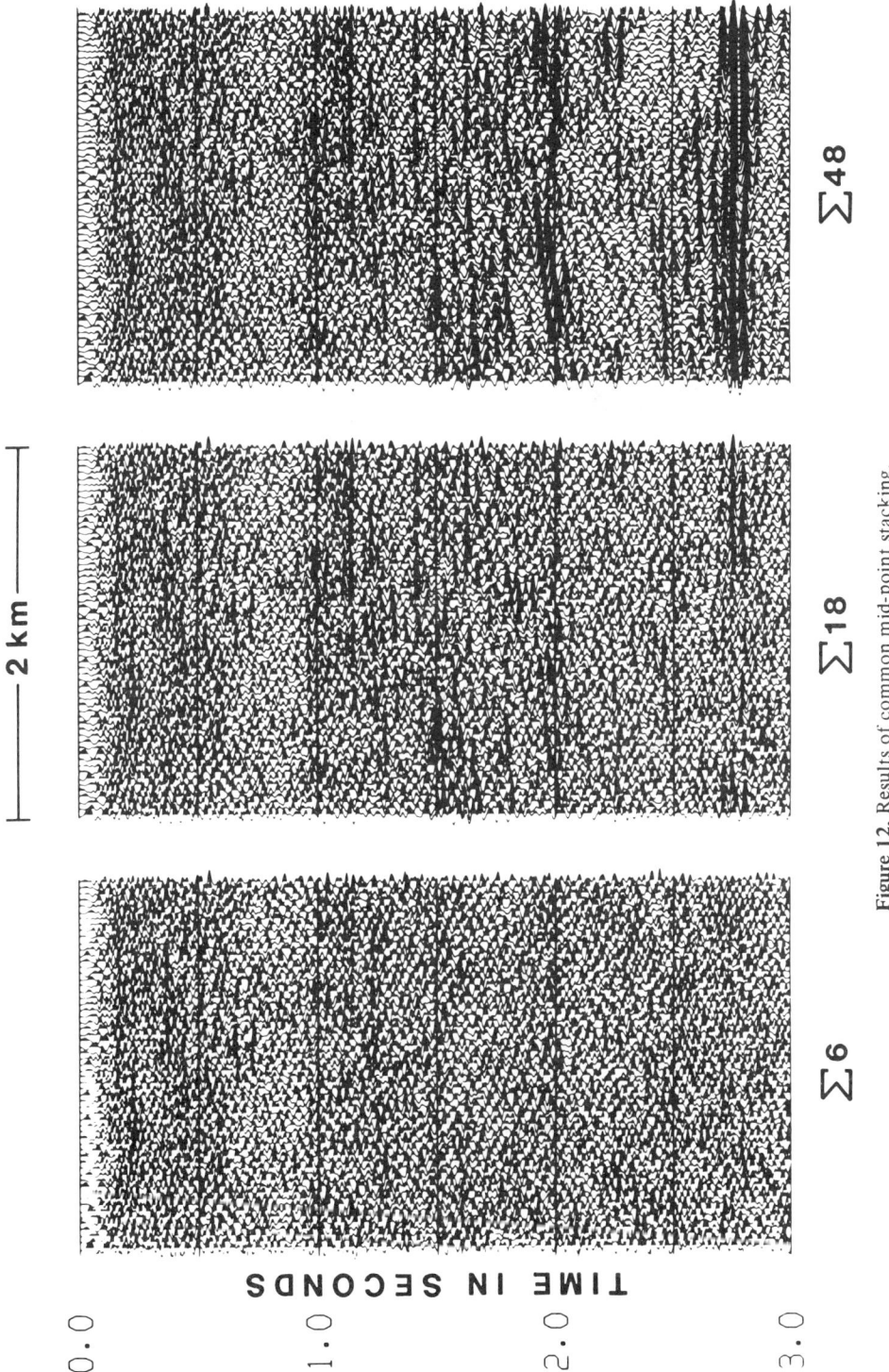

Figure 12. Results of common mid-point stacking.

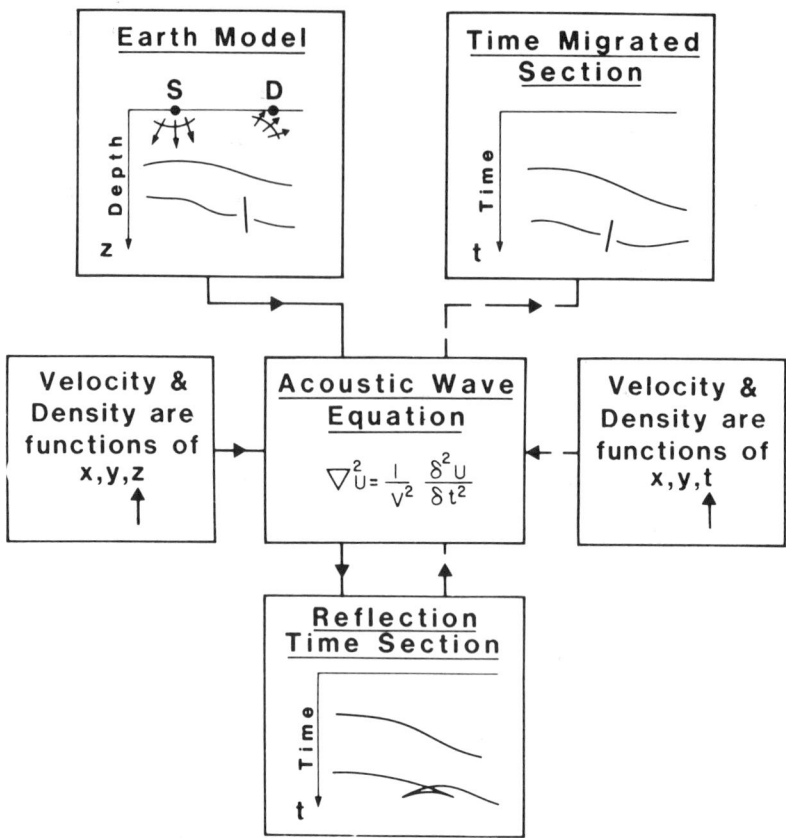

Figure 13. Seismic modelling. Down the solid line – forward modelling, depth-to-time. Up the dashed line – time migration as inverse modelling, time-to-time.

ing out a seismic reflection survey since the spatial variations in velocity and density are all that can be deduced from measurements of mechanical reflections. In practice, we make an estimate of the gross features of the velocity variations as a function of travel time (an observable) from prior data analysis or by extrapolation from nearby borehole information, ignoring variations in density, and are agreeably surprised that 9 times out of 10, the images on the migrated section are very much clearer and interpretable than their equivalents on the unmigrated section. Figs 14, 15, 16 and 17 show typical examples. The essence of the improvement is that the geometric forms of the reflections are more correctly presented on the migrated section – migration gives no significant additional information on velocities and densities, nor will it do so in my lifetime, except in the limited situation of extrapolation away from a well – a confident statement which many research workers are endeavouring to prove wrong and which I make with the hope that it may irritate others and so spur them in the attempt. What will be achieved in the next few years is a closed loop, interactive, computer graphics, modelling system which will iterate on a gross *spatial* velocity function and control record section migration to achieve clearer and more correctly positioned geometrical images of the subsurface. The reason I doubt that development of the method will allow significant velocity/density information may be summed up in two phrases – 'noise' and 'signal waveform'. If more research was carried out on the properties of source-generated noise instead of making the ludicrous assumptions that it is white, random

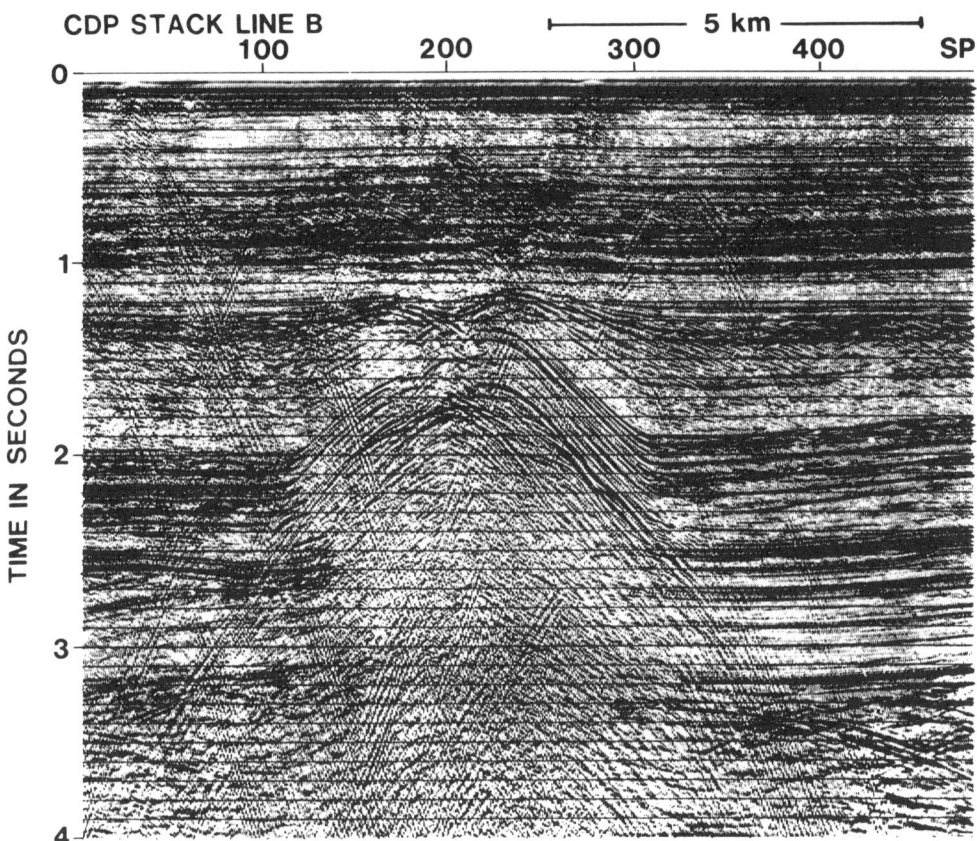

Figure 14. A seismic section before migration.

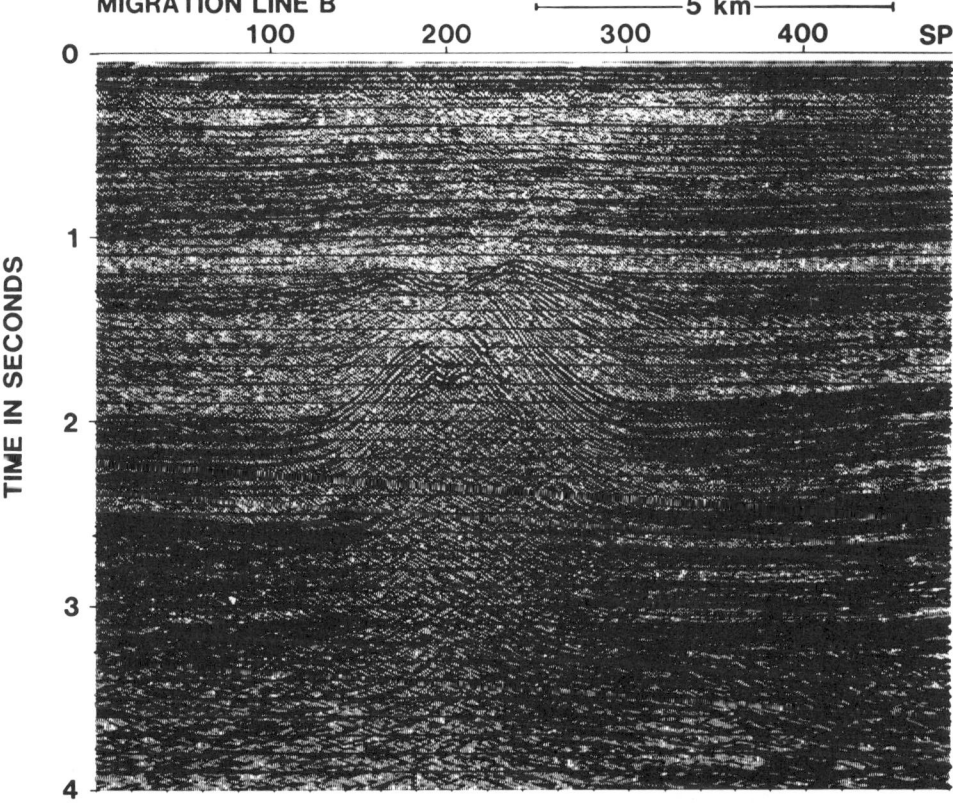

Figure 15. The section in Fig. 14 after migration. Note the narrowing of the salt dome and the clarification of the collapse structures on its peak.

and Gaussian, then faster progress might be made. In principle, the waveform radiated by the source may be precisely measured, and at sea this will 'soon' be achieved. But the waveform required for detailed unravelling of the velocity and density functions is that of the recorded echo, and in the absence of a well this can be estimated only by (spectral) analysis of the recorded traces; and 2–3 s of data, even if noise free, is not enough for a sufficiently accurate estimate.

All current algorithms are based on solutions of the *acoustic* wave equation, so shear waves are ignored. It is also assumed that density is constant. Further, the records to be migrated are assumed to contain no echoes which have been reflected more than once. Since the original field records are often full of multiply-reflected echoes this means that the multiples must have been considerably reduced by pre-migration processing (deconvolution and CMP stacking). It is also normally assumed that the CMP stacked trace (Fig. 11) is equivalent to the trace which would have been recorded if the source and detector had been co-incident (zero-offset).

Most migration procedures then make use of the 'exploding reflector' hypothesis. This hypothesis may be understood by reference to Fig. 18. Fig. 18(a) illustrates a physically possible experiment in which a coincident source-detector pair are moved incrementally along a line, recording a single trace at each location. Plotting these traces side-by-side will produce the seismic section illustrated in Fig. 18(c). Consider now the result of the hypothetical experiment in Fig. 18(b) where a line source (a sheet in 3D) is initiated at time zero. Suppose that at each point along the line the source strength is proportional to the amplitude and polarity of the reflection coefficient at the corresponding position in the real Earth. The wave travelling up from the 'exploding reflector' will reach the ground surface at times equal to exactly one-half of those recorded on the surface-to-surface CMP record section. If we have a lot of reflections, as we do, we may solve the wave equation for each in turn and, since it is a linear equation, superpose the solutions and double all times to obtain the recorded section. So, if we can obtain the appropriate exploding reflector model this will be identical to the actual distribution of reflectors (and diffractors) in the real Earth.

This is done by making use of the Kirchoff Boundary Integral to carry out inverse modelling — that is, to proceed from the zero-offset (CMP) time section to the sought-for exploding reflection model. Refer to Fig. 19. P is an elementary point source at a distance R from an observation point on the free surface. It may be shown (e.g. Hosken 1981) that the strength of the source, u, at position $P(x_1, y_1)$ is given by

$$u(x_1, y_1, Vt_0) \simeq -\frac{1}{2\pi} \cdot \frac{t_0}{V^2} \int_S \frac{1}{t_1^2} (\partial u/\partial t)_{S, t_1} dS \tag{1}$$

where x, y and z are rectangular co-ordinates with z vertically downwards, V is the acoustic wave velocity, t_0 is Z_1/V, t_1 is R/V, S is the surface defined by the ground and an infinite hemisphere.

Since values on the infinitely distant hemisphere make a negligible contribution to the integral, u_P may be calculated from the ground surface observations — provided V is known.

In practice, it is assumed that the earth section is invariant in the y-direction and the surface integral then reduces to the line integral:

$$u(x_1, y_1 = 0, z_1) \simeq \frac{1}{\pi} \int_0^\infty \frac{t_0}{V^2 t_x^{3/2}} [f(t_x) * u(x, 0, 0, t_x)] \, dx, \tag{2}$$

where $*$ denotes convolution and

$$f(t_x) = (2|t_x|)^{-1/2} \delta(t) - (2|t_x|)^{-3/2} [1 - H(t)],$$

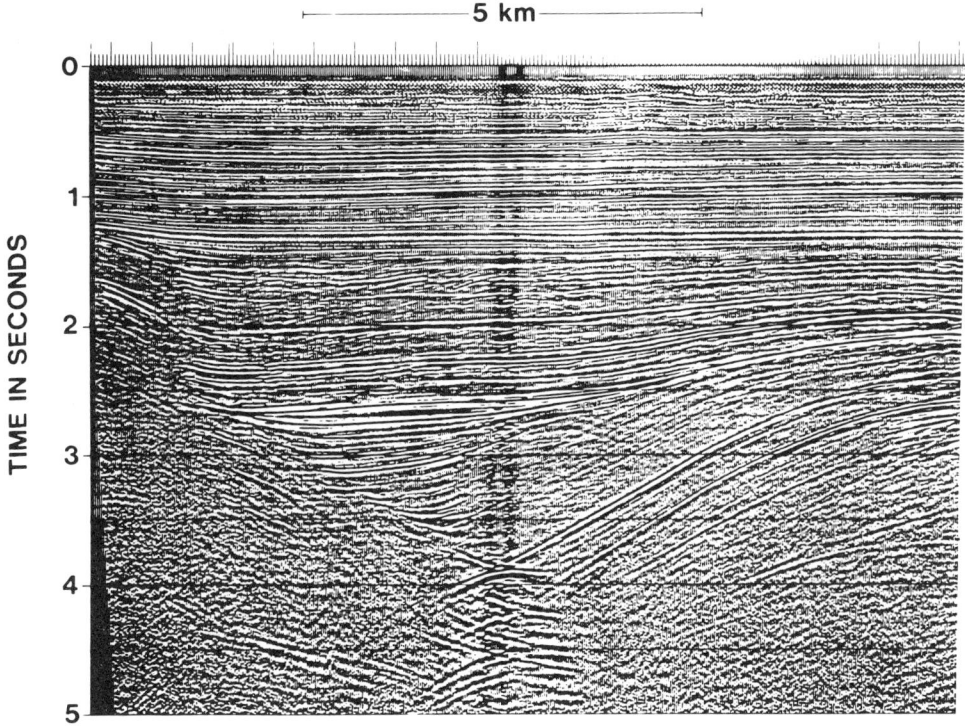

Figure 16. A seismic section before migration.

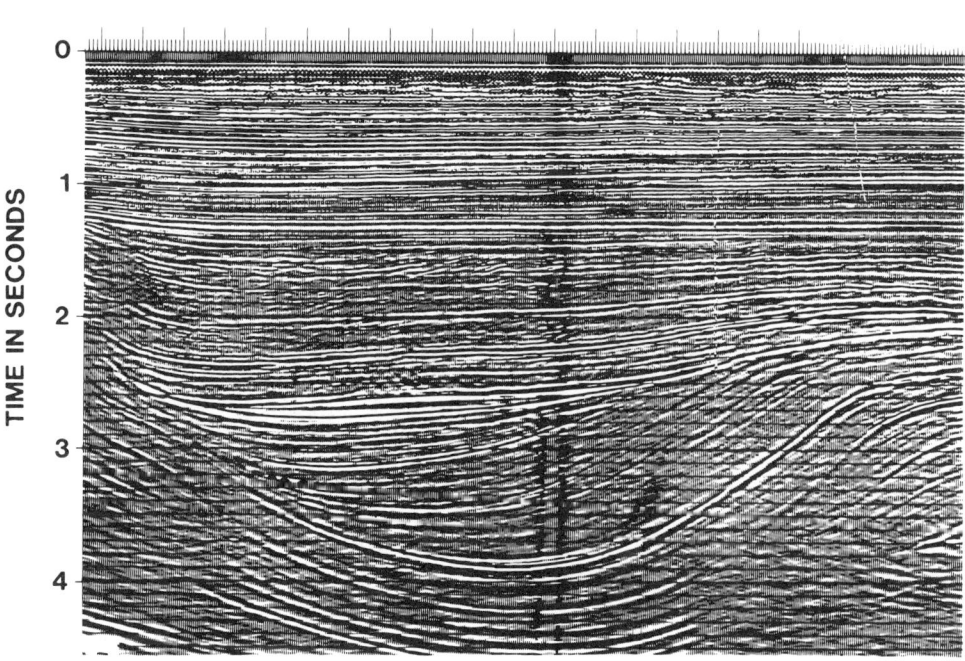

Figure 17. The section in Fig. 16 after migration. Note the broadening of the syncline and the improved clarity of the reflection terminations.

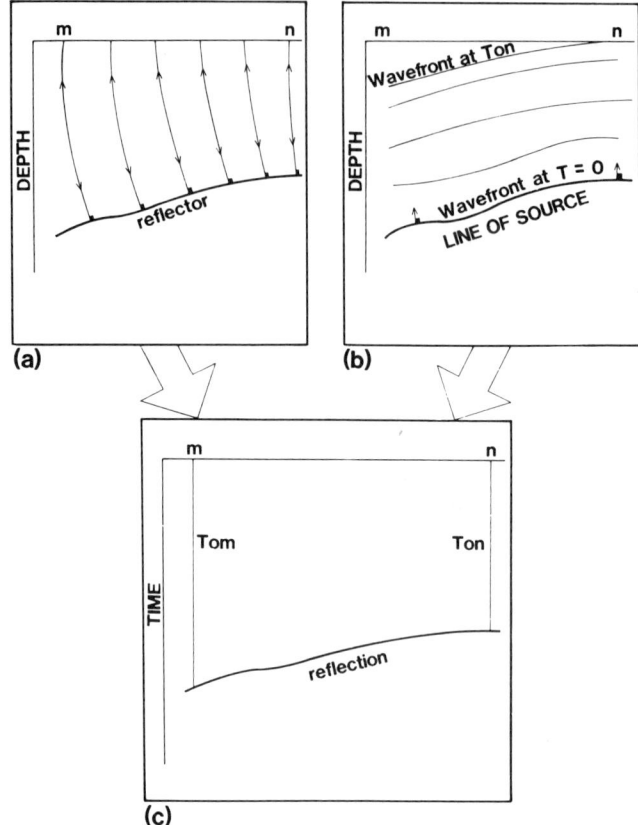

Figure 18. Imaging principle – the exploding reflector hypothesis. (a) The real experiment with coincident source and detector. (b) The hypothetical experiment. The medium is identical to that in (a) except that all velocities are halved. (c) The zero-offset section. Note that (a) and (b) are not equivalent if there are extreme lateral variations in the medium.

where $\delta(t)$ is the Dirac impulse and $H(t)$ is the Heaviside unit step. $f(t_x)$ is a maximum-phase, half-differentiating filter operator, giving 45° phase lag and a 3 dB per octave increasing amplitude with frequency. The integral in equation (2) is approximated by a summation and carried out in the computer as indicated diagrammatically in Fig. 20. The curve along

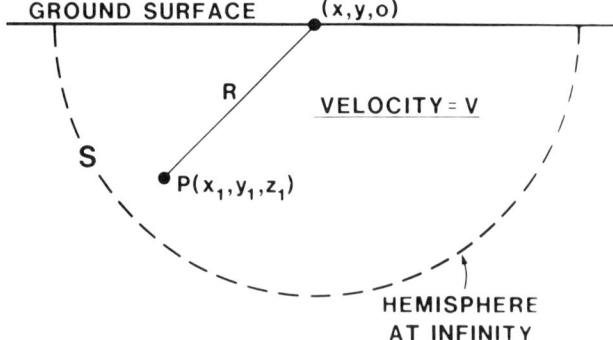

Figure 19. The Kirchoff integral. The radiation from a buried source at P is recorded on the ground surface at $(x, y, 0)$.

which the samples are 'gathered and summed' is that relating t_x to t_0, i.e.

$$t_x^2 = t_0^2 + (x - x_1)^2/V^2$$

and is therefore a hyperbola. If V varies with depth only equation (2) may still be used, and the summation curves remain closely hyperbolic, though V is now replaced with its time-averaged, root mean square value $V_{\text{rms},t}$. If V varies only slowly with x, then summation curves which are easy to implement in the computer can still be defined, but as soon as lateral variations in V become large the procedure, while still being useful in that it clarifies the image, will position that image incorrectly. This mislocation of the image was first discussed by Hubral (1977) who introduced the concept of image rays and indicated how they might be used to remove the mislocation error, a subject elaborated upon by Hatton (1980) and Larner et al. (1981). Image rays are used extensively when accurate positioning is required, such as when locating a well, but normally this is done only after the number of samples on the record section has been drastically reduced (by up to a factor of 1000) by identifying a few key reflections on each trace, representing them by a single sample at the appropriate time and replacing all other samples by zeros.

It has been implied in the previous paragraph that migration will appreciably clarify the image even if the velocity field is not known very accurately. This is an advantage in that it means that worthwhile signal processing can proceed without detailed knowledge of the velocity. On the other hand, it means that measurement of image clarity will not lead to accurate velocity estimates and accurate velocities are in many cases what are needed for the proper location of wells and the detailed mapping of potential hydrocarbon reservoirs, even when the image ray distortion mentioned above is negligible. As a simple example consider the problem illustrated in Fig. 21. It is desired to test the indicated fault block. Referrring to Fig. 26 and putting $V_1 = V_2 = V_3 = V$ it may be seen that the lateral migration shift of a reflection segment is given by $VT_n \sin \theta/2$, which is proportional to V^2. A 10 per cent error in V will therefore give a 20 per cent error in lateral shift, which could result in the well being located outside of the edges of the fault block.

The most popular migration algorithms are based on finite difference schemes for the solution of the wave equation, not the boundary integral method outlined above. Their

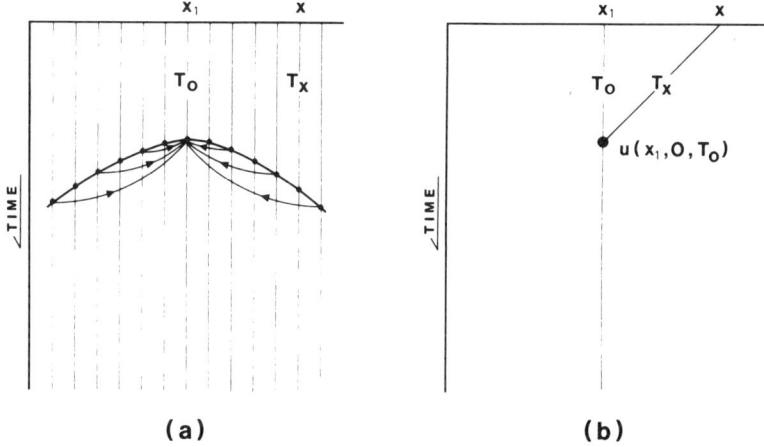

Figure 20. Migration by Kirchoff summation. (a) Zero-offset field (CMP stacked section). Integration of the integral is carried out by summing together, after scaling and filtering according to equation (2), all the u values which lie along the hyperbola. This summation, which takes place over a few hundreds of traces, is output at x_1, T_0 as indicated in (b).

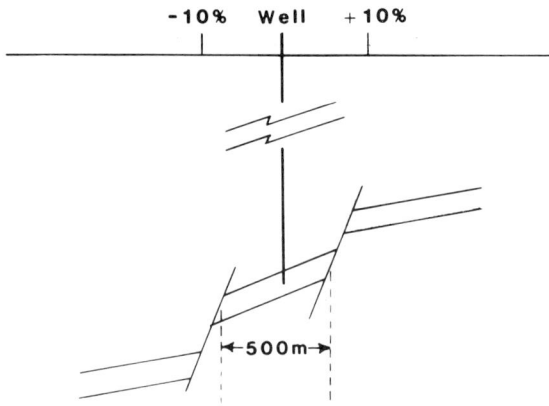

Figure 21. Migration positioning error. Errors of ±10 per cent in velocity will not strongly affect clarity of the image on the seismic section but may result in a well missing its target.

introduction was due to Claerbout (1971, 1976), who also introduced a frame of reference which moved upwards with the average wave velocity with the consequence that in the calculations any downward moving energy (i.e. multiples) was rapidly attenuated. The scheme is less demanding of computer time than the Kirchoff method and is therefore cheaper, it also has the surprising advantage of being more effective because it is less accurate! This is because of the presence of noise. The Kirchoff method is accurate even at very large dips so that noise spikes – which, of course, the algorithm treats the same as signal – are smeared out over large circular arcs producing the well known 'smiles', much as may be seen within the salt of Fig. 15. The finite difference operator does not deal so well with large dips and therefore does not organize the noise to nearly the same extent. Of course, if reflectors with large dips are present then Kirchhoff Summation may be preferable to the Finite Difference method.

The fastest algorithm of all makes use of the Fourier transform to carry out its manipulations in the time-frequency spatial wavenumber domain. As introduced by Stolt (1978), it is exemplary if velocity is constant (it never is), adequate if velocity varies only with depth and poor if the velocity varies laterally. Because of its intrinsic speed, many researchers are endeavouring to modify it to handle lateral velocity variations; if they succeed it will no doubt become the preferred method. In passing, it may be noted that migration is no longer reserved for sections which exhibit severe geometrical complexity but is being used more and more to remove scattered energy and to clarify reflector terminations at unconformities and faults. This is due in large measure to the impetus given to the introduction of migration as a routine procedure as a result of the *cheapness* of $f-k$ migration (Stolt 1978).

Migration as described above starts with a zero-offset reflection section. Such a section could be obtained by surveying with source and receiver placed very close together but the signal-to-noise ratio would normally be so poor that reflections would be largely invisible. A CMP stack is required to raise the SNR to an acceptable level. Fortunately such a stack normally approximates closely to a zero offset section which, via the exploding reflection hypothesis, is taken to be a solution of the wave equation. Nevertheless, when reflector structure is complex a CMP stack does not approximate a zero offset section, nor is it a solution of the wave equation since it results from a superposition after the *non-linear* CMP processing of the individual solutions (the input shot records). Consequently an inverse modelling procedure (migration) based on the wave equation will not work. There are various ways of 'fudging' the migration procedure to handle such data but the only satisfactory procedure is to go back to first principles, keeping the high degree of redundancy necessary for signal-to-noise enhancement, while applying wave-equation processing to individual shot records, each of which is obviously a solution to the wave equation since it is the result of an actual experiment. Such a scheme has been described by Schultz & Sherwood (1980). In their procedure each field record is taken separately and the wave equation used to compute what the record would have been if the detectors had been placed a small distance (say 50 m) below the ground surface. This is illustrated diagrammatically in Fig. 22. The data are then sorted into records, each of which relates to a single detector but all the sources. Reciprocity is invoked to say that detectors and sources may be interchanged, with the result that the record may be treated as if it were the result of a single shot (replacing the single detector at 50 m depth) and a number of detectors (replacing the several sources on the ground surface). The wave equation is then used to compute the record which would have been obtained with the 'reciprocal detectors' placed at the −50 m level. These two steps result in obtaining a set of records which would have been obtained if both the sources and detectors had been 50 m beneath ground surface. This two-stage procedure is continuously repeated and on each occasion the reflection times with respect to the lowered recording level become earlier and the reflected energy moves inward from the larger offsets (source-detector distances) toward zero-offset. As the revised reflection time reaches zero, so does the reflected energy reach the zero-offset trace; the procedure is stopped for that (time, energy) pair and the sample is output to form part of the output migrated record section (Fig. 23). Assuming that reciprocity applies sufficiently well, which is somewhat uncertain, each 'downward continued' record is a true solution to the wave equation which, in the case of Schultz and Sherwood, is obtained by means of a finite difference scheme. Although their paper was called 'Depth Migration before Stack' no

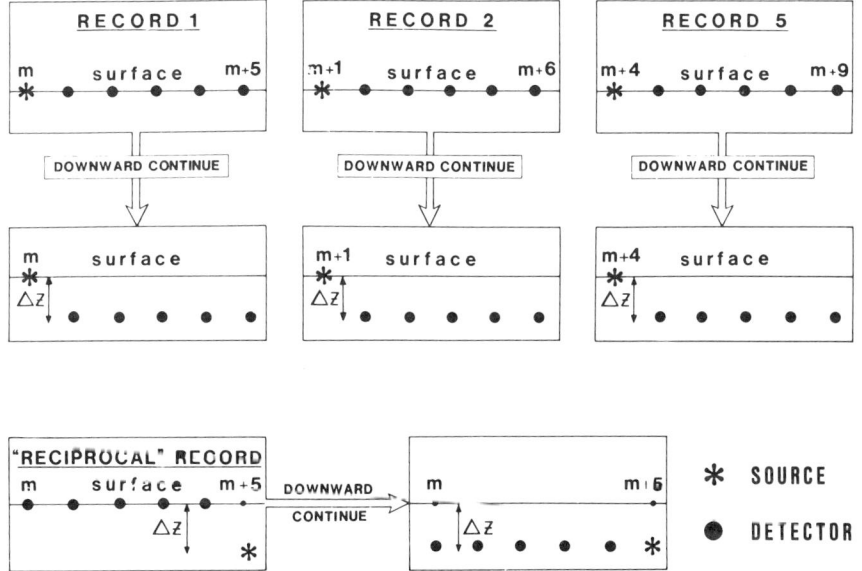

Figure 22. Downward continuation. By invoking reciprocity it is possible to calculate the records which would have been obtained at depths of $\Delta z, 2\Delta z \ldots n\Delta z$. In this illustration the source and five detectors are equally spaced and successive records are taken after moving them along the line by increments of one spacing interval.

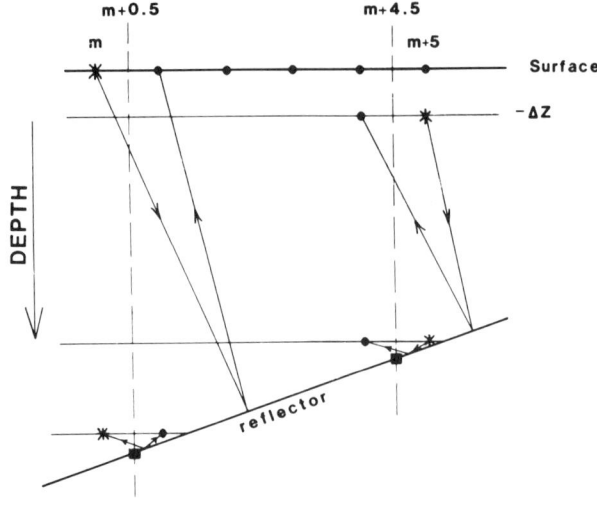

Figure 23. Downward continuation. As the hypothetical recording level is lowered the reflection (and diffraction) times become less and less. When an event time reaches zero the 'recording' level has been lowered to the reflector (or, diffractor) position. Note that the output is a depth section and this requires the velocity field to be specified as a function of space (2 or 3D), not time.

subsequent stacking procedure is required since the technique implicity allows for moveout correction and stacking at the same time as it migrates.

This downward continuation procedure is predicated upon a velocity–*depth* model and hence is called depth migration. The procedures outlined previously require velocity–*time* models as input and hence are called time migrations. In depth migration the initial velocity–depth model is obtained from interpretation of a CMP stacked section or from a time-migrated version of it. This initial model is then changed interactively and iteratively until the resulting depth migration is judged by some set of criteria to be an optimum. Since depth migration works on the individual field records, it has to process 20–100 times as much data as the conventional time migration, for which the input is the CMP-stacked section. Also, because it honours a velocity–depth model of arbitrary complexity, it uses a more complex algorithm. Consequently, it is 100 times or more as costly as time-migration and is rarely used. Hosken & Deregowski (1982) discuss the principles of decision making when moving from less accurate to more accurate migration procedures — there are many more choices than I have mentioned — and give a practical example.

The first paper on seismic record section migration was probably that by Rockwell (1971). It rapidly became the most active area of research into seismic data processing and continues to be so. But we need a lot more attention paid to ensuring that the earth models used and noise properties assumed are reliably accurate idealisations and a lot less (or, at any rate, a little less) attention paid to purely algorithmic development. Of the many post-war improvements in seismic reflection data processing, record section migration ranks third after CMP stacking and statistical deconvolution.

4 Interpretation technique

Interpretation is concerned with turning seismic record sections into geological ones and then using these to deduce basin history as it relates to petroleum generation, migration and accumulation. Stage one, therefore, is concerned with getting as good a description of today's geology as possible. In decreasing order of reliability one obtains structure including faults, unconformities, lithology, depositional environment and stratigraphy. The last three items are particularly dependent on well information and may be put in different orders by different geophysicsists. No mention has been made of direct detection of hydrocarbons: it can be done — but rarely. Stage two, obtaining the petroleum-related history of the sedimentary basin obviously moves into the area of mainstream geology and so falls outside the scope of this article.

Just as there are chrono-stratigraphy, bio-stratigraphy and litho-stratigraphy so too there is seismo-stratigraphy. This is concerned mainly with mapping unconformities and with identifying and interpreting offlaps, onlaps, bottom laps and related reflector terminations. The principles involved and a self-contained and largely accepted terminology is given in a series of papers by Vail *et al.* of the Exxon group (e.g. Vail *et al.* 1977). Although the industry concerned itself with these matters before the publications of the Exxon group, there is no doubt that they raised seismic stratigraphy almost to a separate discipline by their series of outstandingly innovative contributions. Their 'invention' of seismic stratigraphy certainly rates in importance with that of seismic migration.

But the bulk of seismic interpretation is concerned with structural mapping, an example of which is given in the next section.

4.1 RAY TRACE MODELLING

The CMP stacked section in Fig. 24 shows a series of reflectors terminating against the flank of a salt wall on the left hand side of the figure. Fig. 25 shows a contour map at potential reservoir level prepared from a number of such sections. The next step is to choose a well location to test the reservoir for the presence of oil. It must be located so that the well meets the reservoir sand just outside the salt wall but remains inside the shaded area which indicates the maximum possible lateral extent (closure) of any oil accumulation. Reflect*ion* (x, t) segments measured on Fig. 24 must be accurately migrated to reflect*or* (x_m, z_m) pairs as indicated in Fig. 26. Record section migration as described in Section 3.2 is as yet neither sufficiently accurate nor sufficiently flexible to 'solve' the problem. Instead, we turn to a much simpler technique, ray trace migration. Referring to Fig. 26, and remembering that Fig. 24 shows a *zero-offset* section, it may easily be seen that at the surface the angle of approach of the zero offset ray, θ_1, corresponding to the reflection recorded at time t at location x is given by the equation

$$\sin \theta_1 = V_1 \cdot \partial t / (2 \partial x);$$

∂t and ∂x may be measured off the seismic section so that, if V is known, the ray may be traced backwards into the earth model, refracting it at each velocity contrast, calculating the travel time $\Sigma(L/V)$ and terminating the procedure when it equals one-half of the observed reflection time. The reflecting element may then be drawn as a small segment perpendicular to the ray end. This procedure starts with the shallowest reflector, in this case the sea-bed, and maps in successively deeper reflectors until all the reflections have been migrated. Fig. 27 shows the reflection segments which were 'picked' on the CMP stacked section shown in Fig. 24 and Fig. 28 shows a near final result of ray trace migration.

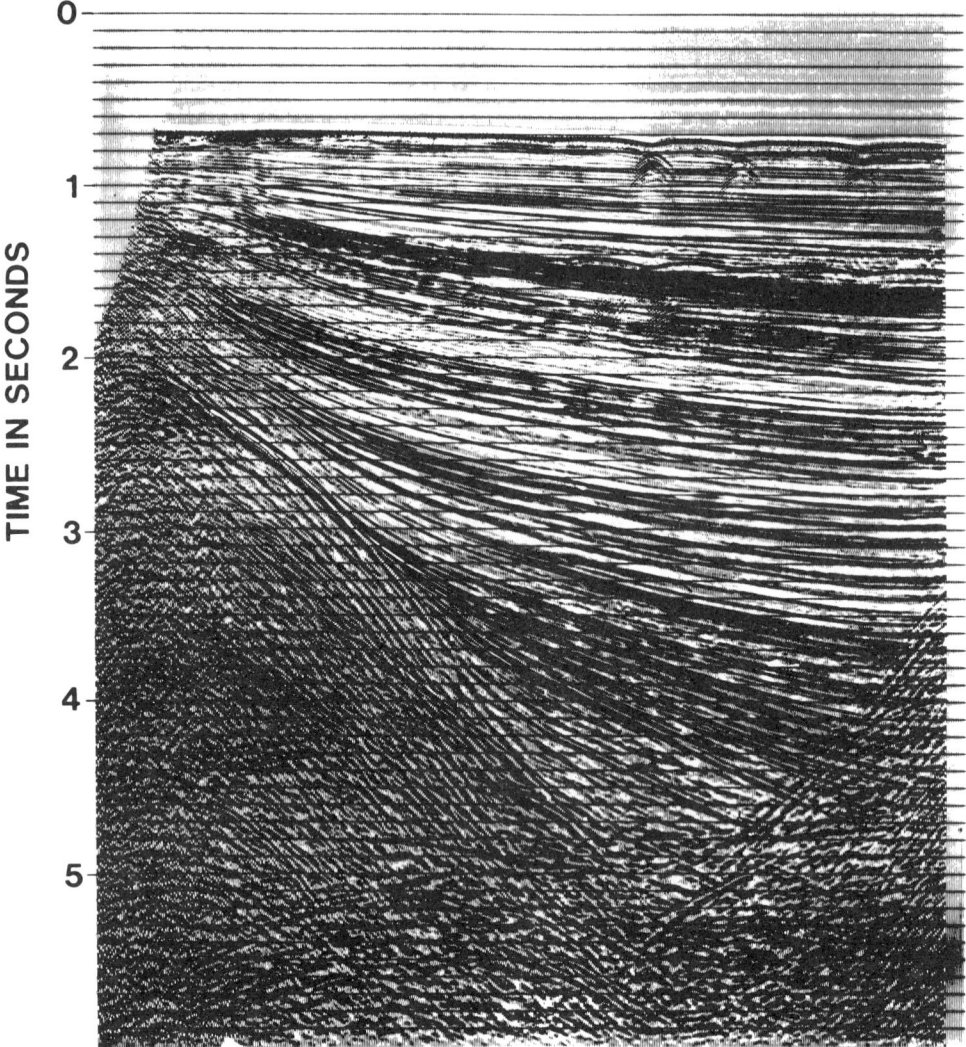

Figure 24. CMP stacked section in a salt province. Note reflections terminating against a salt wall on the left. There is also a deeper salt feature on the right.

Segment numbering informs the interpreter which reflection segment defines which part of the structure so that if he does not like the migrated result he can assess whether or not it is permissible to alter his initial x, t 'picks'. Note in this case that segment 11 is decreased in spatial extent and moved laterally by about 4.0 km, while segment 10 is increased in spatial extent and reversed in curvature. Note also that in spite of continuous dense recording on the ground surface significant gaps occur in coverage on the reflector.

An exact depth model depends upon exact velocities. These mainly come from analyses of CMP gathers controlled, whenever possible, by measurements in wells. Reference to

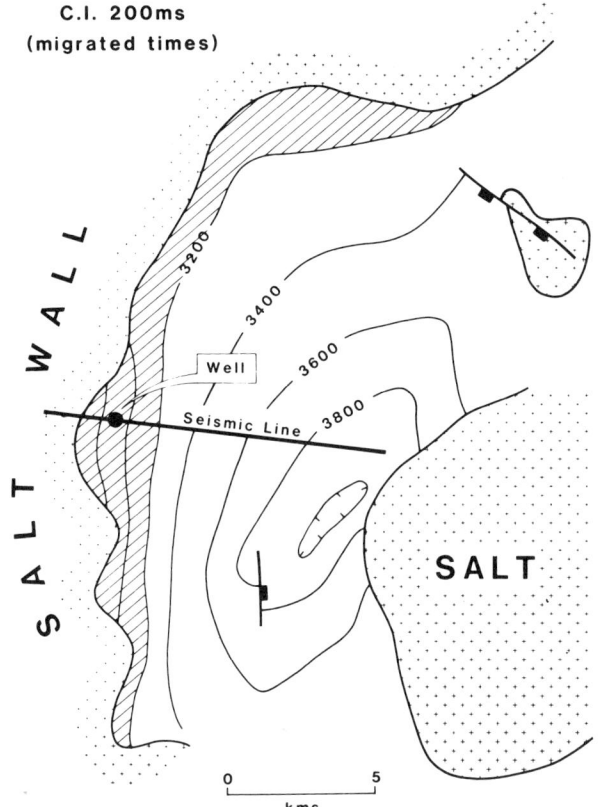

Figure 25. Isochrons at reservoir level.

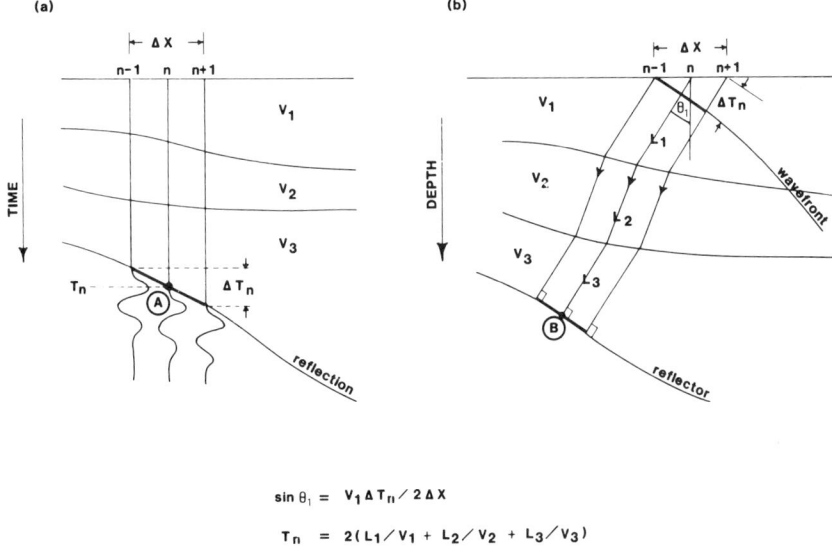

$$\sin \theta_1 = V_1 \Delta T_n / 2 \Delta X$$

$$T_n = 2(L_1/V_1 + L_2/V_2 + L_3/V_3)$$

Figure 26. Ray trace migration. (a) Zero-offset (CMP stack) section showing reflec*tion* segment A. (b) Depth section. Reflec*tor* segment B is the origin of A, its accurate location demands accurate knowledge of the velocity field.

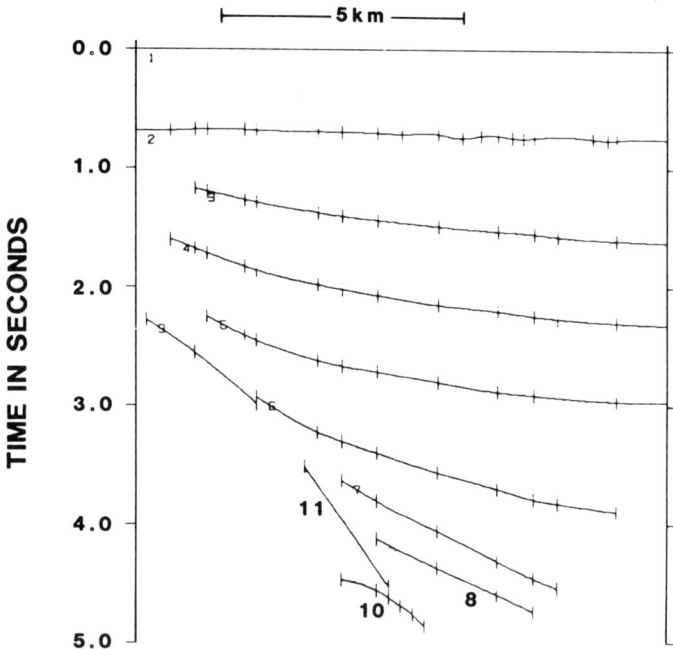

Figure 27. 'Picked' reflections. This is an interpreted overlay to the CMP stacked section of Fig. 24.

Section 3.1 and Fig. 11 makes it obvious that the reflection moveout measured on a CMP gather is directly related to velocity. In fact, for the ideal case of a constant velocity, V,

$$t_x^2 = t_0^2 + x^2/V^2.$$

So, by analysing observed moveout times layer velocities may be estimated. Note that an estimate may be obtained at each common mid-point that is, at 25 m intervals in this case.

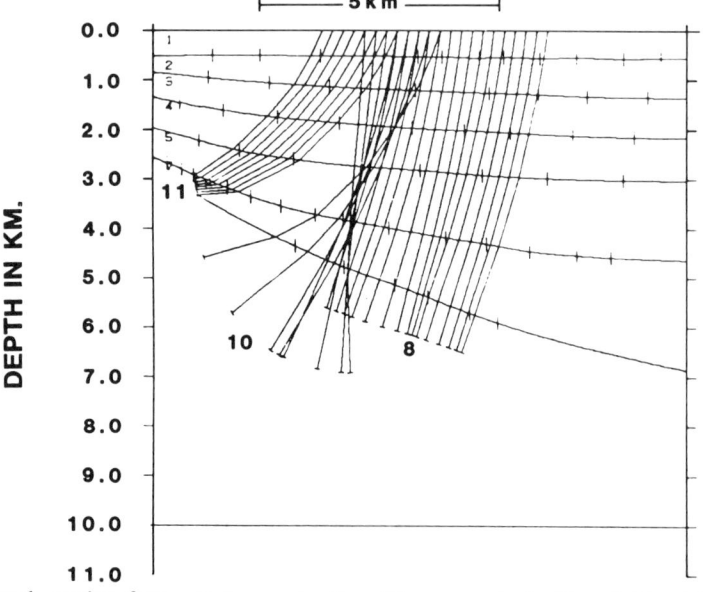

Figure 28. Depth section from ray trace migration. The ray ends 8, 10 and 11 originate from the correspondingly numbered reflection segments in Fig. 27.

Since layer velocities vary laterally due to lithologic changes and vertically due to changing overburden pressure, this essentially continuous velocity estimation procedure is essential. Constant velocity layers do not exist or, at least, only very rarely.

So the final depth model, obtained after several iterations, satisfies not only the t, x pairs on the CMP stacked section but also the moveout functions from each primary reflection on the individual CMP records. The result, in this case, was a mean depth error of around 0.75 per cent. Unfortunately, as with the results from 54 other wells drilled in the same sedimentary basin, no commercial oil was found!

Note that this graphically interactive ray trace procedure starts from the CMP stacked section, which is the basic result of a seismic survey, and ends with a depth model. It is also possible to start with a depth model and compute zero offset section by ray tracing. This would then be compared with the observed CMP section and the depth section altered iteratively until the simulated and real sections agree. This latter procedure founders on the difficulty of generating the initial depth model so, while many programs have been written which start from a depth model, few have had significant use.

5 The future

Though the principles remain essentially the same, the pace of technological change advances the practice of seismic reflection survey with unabated speed. Mainly this relates to the gathering of increased data volumes and the use of digital computers to handle them. The increased data volume is due partly to an increase in the data redundancy factor as a means of combating shot-generated noise and partly to decreasing the line-spacing in the survey. The latter is necessary in order to enable us to drop the over-simplifying assumption of a 'two-dimensional' subsurface, which currently underlies the vast majority of seismic data processing. Digital computers and related equipment continue their rapid improvement in cost effectiveness but the day is still far off when their power will be cheap enough to warrant application of signal enhancement and imaging algorithms which are already definable. In fact, it will still be a long time before the computers are large enough, much less cheap enough.

1983 is the year of the colour graphics work station. These are of use when interpreting seismic sections (such as those shown in Figs 2 and 3) after they have undergone the full range of signal enhancement procedures. Current work stations can neither store, nor call-up, nor display, nor provide hard copy of enough data fast enough, nor cheaply enough, for ubiquitous use. But in five years they will — and then the seismic interpreter will no longer need to handle paper?

In 1982 the first commercial service was offered for satellite transmission of seismic data from the survey area to HQ. The maximum transmittal rate is 56 kbaud, which is sufficient for many quality control purposes. A rate of 1.544 Mbaud is promised which would enable *current* data acquisition results to be transmitted in real time. HQ processing and interpretation will then have the challenge of reacting fast enough to modify the survey specification during the data acquisition period, even for fast moving marine surveys. It will not happen.

From a purely technical viewpoint it might be better to aim at moving the computer power to the acquisition ship or base camp — something which has been talked about for a long time and is gradually being achieved — but, even if computer developments make this worthwhile, the desire to get data back to the safety of HQ, where it can be more easily integrated with all the other exploration and commercial considerations, will always be a strong counterforce.

There are a number of peripheral seismic techniques and measurements which have been in R&D for many years and one day may survive in the harsh environment of the real world. Shear wave reflections may give added useful information, particularly if we can devise worthwhile numerical models for those rock properties which tell us something we need to know; attenuation — a very blunt tool indeed — may some day be worth measuring; and more measurements with detectors and/or sources down deep holes may occasionally be of benefit. One of the great unsolved problems in seismic prospecting is mapping the near surface sufficiently well to reduce significantly its masking effect on the deep subsurface. Perhaps this problem may be alleviated by making better use of interface modes of propagation, though we shall certainly have to drop the laughable assumption that the Earth consists of plane, parallel, layers.

At present the earth models underlying acquisition, processing and interpretation form an hierarchy of increasing complexity. Attempts are underway to dispense with the simpler models and to use, iteratively and interactively, the most complex of these models to control acquisition and processing parameters in addition to its current role in providing a drilling location. We know what we want to do — we merely need cheaper and bigger computers and display devices in order to do it.

Whatever may be thought of it in other fields, Marshall Macluan's dictum 'The medium is the message' certainly applies to seismic prospecting. To unravel the message we must study the medium, or at any rate our model of it. If we do so it is just possible that eventually, and before the oil and gas 'run-out', we will be able to detect hydrocarbon accumulations from surface seismic measurements as a matter of routine, though not in my lifetime.

Acknowledgments

I wish to thank those of my colleagues who commented on the first draft of this paper and the Chairman and Board of Directors of The British Petroleum Company plc for their permission to publish it.

References

Berkhout, A. J., 1980. *Seismic Migration,* Elsevier, Amsterdam.
Blake, F. G., 1952. Spherical wave propagation in solid media, *J. Ac. Soc. Am.,* **24,** 211–215.
Bruel & Kjaer, *Hydrophones – their characteristics and applications,* Bruel & Kjaer (UK) Ltd., Hounslow, England.
Claerbout, J. F., 1971. Toward a unified theory of reflector mapping, *Geophysics,* **36,** 467–481.
Claerbout, J. F., 1976. *Fundamentals of Geophysical Data Processing,* McGraw-Hill, New York.
Dennison, A. T., 1953. The design of electromagnetic geophones, *Geophys. Prosp.,* **1,** 3–28.
Edelmann, H. A. K., 1975. *Improved Airgun Arrays,* Prakla-Seismos Report, Prakla-Seismos GMBH, Hannover.
Evenden, B. S. & Stone, D. R., 1971. *Seismic Prospecting Instruments,* Volume 2, Instrument Performance and Testing, Borntraeger, Berlin (under revision).
Giles, B. F. & Johnston, R. C., 1973. System approach to air-gun array design, *Geophys. Pros.,* **21,** 77–101.
Hatton, L., 1980. Depth migration of seismic data from inhomogeneous media, in *Seapex Proc.,* **5,** 11–22.
Hood, P. 1981. Migration, in *Developments in Geophysical Exploration Methods,* **2,** pp. 151–230. ed. Fitch, A. A., Applied Science Publishers.
Hosken, J. W. J. 1981. Imaging the earth's subsurface with seismic reflections, in *The Solution of the Inverse Problem in Geophysical Interpretation,* pp. 179–210, ed. Cassinis, R., Plenum Press, New York.
Hosken, J. W. J. & Deregowski, S. M., 1982. Migration processing: a strategy, in *Proc. of Second ASCOPE Conf. 1981,* pp. 721–744, ed. Saldivar-Sali, A.

Hubral, P., 1977. Time migration – some ray theoretical aspects, *Geophys. Pros.,* **25**, 738–745.

Jeffreys, H., 1931. On the cause of oscillatory movement in seismograms, *MNRAS–Geophys. Suppl.,* 407–416.

Klassen, K. B. & van Peppen, J. C. L., 1982. A novel high-resolution geophone, presented at the *44th Meeting of E.A.E.G.,* Cannes, June, pp. 8–11.

Larner, K. L., Hatton, L., Gibson, B. S. & Hsu, I-C., 1981. Depth migration of imaged time sections, *Geophysics,* **46**, 734–750.

Lerwell, W. E., 1979. Seismic sources on land, in *Developments in Geophysical Prospecting Methods,* **1**, pp. 115–141, ed. Fitch, A. A., Applied Science Publishers, London.

Lerwill, W. E., 1981. The amplitude and phase response of a seismic vibrator, *Geophys. Pros.,* **29**, 503–528.

Leuhrmann, W. H., 1972. Seismic tool boosts S/N ratio of sea streamers, *Oil and Gas J.,* **70** (48), 102–109.

Lugg, R., 1979. Marine seismic source, in *Developments in Geophysical Prospecting Methods,* **1**, pp. 143–203, ed. Fitch, A. A., Applied Science Publishers, London.

Mayne, W. H., 1962. Common reflection point horizontal stacking techniques, *Geophysics,* **27**, 927–938.

Miller, G. F. & Pursey, H., 1954. The field and radiation impedance of mechanical radiators on the free surface of a semi-infinite isotropic solid, *Proc. R. Soc. (London), Ser. A,* **223**, 521–541.

Nooteboom, J. J., 1978. Signature and amplitude of linear airgun arrays, *Geophys. Pros.,* **26**, 194–201.

O'Brien, J. T., Kamp, W. P. & Hoover, G. M., 1982. Sign-bit amplitude recovery with applications to seismic data, *Geophysics,* **47**, 1527–1539.

O'Brien, P. N. S., 1969. Some experiments concerning the primary seismic pulse, *Geophys. Pros.,* **17**, 511–547.

Pursey, H., 1956. The power radiated by an electromechanical wave source, *Proc. Phys. Soc., B,* **69**, 139–144.

Rockwell, D. V., 1971. Migration stack aids interpretation, *Oil and Gas J.,* April 19, 202–218.

Safar, M. H., 1976. The radiation of acoustic waves from an airgun, *Geophys. Pros.,* **24**, 756–772.

Schneider, W. A. & Backus, M. M., 1968. Dynamic correlation analysis, *Geophysics,* **33**, 106–126.

Schultz, P. S. & Sherwood, J. W. C., 1980. Depth migration before stack, *Geophysics,* **45**, 376–393.

Sharpe, J. A., 1942a. The production of elastic waves by explosion pressures, 1, *Geophysics,* **7**, 144–155.

Sharpe, J. A., 1942b. The production of elastic waves by explosion pressures, 2, *Geophysics,* **7**, 311–322.

Sixta, D. P., 1982. Comparison and analysis of downgoing waveforms from land seismic sources, *M.Sc. thesis,* Geophys. Dept., Colorado School of Mines, U.S.A.

Stolt, R. H., 1978. Migration by Fourier transform, *Geophys.,* **43**, 23–48.

Telford, W. M., Geldart, L. P., Sheriff, R. E. & Keys, D. A., 1976. *Applied Geophysics,* C. U. P., Cambridge.

Vail, P. R., Mitchum, Jr., R. M. *et al.* 1977. Seismic stratigraphy and global changes of sea level, in *Seismic Stratigraphy – Applications to Hydrocarbon Exploration,* AAPG Memoir 26, Tulsa.

Waters, K. H., 1978. *Reflection Seismology,* Wiley-Interscience, New York.

White, R. A. & O'Brien, P. N. S., 1974. Estimation of the primary seismic pulse, *Geophys. Pros.,* **22**, 627–651.

Ziolkowski, A., Parkes, G. E., Hatton, L. & Haugland, T., 1982. The signature of an Airgun array: computation from near-field measurements including interaction, *Geophysics,* **47**, 1413–1421.

Predictive Isopach Mapping of Gas Sands from Seismic Impedance: Modeled and Empirical Cases from Ship Shoal Block 134 Field[1]

ROBERT D. WOOCK[2] and ALAN R. KIN[3]

ABSTRACT

Gulf of Mexico Miocene through Holocene stratigraphy exhibits general physical properties conducive to detailed geophysical investigation. Selected seismic attributes may be used directly to produce net gas sand thickness. Synthetic examples demonstrate the relationship between the energy of a sand-top reflected wavelet and the associated gas pay. Reflected seismic energy displayed as acoustic impedance sections are used to produce pseudo "net pay" isopach maps, which are used to predict gas sand thickness, gas distribution, and gas reserves.

The techniques and concepts used in this form of geologic analysis are demonstrated in the rediscovery and development of the Ship Shoal Block 134 gas field of offshore Louisiana. Using the properties of amplitude, acoustic impedance, and reflected energy to produce net pay isopach maps has led to 95% successful prediction of gas reserves within this study area and other areas. The thickness of gas sands estimated by impedance methods is substantially less than the accepted minimum thickness interpretable from conventional seismic sections. In this case study, gas sand thicknesses ranging from 1.5 m to 10 m have been reliably mapped.

INTRODUCTION

Offshore Louisiana, Mississippi, and Alabama sediment densities range from 2.0 to 2.5 g/cm^3; interval velocities range from 1,189 m/sec (shallow gas sands) to 4,267 m/sec (salt); and Poisson's ratios range from 0.1 to 0.5. In a sand and shale depositional system, median values of density, acoustic velocity, and Poisson's ratio result in excellent reflectivity contrasts, which are locally facies dependent. Modern seismic acquisition and processing technology can reveal detailed and accurate data on these lithologic properties. Interval velocity and impedance values can be determined and their distribution mapped in sandstone reservoirs to depths of 3,600 m. An integrated impedance or relative energy scale, tied either to nearby well control or to modeled responses, can yield extent and thickness of gas reservoirs through construction of synthetic net-pay isopachs. The use of impedance-based, synthetic, net-pay isopach maps in planning exploratory and production drilling is a cost-effective means of accelerating the phases of field identification and development.

DISCUSSION

Areas of subtle dip have presented problems of structural delineation to explorationists charged with designing efficient exploration and development programs. Standard seismic mapping techniques yield time-structure maps that may be misleading. The additive effect of slow velocities due to gas-charged or undercompacted sediments may result in apparent time-depth structural inversion. Horizons beneath such zones appear and, thus, are mapped deeper than adjacent areas.

Time-structure maps are devoid of important facies information. The explorationist may attempt to fill this void by constructing a regional net-sand map or porosity-isopach map from available well logs. Outlining seismic anomalies is another approach that can select good well locations.

Recent advances in the use of acoustic impedance data, derived from reflection seismic data, have given the explorationist facies-specific information. Acoustic impedance or relative energy data may be mapped to yield a superior criterion for locating wells and for estimating reserves. In fact, wells drilled on the basis of an impedance-derived facies isopach map (synthetic "net pay") often are later determined to be at either the highest structural position or in the optimal facies development position. Development of the techniques described in this paper paralleled the appraisal, discovery, and delineation of the Ship Shoal Block 134 field.

SYNTHETIC EXAMPLES

The relationship between gas-sand thickness and the impedance of relative energy of a bright spot on a seismic

---

©Copyright 1987. The American Association of Petroleum Geologists. All rights reserved.

[1]Manuscript received, February 27, 1987; accepted, August 5, 1987.
[2]Odeco Oil & Gas Company, 1600 Canal Street, P.O. Box 61780, New Orleans, Louisiana 70161.
[3]Seismograph Service Corporation. Presently K.G.S., Inc., 8203 E. 60th Street, Apartment 1323, Tulsa, Oklahoma 74145.

The writers thank Odeco Oil and Gas Company for permission to publish this paper, and for generous support during its preparation. Also, we thank Seismograph Service Corporation, Tulsa, Oklahoma, for generous technical support, our typist Angie Bova, our draftsman Art Davis, and David L. Godard for his assistance in grammatical matters.

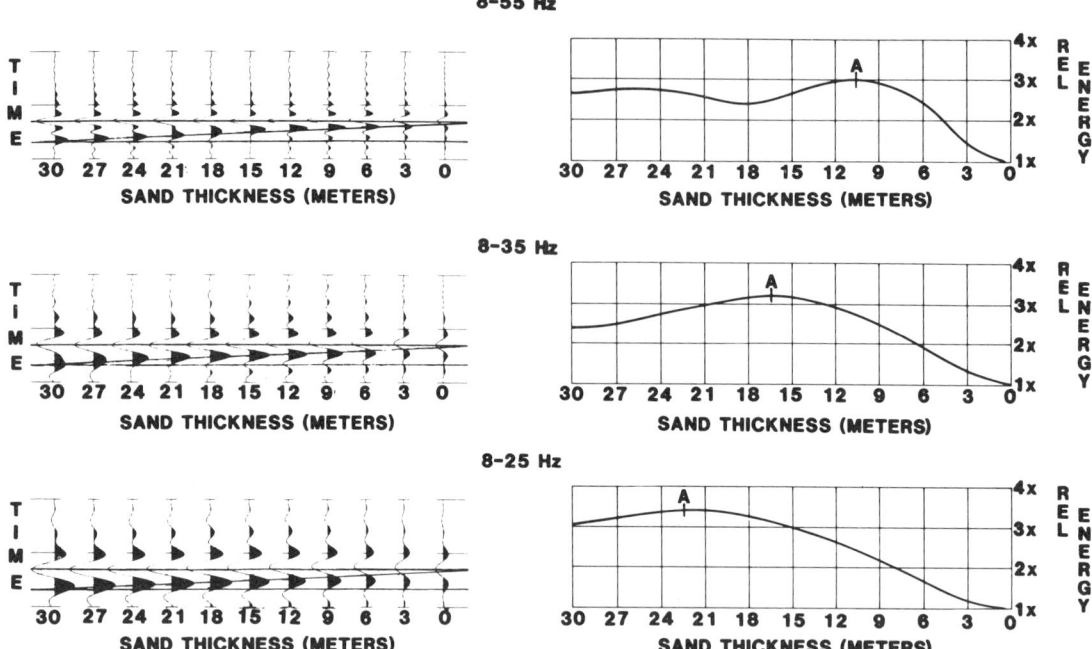

Figure 1—Synthetic responses, with relative-energy vs. gas-pay curves, for a modeled gas sand. Temporal response for top and bottom occurs at and above 24 m in 8-55 Hz band, but not below 30 m for either of other two bands. Areas of relative-energy curves at right of A are useful zones for correlating gas-sand thickness to relative (reflected) energy. Impedance data will (in theory) avoid these limitations.

time section is controlled by the local environment. The parameters affecting the wavelet energy to be considered are the overburden interval velocity, the gas-sand interval velocity, the gas-free sand interval velocity, and the seismic bandwidth.

The limit of temporal resolution and the tuning limit (defining the lower limit of bed thickness resolution) have been discussed by Kallweit and Wood (1982). Temporal resolution (t_r) is determined by

$$t_r = 1.0/(1.509 \times Fu), \tag{1}$$

where Fu = the highest resolved frequency. The timing limit (b/2), the approximate peak-to-trough distance on a wavelet, is

$$b/2 = 1.0/(1.4 \times Fu). \tag{2}$$

For a wavelet with an upper frequency limit of 55 Hz, the limit of resolution will be approximately 12.1 msec. At a reasonable sand velocity of 1.967 m/msec, this yields a thickness resolution of approximately 24 m: $1.0/1.509 \times 55$ cycles/sec = 0.0121 sec/cycle, 1.967 m/msec × 0.0121 sec = 24 meters. Using the same equations for 35 Hz, the thickness will be approximately 40 m, and at 25 Hz approximately 55 m.

However, in the Miocene through Holocene of the Gulf of Mexico, a typical gas-sand thickness is much less than 24 m. Thus, the gas-sand thickness cannot be measured by temporal resolution on a standard seismic time section.

We can examine this apparent limitation with the aid of a simple synthetic seismogram illustrating a total sand thickness of 30.5 m and a gas-sand thickness that ranges

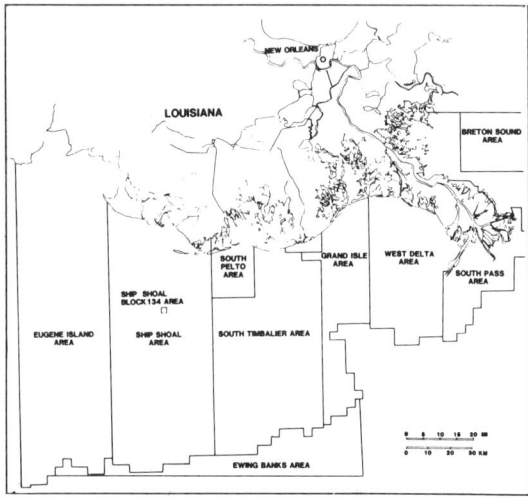

Figure 2—Ship Shoal Block 134 gas field location plat. Field is part of Ship Shoal 113A complex.

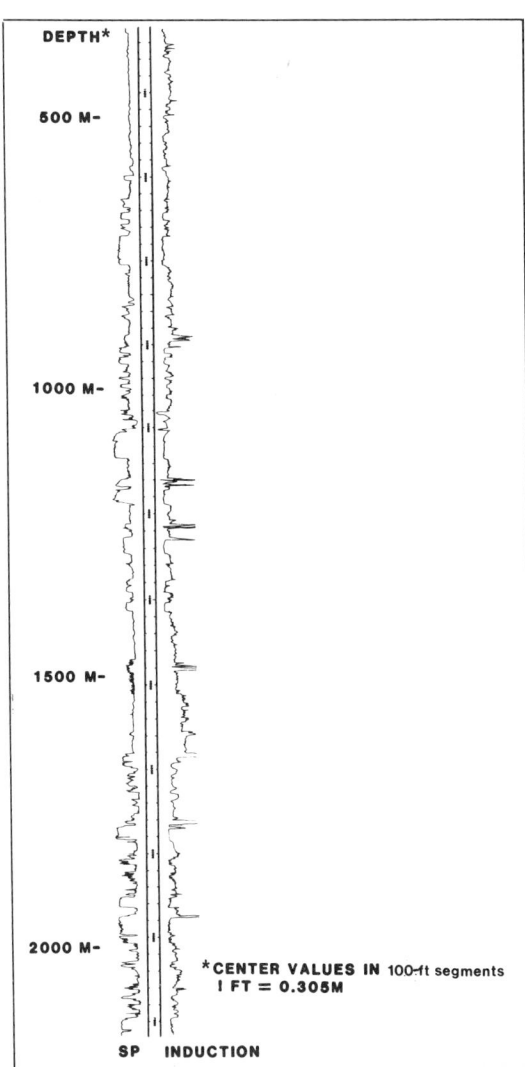

Figure 3—Composite type log, Ship Shoal Block 134 field. Resistive zones are gas sands.

from 0 to 30.5 m (Figure 1). On the left in Figure 1 is the seismic response of a monotonically thickening gas sand modeled with three zero-phase seismic wavelets of different bandwidths (8-55 Hz, 8-35 Hz, 8-25 Hz). The gas-sand thickness varies from zero on the right side to 30.5 m on the left. On the 8-55 Hz bandwidth section, the base of the gas can be differentiated seismically for thicknesses greater than 24 m, but not for lesser thicknesses. For the 8-35 Hz bandwidth, the base of the gas sand is difficult to differentiate even at 30.5 m thicknesses; in the 8-25 Hz bandwidth, the base of the gas sand is not differentiable at any thickness modeled.

The right side of Figure 1 shows the measured relative energy of the seismic wavelet at the top of the sand, plotted to correspond to the structure seen on the left. At 8-55 Hz, an increase in relative energy is seen where thicknesses are 0-10.7 m. For 8-35 Hz, an increase in energy is seen for thicknesses 0-16.8 m, and the same is true at 8-25 Hz for 0.23 m. Beyond these thicknesses, the energy decreases and then oscillates, becoming constant where the sand is totally gas saturated. This oscillation is caused by phase interference between the reflected wavelets of the top and bottom gas sands as the gas sand thickness varies.

If any of the velocity attributes (e.g., overburden, gas sand, or gas-free sand) vary, the general shape of the energy vs. gas-pay curve does not change for each bandwidth. Rather, the amplitude of the curve is modified. The same effect can be seen if the phase of the seismic data is changed. Thus, the response curves can be calibrated, using data from any well in the area being studied, and valid results obtained even from data that have not been adjusted to zero phase, the optimum phase for seismic resolution (Aki and Richards, 1980), or have not been restacked to use only the low angle "zero offset stack" (Gelfand and Larner, 1984).

Thus, the energy of the seismic reflector at the top of the sand in a gas-charged environment can be measured relative to the same reflector in a region that is not gas charged, and, with a knowledge of the seismic bandwidth, the net gas-sand thickness can be predicted.

CASE STUDY

Ship Shoal Block 134 field is located 48 km off the Louisiana coast in less than 15 m of water (see Figure 2). Thirty-three commercial wells are capable of a combined delivery of over 200 MMCFGD. Original proved reserves for the field were 191 bcf. Cumulative gross production as of July 1987 has been 115 bcf. Recent reserve revisions have added about 20 bcf to the expected ultimate recovery.

The gas is produced from 14 Pliocene and Pleistocene shelf sands (Figure 3) interpreted by the writers as deltaic, including channel fill, overbank splays, distributary mouth bars, and reworked longshore bars from 793-1,967 m subsea. Support for this depositional environment interpretation is gained from the well-log morphology, sand grain-size distribution, mapped distribution patterns, and core contents. Several lignite zones occurring in this sequence serve as criteria for identifying a lower delta plain environment (Saxena, 1982).

Figure 3 is a composite type log from this field showing most of the gas-producing sands developed in this deltaic complex. Resistive zones, as indicated by kicks to the right on the induction log, are the gas sands. Only three of the upper pays are discussed in this study.

Figure 4 is a structural cross section along the seismic line modeled in this paper. Although some inferred minor faults are not presented, the cross section illustrates the key role stratigraphic control, gained through seismic coverage, plays in delineating this field.

Interest in the field (Tract A) stemmed from pursuing the traditional industry methods of studying trends and shows in old wells. Area gas production was known to be

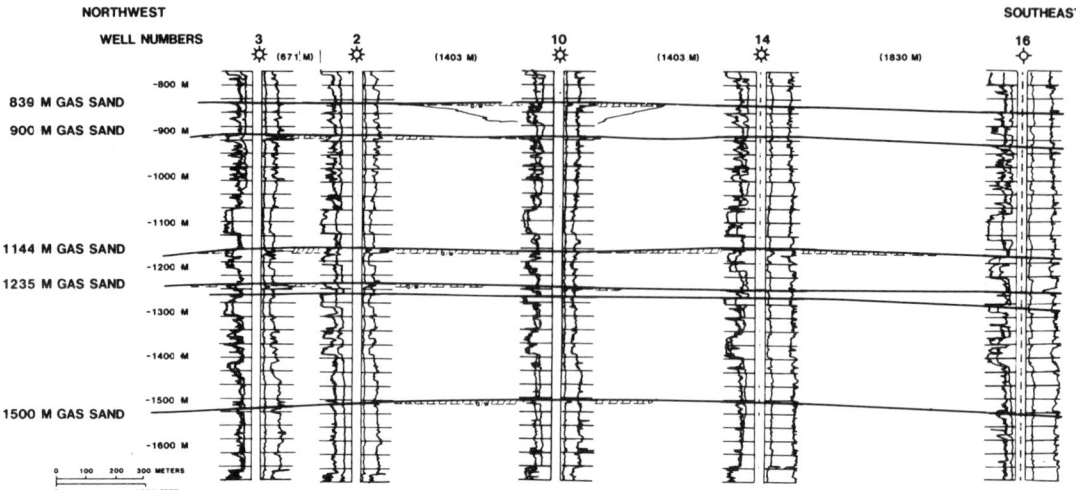

Figure 4—Structural cross section through Ship Shoal Block 134 field (Odeco wells). Cross section underlies seismic line modeled in this paper. All wells are hung on true depth. Gas accumulations (hatched lines) are controlled largely by stratigraphic changes.

"bright-spot associated" as well as high on mapped structure. Two wells, No. 8 and No. 11 (one a dry hole and the other with minor shows), were drilled in the early 1960s. Both were mapped downdip from an adjacent tract (Ship Shoal Block 134), which was 2 mi from the nearest production at Ship Shoal Block 135. Block 134 was available for bidding in an upcoming lease sale, so an assessment was needed.

Strong seismic "events" crossed the tract at shallow levels. Time-structure mapping placed these extensive features in lows in a generally flat, local setting. Work proceeded on the assumption that these "events" were bright spots without the "spots" (i.e., areally large, gas-filled sands).

Two seismic lines (one proprietary and the second speculative data) that tied to known production were used for direct analysis and to provide cross calibration. The "true-amplitude" sections were reprocessed to zero phase with a frequency bandwidth of 8-55 Hz.

As shown earlier, sands thinner than approximately 24 m (with a seismic upper frequency limit of 55 Hz) cannot be distinguished by temporal methods (i.e., top and bottom cannot be resolved by the highest frequency available). Thicker sands should yield to temporal methods of determining thickness by using standard reflectivity sections. Applying these guidelines led to the conclusion that at least some of the bright spots on the true-amplitude sections resulted from gas-filled sands, some with gas-water contacts.

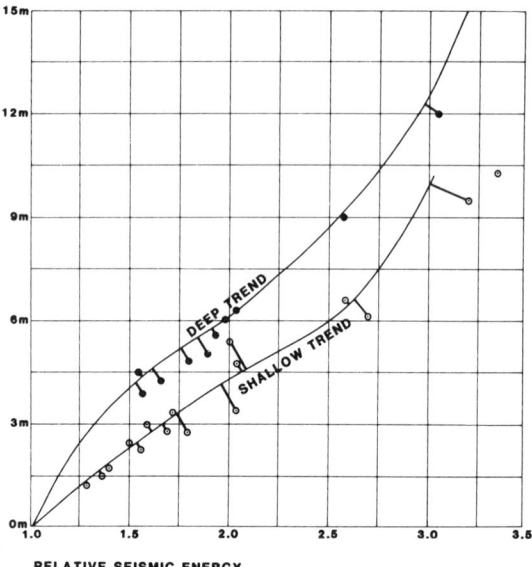

Figure 5—Energy vs. gas-pay trend graph displays empirical relationship between reflected seismic energy and gas column for northeastern Ship Shoal area. Shallow sands are those above 1,350 m. Such trends are used to calibrate relative reflected energy or impedance data.

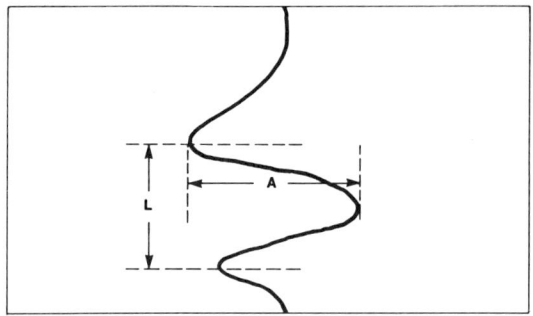

Figure 6—Energy approximation determination equals area of rectangle whose sides are L (function of wavelength) and A (peak to trough amplitude).

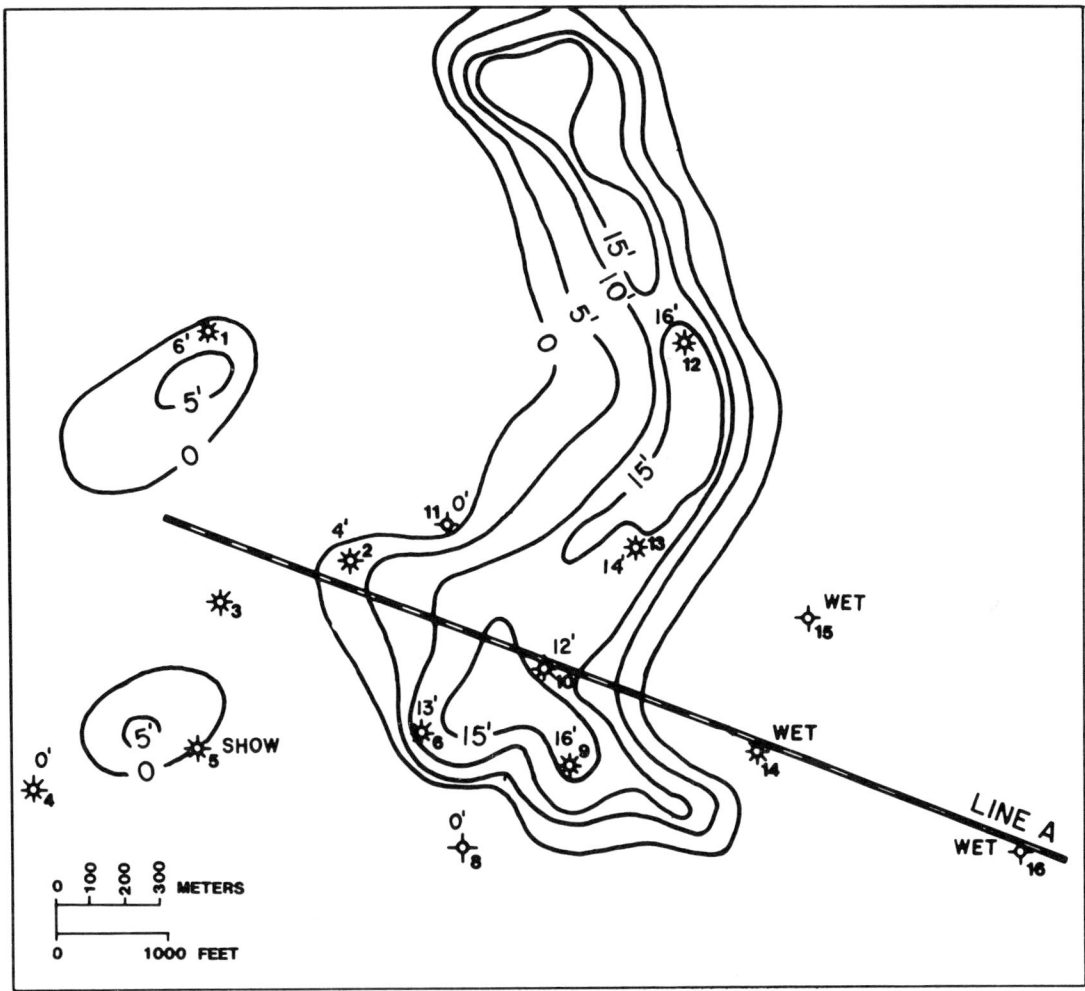

Figure 7—Gas-pay isopach map from well data and impedance for 839-m level gas sand. C.I. = 5 ft (1.52 m).

Obviously, traditional seismic analysis would not give quantitative measurements of net-gas thickness since well control indicated the gas pay would be thinner than could be distinguished from conventional seismic displays.

The first step in analyzing and quantifying the gas pay was to construct a trend graph showing gas-pay thickness from 26 points of well control vs. reflected seismic energy. Subsequent reevaluation of this trend graph led to the set of curves shown in Figure 5. Two distinct trends are visible, one for the shallow reservoirs and one for deeper reservoirs. These trends correspond to predictive trends for different bandwidths, with the deeper data having lower frequency content.

Initial energy measurements were performed manually using the assumption that the energy of a wavelet can be represented by the area of a rectangle whose sides are L (a function of wavelength) and A (the peak to trough amplitude) (see Figure 6).

Analysis of the data, combined with area extent of the seismic anomalies, led to the production of possible reserve estimates. These reserve estimates were used to appraise the value of the tract for setting bid levels. The reserves estimated for the levels illustrated in Figures 7-9 amounted to 104 bcf, unrisked. The shallower gas sands effectively hid several deeper gas sands with substantial reserves, which were discovered by drilling.

Manually made measurements, although yielding reasonable results, proved somewhat tedious and time consuming. A more convenient and accurate approach was available using a computer to integrate the wavelength to obtain the area under the curve or, in layman's terms, a direct measurement of the wavelet energy. Using the assumption that the majority of the seismic data is not associated with gas production, an analysis of the measured energies in an anomalous zone can be compared to relative measurements of the average energy of the seismic wavelet for the equivalent water-wet sand. Further,

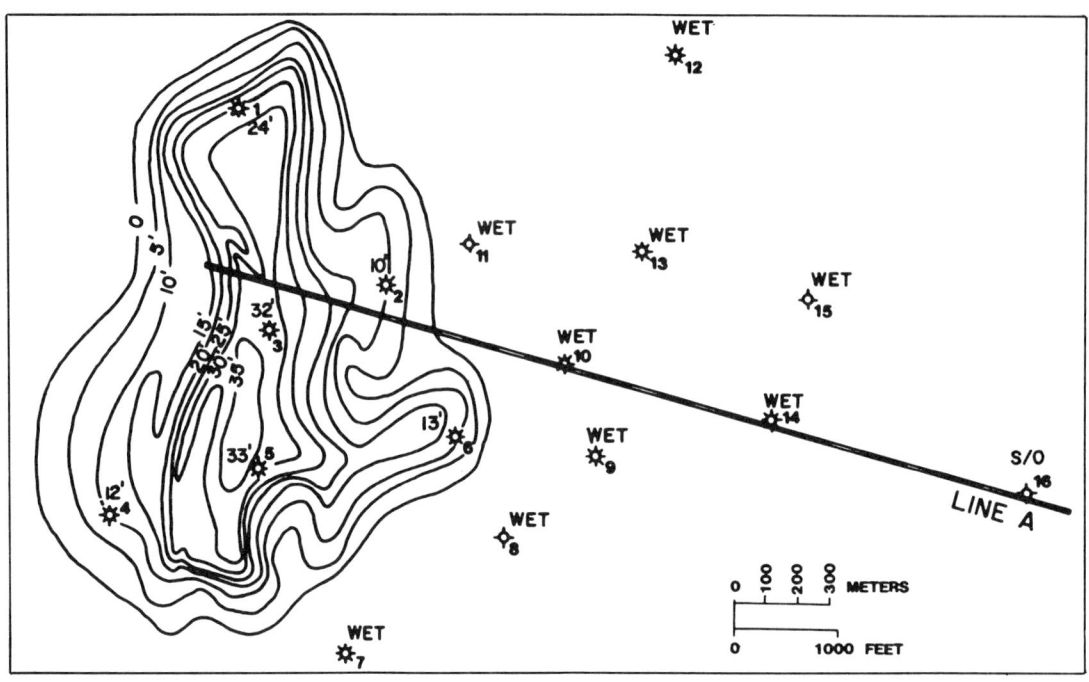

Figure 8—Gas-pay isopach map from well data and impedance for 900-m level gas sand. C.I. = 5 ft (1.52 m).

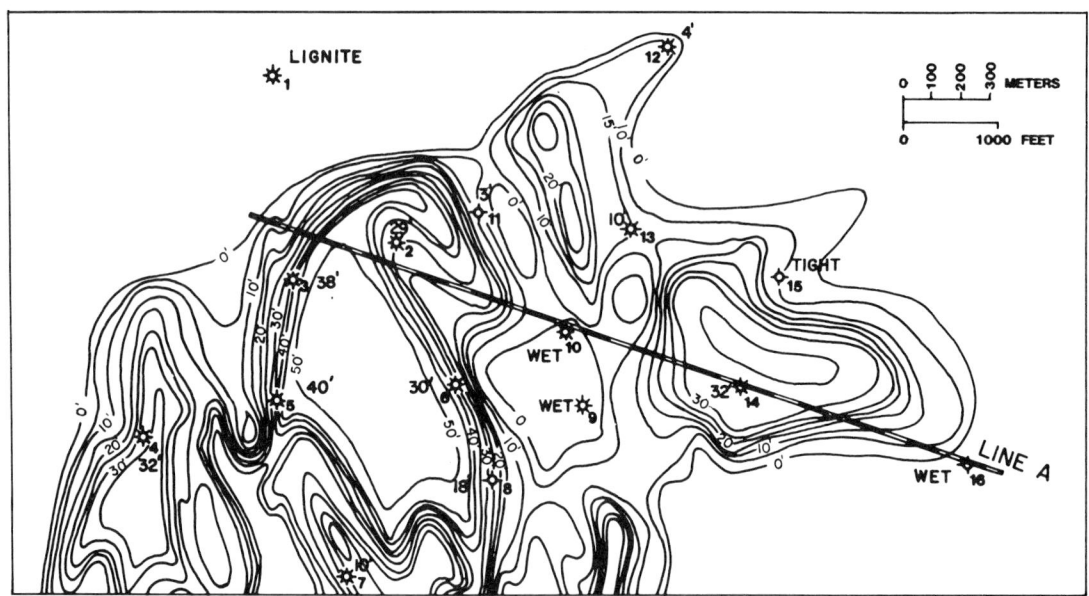

Figure 9—Gas-pay isopach map from well data and impedance for 1,144-m gas sand. Note lignite at this horizon near No. 1 well. Nearby zero contour is lignite-sand contact, indicative of a lower delta-plain environment. C.I. = 5 ft (1.52 m).

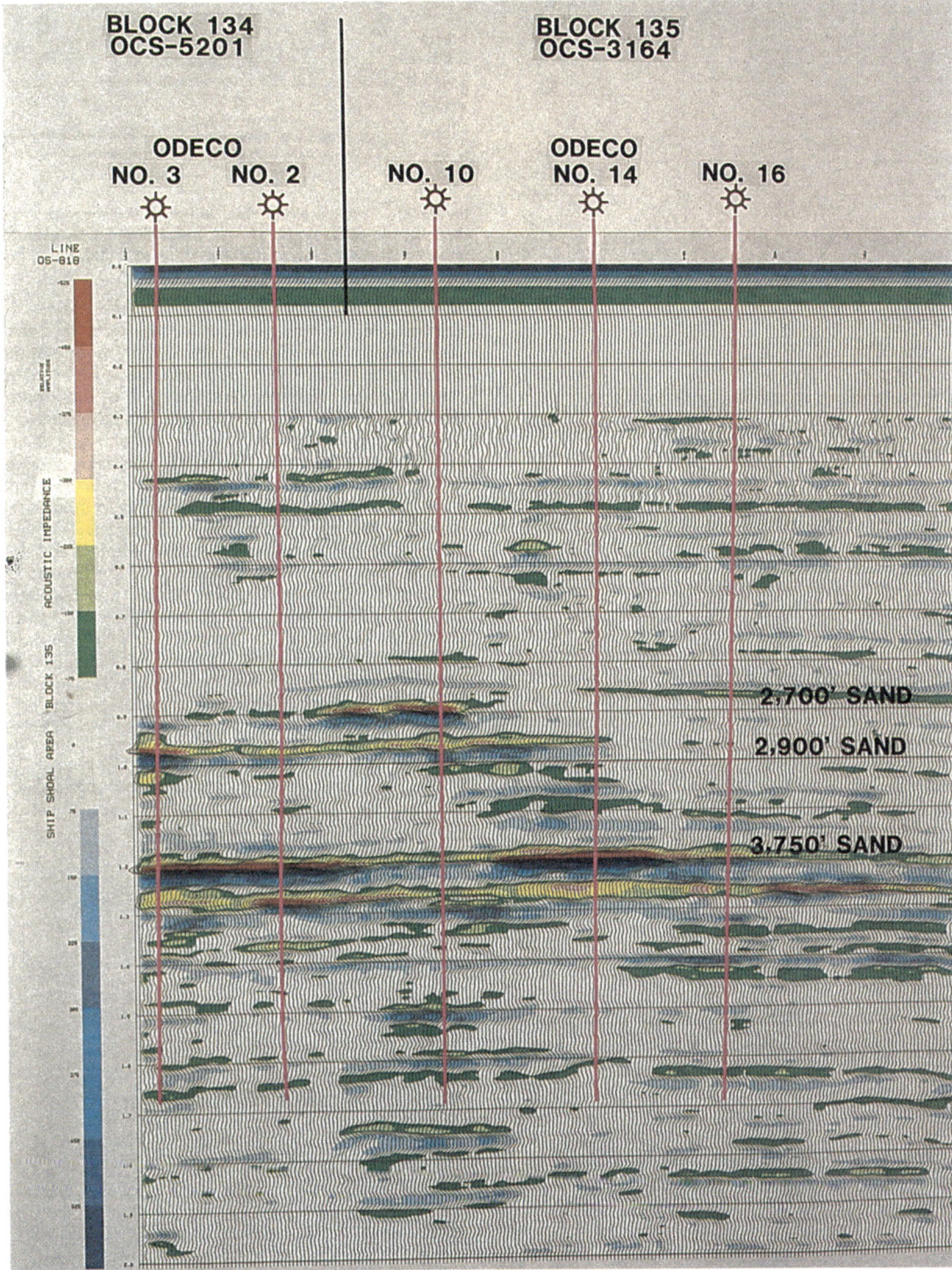

Figure 10—Type line-impedance display. These data are typical of basic measurements used in generating isopachs of gas sands. Wells are same as those used in other figures.

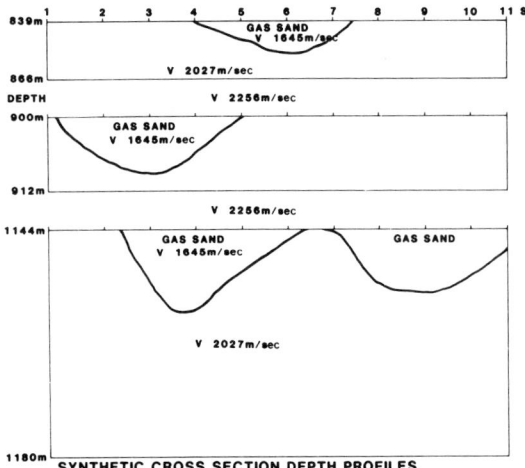

Figure 11—Synthetic cross-section model used to generate synthetic time section in Figure 12.

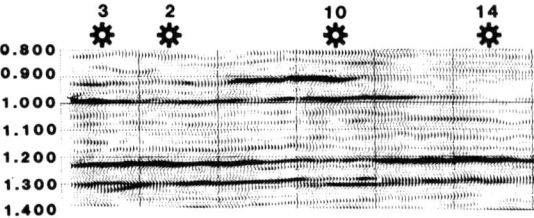

Figure 13—True-amplitude time section. Compare with synthetic section of Figure 12. These sections demonstrate quantitative effect of gas-column thickness on seismic data.

with knowledge of the seismic bandwidth and with well control for calibration, direct reserve estimates may be made from these relative-energy measurements.

In the area under study, a mixture of seismic data was used, which consisted of new proprietary shooting, older proprietary shooting, and old speculative data. All the relevant data were reprocessed for phase and frequency-range compatibility. Probably the most important factors in the reprocessing were the closely spaced velocity analyses and the zero-phase deconvolution. The deconvolution should be designed so that a close match exists between the seismic profile and the appropriate synthetic well profile, or preferably, a vertical seismic profile. The data were migrated to collapse diffracted energy. The seismic data were then converted to acoustic impedance through a typical inversion program and displayed in color (Figure 10). The presence of gas sand is indicated by the relative low-velocity side of the scale, which is the top, or red end, of the spectrum. The scale consists of arbitrary units of relative acoustic impedance, held constant through the study area. This display mode permits quantitative measurements of anomalous zones based on the strength of impedance and thickness of such zones. Such data are the mappable features that have been correlated to gas-sand thickness in this study.

Analyzing the seismic data in the area using relative-energy and impedance measurements, such as the acoustic impedance display for the type line (Figure 10) and the energy vs. gas-pay trend graphs for each zone, led to the series of net-pay isopachs used for well locations and estimating production. Isopach maps were constructed by plotting the thickness and color value of the associated impedance anomaly in much the same way as structure mapping. The results are shown in final versions with well control (Figures 7-9). Figure 7 is an impedance-derived net-pay isopach map for a sand layer at approximately 834 m depth. Figure 8, from the same area, is an interpretation at approximately 900 m, and Figure 9 is an interpretation at 1,144 m.

Although no structural control was incorporated in the construction of these isopachs, subsequent in-fill drilling permitted detailed structure mapping. These structure maps are not available for publication, but they closely agree with the isopach maps shown.

To understand how and why structural and stratigraphic detail can be extracted from impedance data, we

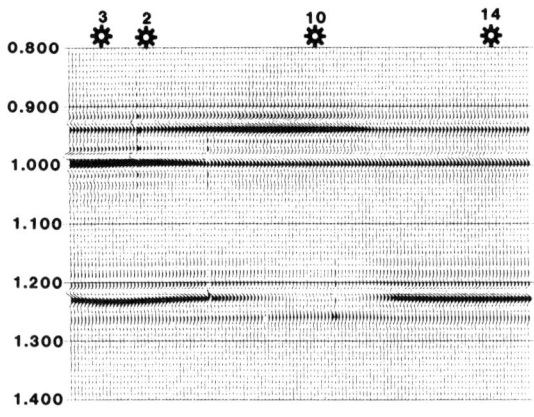

Figure 12—Synthetic time section generated from model using interval velocities derived from sonic logs. Wells are same as those in other figures.

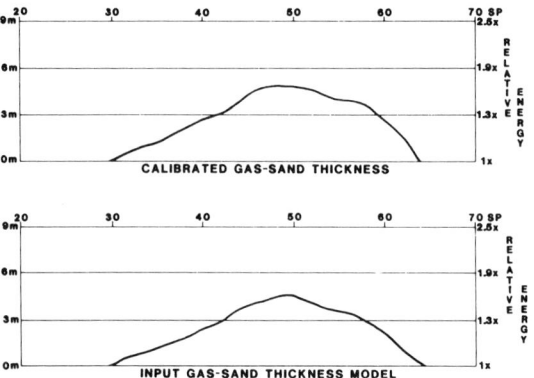

Figure 14—Calibrated gas-sand thickness from synthetic section.

have modeled the results of impedance-mapped data from the case study area along a known seismic line visible on the isopach maps. Here, the modeling was performed using the AGS-II modeling system.

The seismic line superimposed on the maps runs northwest-southeast. The depth profiles for this line (Figure 11) are taken from these interpretations. (See Figure 4 for cross-section details.) We have not included any structure in the model. In fact, the structure in this area is, for all intents, flat.

Using interval velocities derived from sonic logs in the same area, we construct synthetic time section (Figure 12) with an 8-35 Hz wavelet from the modeling package.

The true-amplitude processed seismic section corresponding to this cross section is displayed in Figure 13. Strong similarities are apparent in each of the modeled horizons. Note also, the apparent time structure on the deepest layer. The apparent syncline in the base sand is a function of the low gas-sand velocity and is not a geologic structure.

On the synthetic time cross section (Figure 12), the top sand shows no evidence of a base gas-sand reflector; however, a strong amplitude anomaly exists on the top sand reflector. At the bandwidth examined here, apart from the amplitude anomaly, the top sand reflector has the same character across the entire horizon. From the viewpoint of a conventional interpretation, all that can be said of the amplitude anomaly is that gas is likely to be present and, from our model studies discussed earlier, the gas-sand thickness is less than 30.5 m.

Figure 14 shows the calibrated relative energy for the top sand reflector. A marked increase in energy is apparent in the area of the "bright spot." Since the velocity and bandwidth properties of the sand reflector are the same as those in the 8-35 Hz synthetic section shown in Figure 1, the energy to gas-sand thickness relationship should be the same as the 8-35 Hz response curve. Using the empirically derived relationship from Figure 1, we arrived at the gas-sand thickness profile shown in the top frame of Figure 14. Note the strong similarity between the input model thickness and the final computed thickness.

The maximum error in gas-sand thickness evaluation for this model is approximately 1.2 m over a limited range of the anomaly. For the majority of the anomaly, we have accurately predicted, from seismic data, gas-sand thicknesses of 0-5 m. Normally, seismic data have an apparent resolution limit of approximately 40 m when viewed in the conventional sense.

At the time of this writing, the impedance mapping technique has been applied by the writers in 18 other areas for 29 horizons. Forty-three wells, including 16 in the Ship Shoal 134 field, have been affected by this procedure. These include 32 new discoveries, delineation of seven preexisting discoveries, delineation of two new targets near preexisting dry holes, and two unsuccessful wells. One dry hole encountered enough zeolite minerals in the target sand to generate a velocity situation similar to gas saturation. The second dry hole occurred within a proven reservoir and apparently represents a shale-out zone of less than 10 ac, below seismic resolution. Pay thickness prediction improves with the amount of control. Drilling experience shows the predicted thickness in wildcat situations to be conservative due to pay sands hidden under the target by seismic attenuation. Although greater accuracy may be achievable with the technique presented here, net-pay thickness prediction has usually been expressed to the nearest 1-2 m with the same variability, typically representing a 10 to 20% range. The variability of found net-pay thickness vs. predicted net-pay thickness is in the same 10-20% range.

CONCLUSION

Impedance processing is conducted on most Miocene, Pliocene, and Pleistocene gas prospects less than 3,600 m deep in the Gulf of Mexico by those companies familiar with the benefits. Impedance mapping, however, has been generally restricted to computer-assisted three-dimensional seismic projects. The reflection-energy attribute is a concept rarely seen in geophysical literature. This paper quantifies the properties of amplitude, impedance, and reflected energy, and the suitability of each for constructing meaningful seismic facies maps with the goal of predicting gas-sand thickness.

The impedance mapping technique has been successfully employed by the writers in 19 areas of the Gulf of Mexico from the Mobile area to South Marsh Island. In 95% of the drilled prospects, commercial gas was encountered with pay thickness typically falling within a 20% range of that predicted. Pay-thickness prediction improves with increased well control and, in stratigraphically complex areas, with increased seismic resolution and grid density.

SELECTED REFERENCES

Aki, K., and P. G. Richards, 1980, Quantitative seismology: theory and methods, v. 1: New York, W. H. Freeman, 573 p.

Bodine, J. H., 1986, Wave-form analysis with seismic attributes: Oil and Gas Journal, v. 84, (June 9), p. 59-63.

Coen, S., and M. Meadows, 1986, Exact inversion of plane-layered isotropic and anisotropic elastic media by the state-space approach: Geophysics, v. 51, p. 2031-2050.

Dix, C. H., 1955, Seismic velocities from surface measurements: Geophysics, v. 20, p. 68-86.

Gassaway, G. S., and H. J. Richgels, 1984, Seismic amplitude measurement for primary lithology estimation (SAMPLE): case histories from Tertiary western basins: Offshore Technology Conference, OTC 4784, p. 85-87.

Gelfand, V. A., and K. L. Larner, 1984, Seismic lithologic modeling: The Leading Edge, v. 3, p. 30-35.

Hubral, P., and T. Krey, 1980, Interval velocities from seismic reflection time measurements: Tulsa, Oklahoma, Society of Exploration Geophysicists, 203 p.

Kallweit, R. S., and L. C. Wood, 1982, The limits of resolution of zero-phase wavelets: Geophysics, v. 47, p. 1035-1046.

Oldenburg, D. W., T. Scheuer, and S. Levy, 1983, Recovery of acoustic impedance from reflection seismograms: Geophysics, v. 48, p. 1318-1337.

Saxena, R. S., 1982, Exploration models and recognition criteria for deltaic sandbodies, in Deltas, GSA Short Course, p. 19-71.

Walker, C., and T. J. Ulrych, 1983, Auto-regressive recovery of the acoustic impedance: Geophysics, v. 48, p. 1338-1350.

Widess, M. B., 1985, How thin is a thin bed?: Geophysics, v. 50, p. 2061-2065.

New seismic technology can guide field development

J. P. Lindsey, M. W. Schramm, Jr., and **L. K. Nemeth,** consultants, GeoQuest International, Ltd., and J. R. Butler and Co.

10-second summary

Recent advances in seismic technology, combined with preliminary production and engineering data from initial wells drilled in a new field, can be used to optimize field development. By incorporating existing seismic data at time of discovery with new data generated as the field grows, it is possible to continually refine field limits and significantly reduce the possibility of drilling dry or uneconomic wells.

REFINEMENTS in the use of seismic data are helping to improve drilling success ratios, both in finding new fields and in developing them once they have been discovered. As a result, progress is being made towards the extremely desirable goal of maximizing recovery from a new hydrocarbon deposit at minimum development cost.

Three years ago, the so-called "bright spot" technology appeared as a qualitative method of direct hydrocarbon detection. At that time, the method was used mainly to locate gas reserves on the U.S. Gulf Coast. Since then, technology has been refined and now various quantitative methods exist which not only serve as an indicator of gas reserves, but oil reserves as well. Further, these techniques can provide insight into actual stratigraphy.

Development of this technology requires integration of geophysical and geological techniques, and interpretative skills, to a degree which previously was not generally practiced. Evidence of this factor was seen at last year's annual AAPG meeting in Dallas, Texas, where two sessions were held on stratigraphic analysis techniques using seismic methodology.[1-12]

These techniques, initially applied in an exploration context in the search for stratigraphic fields, now are being extended to guide development once a field has been discovered.[13,14] Reason for this new trend is that most of the newer methods used to find stratigraphic traps can be further calibrated and refined, once data are available from one or more wells in a field, to provide better insight into field exploitation.

These new techniques have a variety of uses. For example, they can be used to plan delineation and development wells and to make more reliable estimates of reserves early in the field life. They also provide more reliable information for reservoir model studies, particularly in the area of lithology, net pay thickness, field geometry and gas or other liquid contacts.

Technical details of this methodology are beyond the scope of this paper. However, basic concepts can be described. In essence, much of this work is based on the simple idea of using seismic information—existing at the time of field discovery or acquired after discovery—in an integrated fashion all through the development life of the reservoir. Traditionally, seismic information is rarely used once a field is discovered. From that point, major reliance is usually placed on well, production and engineering data. However, many relationships exist between these classes of data that previously have been ignored or not understood.

An article in the June 1975 issue of WORLD OIL described basic principles of stratigraphic modeling and the importance of understanding the propagating wave shape.[15] This presentation illustrated basic techniques required to process seismic data in such a way that resulting sections were more useful in a stratigraphic interpretation context.

Recent advances have extended this technology from a quantitative viewpoint to consider analysis of thin-bed sequences, gradational boundaries, lateral changes in stratigraphy and identification of gas/fluid and fluid/fluid contacts.

The remainder of this article schematically illustrates possible uses of new seismic/geological techniques.

Effects of shale on seismic responses

An acoustically thin sand (less than one-quarter wavelength) gives a seismic response that varies in amplitude, but not in shape or character, with thickness. Consequently, thickness of a thin sand can be measured from amplitude of its seismic response if a suitable calibration can be established. Figs. 1 and 2 show how seismic response depends on a contaminant in the sand, such as shale. Vertical distribution of shale and relative thickness are varied to observe consequences on seismic response.

The first model (Fig. 1) has a 17-foot shale stringer embedded in a 50-foot sand that changes in its vertical placement. Seismic response shows no amplitude variation corresponding to the stringer. Trough-to-peak amplitude and time difference are measured and plotted at the bottom of the figure. As shown the amplitude variation correlates very well with net sand thickness measured from the model.

The second model (Fig. 2) varies both thickness and vertical position of the embedded shale. Once again the amplitude measure correlates with net sand thickness, whereas the timing measure is uniform and provides no information about thickness.

The conclusion drawn from these models is that when a rock layer is thin, presence of host-rock contaminants reduces seismic response amplitude proportionate to the amount of contaminant. The amplitude measure then becomes proportional to net rock thickness and, when properly calibrated, provides a degree of net thickness resolution that cannot be obtained from a timing measurement. This indicates seismic reflection amplitudes may be used to measure thickness and areal extent of reservoirs that are too thin to resolve by time-difference measurement.

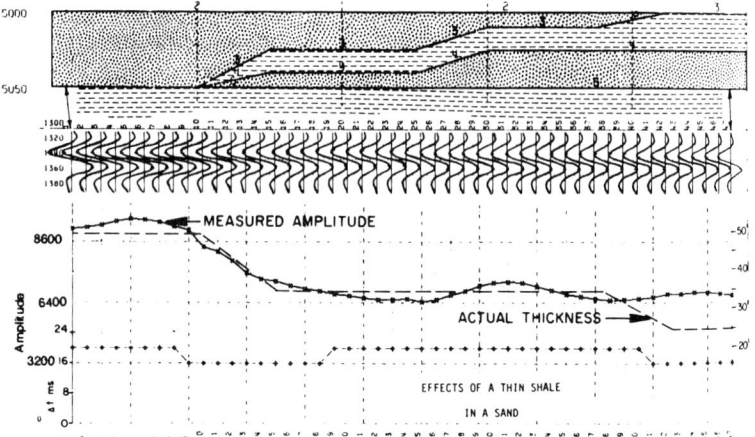

Fig. 1—Effects of a shale stringer in a thin sand on seismic reflection from the sand body. The seismic wavelet shape does not vary with shale thickness or distribution in the sand. The amplitude, however, is directly related to the net sand amount.

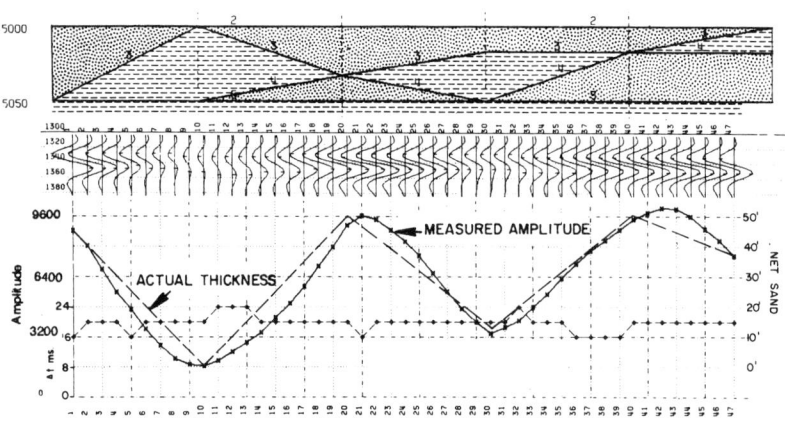

Fig. 2—Effect of shale thickness and distribution in a thin sand on the seismic reflection from the sand unit. Seismic wave shape is invariant, but seismic amplitude is directly correlated with net sand thickness.

Stratigraphic gas sand model

Fig. 3 shows a gas-filled sand which loses porosity and permeability in the updip direction over a mild relief structure. Closure is less than 200 feet in about three miles, and pay is not at the crest of the structure. Seismic response to this primarily stratigraphic situation shows the typical "bright spot" on the flank of structure.

Seismic amplitude and time difference plots were made with the computer as in Figs. 1 and 2. Because the sand is thin, time difference has no change moving across structure. But amplitude is related to the gas zone. The manner in which amplitude reduces downdip (as shown in Time Model in the lower part of Fig. 3) indicates a gas-water contact, even though an expected "flat spot" is not visible because the sand is too thin. Updip amplitude behavior is more characteristic of loss of porosity (actual loss of sand, loss of sand quality, permeability barrier, etc.).

Correspondence between net gas sand thickness and seismic amplitude is shown in the top panel of the figure. If this seismic profile represents one of several which define the gas sand, it is apparent that the calibrated measure of amplitude for each profile would make it possible to contour, in feet, the net gas sand pocket on the base map. Volumetric measurement of the reservoir rock in place then is made. Typically, net reservoir rock thickness is contoured at 10-foot levels for Gulf Co gas sands.

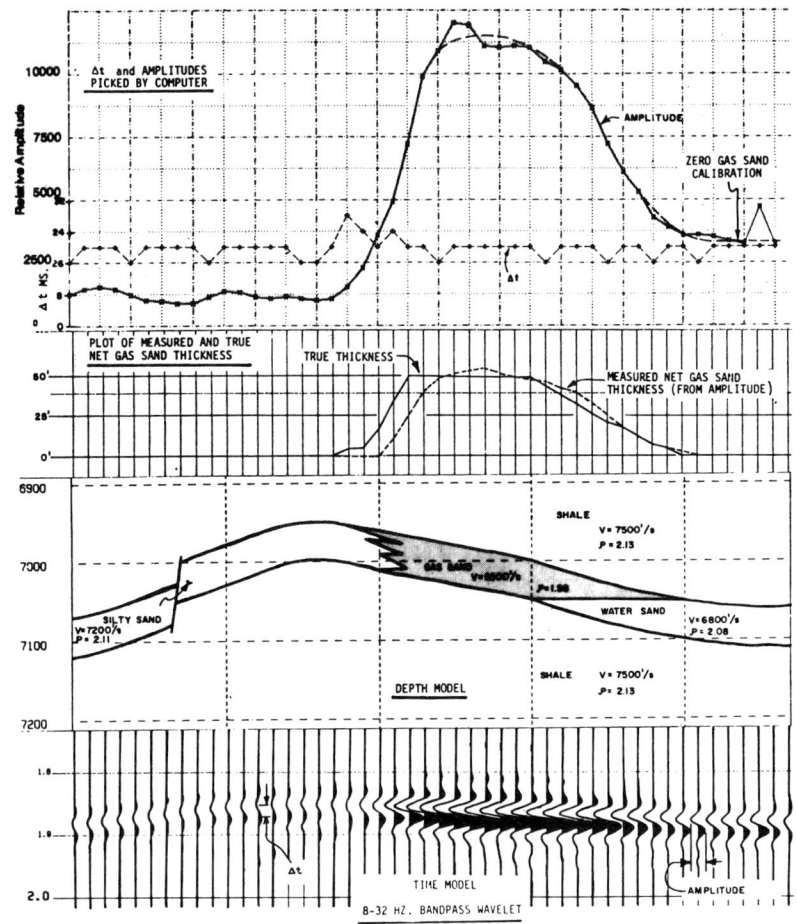

Fig. 3—Seismic amplitude as an indicator of net hydrocarbon thickness, loss of porosity, and gas-fluid contact.

Sand-shale interfingering model

Lithologic boundaries are not always mirror smooth in the subsurface. Depositional characteristics of plastic flowage, interbedding or lithologic gradation can cause a diffuse boundary in the vertical direction and lack of a single boundary influences seismic reflection. The seismic wave averages properties over a near-circular subsurface area when reflecting. This area is called a "Fresnel Zone" and is usually 600-2,000 feet in diameter, depending upon depth of the reflecting boundary and wave length of the wavefront.

When a diffuse boundary is involved, the seismic reflection is weakened and loses some degree of lateral continuity. Moreover, seismic response may differ from that predicted from a well log. Logs measure the vertical lithologic sequence using a few inches or feet of lateral penetration at the most. Seismic waves average

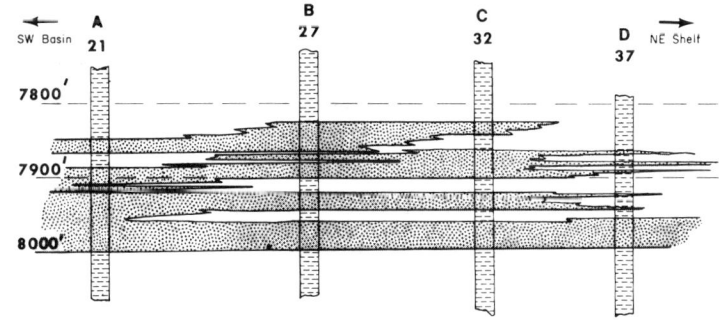

Fig. 4—Typical cross section in depth of sand imbedded in shale.

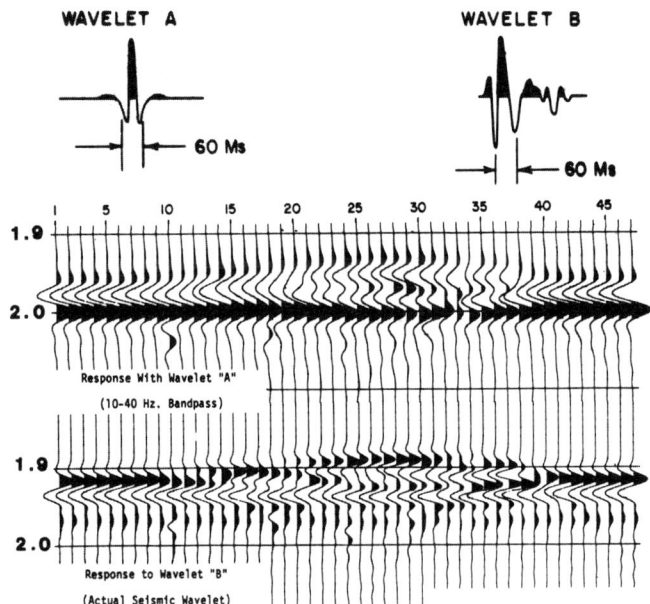

Fig. 5—Seismic model response to the sand unit of Fig. 4 using (top) a symmetric zero-phase wave shape, and (bottom) a more complex wavelet extracted from marine seismic data.

shown at location A, B, C and D, which will correspond in subsequent figures to shot point locations 21, 27, 32 and 37, respectively (in a seismic model).

This schematic situation was modeled using AIMS (Advanced Interpretive Modeling System) to simulate several types of seismic responses that could be compared in time on a corresponding basis with synthetic seismograms derived from reflectivity sequences at the various well locations.

Seismic response to the complete sand unit is shown in Fig. 5. The upper version uses an ideal seismic wavelet that is symmetrical and polarized such that a positive reflection (transition from a porous sand to a harder shale) will be displayed as a black or right-hand reflection. This is an ideal wavelet for visual detection of detailed stratigraphy and acoustic lithology.

The lower seismic response is the same model using a propagating wavelet actually found for a marine seismic line. The wavelet is not symmetrical or regular in any sense and is, therefore, not suited for stratigraphic interpretation. Comparison of these two responses clearly shows that a simple representation is possible using the ideal wavelet.

Fig. 6 shows the synthetic trace (under color blocks) generated from the acoustic

several hundred feet laterally and thus reveal different boundary properties than do the log data.

Fig. 4 illustrates schematically a stratigraphic situation involving diffuse boundaries. A sand of about 100 feet thickness is embedded in a shale. The sand base is very uniform over a large area, but the sand top is locally interfingered with the overlying shale. Four hypothetical well locations are

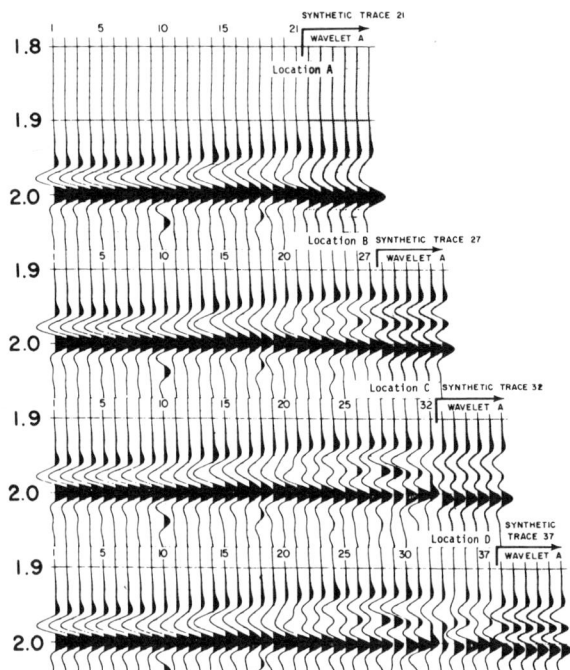

Fig. 6—Comparison of seismic response of the sand unit of Fig. 4 with synthetic seismogram derived from well logs using ideal wavelets.

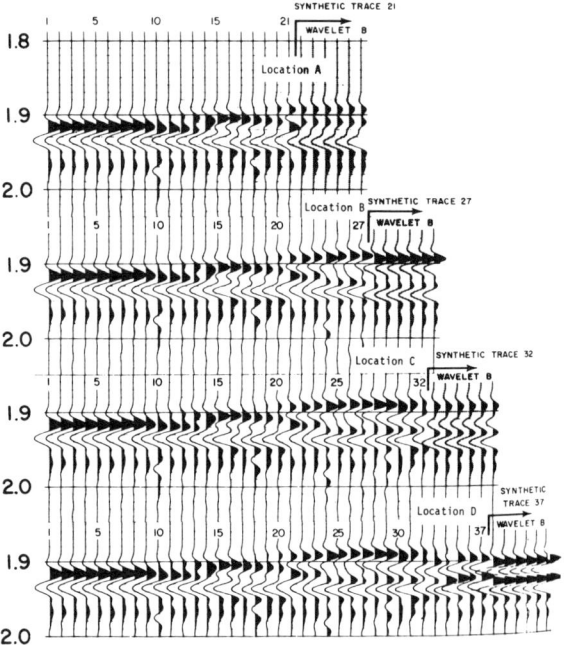

Fig. 7—Same comparison as Fig. 6, except that seismic response and synthetics used complex marine seismic wavelet.

log run at each well location compared to the seismic model response at the same location using the ideal wavelet. Remember that the model response is an *average* response over Fresnel Zone, whereas the well log synthetic is a simulation that *assumes* the lithologic sequence seen at the well extends laterally in all directions.

Fig. 7 is a similar presentation using the more complex marine wavelet. Comparison with Fig. 6 shows the significantly increased difficulty of accurately interpreting the stratigraphic sequence.

Fig. 8 illustrates a common situation when comparing a synthetic trace derived from an acoustic well log using an ideal wavelet, to a seismic section which has a different wavelet.

The main point is that seismic data and synthetics generated from well logs should not be expected to agree if the synthetic was generated using a wavelet that differs from the wavelet of the seismic section.

The following conclusions can be drawn, from this brief model study:

▶ Synthetic seismograms must be generated with the same wavelet that exists in the seismic section if a meaningful comparison is to be achieved.

▶ The seismic response and the synthetic seismogram should be expected to agree to the extent that the log sequence extrapolates laterally for a Fresnel Zone. The Fresnel Zone for this model has a 1,780-foot diameter for 20-Hz energy.

▶ When the same wavelet has been used to make the synthetic seismogram that exists in the seismic section, a lack of agreement between the two is indicative of depositional and stratigraphic properties seen differently by the log and seismic waves, rather than an error in the measurement of either.

These models are intended to illustrate some of the key principles that now can be applied to stratigraphic interpretation. Modeling and knowledge of wavelet shape will play an increasingly more prominent role in defining calibration techniques, particularly in relating well data to seismic data. Acoustically meaningful parameters and their translation into reservoir descriptive parameters are complex processes which will require careful integration of geophysical, geological and engineering disciplines, and analytical techniques which facilitate such integration.

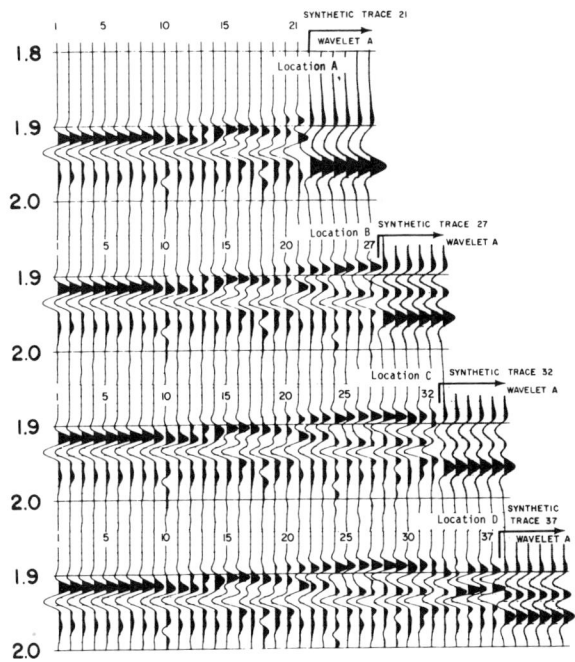

Fig. 8—Comparison of seismic response (complex wavelet) with synthetic response derived from well logs (ideal wavelet).

The reader can extrapolate these concepts to see how they might be used to guide field development once a field has been found. The key, of course, is calibration. The more well data and engineering data that exist, the more refined the comparison between seismic response and reservoir definition.

About the authors

J. P. LINDSEY, *a founder and principal of GeoQuest, received B.S. and M.S. degrees in electrical engineering from Oklahoma State University. His past experience includes work as a research engineer with Phillips Petroleum Co. and as chief research geophysicist and vice president of research and development of Geocom, Inc.*

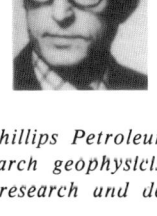

MARTIN W. SCHRAMM, JR., *holds a B.S. and M.S. in petroleum engineering and geology from the University of Pittsburgh and a Ph.D. from Oklahoma University. He has worked as an independent consultant and in geological and management positions with Gulf Oil*

Corp., Cities Service Oil Co., Cities Service International Co. and White Shield Exploration Corp. Dr. Schramm is a founder and principal of GeoQuest and is a vice president of J. R. Butler and Co.

L. K. (LES) NEMETH, *holds a B.S. in geology from Rensselaer Polytechnic Institute, an M.S. in petroleum and natural gas engineering from Pennsylvania State University and a Ph.D. in petroleum engineering from Texas A & M University. His experience in reservoir, production, drilling and general field engineering was acquired with Texaco, Tenneco and as consultant and technical supervisor for secondary recovery projects for the national oil company of Argentina (YPF). During the past four years, he was associated with Butler, Miller and Lents, Ltd., an independent consulting firm. Currently, Dr. Nemeth is head of J. R. Butler and Co., oil and gas consulting firm affiliated with GeoQuest and specializing in advancing the state of the art in reservoir definition.*

LITERATURE CITED

[1] Vail, P. R.; Mitchum, R. M., Jr.; Sangree, J. B., and Thompson, S., III, "Stratigraphic Framework and Eustatic Cycles

[1] from Seismic Stratigraphic Analysis," AAPG, Dallas, Texas, April 7-9, 1975.
[2] Todd, R. G., and Mitchum, R. M., Jr., "Seismic Stratigraphic Identification of Eustatic Cycles in Late Triassic, Jurassic and Early Cretaceous Rocks Gulf of Mexico and West Africa," AAPG, Dallas, Texas, April 7-9, 1975.
[3] Widmier, J. M., and Sangree, J. R., "Depositional Environments from Seismic-Facies Analysis," AAPG, Dallas, Texas, April 7-9, 1975.
[4] Galloway, W. E.; Yances, M. S., and Whipple, A. P., "Seismic Stratigraphic Model of Depositional Platform Margin, Eastern Anadarko Basin," AAPG, Dallas, Texas, April 7-9, 1976.
[5] Bryant, W. R., and Antoine, J. W., "Seismic Stratigraphy of Texas and Louisiana Shelf Sediments," AAPG, Dallas, Texas, April 7-9, 1975.
[6] Stuart, C. J., "Sedimentologic and Stratigraphic Interpretation of Seismic Data in Gulf of Mexico," AAPG, Dallas, Texas, April 7-9, 1975.
[7] Wu, C., and Mateker, E. J., Jr., "Lithology Determination from Reflection Amplitude," AAPG, Dallas, Texas, April 7-9, 1975.
[8] Crow, C., and Alhilali, K., "Attenuation of Seismic Reflections as Key to Lithology and Pore Filter," AAPG, Dallas, Texas, April 7-9, 1975.
[9] Davis, T. L., "Seismic Evidence of Late Cretaceous Growth Faulting, Denver Basin, Colorado," AAPG, Dallas, Texas, April 7-9, 1975.
[10] Kim, D., "Stratigraphic Mapping of Cambrian Erosional Feature in Central Ohio," AAPG, Dallas, Texas, April 7-9, 1975.
[11] Sheriff, R. E., "Physical Principles for Identifying Lithology and Stratigraphy from Seismic Data," AAPG, Dallas, Texas, April 7-9, 1975.
[12] Dedman, E. V.; Lindsey, J. P., and Schramm, M. W., "Extracting Geologic Parameters from Seismic Data Stratigraphic Technique," AAPG, Dallas, Texas, April 7-9, 1975.
[13] Lindseth, Roy O., "Lithology Determination (Sands, Shales, Reefs, etc.)," Denver Geophysical Society Continuing Education Seminar, April 17-18, 1975, Denver, Colo.
[14] Domenico, Norman S., "Rock Characteristics and Pore Fluid Content," Denver Geophysical Society Continuing Education Seminar, April 17-18, 1975, Denver, Colo.
[15] Lindsey, J. P.; Dedman, E. V., and Schramm, M. W., "Stratigraphic Modeling: A Step Beyond Bright Spot," *World Oil*, May 1975.

How hydrocarbon reserves are estimated from seismic data

J. P. Lindsey, Geophysical Consultant, and **C. I. Craft,** President, Geocom, Houston

15-second summary

Extent of industry use of the "hot spot" seismic technique for locating offshore hydrocarbons was evidenced by intensive competitive bidding for offshore tracts offered in the recent federal lease sale off Texas. This article discusses the next step after "hot spot"—estimating limits of seismic-indicated reservoirs and calculating reserves contained in them.

DETECTION of hydrocarbons from seismic data is now a reality, as a result of routine recording of amplitude information on digital instrumentation. A processing center can acquire this information from field tapes and design a gain control function that simultaneously removes spherical divergence and regional attenuation effects, and preserves local amplitude anomalies which can be hydrocarbon indicators. The procedure works equally well for structural and stratigraphic traps.

These amplitude anomalies are produced by the increased reflection coefficients of low velocity reservoir rocks (frequently less than 5,000 fps for Gulf Coast gas sands). To some extent, "tuning" of the seismic wavelet to gas sand thickness is involved as well.

With faithful preservation of reflection amplitudes provided by this new processing technology, estimation of lithology has become a possibility. To understand how this is possible involves the basic physics of sound propagation in solid materials.

When rocks of different types are interfaced in the subsurface, some of the compressional wave energy passing through the interface is reflected. These reflections are detected at the surface and interpreted to define the subsurface structure.

Strength of this reflected energy, after removing near-surface effects, depends on how different the "acoustic impedances" of the interfaced rocks are. Acoustic impedance is a measure of how much resistance to particle motion is offered by the rock to sound energy moving through it. It is the product of rock density and sonic velocity. A soft, low density rock adjacent to a hard, high density rock will produce a strong reflected wave amplitude. A rock interface at which velocity increases slightly and density decreases in the same ratio will be totally transparent to sound, since acoustic impedances are the same.

Rock velocity and density are basic clues to lithology. Considerable documentation exists of the velocities and densities of the many rock types associated with the world's hydrocarbon provinces. Consequently, knowledge of velocity and density of rocks associated with a likely reservoir is sufficient to establish basic lithologic parameters. Then the presence and quantity of hydrocarbon in-place can be estimated based on the relationship between velocity, density, porosity, and pore volume content. This relationship must be estimated initially for wildcat provinces and subsequently refined by drilling experience and bore hole measurements.

Starting with an appropriately processed seismic section having an indicated hydrocarbon reservoir, lithology is estimated using the following basic information:

1. A catalog of rock types to which the reservoir composition is limited

BASIC INFORMATION required to estimate lithology from a seismic section with an indicated hydrocarbon-bearing reservoir includes a catalog of rock types to which reservoir composition is limited; density and interval velocity values of each rock type; time structure of indicated interfaces picked from the seismic section; and the characteristic of the "basic" seismic wavelet that propagated through the reservoir.

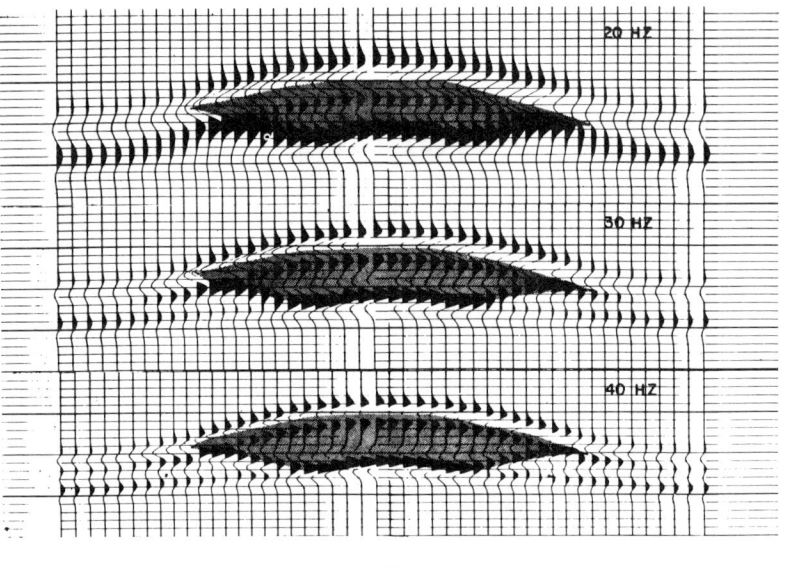
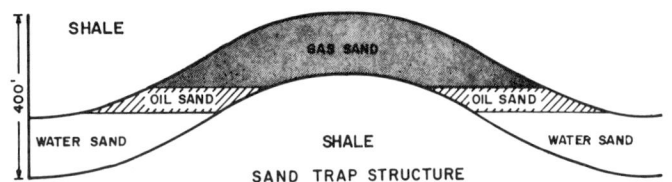

FIG. 1—Model of 150-foot-thick gas sand with 250 feet of vertical relief over 1.5 miles. Three seismic solutions are shown for the model. Low seismic velocity in the gas sand distorts the time section relative to the depth section.

About the authors

J. P. LINDSEY *began his career in geophysics in 1953, with the Research and Development Department of Phillips Petroleum Co. in Bartlesville, Okla. He participated in the early use of magnetic recording, common depth point technique and digital technology applied to seismic exploration for oil and gas. After sixteen years as a researcher, he joined Geocom, Inc. as chief research geophysicist, later becoming vice president of research and development. In this capacity he was a participant in the continuing development of seismic processing technology, including automatic migration, multiple reflection elimination, direct hydrocarbon detection on seismic data and the modeling of seismic-indicated reservoirs for lithology. He now is a consultant in geophysics, working on problems in data acquisition, processing, amplitude interpretation, and special projects. Mr. Lindsey holds an M.S. in electrical engineering from Oklahoma State University and is a member of the Houston Geophysical Society, SEG and IEEE.*

CECIL I. CRAFT *graduated from the Colorado School of Mines in 1957 in geophysical engineering. Upon graduation, Mr. Craft worked for Shell Oil Co. in Denver, New Orleans and Houston. He worked in various capacities, as seismologist, party chief and senior geophysicist. In 1964, he became head of the Seismic Processing Division in Houston, with the title of senior geophysicist. Geocom was founded in 1965 by Mr. Craft, who has been its president since its inception.*

2. Density and interval velocity values of each rock type

3. Time structure of indicated interfaces picked from the seismic section

4. The characteristic of the "basic" seismic wavelet that propagated through the reservoir strata.

Picking rock interfaces from the seismic section requires considerable expertise to insure that false indications are eliminated and that all the real ones are picked. Errors at this point are largely eliminated by limitations imposed by available rocks in the catalog and physical properties of hydrocarbon traps.

For example, in Gulf Coast reservoirs only sands and shales comprise the catalog. Additionally, shale must overlie gas sands and water sands cannot. A further constraint is the requirement that polarity of interface picks on the seismic section match polarity of corresponding reflection coefficients established by the rock properties of each layer.

This latter requirement is perhaps the most stringent and difficult one to meet. But it is an advantage in disguise, eliminating many false rock sequences and improving accuracy of the ultimate solution.

When the "puzzle" of rock sequence has been solved, delineation of the reservoir depth structure and computation of its synthetic seismic response can be accomplished. *The ultimate objective is to have the synthetic response match basic character and amplitude properties of the real seismic section and to produce a realistic depth structure.* Failure to satisfactorily "model" the real seismic reservoir response points to errors in either interface picks or rock layer properties.

Identifying interfaces from the seismic data is not simple. It involves experience with how seismic reflections merge for thin layers, and knowledge of the basic shape of the reflection wavelet. Not all legs of the seismic complex are associated with a rock interface. Conversely, not all rock interfaces produce an identifiable pick on the record. Making interface identification from the seismic response involves a trial and error approach to achieve an answer consistent with the data.

A further complexity is that determination of key rock layer parameters (those of the porous zones) is only partially independent of the interface time picks. However, solutions are obtained with benefits that far outweigh the cost, which is on the order of $500-$1,000 per case.

Total hydrocarbon content is determined from pay zone thickness, areal extent and porosity derived from the modeling process. Accuracy is a function of the degree to which the relationship between velocity and porosity is known. For wildcat efforts, reasonably good general relationships can be employed which will adequately rank prospects for indicated hydrocarbon content. Quantitative accuracy is improved with drilling experience by using well logs to calibrate constants in the velocity-porosity equation. In well developed areas, very satisfactory results can be obtained.

Some of the relationships between

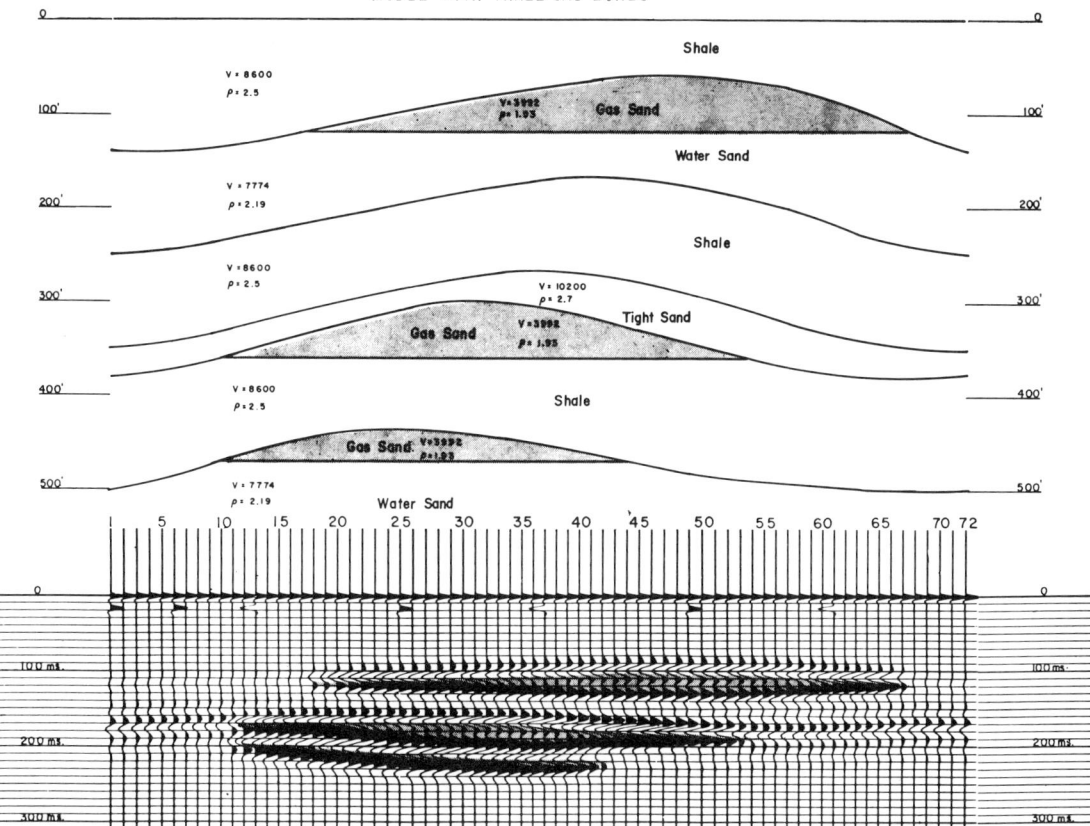

FIG. 2—Model of three gas sands. High amplitude zones clearly indicate gas in the two upper sections, while the lower, thinner sand is at the limit of resolution.

reservoir structure and the seismic section are illustrated in Figs. 1 and 2. Fig. 1 represents a sand 150 feet thick, with 250 feet of vertical relief over 1.5 miles. Gas is trapped at the top of the reservoir, with oil overlying water on the flanks. Shale encompasses the sand at both top and bottom.

Three seismic solutions are shown for this model, differing in resolution of fine detail because of the frequencies used. The 20 hertz solution represents typical marine data with no deconvolution processing. If data quality is good, processing can improve this resolution to that shown in the 30 hertz and 40 hertz solutions. The 40 hertz solution represents the upper limit of resolution obtainable with the seismic method for deep returns.

The gas-shale and gas-oil contacts provide the strongest reflections from the reservoir structure. Low velocity in the gas sand distorts the time section relative to the depth section. Conflict between time "pushdown" caused by low gas sand velocity and time "pull-up," caused by structure, causes the relatively flat bottom of the gas zone to appear otherwise. The reflection at the bottom of the gas zone is positive and causes the black deflections. These are visually dominant on the seismic display, the gas top being less evident.

Fig. 2 is a three gas sand model with two 60-foot-thick sands and one 40-foot sand. High amplitude zones clearly mark presence of gas in the

> **Where "hot" spot and modeling techniques have been used to detect hydrocarbons and estimate reserves**
> ► Gulf Coast, from Florida to Mexico
> ► Sacramento Valley, California
> ► Nigeria
> ► Indonesia
>
> Although no public information has been released, the techniques will likely work equally well in parts of the North Sea, the interior portion of the United States, Canada and Venezuela.

two 60-foot sands. The 40-foot sand is just at the limit of resolution.

The general anticlinal character of the reservoir structure is not so apparent in the seismic solution. This is primarily because black deflections mark gas bottoms. A reversal in plotting polarity would change this significantly. It is becoming common practice to require plots of both polarities for interpretation of hydrocarbon indicators on controlled gain seismic sections.

The objective of modeling is to deduce the depth structures of Figs. 1 and 2 from the seismic response.

It is important to emphasize that modeling alone does not determine presence or absence of hydrocarbons. This is an interpretation function, which is significantly improved when amplitude information is preserved.

The new combination of amplitude interpretation and reservoir modeling significantly extends the power of the seismic tool beyond determination of structure into detection of hydrocarbons in-place and an estimate of quantities involved. ■

Distinguished Author Series

Neidell **Beard**

Progress in Stratigraphic Seismic Exploration and the Definition of Reservoirs

by Norman S. Neidell and John H. Beard

Norman S. Neidell is president and chief operating officer of Zenith Exploration Co. in Houston. After working with Gulf Oil Corp. and Seiscom Delta Inc., he helped found GeoQuest Intl. Inc., an exploration consulting firm and seismic contractor. He cofounded Zenith Exploration in 1976 and founded Delphian Signals in 1979 to exploit applications of his research on dolphin echolocation. Neidell holds many patents and has taught geophysical courses for SPE, SEG, and AAPG. He received a BA degree from New York U., an applied geophysics diploma from Imperial C. (London), and a PhD degree in geodesy and geophysics from Cambridge U.

John H. Beard is senior vice president of exploration with Zenith. Before joining the company in mid-1981 he worked more than 22 years with Exxon as a senior exploration and research geologist and geophysicist in the U.S., South America, West Africa, the Mediterranean, Indonesia, and the North Sea. Beard has written many articles for scientific journals during the past 12 years.

Summary

The blending of exploration tools calling upon geology and geophysics has produced an approach to stratigraphic seismic exploration of great power in the identification and definition of hydrocarbon reservoirs. From the geologic side, facies analysis used conventionally and with color-derived seismic displays scaled in velocity can suggest the most likely depositional models and the associated lithology. Next, very detailed and precise velocity measurements from the seismic data by means of the moveout relationship and independently from the reflection strength or seismic amplitude measure can confirm or contradict a proposed model.

When full consistency is established between these components and all other available data, then the exploration data have been utilized fully. Furthermore, maps of velocity variation within proposed reservoir units can be developed to guide initial drilling, and after calibration with wells, to develop and operate the field. The full procedure also surmounts certain fundamental limitations inherent in conventional exploration technology and leads to new and exciting exploration plays. This paper illustrates all concepts and procedures with case studies and examples.

Understanding Seismic Stratigraphic Exploration

Stratigraphic and Structural Traps. Many knowledgeable explorationists have recognized the current shift in exploration emphasis from the quest for the simple structural trap to stratigraphic traps. The special volume put together by Halbouty[1] exemplifies the collective current view. The difference between stratigraphic and structural traps can be as clear as the difference between an anticline and a pinchout, or it may be quite subtle where both elements play a role.

Structural traps may be complex in geometric terms or of gentle expression, and so can become difficult to recognize and find. Stratigraphic traps, on the other hand, with or without the role of structure, by definition entail greater subtlety. We can understand this point best by looking more closely at some of the mechanisms by which such traps are formed. Changes in lithology and porosity development by fracturing or diagenetic activity are just some of the ways stratigraphic traps may develop.

Conventional exploration tools often are taxed beyond their capability by the task of defining certain stratigraphic traps. The following discussion explains the nature of such limitations and formulates tools and procedures that can be used effectively to define these exploration objectives. We also note that the technology used to attain this goal also has bearing on the delineation of hydrocarbon reservoirs.

Limits of Seismic Visibility. In recent times it has been fashionable to discuss how thin a bed may be measured by the seismic method.[2] More recently in a rather comprehensive discussion, Neidell[3] showed that very thin beds indeed may be defined with conventional seismic data, and several case studies have since been published. The key to making such determinations is the availability of reliable amplitude information in the seismic data. While it is not truly relevant to this discussion to describe the theory of thin bed determination—which entails a good measure of sophistication—it is important to mention it because of the implications of such technology.

If we can use conventionally acquired seismic data to map thin beds, say with 5-ft [1.5-m] contour intervals at depths of 12,000 ft [3658 m], and to do so reliably, then little of exploration significance should escape the scrutiny of the modern seismic method! Unfortunately this is not at all the case.

Presumptions in the methods mentioned can lead explorationists astray. Perhaps the best way to make this point is with Fig. 1. Note here in the upper portion a more or less conventional black and white seismic display dipping southeast into the Gulf of Mexico. The profile is in Houston County, TX, and the prominent sequence of reflectors in the middle region of the data is from the Cretaceous sequence of beds.

Directly below is a portrayal of the same data in another format and in color. The colors are keyed to velocity estimates, in this instance 1,000 ft/sec [304 m/s] steps. A transformation has been applied which in approximation seeks to turn each seismic trace of the profile into a sonic log, but in time rather than depth. More discussion of this figure is provided later, both in terms of the exploration problem and of the methods by which such a display can be developed. Comment also must be made regarding reliability of such a transformation.

The color display shows with startling clarity an *en echelon* sequence of carbonate banks. Their structure is classic in terms of the velocities we measure. The two most prominent members of the sequence, upper and lower Edwards cycles, have much in common with the bullseye of a target in terms of visibility.

In fact, the banks are shelfal developments which relate to the East Texas basin north of the profile.

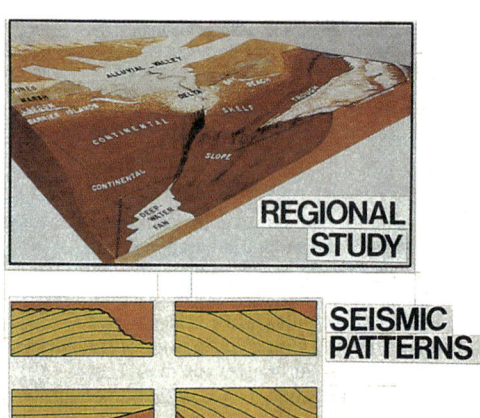

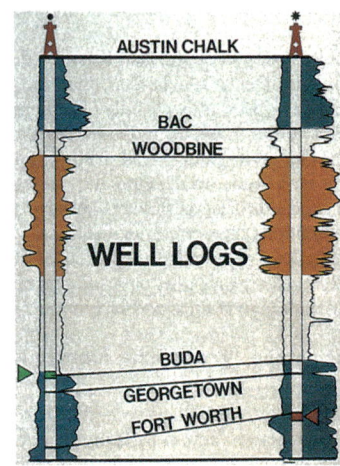

Fig. 1—A lesson in seismic visibility.

Fig. 2—Geologic foundations.

They formed at a time when the section dipped into this basin. The big Edwards member is at least 300 ft [91 m] thick and is some 12,000 ft [3658 m] deep.

While we can map units with 5-ft [1.5-m] accuracy *once they are recognized,* they first must be found. The entire bank sequence is not seismically visible on the conventional presentation. It would have been missed entirely by conventional approaches.

Clearly, then, the color display of seismic data scaled in estimated interval velocity must be at least a part of the total package by which stratigraphic traps can be defined. The following describes the full technology and illustrates its use.

An Integrated Exploration Framework for Stratigraphic Traps

Building on Classical Approaches. Fig. 2 presents the geological components that form the basis of an integrated approach. We want to make use of all available information and in the figure, the regional model and the information from well logs is most important. The third element of the figure, suggested only in general terms, is a more recent development.

Vail *et al.*[4] in a fundamental series of works clarified and classified the use of background patterns in seismic data as a means of recognizing depositional environments and even lithology. This "geologic" approach to the seismic data is incorporated also with the more traditional sources of intelligence. The

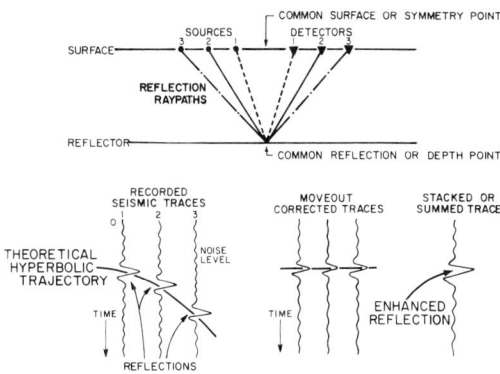

Fig. 3—The CDP imaging method and the role of the moveout curve: the basis for seismic velocity analysis.

application of facies analysis using the color display scaled in velocity proves a significant enhancement for both tools.

From the geophysical side, and particularly in regard to the seismic data, we seek to include something beyond the usual. The color display of velocity information is derived largely from the amplitude dimension of the seismic data. Yet another way of

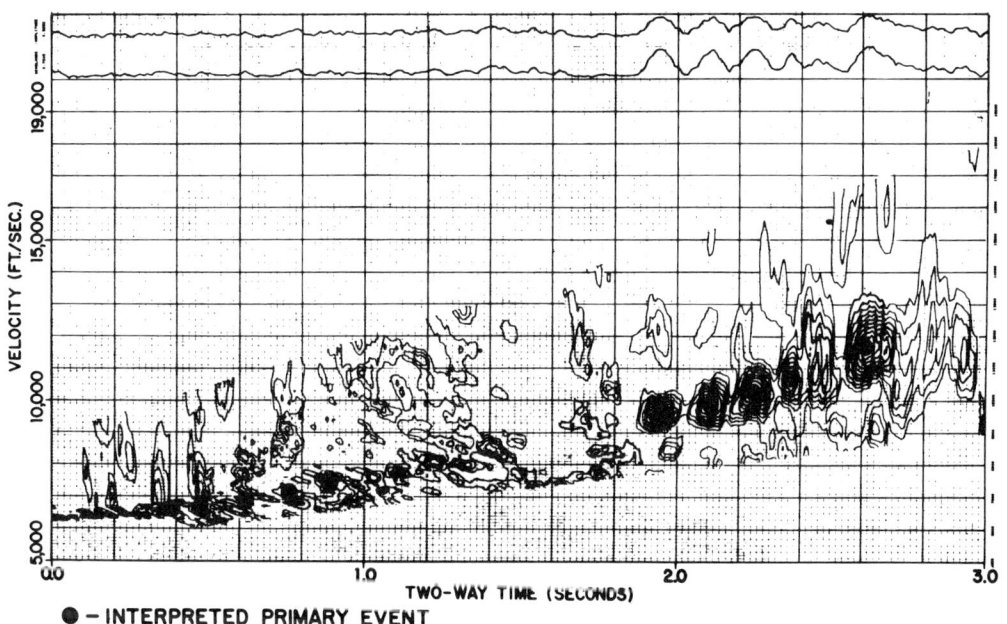

Fig. 4—A seismic velocity analysis.

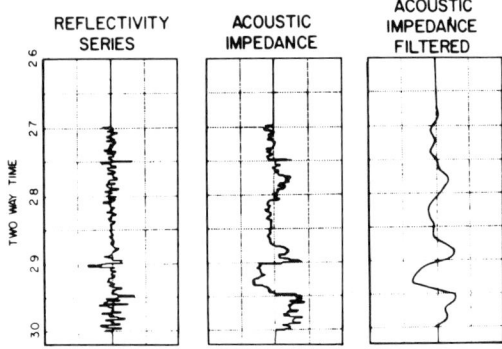

Fig. 5—Frequency considerations concerning well log information.

measuring velocity from seismic surveys relates more to the geometry of the survey. This latter method also can be made with greater precision, which is needed in defining many stratigraphic hydrocarbon accumulations.

At this point, one may present philosophy and pick up details later. Measuring seismic velocity more often and more precisely from the best possible processed seismic data, and according to both methods by which this may be done, provides a powerful base for establishing consistency of exploration hypothesis. An exploration "model" or hypothesis can be conceived from the seismic data and geological inputs, but it also must be consistent with both types of velocity measurements. When such consistency is established, it becomes most difficult to imagine other ways to test the model, short of drilling.

Seismic Velocity Measurements. Velocity information can be derived from seismic data in two different ways: from the moveout relation and from seismic amplitudes. Since the use of amplitudes already has been mentioned, we look first at what the moveout curve appears to offer.

Moveout-Derived Velocity. Fig. 3[5] clarifies the role of the moveout curve in the common depth point (CDP) imaging process.[6] By noting the arrival time of a reflection from a more or less common subsurface (depth) point, a horizontal component of travel effectively can be measured. This is, of course, the conceptual key to the velocity analysis. The most effective analysis and display format in our opinion is the one developed by Taner and Koehler,[7] one of the earliest analyses suitable for routine use. This analysis, however, is not used correctly in industry standard practice, as we will show.

The basic calculation of seismic velocity analysis involves the systematic fitting of hyperbolae to moveout curves. The resulting display requires a great element of interpretation. One of these is shown in Fig. 4. Owing to the large scale of computing, it is traditional to make one such calculation per mile with a few extra ones at key places along the profile.

Returning to Fig. 4, note that several of the contour closures, which an interpreter has deemed primary events, have been marked. It is important to realize that one analysis per mile provides only sporadic information. Nor can we usually attain resolution in velocity of beds thinner than 100 milliseconds in two-way time. For an 8,000 ft/sec [2438 m/s] velocity this corresponds to a unit thickness of some 400 ft [122 m]. Hence the information would appear rather coarse. Nevertheless, certain classic studies used this type of velocity information for lithologic work with remarkable accuracy. See, for example, the work of Cook and Taner.[8] The techniques discussed herein follow in good measure from these early efforts.

Seismic Amplitude-Derived Velocity. Next, means to obtain seismic velocities from seismic amplitude information must be discussed in some detail. First, converting single traces to a log-like format will not recover the low-frequency trend or the high-frequency data above the limits of the seismic sampling interval. The nature of this information loss can be appreciated more fully by looking at logs from a North Sea well. The panels of Fig. 5 show comparable portions of both a reflectivity series and an acoustic impedance log developed in two-way travel time from the North Sea velocity and density logs. Of course, log editing procedures have been applied first. A filtered version of the acoustic impedance log also is shown with a frequency content comparable to what one might expect to recover from a single seismic trace.

After appropriate treatment of the seismic waveform, by integrating seismic traces one can estimate the acoustic impedance trace by trace. Fig. 6 provides a close-up display of the carbonate bank sequence in the Cretaceous section in Houston County where such an analysis has been performed. We have already seen this section in Fig. 1. A trend function developed from the seismic data through the moveout velocity estimation has been applied to produce the background color display. The trend supplies the low frequencies not recorded in the single seismic traces. The works of Lavergne and Willm[9] and Lindseth,[10] pioneers in the development of this technology, must be mentioned here. Lindseth further recognized the value of color displays in presenting subsurface parameter estimates, which is rapidly becoming an industry standard. We fully subscribe to this philosophy.

Note in Fig. 6 that the integrated trace nearest the wellbore agrees very well with the filtered acoustic impedance function as scaled in time. The treatment for the log data here is very analogous to that described for the North Sea single-trace data of Fig. 5.

Note also that if one wishes to recovery velocity information rather than acoustic impedance, some reasonable assumptions about the relation of velocity

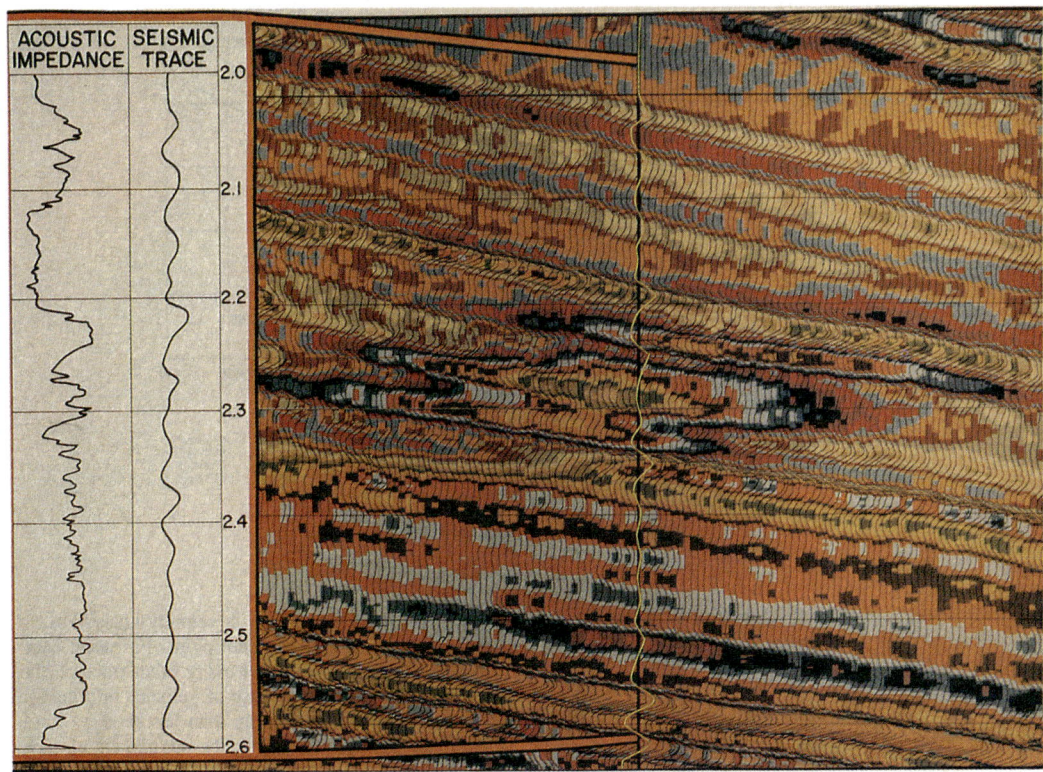

Fig. 6—Acoustic impedance estimates trace by trace, Houston County, TX.

(V) and density (ρ) must be made. One usual approach is to invoke Gardner's relation,[11] which leads to $\rho \propto V^{1/4}$.

In Fig. 6 this assumption has been used and the velocity increment is 400 ft/sec [122 m/s]. The color acoustic impedance or better "velocity" display is an important component in the stratigraphic approaches to land data revealed in this text. At all times, however, velocity information for land data derived from seismic trace amplitudes must embody an unknown but significant component of error and uncertainty, and here we treat it accordingly. The independent but more reliable coarser view of seismic velocities offered by the velocity analysis also will be used.

Using Seismic Velocity Measurements

Velocity and Lithology. There is a long history and tradition in exploration and related disciplines that relate velocity measures to other desired information. Fig. 7 embodies this precept quite clearly. It is adopted from a handbook of well log interpretation charts and indicates how sonic travel-time measurements or equivalently velocities (their reciprocals) may be related to porosity with minimal subsurface calibration information.

If correlations such as implied by Fig. 7 can be made reliably, then we can go still further. Returning to Figs. 1 and 6 which treat the Houston County carbonate bank sequence, we can appreciate that amplitude-derived velocity information is available in detail for each profile of the survey grid. Reliable velocity information may also be available from moveout curves.

Hence we can use such information in an integrated fashion over the entire survey to produce maps such as the one in Fig. 8 for the Glen Rose C member of the bank sequence. Estimated maps of unit thickness have been available for some time from seismic data. Maps of velocity variation add an important new kind of information. Since all seismic-derived maps may be prepared before any drilling, a thickness map in conjunction with velocity variations can suggest the best well locations. Once calibration factors are available from well control, maps incorporating such factors become the basis for operating the reservoir.

It is interesting that progress in exploration toward

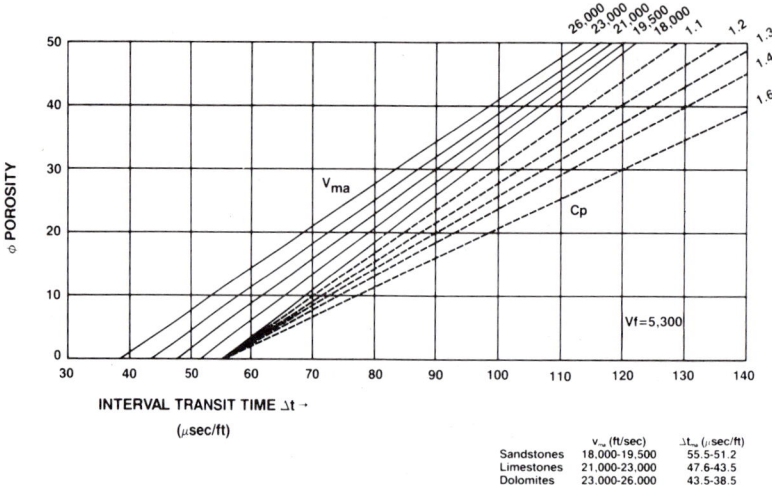

	V_{ma} (ft/sec)	Δt_{ma} (μsec/ft)
Sandstones	18,000-19,500	55.5-51.2
Limestones	21,000-23,000	47.6-43.5
Dolomites	23,000-26,000	43.5-38.5

Fig. 7—Porosity from sonic log.

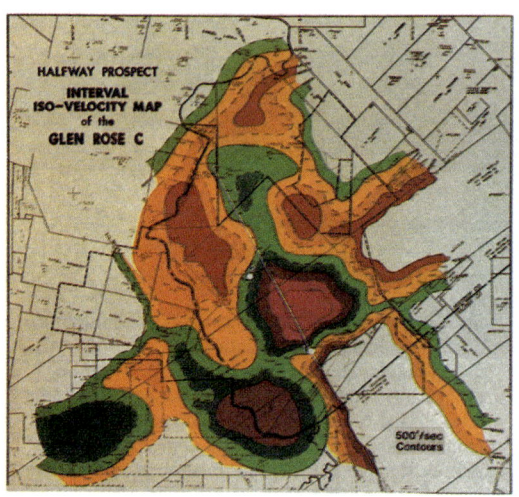

Fig. 8—Glen Rose C interval isovelocity map.

defining stratigraphic traps has also acted to close the gap between exploration and production technology. Since we wish to make optimal use of the moveout-derived velocities, we now describe how this may be accomplished.

Using Moveout Velocities. To utilize fully the velocity information inherent in moveout curves and to minimize coarseness, all parameters and procedures must be optimized wherever possible. Starting with wavelet-processed land data, a velocity analysis is computed at least at every shot point. For usual data this represents an analysis of every other trace on the stacked section, and a density of velocity information of the order of 50 times more than has been customary. At the same time, the parameters of each analysis are set to reflect the known waveform duration with small steps taken in both the velocity and two-way time-search variables. While such methods raise the price of each calculation considerably, the certainty and stability of the results improves in direct relation.

Further refinements and enhancements are used in the velocity analysis, but keep in mind that localized information in spatial terms is desired. The goal, therefore, is always to follow specific and key reflecting horizons.

Fig. 4, as noted previously, shows one typical velocity analysis of the type just described. A procedure now is shown that brings out fully the velocity information from this source.

It is well known that abrupt lateral changes in subsurface lithology distort stacking velocities as determined for reflections at depths below them, making the relation of stacking velocities to interval velocities quite complex. Zones of greater porosity (possibly hydrocarbon-filled porosity) are likely to have a slower stacking velocity, as picked from such an analysis for the reflectors below such zone, than would be seen if little or no lateral velocity change occurs. For subsurface geometries that are relatively simple (flat layers, conformable dipping surfaces, etc.), stacking velocities bear fairly straightforward relationships to interval velocities. In fact, the magnitudes of stacking velocities usually relate quite

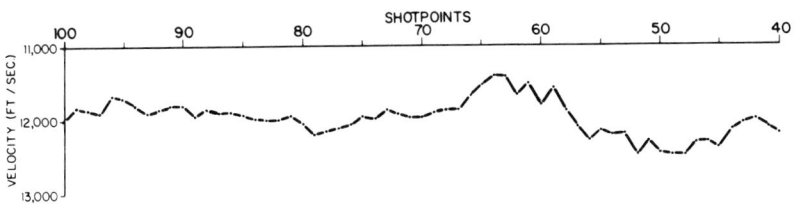

Fig. 9—Velocity plot for a key reflector.

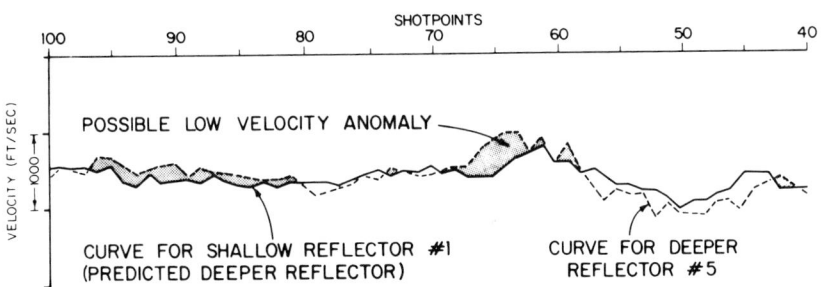

Fig. 10—Calibrated velocity-curve pair bracketing an interval of interest.

directly to the lithology. Typically these velocities increase as one goes from shales and sands to sandstones and limestones and then to dolomite. In all circumstances, however, a rock of reservoir quality shows a reduced velocity compared to that same rock without appropriate porosity. Where gas is the pore fill, the velocity reduction is emphasized further.

According to our method, Differential Interformational Velocity Analysis® (DIVA), the stacking velocities of the reflections of interest are tracked laterally on an analysis-by-analysis basis by using the times for the reflectors from the seismic profile to aid in event identification on the velocity analyses. For each recorded event time and for each analysis, a velocity is picked.

Referring to Fig. 9, the velocities so obtained are first plotted separately for each reflection according to the analysis location. The horizontal axis of this plot consists of velocity analysis locations along the line of the survey. Velocity, in feet per second, is plotted vertically along the profile with velocity increasing in a downward direction. This same procedure is undertaken for each reflection tracked on the profile. Plots thus obtained present stacking velocity trends as picked for each reflector of interest.

Looking at Fig. 10, each of the individual velocity plots is then "overlaid" in pairs with all other velocity plots to "bracket" effectively all zones between reflections. This process may be viewed as predicting a deeper stacking velocity from the velocity curve of the shallower reflection, assuming the interval between the two particular reflectors undergoing comparison is uniform or varies only regionally in interval velocity. A semiempirical alignment compensates for the possible development of discordant dip between the particular reflector pair. Plotted velocity values for the deeper of the two reflectors are connected with dashed straight lines for easy identification.

Any overlaid curve pair under study should approximately follow one another if the lithology between them is uniform. Local change in the intervening layers results in their divergence or crossover. When the dashed velocity curve for the deeper reflector approaches or crosses over the prediction curve for the shallower event, this may be recognized readily on such display as a possible zone of anomalously low velocity. The word *possible* is necessary since the basic underlying data have a noise component that is rarely insignificant. Anomalous low-velocity zones could indicate possible porosity and even hydrocarbons in appropriate circumstances.

The basic mathematical relations for a calibration procedure for the case of simple geometry start with the Dix equation[12]:

$$V_{int} = \sqrt{\frac{V_d^2 t_d - V_s^2 t_s}{t_d - t_s}},$$

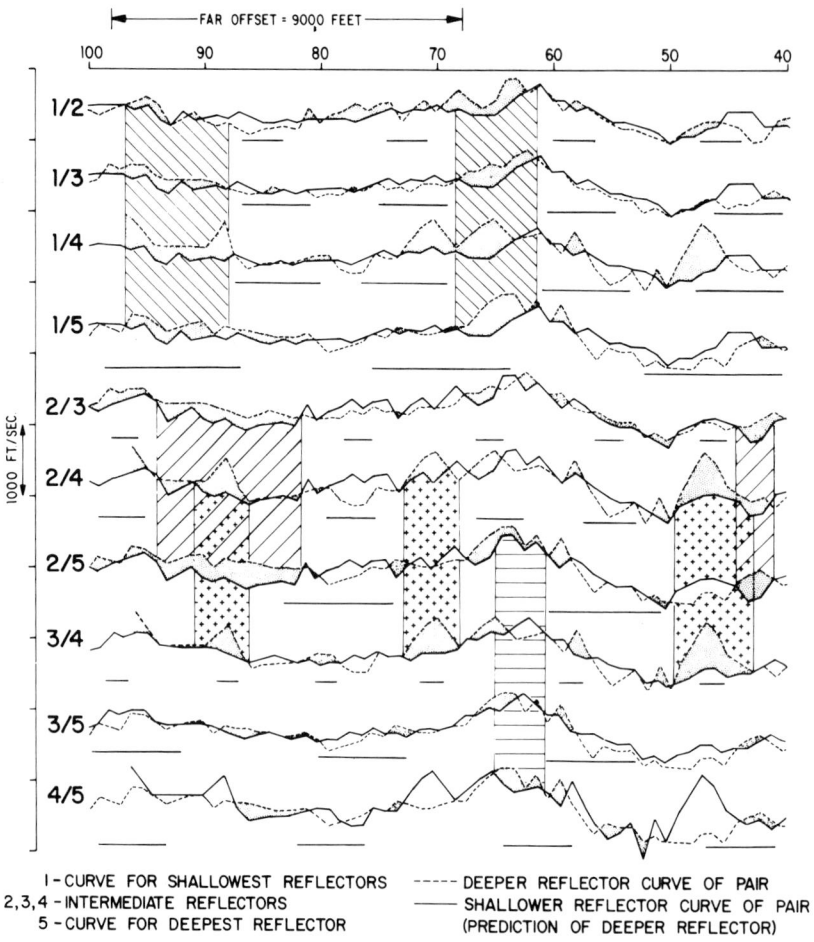

Fig. 11—Differential interformational velocity analysis.

where

V_d = velocity of the deeper reflection of a pair,
V_s = velocity of the shallower reflection,
t_d = time of the deeper reflection,
t_s = time of the shallower reflection, and
V_{int} = interval velocity between pair of reflection boundaries.

This calculation is applied for any two reflections that are closely spaced [about 1,000 ft (300 m) or less]. A velocity estimate for the interval between the two reflections compared according to the Dix equation is made for every spatial velocity analysis location. The result is a set of "noisy" interval velocity values.

When all possible combinations of velocity curves by key reflectors are overlaid according to the method described, a final display showing them all is constructed as shown in Fig. 11. The horizontal and vertical scales for each curve pair are the same as described for Figs. 9 and 10. All curve combination pairs are aligned beneath the proper surface reference positions on the display. The shallowest reflections are placed at the top of the presentation with comparisons in turn with deeper reflections taken in depth order. Then the next shallowest reflector is treated in similar fashion, and so on. Places along the profile where the dashed line (deeper velocity values) rises above the solid line (shallower or predicted velocity values) denoting possible low velocities are often colored pink or red for visual emphasis.

Where low-velocity anomalies line up vertically and bracket a suspected low-velocity zone in common, this is interpreted as strong corroboration of the reality of such anomaly. Vertical lines now are drawn to delineate the anomalous zone. The vertical band between such lines is color coded or coded graphically as shown in Fig. 11, according to the geologic interval in which the anomaly is thought to occur. The specific interval is localized by the curve pair bracketing the narrowest interval within the vertical band of consistent low-velocity indications. Vertical zones showing an anomaly in more than one geologic interval are coded with stripes corresponding to the colors of each interval in which an anomaly is interpreted. For reference, the physical dimension of the far source/receiver offset for the particular seismic profile under analysis is placed to scale on the upper part of the display.

From this display, then, it is possible to view very precise moveout-derived velocity information that is localized very well in terms of position along the profile but not particularly well in the time variable. Information from another source will assist with localizing low-velocity anomalies in reflection time.

It is worth remarking that interpretation of this display is learned rather quickly. One reason for the familiarity that many explorationists feel toward it is that each velocity curve pair resembles a porosity log display turned sideways. The resemblance is quite intentional.

Integration of Amplitude-Derived Velocities.
Velocity information from seismic amplitudes, though not fully reliable, is used in our procedures by means of the color acoustic impedance section display. These displays play a dual role. They can show velocity anomalies quite clearly in relation to their reflection time in finer detail than the DIVA. Hence these displays are most important in the time localization of detected anomalies. Even though the anomaly magnitudes may not be correct on these displays, the sense (high or low velocity) is usually right. Between the DIVA and this display, we have been able to localize with great accuracy low-velocity anomalies in spatial position along the profile and in reflection time. At the same time, a precise determination of the velocity is also available.

Implications for the definition of hydrocarbon reservoirs are quite evident. Between both techniques it becomes clear that accuracies approaching 200 ft/sec [61 m/s] in velocity variation within a lithologic unit of appropriate thickness may be determined reliably.

There is another important function that the color acoustic impedance section must fulfill. Even if low velocity is indicated, porosity represents only one possible explanation for its occurrence. Lithologic changes also can cause lowered velocities; these are most troublesome, particularly if unanticipated. Major changes in lithology usually are dramatically visible on the color acoustic impedance display, often both by contrast and by their geometry. The explorationist thus is presented with what one may regard as a geologic perspective. We have already noted the value of such displays in regard to the use of seismic facies analysis.

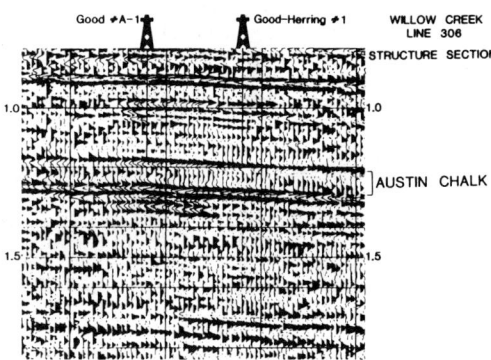

Fig. 12—Close-up of Section 306, Good A-1 well.

Case Studies and the Consistency Criterion

Fracture Porosity Detection in the Austin Chalk. In a unique scientific experiment near Dilley, Frio County, TX, ten 7,000-ft [2134-m] wells were drilled to "prove" the validity of the DIVA-based stratigraphic approach. Six wells encountered natural fracture porosity in the Austin Chalk (such fracturing being necessary for its production) in an area distinguished by no faulting and only the most subtle of structures. One well, a "throwaway" drilled only to hold a valuable lease, balanced the experimental statistics. If previous work in this same area were regarded as random drilling then one well in 20 encountering fractures should serve as a simple standard for comparison. More than $4 million worth of oil and gas were produced in just under 12 months from initiation of the $6 million program. The scientific project obviously was a commercial success. At the same time, carefully monitored and controlled operational procedures allowed further refinement of the technology in terms of calibration, form of presentation, and interpretational method.

Fig. 12 shows close-up seismic Section 306 of the project passing through the best well, the Good A-1. This well had an initial potential flow of 600 B/D [95 m^3/d] of oil and is on its way to producing 100,000 bbl [15 890 m^3] from fractures with no stimulation. Note that this well may be said to be on a structure. On this section the chalk is bracketed by the Pecan Gap and the Eagleford shale, just above the Buda limestone. Note that on these wavelet-processed sections each boundary is marked by peaks and troughs as appropriate to the reflection coefficient polarity. That is, the Pecan Gap upper boundary is a peak, while a trough marks the top of the Eagleford shale.

The display for Line 306 (Fig. 13) shows a prominent anomaly for the interval containing the chalk below the Good A-1 location. Experience with

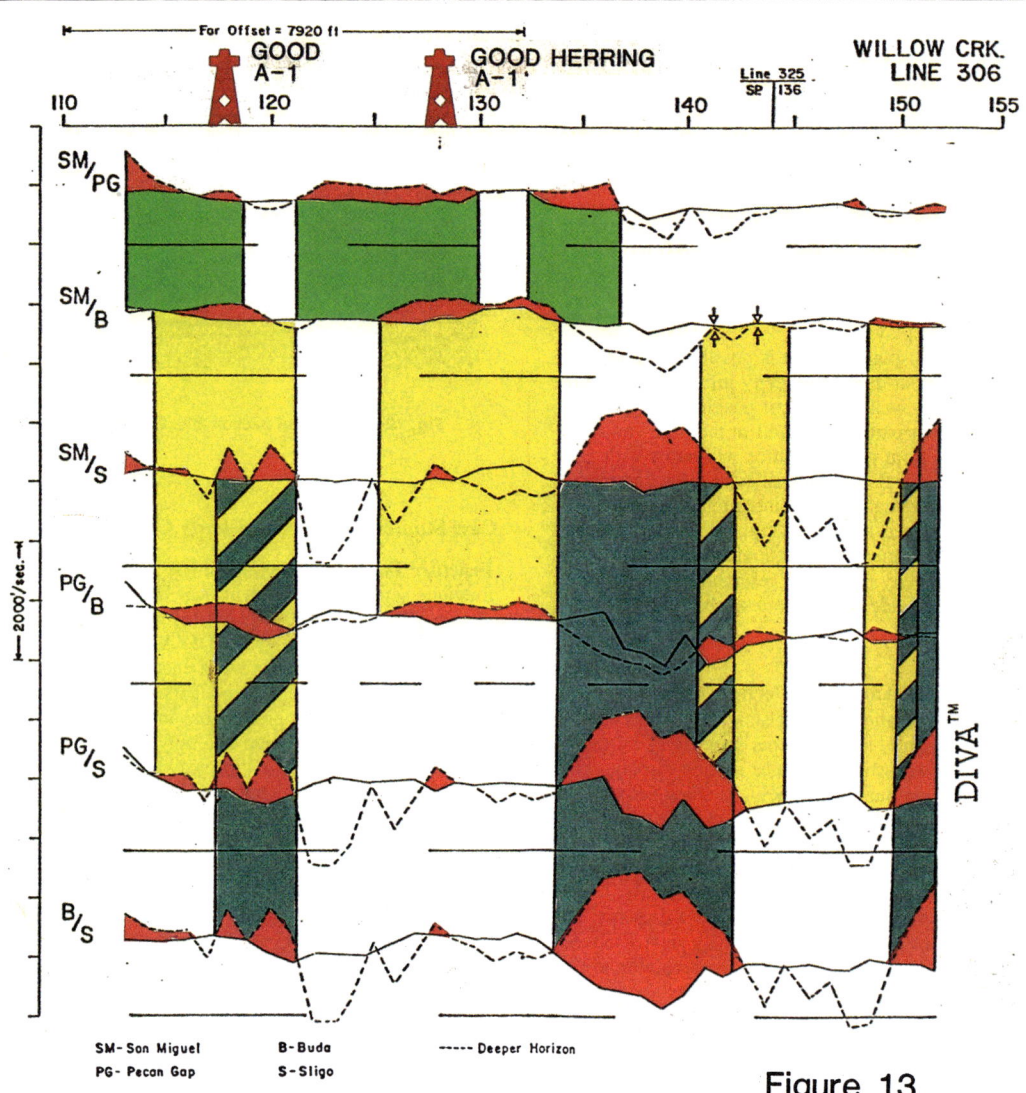

Fig. 13—DIVA™ display for Line 306.

the program and other work suggests that velocity drops of more than 400 ft/sec [122 m/s] can be strong evidence of gas presence. This well, in fact, had a far higher gas/oil ratio than the other wells. Added information was derived from the color acoustic impedance section for Line 306, shown in Fig. 14. A prominent low-velocity anomaly is notable in the chalk below the well location. This type of display, while unreliable in quantitative terms, in most cases does provide a correct polarity or "sense" for anomalies and a precise position in time. By contrast, the DIVA has excellent resolution in velocity but poor resolution in time. The acoustic impedance section while labeled in velocity is in truth only a presentation of amplitude information. The principle used for the color display here in relation to the full procedure is to establish anomaly time position and to monitor regional lithologic changes. In other cases, the color display can play a much expanded role.

As a standard completion practice, all Dilley wells were drilled to top of chalk with subsequent air drilling into the chalk. Only 4 of the 10 wells were acidized and lightly fractured to enhance production. Although the experiment began in late 1981, it is still

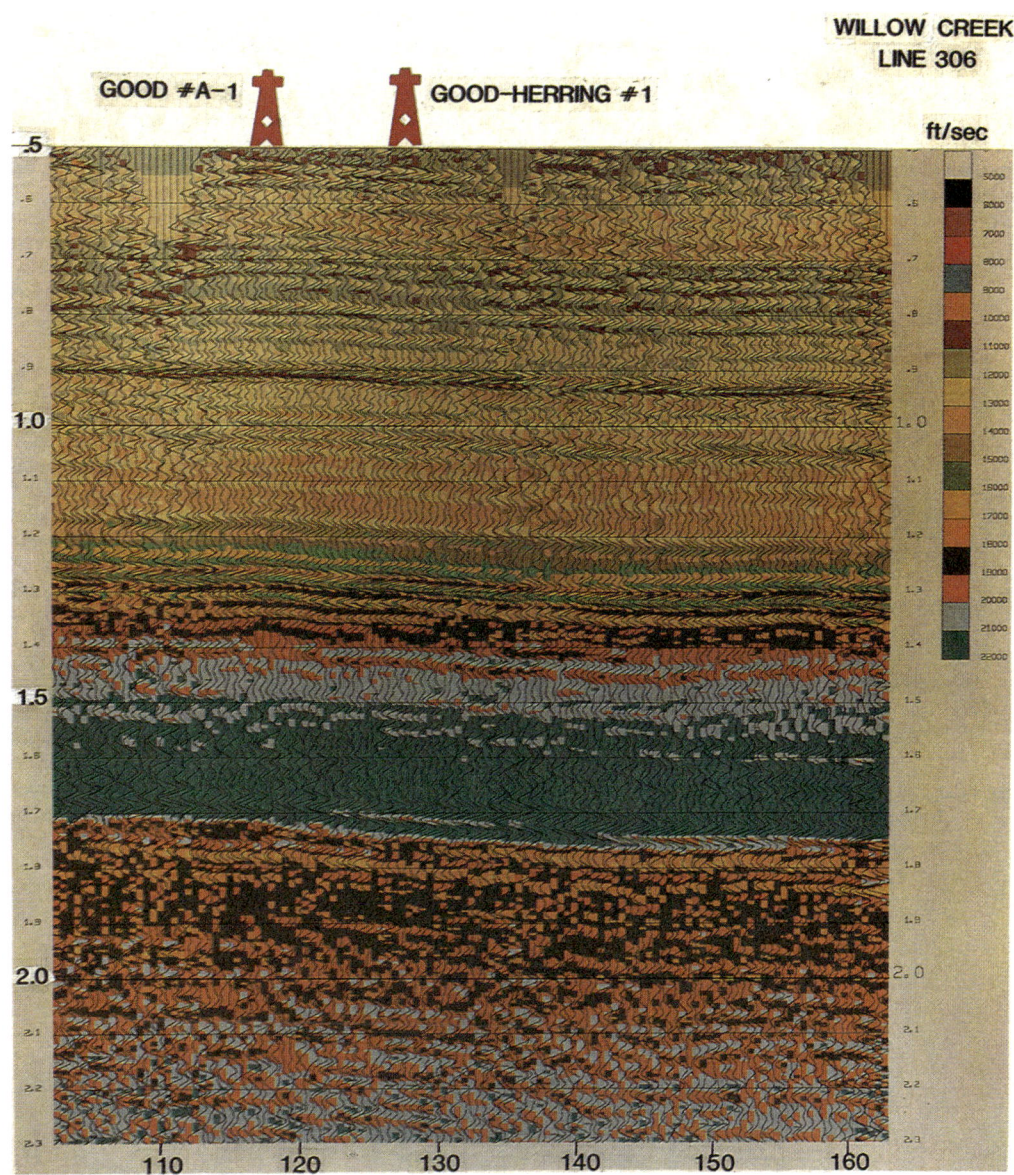

Fig. 14—Total acoustic impedance, velocity scaled.

ongoing with some wells not yet completed. More than 200,000 bbl [31 780 m³] oil has been produced to date.

During the experiment, it was confirmed that porosity, as further corroborated by the drilling rate, correlated well with the velocity reduction and that the presence of gas substantially enhanced the anomaly magnitude. Velocity anomalies, as identified on color acoustic impedance, were found unreliable in terms of absolute magnitude, properly relegating this tool to a subordinate or supportive role in this problem context.

As a final note, certain DIVA anomalies that proved to be oil in fracture porosity were of magnitudes as small as 150 ft/sec [46 m/s]. For an average chalk matrix velocity of 17,000 ft/sec [5182 m/s], this represents detection at better than a 1% level. While the sensitivity of the method obviously depends on data quality, data fold, surface conditions, noise backgrounds, and reflection contrasts, these results are nevertheless remarkable.

Edwards, Glen Rose, and Related Carbonates. A close-up of the acoustic impedance plot for a dip line in Houston County was noted in Fig. 6. The counterpart wavelet-processed seismic section was noted in Fig. 1 along with a color plot of acoustic impedance or velocity at coarser intervals. We discussed previously the *en echelon* carbonate buildup on the shelf of the East Texas basin.[13]

In this case, it was necessary to develop the DIVA display to bracket reservoir members separately, as overlying banks within a single interval have a signature that is quite difficult to unravel. The more and less porous zones of successive banks and related low-velocity shale zones interfered to produce this confusion and ambiguity. Portions of an appropriately computed display for this profile, shown in Fig. 15, clearly show several of the porosity zones.

The search for carbonate developments including patch reefs and pinnacles can be addressed much more generally. Owing to pronounced differences of depositional bedding characteristics, reflection seismic data have been used to differentiate carbonate buildups such as reefs and banks (including pinnacle reefs) from the enveloping strata. Geophysical criteria that allow recognition of buildups can be either *direct* or *indirect*.

Seismic parameters that directly outline buildups are reflections from the boundaries of the buildups, onlap of overlying cycles, or seismic facies changes between the buildups and the enveloping beds. Seismic parameters that indirectly indicate the presence of buildups or reefs are drape, spurious, or stray events, and more recently velocity anomalies both as indicated in color acoustic impedance plots scaled in velocity and in detailed studies which use moveout velocities. Of course the indirect indications always entail more subtlety. Types of carbonate buildups recognized most

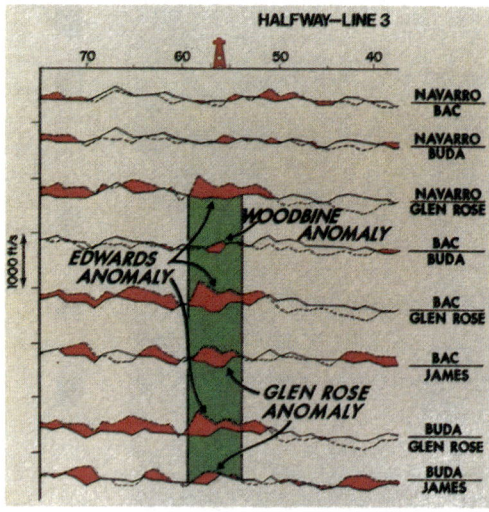

Fig. 15—Close-up of DIVA' showing key anomalies in carbonate bank sequence.

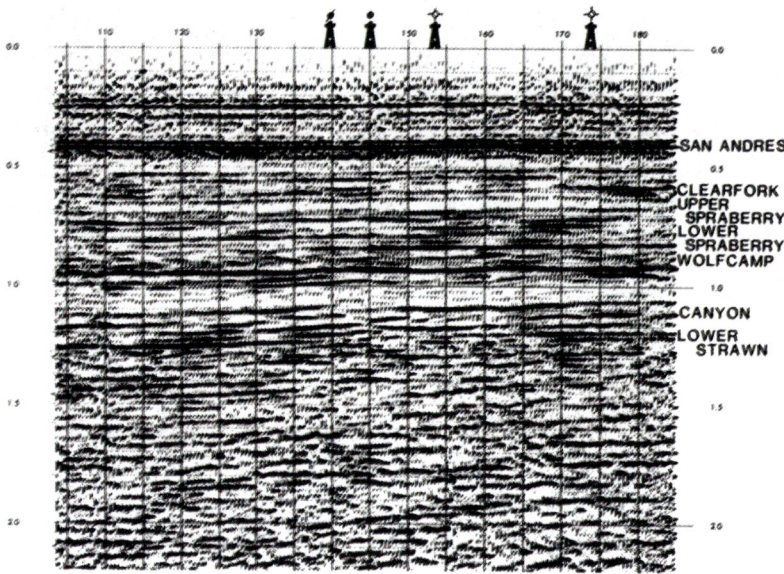

Fig. 16—Structural section for Line 12, Permian basin.

easily from seismic data are shelf-margin and barrier developments and those of larger areal extent. Pinnacle and patch reefs, particularly of smaller areal extent, are substantially more difficult to recognize using conventional seismic-reflection character changes, and more frequently require use of the indirect clues.

Shelf-margin buildups mostly are linear with deep water on one side and shallow water on the other. Barriers are also linear but with relatively deep water on both sides. Pinnacle buildups are roughly equidimensional and were surrounded by deep water, whereas patch reefs form in shallow water in broad shallow seas. Use of basin architecture is an additional and important geological line of evidence to infer locations of buildups. Nevertheless, the difficulty of recognizing the various carbonate buildups or reefs from conventional seismic data using direct-reflection criterion from diverse seismic sources is readily illustrated. Bubb and Hatlelid[14] in a classic discussion treat the state of the art in conventional terms.

An illustrative study, the search for patch reefs in the Canyon formation south of Horseshoe atoll in the Permian basin, was cited briefly in earlier papers.[13,15] The study was undertaken where substantial subsurface control existed with the specific aim of using the velocity-related detection tools to predict reefing on other seismic lines in the prospective area. This was clearly a circumstance where only indirect evidence could be used.

Note one of several wavelet-processed seismic profiles (Line 12) in Fig. 16 and the related detected moveout-derived velocity anomalies (Fig. 17) for this line. On the black and white section there is a lack of positive reflection energy at the top of the Canyon in the area of the reef. This probably is due to the buildup of carbonaceous material within the shaley section above the reef, frequently associated with such reefing. However, this characteristic in itself would hardly be sufficient evidence for drilling a well, as numerous dry holes surrounding some of the smaller reefs attest. The display noted in Fig. 17 for this case shows one good anomaly with lowered velocity of the order of 500 ft/sec [152 m/s] under ground location 146, which covers only the well producing from the reef and not the two wells on either side of it that did not produce from the reef.

The most diagnostic indication of the anomaly is given by the narrowest bracketing of the Canyon formation—here, the curve pair labeled Canyon to Lower Strawn. Confirmation of the anomaly is provided by the broader intervals. We also must consider the second anomaly under ground location 174, the location of a dry hole in the Upper Canyon formation. The color acoustic impedance section in Fig. 18 under ground location 146 shows an interesting and typical reefal "fisheye" signature with hard high-velocity carbonates of the Upper Canyon lime on either side of the reef and with a lower-velocity central area containing the reef. Immediately above the reef the low-velocity carbonaceous shale on either flank is interrupted by a higher-velocity transitional carbonaceous material associated with the reef. This provides a very distinctive signature that we have observed in many other areas, such as the Lockport in eastern Ohio.

A close-up of the particular reef and its signature from the color plot (Fig. 19) makes clear that such subtlety is best treated using approaches that draw

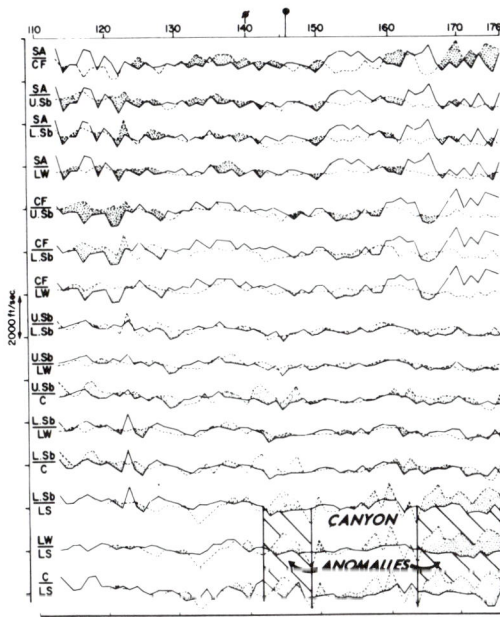

Fig. 17—DIVA™ analysis for Line 12, Permian basin.

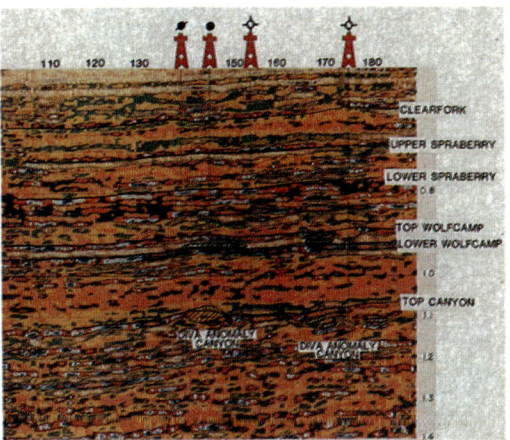

Fig. 18—Color acoustic-impedance section (velocity display) for Line 12, Permian basin.

time but only approximately in the horizontal dimension (and magnitude). The effectiveness of the complete procedure illustrated by this study, in which the geologic concepts starting with the basin in the large are integrated with velocity determinations for consistency and verification.

King County, Tannehill Sands. We next look at an area in King County, TX, where primary exploration objectives are quite different: the Tannehill, Strawn, and Ellenburger formations. Ashland Oil Co. originally started exploring in this area during the late 1970's. Lear Petroleum Corp. subsequently acquired this acreage as part of a larger package of prospects and data. To further develop the prospect with its suspected structural high at the Ellenburger horizon, Lear collected three seismic lines across the leases.

These three seismic lines had been processed very poorly by current standards, so the identification of the Ellenburger was at best speculative. Since the tapes for these lines could not be found, Zenith Exploration Co. Inc. opted to execute a four-line, 18-mile [29-km]

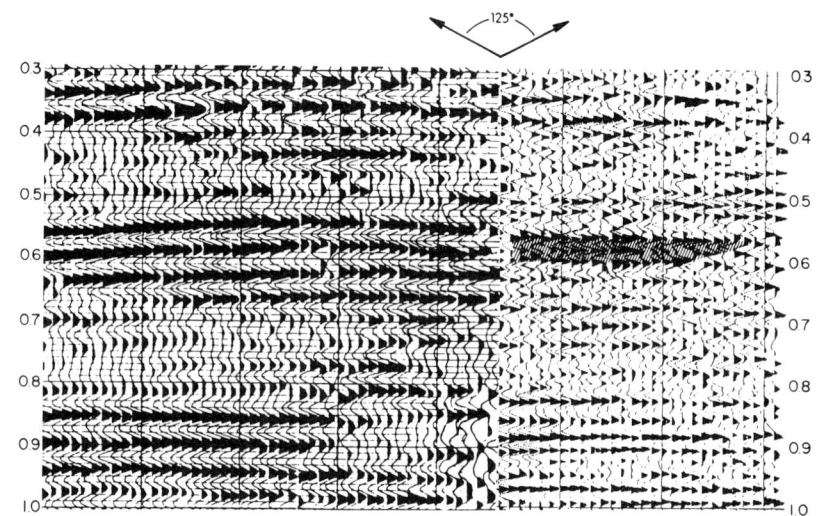

Fig. 21—Seismic resolution of Tannehill sand channel using high-frequency seismic data.

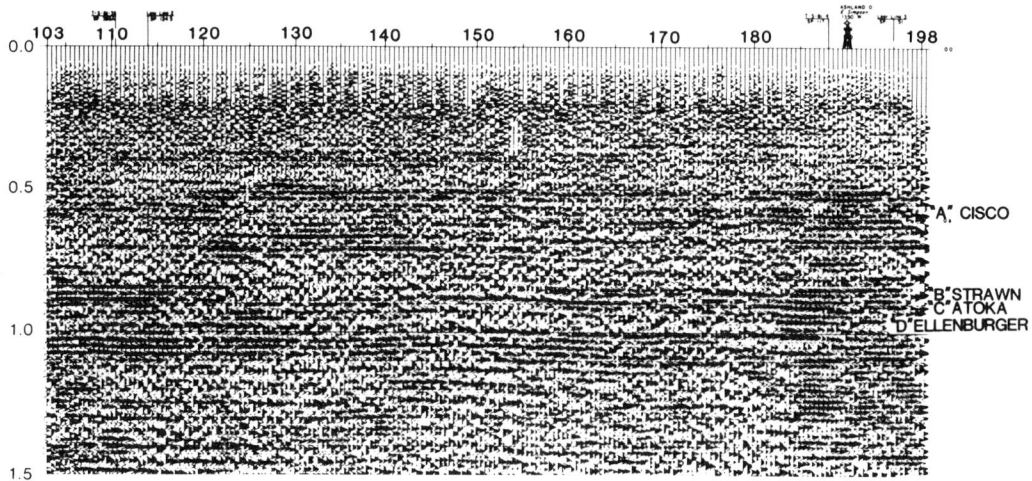

Fig. 22—Simpson prospect (King County, TX) Line T-S-81-1, structural section.

upon the velocity information. In this case, the large anomaly seen from the moveout-derived velocities (Fig. 17) arises from the unusually high porosity (more than 20%), since no gas was present here. Hence, statements of general nature concerning the lithology enable one to attempt quantitative estimation of porosity or gas presence from the velocity anomaly magnitudes. In philosophy, such approach is much like that applied to borehole measurements as, for example, embodied in published charts that relate porosity to sonic travel times for specified lithologies (Fig. 7).

Fig. 20 shows yet another such reef close-up (Line 4) and its color acoustic impedance signature. The subtleties involved again speak eloquently for the introduction of tools beyond the most elementary use of the seismic data. As before, we suggest fully utilizing the velocity information and incorporating appropriate geological models.

A similar pattern of moveout-based velocity anomalies and acoustic impedance signatures was seen over all the known reefs in the study. Drilling results indicate that moveout-derived velocities probably provide a more exact spatial location of the reefal porosity than do the acoustic impedance signatures taken alone. On the other hand, as evidenced by the Canyon anomaly under ground location 174 (Fig. 18), the moveout-derived velocity anomalies alone do not indicate the presence of reefal porosity in the absence of corroborating acoustic impedance evidence. This is because moveout-derived results may cover a broad gate in reflection time (as they do here within the Canyon and Strawn section), and thus can indicate lowered velocities for anomalies anywhere within that section. We also must be alert to anomalies that lie off the line to one side or the other. The nature of seismic reflections and the role of the Fresnel zone[3] makes clear how these off-line features can make apparent contributions. Acoustic impedance sections, on the other hand, pinpoint the anomalies better in reflection

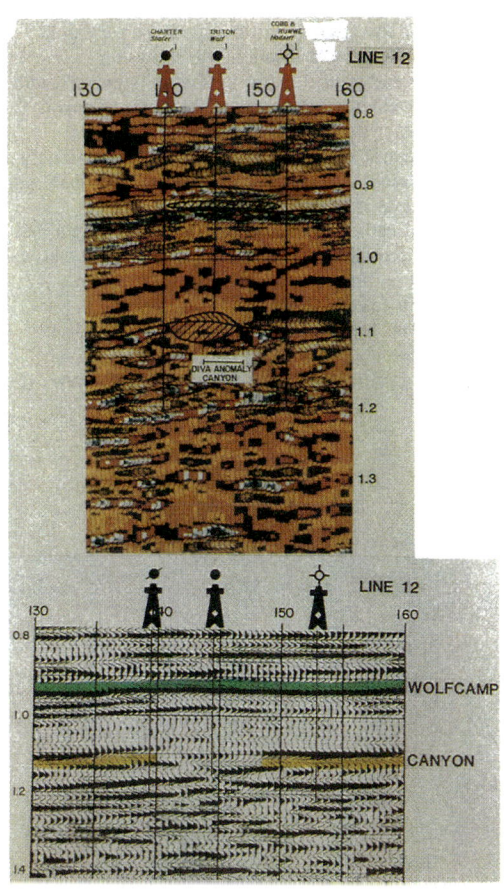

Fig. 19—Close-up of reef seismic signatures on conventional structural section and color impedance, Line 12.

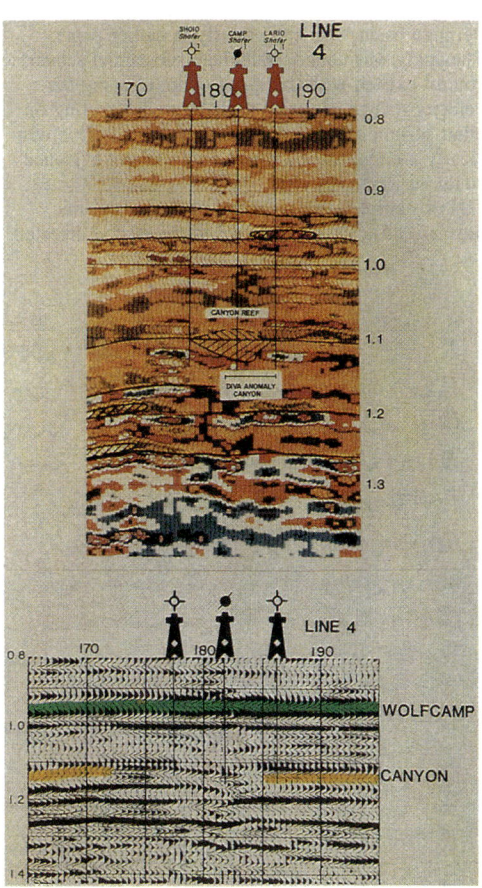

Fig. 20—Close-up of reef seismic signatures on conventional structural section and color impedance, Line 4.

seismic program to refine and further delineate the structure as suggested by the original Ashland mapping, as well as to define stratigraphic elements. The new data, with their broader frequency band and superior processing including wavelet treatment, made it possible to recognize and follow the subtle expression of the Tannehill sand channels, as the comparison of old and new data intersecting in Fig. 21 shows. Note that with wavelet processing, the negative event is indicated by a trough which defines the top of the channel and tells us that the channel fill is low-velocity compared to its surrounding. The seismic expression is directly comparable to the geologic model of a typical channel offered recently by Reneer.[16]

As before, an integral part of the interpretive approach involves using acoustic impedance color sections to correlate with the available geologic and seismic study. Logs from wells located on the seismic lines and a velocity survey and synthetic seismogram from Ashland's Eunice Simpson No. 1 were used to correlate lithologic units identified on the seismic color acoustic impedance sections and also the wavelet-processed seismic time sections.

In this case, owing to noise conditions and acquisition problems, the moveout-based analysis

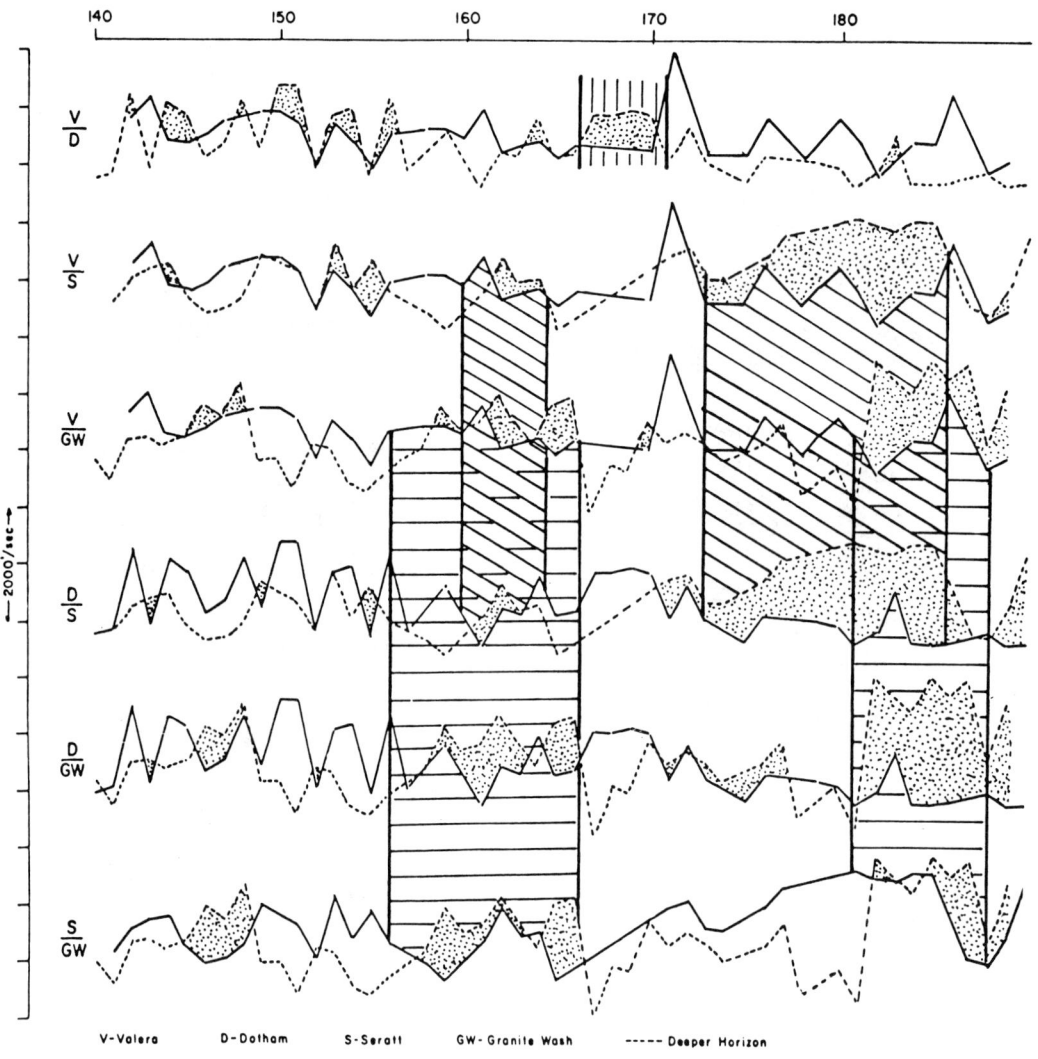

Fig. 23—Simpson prospect (King County, TX) Line T-S-81-1, DIVA™ display.

could be run on one line (Line T-S-81-1, shown in Fig. 22) and only as a test (Fig. 23 shows the velocity anomalies). It was uncertain from the outset of the project whether the velocity analyses would be of sufficiently high quality in this area, owing to use of a different seismic processor to speed up turnaround, the nature of the west Texas surface condition and rapid subsurface changes, and a number of operational problems during the seismic acquisition operation.

In spite of these obstacles, fairly good velocity picks were obtained and the anomaly display was produced for Line T-S-81-1. It showed low velocity anomalies that were largely in agreement with those seen on the color acoustic impedance section (Fig. 24). Of course, the color section amplitude derived velocity anomalies previously deemed credible also were corroborated on a structural basis using a regional picture developed from the geologic view. For example, it was necessary for the channel developments to map conformably around the structural highs. Agreements in such regard proved quite remarkable.

Color seismic acoustic impedance displays as in Fig. 24 provided basic information used to delineate and map facies changes in a manner that allowed interpretation in terms of rock type and porosity, and therefore, to some degree, the depositional environment. It was then possible to interpret the probable existence of favorably located reservoirs.

For this case, the acoustic impedance plots were produced by Teknica. Maps were made using both the color sections and the basic seismic sections to delineate interpreted porosity zones. Seismic lines away from wells were converted to depth by using velocities from the stacking velocity analyses along the particular line and correcting these to agree, where possible, with velocities derived from the well. This correctional conversion procedure was necessary to remove possible distortions in the geometry of the seismic data caused by differing thicknesses of the high-velocity carbonate section in this area as evidenced by well data.

Use of these new techniques not only enabled us to define areas of likely porosity but also made it possible to correlate lithologic units delineated by velocity slowdowns on the color sections with structural components and to map these areas with great confidence. Fig. 25 partially illustrates the results and successes of the approach. Note that the mapped Tannehill channels indeed do skirt the Top Cisco structural highs. Several significant Tannehill discovery wells, indicated by the black arrows, were drilled after this study and appear to confirm the channel-system map quite well.

Conclusions and Future Directions

It should be clear from the discussion and case studies that new exploration tools and methods designed to seek out stratigraphic traps also have great bearing on the definition of hydrocarbon reservoirs and their properties. The approach outlined couples geological inputs, including the recently developed seismic facies analysis, with geophysical directions. In particular, velocity measurements from seismic data are featured. At the same time, the resolving power of seismic data in structural terms and in defining unit sizes and thicknesses also has improved greatly.

Thus, we should be able to define reservoirs with

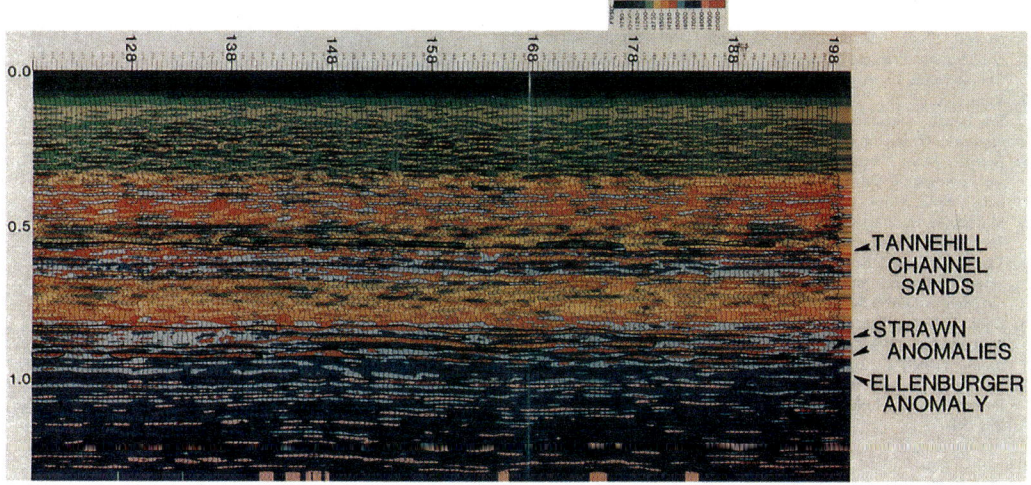

Fig. 24—Simpson prospect (King County, TX) Line T-S-81-1, color acoustic impedance.

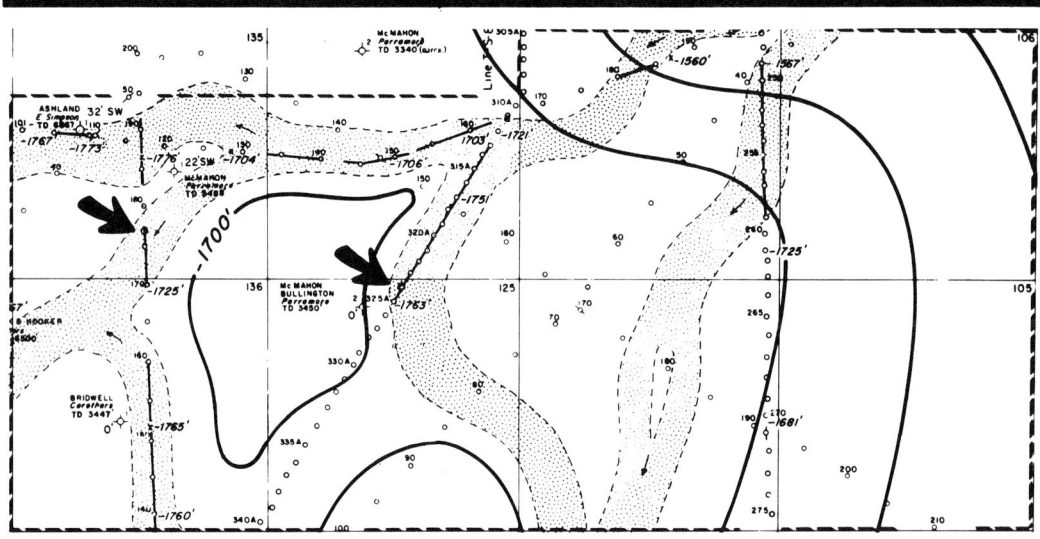

Fig. 25—Seismic definition of Tannehill sand channel and recent discoveries.

greater precision than ever before and also to suggest in advance of drilling their properties such as porosity. With subsurface calibration information, gas presence usually may be recognized. As more subsurface information becomes available, when used in conjunction with exploration maps, these maps appropriately scaled and modified should become the basis for producing the reservoir in an optimal way.

References

1. *The Deliberate Search for the Subtle Trap,* M.T. Halbouty (ed.), AAPG Memoir 32 (1982).
2. Widdess, M.: "How Thin is a Thin Bed?" *Geophysics* (1973) **38**, 6, 1176-80.
3. Neidell, N.S.: *Stratigraphic Modeling and Interpretation: Geophysical Principles and Techniques,* Continuing Education Course Notes 13, AAPG (1979).
4. Vail, P.R. *et al.*: "Seismic Stratigraphy and Global Changes in Sea Level," *Seismic Stratigraphy in Applied Hydrocarbon Exploration,* C. Payton (ed.), AAPG Memoir 26 (1977) 1149-1212.
5. Neidell, N.S.: "Technical Impact of Geophysics," *Techniques and Concepts in Oil and Gas Exploration,* de Figuerida and Jain (eds.), SEG, Tulsa (1982).
6. Mayne, W.H.: "Common Reflection Point Horizontal Stacking Techniques," *Geophysics* (1962) **27**, 6, 927-38.
7. Taner, M.T. and Koehler, F.: "Velocity Spectra-Digital Computer Derivation and Applications of Velocity Functions," *Geophysics* (1969) **34**, 6, 859-81.
8. Cook, E.E. and Taner, M.T.: "Velocity Spectra and Their Use in Stratigraphic and Lithologic Differentiation," *Geophysical Prospecting* (1969) **17**, 1, 433-48.
9. Lavergne, M. and Willm, C.: "Inversion of Seismograms and Pseudo Velocity Logs," *Geophysical Prospecting* (1977) **25**, 2, 231-50.
10. Lindseth, R.O.: "Seislogs," AAPG-SEG School on Stratigraphic Interpretation of Seismic Data, Houston, Sept. 20-25, 1976.
11. Gardner, G.H.F., Gardner, L.W., and Gregory, A.R.: "Formation Velocity and Density—The Diagnostic Basics for Stratigraphic Traps," *Geophysics* (1970) **39**, 6, 770-80.
12. Sheriff, R.E.: *Encyclopedic Dictionary of Exploration Geophysics,* SEG, Tulsa (1973) 266.
13. Neidell, N. *et al.*: "Improve Prospect Picks with Moveout Velocity Analysis," *World Oil* (Jan. 1984) 129-42.
14. Bubb, J.N. and Hatlelid, W.G.: "Seismic Recognition of Carbonate Buildups," *Seismic Stratigraphy—Applications to Hydrocarbon Exploration,* AAPG Memoir 26 (1977) 185-204.
15. Cook, E.E. and Neidell, N.S.: "Project Dilley: A Drilling Program to Evaluate a New Stratigraphic Technique," paper presented at 52nd SEG Meeting, Dallas, Oct. 17-21, 1982.
16. Reneer, B.: "Depositional History, Potential of Permian Tannehill Sandstone, King, Knox Counties, Texas," *Oil & Gas J.* (Nov. 21, 1983) 85-87.

SI Metric Conversion Factors

bbl × 1.589 873   E−01 = m^3
cu ft × 2.831 685   E−02 = m^3
ft × 3.048*   E−01 = m

*Conversion factor is exact.

Distinguished Author Series articles are general, descriptive presentations that summarize the state of the art in an area of technology by describing recent developments for readers who are not specialists in the topics discussed. Written by individuals recognized as experts in the areas, these articles provide key references to more definitive work and present specific details only to illustrate the technology. **Purpose:** To inform the general readership of recent advances in various areas of petroleum engineering.

Interpretation of depositional facies from seismic data

J. B. Sangree* and J. M. Widmier‡

Depositional environments can be predicted from seismic data through an orderly approach to the interpretation of seismic reflections. One keystone to this approach is an understanding of the effects of lithology and bed spacing on reflection parameters. Amplitude, frequency, and continuity are some of the parameters most useful for interpreting environments. Reflection amplitude contains information on the velocity and density contrasts at individual interfaces and on the extent of interbedding. Frequency is primarily a characteristic of the nature of the seismic pulse, but it is also related to such geologic factors as the spacing of reflectors or lateral changes in interval velocity. Continuity of reflections is closely associated with continuity of bedding (e.g., continuous reflections suggest widespread, layered deposits).

A second keystone to this interpretive approach is the parallelism of reflection cycles to gross bedding and, therefore, to physical surfaces that separate older from younger sediments. Exceptions to this concept include (1) fluid contact reflections, (2) limitations imposed by seismic resolution, and (3) various nongeologic coherent events. In spite of these exceptions, this concept provides a powerful tool for the analysis of reflection patterns.

Reflection cycle patterns include the configuration of reflections (i.e., layered, chaotic, and reflection-free) and the nature of cycle terminations at the depositional unit boundaries. The external form of the depositional unit can be analyzed from a grid of seismic lines and is valuable in interpreting the depositional processes responsible for the unit. Sheet, sheet drape, wedge, lens, fan, and other forms are described. The areal associations of these forms are often critical to environmental interpretation. Examples of facies interpretation from seismic sections are shown for depositional environments ranging from shelf to basin floor.

INTRODUCTION

Information on reflection amplitude, frequency, continuity, and configuration, as well as the external form and three-dimensional associations of groups of reflections, is available for direct interpretation from seismic sections. Each of these elements of seismic data contains information of stratigraphic significance, as listed in Table 1. Although the discussion in this paper is limited to visual inspection of these elements (Figure 1), computer techniques now provide us with quantitative measurements of many of these parameters.

Analysis of the relationships of Table 1 begins by examining the stratigraphic controls on the reflection process. Other factors that affect reflections are beyond the scope of this paper. These include the effects of hydrocarbon saturation on reflections and of certain factors that vary with increasing traveltime [i.e., intrinsic attenuation (McDonal et al, 1958), intrabed multiples (Schoenberger and Levin, 1974), and amplitude attenuation caused by the reflection process].

STRATIGRAPHIC CONTROLS OF AMPLITUDE AND FREQUENCY

Cycles or events on a seismic section may represent reflection of the down-traveling pulse from a single bedding surface, but much more commonly they represent reflections from several adjacent bedding sur-

---

A version of this paper was presented at the 44th Annual International SEG Meeting, November 12, 1974 in Dallas. A shorter version was published in AAPG's Memoir 26, and the May 1978 AAPG Bulletin. The portions of this paper which duplicate that version published by AAPG appear here by permission of AAPG.
Manuscript received by the Editor June 9, 1977; revised manuscript received July 20, 1978
*Esso Exploration Inc., Europe Africa Region, St. Clements House, Church Street, Surrey, England K12 2QL; formerly with Exxon Production Research Co., Houston, TX.
‡Exxon Production Research Co., P.O. Box 2189, Houston, TX 77001.
0016-8033/79/0201-$03.00. © 1979 Society of Exploration Geophysicists. All rights reserved.

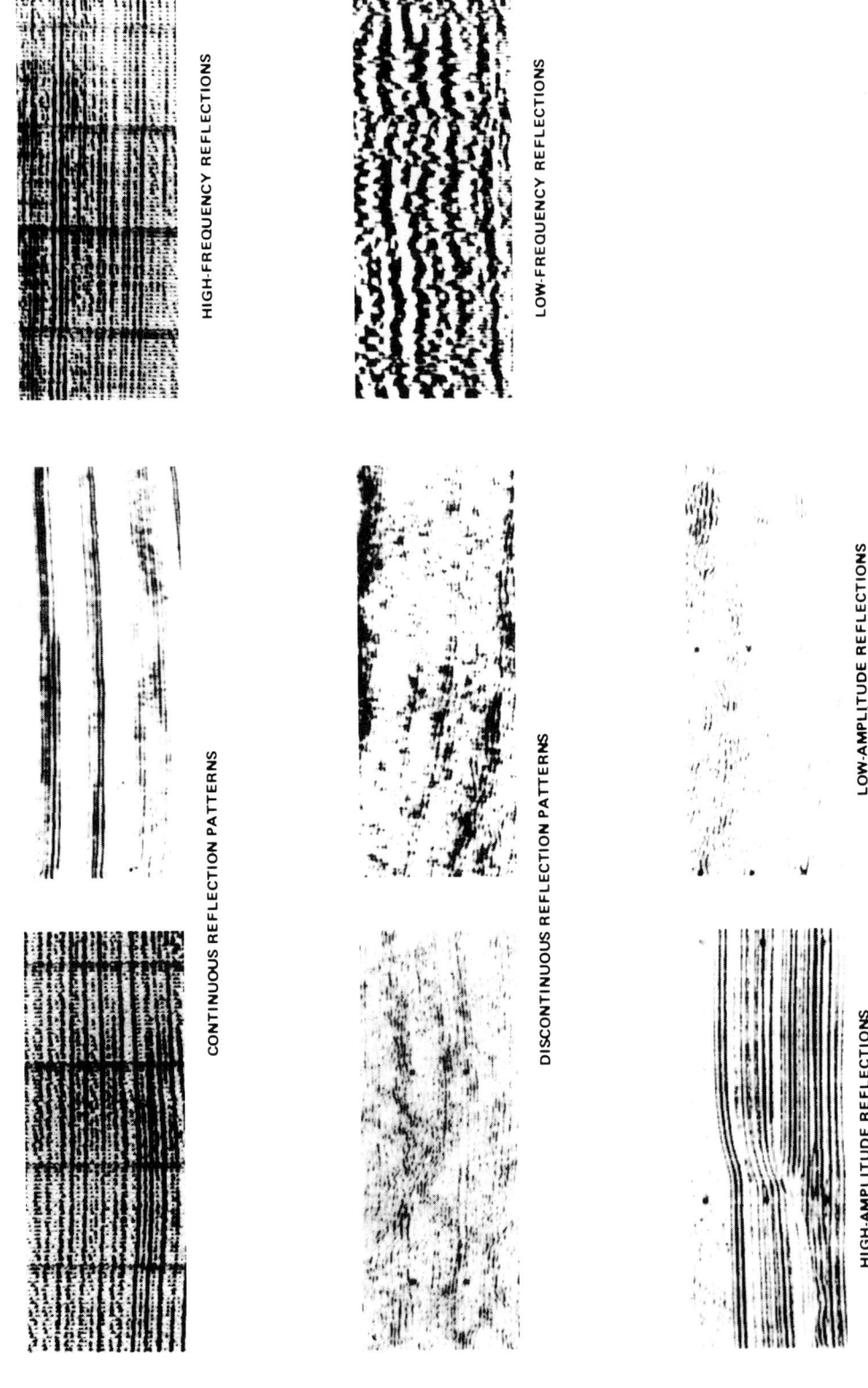

FIG. 1. Continuity, amplitude, and frequency variations in reflection character.

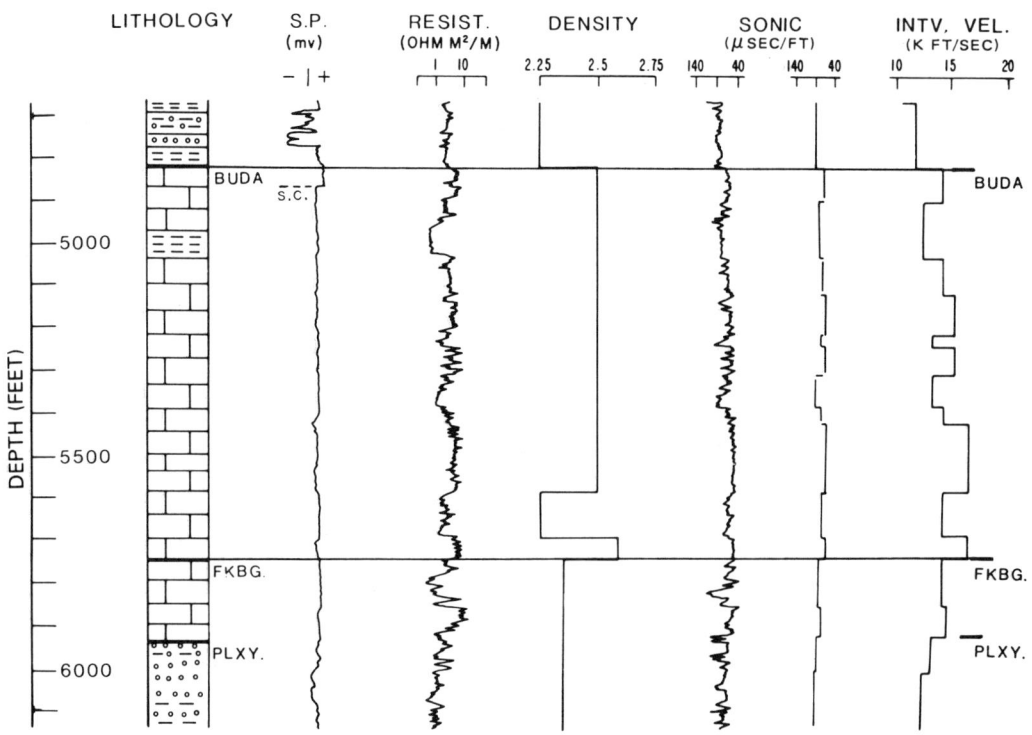

FIG. 2. East Texas well data—depth scale.

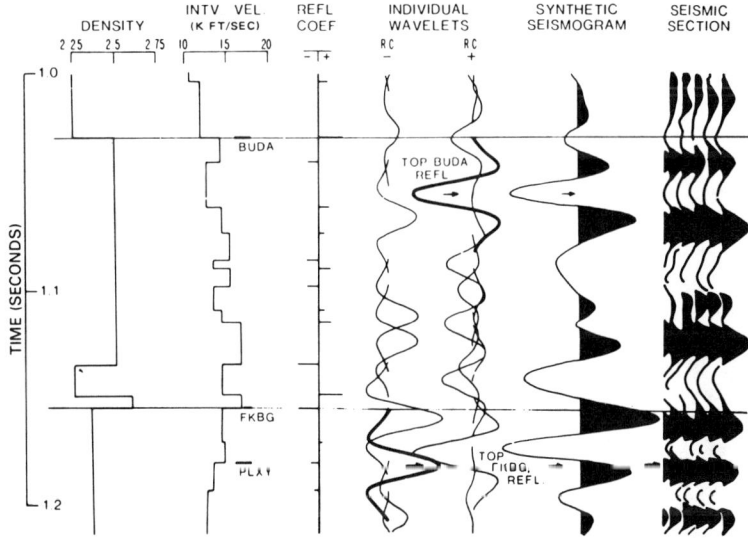

FIG. 3. East Texas synthetic seismogram—time scale.

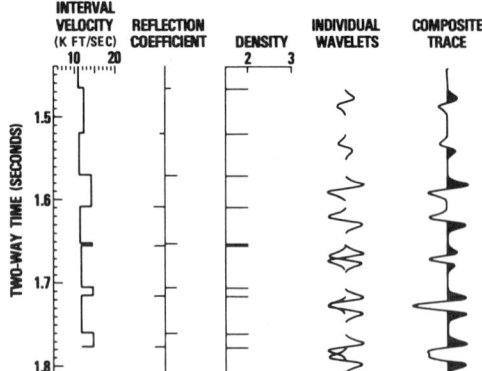

FIG. 4. Composition of reflections, illustrating effect on a composite synthetic seismic trace of varying bed thickness and velocity contrasts.

faces. To understand the geologic relationships shown for amplitude and frequency in Table 1, we need to look at the reflection-composition process.

The composition of reflections and their stratigraphic implications can be illustrated quite easily when one ignores the complexities of intrabed multiples. In our experience, the advantage of this simplification commonly outweighs the limitations. Except for very detailed seismic modeling studies, this approach allows the geologically oriented interpreter to gain a good understanding of the stratigraphic nature of reflections and also provides a sound basis for tying well and seismic data.

Figures 2 and 3 illustrate this simplified approach to the composition of reflections. Figure 2 shows the lithology, electric logs, density plot, and sonic log for an East Texas well. The sonic log has been smoothed and blocked into discrete beds and converted to interval velocity, as shown on the right side of the figure. The vertical scale is linear depth.

Figure 3 repeats the density and blocked interval velocity plots of Figure 2, but the vertical scale is linear time. Also plotted are the reflection coefficients and individual reflection wavelets[1] for each interface. The individual wavelets are separated into positive and negative polarities. The reflection coefficients, and consequently the amplitudes, of each wavelet are calculated for each bedding interface by the familiar relationship,

$$\text{Reflection coefficient} = \frac{V_2 D_2 - V_1 D_1}{V_1 D_1 + V_2 D_2},$$

where V and D are interval velocity and density, respectively, for two adjacent beds, bed 1 lying above bed 2.

In Figure 3 we see the algebraic summation of the individual wavelets in time (stated differently, the convolution of reflections from individual bedding surfaces) to form a composite waveform. On the right of Figure 3 is a portion of a seismic section from a recent common-depth-point (CDP) line close to the well. Similarity of the synthetic waveform and the seismic section confirms the validity of the method.

Amplitude variations are influenced by velocity-density contrasts at bedding interfaces. Further, both amplitude and frequency[2] of the seismic response are strongly controlled by the spacing of reflection surfaces. Figure 4 presents a seismic model of five sand beds, each with velocity higher than the 11,000-ft/sec velocity of the enclosing shale. Densities have been kept constant to simplify the discussion.

The upper two beds are thick enough to show the uncomplicated effect on seismic amplitude of variations in the contrast of velocity and density across a depositional surface. A low degree of contrast, i.e., a weak coefficient, gives a low-amplitude reflection. A high degree of velocity-density contrast results in a much higher amplitude.

The lower three beds are thin enough to show the effects of spacing of the top and base reflecting surfaces (i.e., bed thickness) on the summed (composite) trace. For the thinnest bed, both the amplitude and peak-to-peak cycle breadth of the composite waveform are reduced below those of the individual wavelets. That extremely thin beds lie below the resolution of the seismic system is simply an extension of this effect.

The fourth bed from the top has an optimum thickness for strong constructive addition. Reflection cycle breadth for this bed is the same as that of the wavelet, but composite waveform amplitude is greatly increased over that of the wavelets from either the top or the base of the bed. Thus, the higher amplitudes seen on seismic sections are commonly not from extremely large velocity-density contrasts from individual beds; instead, they represent beds with relatively moderate reflection coefficients whose thickness gives rise to

---

[1]In this example, a Ricker pulse was used for the individual reflection wavelets. It is characterized by a large central peak and two smaller side lobes. In the next example (see Figure 4), a first derivative of a Gaussian pulse was used for the individual wavelets. This pulse is a two-peaked pulse. Normally, the seismic data we are trying to match or analyze will dictate the appropriate pulse to use.

[2]Frequency, as used throughout this paper, refers to the characteristic frequency of a seismic event. It is the inverse of cycle breadth as measured on a seismic section.

Table 1. Geologic interpretation of seismic parameters.

Seismic facies parameters	Geologic interpretation
Reflection configuration	• Stratification patterns • Depositional processes • Erosion and paleotopography
Reflection continuity	• Bedding continuity • Depositional processes
Reflection amplitude	• Velocity-density contrast • Bed spacing • Fluid content
Reflection frequency	• Bed spacing • Fluid content
External form and areal association	• Gross depositional environment • Sediment source • Geologic setting

strong constructive addition. Optimum bed thickness for constructive addition corresponds to ¼ of a reflection cycle breadth on a seismic section (e.g., for a 30-Hz reflection, a 10,000-ft/sec bed has an optimum thickness of 82 ft).

The lowest bed in Figure 4 has a greater thickness than the two beds above it. The cycle breadth is increased, and amplitude is intermediate to that of the two overlying beds. Usually, broad (wide cycle breadth) reflections on a seismic section, interspersed with reflections of less cycle breadth, suggest that bedding is unusually thick and exceeds the optimum thickness for constructive addition. Many offshore Gulf Coast sections show this effect in the upper 2–3 seconds owing to the occurrence of effective bed thicknesses on the order of 200 to 300 ft. Care must be taken to distinguish this effect from the gradual increase in cycle breadth with depth seen on all seismic sections due to the earth's selective attenuation of higher frequencies.

BEDDING AND REFLECTION CONFIGURATION

Underlying the interpretation of reflection continuity, configuration, and external form, as outlined in Table 1, is the basic concept that seismic reflections parallel bedding surfaces within the resolution of a seismic cycle (Vail et al, part 5, 1977). This concept in turn rests on the tendency for stratified sediments to have greater lateral continuity of lithology, and therefore of physical properties, parallel to depositional surfaces than across depositional surfaces. This is not to say that all strata are layers of uniform internal composition; indeed, lateral variations of lithology are commonplace within strata deposited in a given span of geologic time. Although thorough documentation of the parallelism of reflections and bedding is not the subject of this report, we will present one example.

Figures 5 and 6 present a geologic section and a seismic section showing sediments deposited in shallow-marine to nonmarine environments in an intracontinental basin, in this case, the San Juan basin

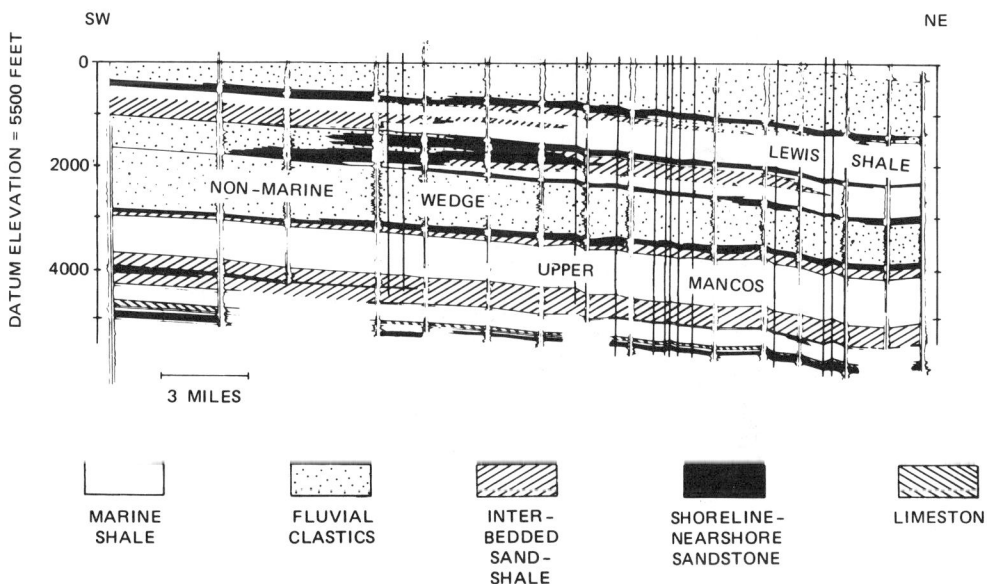

FIG. 5. Geologic cross-section of the San Juan basin.

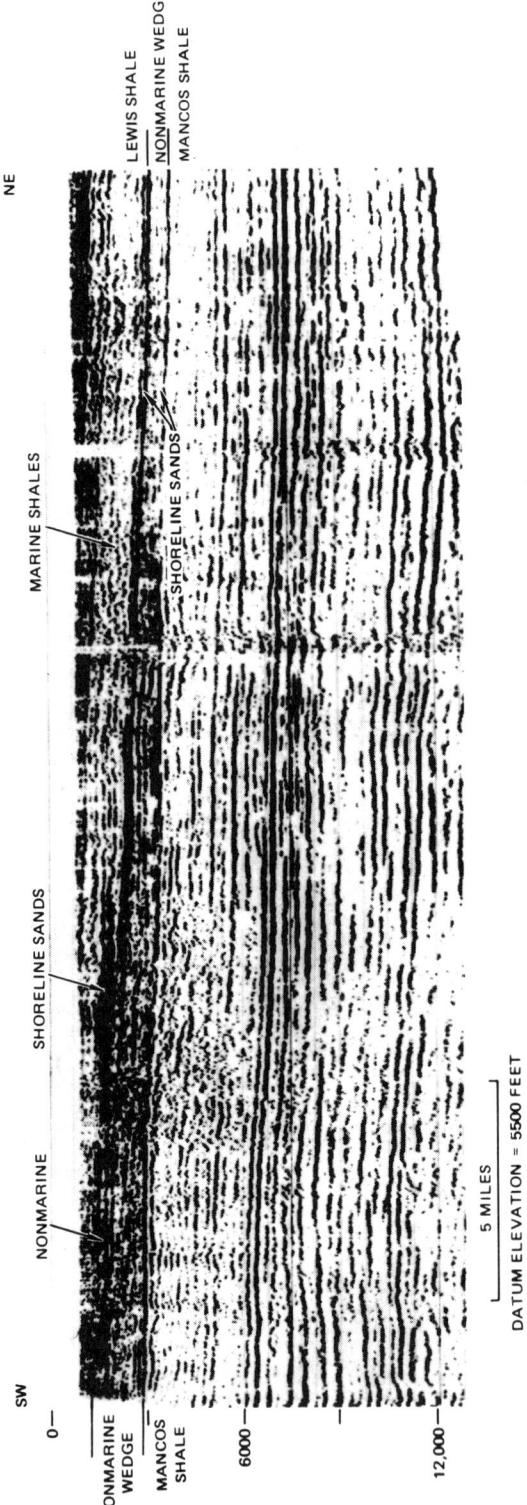

FIG. 6. Residual amplitude seismic section from the San Juan basin. On this section, only amplitudes above a selected threshold are preserved; those lying below this threshold are set equal to zero. This is done to emphasize amplitude variations for analysis.

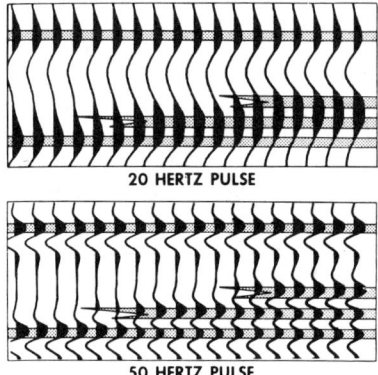

FIG. 7. Parallelism of beds and reflections: Four sand beds are resolved by a 50-Hz pulse, but not by a 20-Hz pulse. On the synthetic seismic section with a 20-Hz pulse, the lower high-amplitude reflection moves up and across the lower sand to the second and third sands.

of New Mexico. The section is about 30 miles long and averages almost one well per mile; the wells are very close to the line of section. The interval of interest is a Cretaceous nonmarine wedge. This unit is overlain and underlain by marine shales; the overlying shale is the Lewis and the underlying shale is the upper Mancos. These marine shales intertongue with the nonmarine wedge, and at the transition between the nonmarine and marine units a series of shoreline beach sands are formed. The pattern of bedding developed from detailed log correlations shows that individual depositional units cross all of the facies patterns shown here. For example, a depositional surface starting in the Lewis shale can be carried into the sandstones of the shoreline and nearshore environment, and finally into the fluvial clastics of the nonmarine wedge.

The seismic line was placed as close as possible to the wells, and closely spaced reflection velocity analyses were used to convert reflection time to depth. High-amplitude reflections mark the top and base of the nonmarine wedge at the zone of interbedding of marine shales and shoreline and nearshore sands.

Following reflections in the marine Lewis shale (Figure 6) from the right across the section and into the nonmarine sediments of the left, we see striking changes in the seismic parameters associated with these reflections. In the Lewis shale the reflection character is one of low amplitude and moderate continuity. As we trace a reflection into the position where the shales interfinger with the shoreline sands, amplitudes increase sharply because of the interbedding of sands and shales as well as the possible presence of gas in some of these sands. Carrying the same reflection farther to the left, we find it ending in a discontinuous pattern of both low and high amplitudes, consistent with the discontinuous bedding of the fluvial sediments.

Considerable experience confirms our confidence in the basic principle of parallelism of bedding and reflections; this is not to say, however, that there are not problems and exceptions to the principle. Problems in the application of this principle can arise from nonstratigraphic, coherent seismic events, such as diffraction patterns or sideswipe reflections originating some distance outside the seismic line of profile (Tucker and Yorston, 1973). Exceptions also include reflections from a fluid contact, such as a gas-water contact in a thick sand.

A more subtle exception originates in the limited resolution of the seismic section. A computer model of several sand bodies interbedded with massive shale illustrates the problem. Figure 7 shows a geologic section with overlays of two noise-free synthetic seismic sections that were made with different simple sine-wave pulses. The reflection lag time has been adjusted so that the reflection peaks superimpose the appropriate sand body to better illustrate the relationship of sands to reflections.

In the lower portion of Figure 7, a high-resolution pulse with a characteristic frequency of 50 Hz faithfully reveals the continuity of the upper and lower sands, and also indicates the lateral facies change from sand to shale of the two middle sands by a loss of amplitude in the shale. The upper portion of the figure shows the same geologic model and a synthetic section made this time with a low-resolution 20-Hz pulse. The continuity of the upper sand is preserved, because this sand is separated from the other sands. The lower three sands, however, interfere in the composite waveform to produce a reflection cycle that drifts across the depositional units, thus appearing to violate the principle of parallelism.

This exception is less of a problem than one might think. First, the staggered geometry of lateral facies transitions must have an unusual degree of an echelon symmetry to produce the indicated effect. Second, abrupt facies transitions may be visible as phase shifts, particularly if reflections above and below are continuous and parallel. Third, improvements in seismic resolution continually reduce the limit of the thinnest bed that can be resolved by a complete seismic cycle, although the inability to resolve thin beds at great depths remains a severe limitation of the seismic system. Bed resolution as a function of reflection frequency is shown in Figure 8.

In summary, it is our experience that reflections do parallel beds within the limitations of seismic resolution. The seismic stratigrapher must learn to handle problems, exceptions, and pitfalls in much the same way that the interpreter of structure from seismic data must cope with geophysical and geologic difficulties.

NATURE OF REFLECTION PATTERNS

A seismic facies unit is defined as a mappable group of reflections whose elements, such as reflection pattern, amplitude, continuity, frequency, and interval velocity, differ from the elements of adjacent units. Reflection pattern includes reflection configuration and the nature of cycle terminations at seismic facies unit boundaries. Since we interpret reflection configurations as gross bedding patterns, we can infer from these patterns considerable information on the transporting processes and the environment of deposition. In Figure 9, three principal types of reflection configuration are shown: (1) reflection-free zones from areas where few reflecting surfaces exist; (2) stratified patterns in which parallel or divergent reflections are present and have a reasonable degree of continuity; and (3) chaotic patterns in which reflections are discontinuous, are often mounded or internally contorted, and frequently are distinguished by a good many diffraction patterns.

Reflection-free configurations are indicative of a uniform, single lithology, possibly of extremely steep dip, or of intense postdepositional homogenization of multiple lithologies. In a clastic sediment regime, this configuration could be from a salt dome or an unusually homogeneous marine shale, whereas in an evaporite-carbonate regime, it could indicate salt or massive carbonate reef-core environments. Massive igneous bodies may also produce reflection-free patterns.

The continuity of reflections varies greatly between layered and chaotic configurations. Continuous reflections, with uniform amplitude and frequency from trace to trace, arise from rock layers that are uniform in their thickness and lithology over the region covered by the section. They represent deposition during periods of stable and uniform depositional conditions.

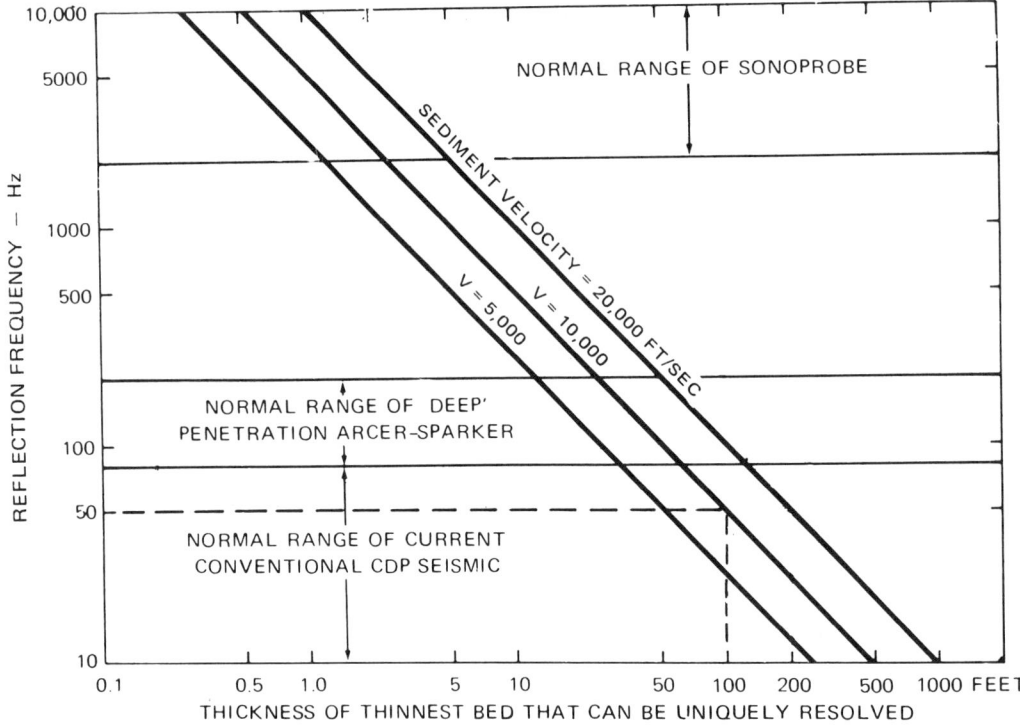

FIG. 8. Resolution as a function of the recorded signal frequency. With a 50-Hz signal used as an example, reflections from the top and bottom of a bed, which is 100-ft thick or greater and has a velocity of 10,000 ft/sec, will not overlap in time and the bed will be uniquely resolved on a recorded seismic trace.

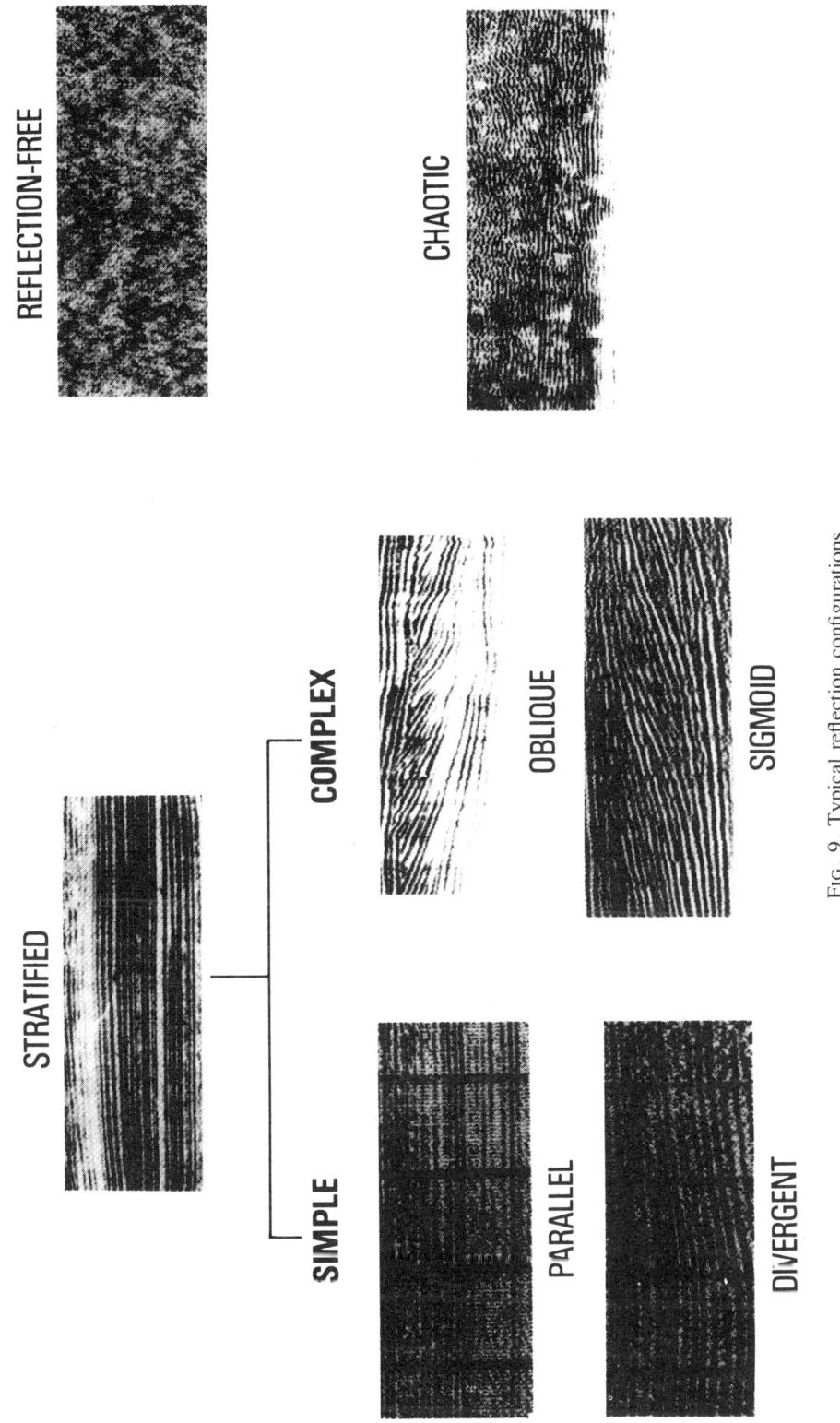

Fig. 9. Typical reflection configurations.

Discontinuous discordant reflections observed in chaotic configurations (Figure 9) suggest a disordered arrangement of reflection surfaces and either a relatively high energy and variability of deposition or a disruption of beds after deposition. When a pattern of discontinuous reflections is associated with closely spaced diffractions, the interpreter should examine the pattern closely to distinguish between sharply folded or broken beds, suggesting disruption, and the occurrence of small, steep-walled channels on unconformity surfaces or as part of submarine cut-and-fill deposition.

Stratified configurations can be simple or complex (Figure 9). Simple stratified configurations include parallel and divergent arrangements of seismic cycles. Parallel arrangements suggest uniform rates of deposition on a stable or uniformly subsiding surface. Divergent arrangements suggest areal variations in the rate of deposition, progressive tilting of the depositional surface, or a combination of the two factors.

Complex stratified configurations (Figure 9) include sigmoid and oblique arrangements. (Differences between these patterns are discussed in the section on shelf-margin and prograded-slope seismic facies.) Both of these arrangements form through progressive development of depositional surfaces that slope from a gently dipping, relatively shallow-water area into deeper water. The resulting reflection configuration can be divided into upper (topset), middle (foreset), and lower (bottomset) zones (Figure 10). If the upper zone represents beds deposited in shallow-water depths subject to wave disturbance, the seismic facies unit may consist of a delta-plain, delta-front, and prodelta sediments. In this case, the upper, mid-

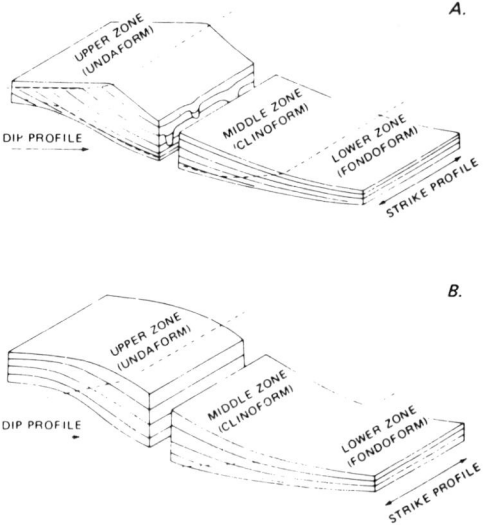

FIG. 10. Diagrammatic oblique- and sigmoid-progradational reflection patterns.

Table 2. Terminology used for objective description of seismic parameters

Interpretational			Objective						
					Reflection configuration		Other seismic facies parameters		
Depositional framework interpretation	Seismic facies unit	Environmental facies interpretation	External form of facies unit	Reflection geometry at boundaries	Principal internal configuration	Configuration modifier	Amplitude	Continuity	Frequency (cycle breadth)
			Sheet Sheet drape Wedge Elongate wedge Lens Elongate lens Mound Elongate mound Fan Trough fill Channel fill Basin fill Slope front fill	Concordant Onlap Downlap Toplap Erosional truncation Structural truncation	Parallel Divergent Sigmoid Oblique Complex sigmoid-oblique Chaotic Mounded Reflection free	Hummocky Lenticular Disrupted Contorted Regular Irregular Uniform	Low Moderate High	Continuous Moderately continuous Discontinuous Uniform—variable	Broad Moderate Narrow

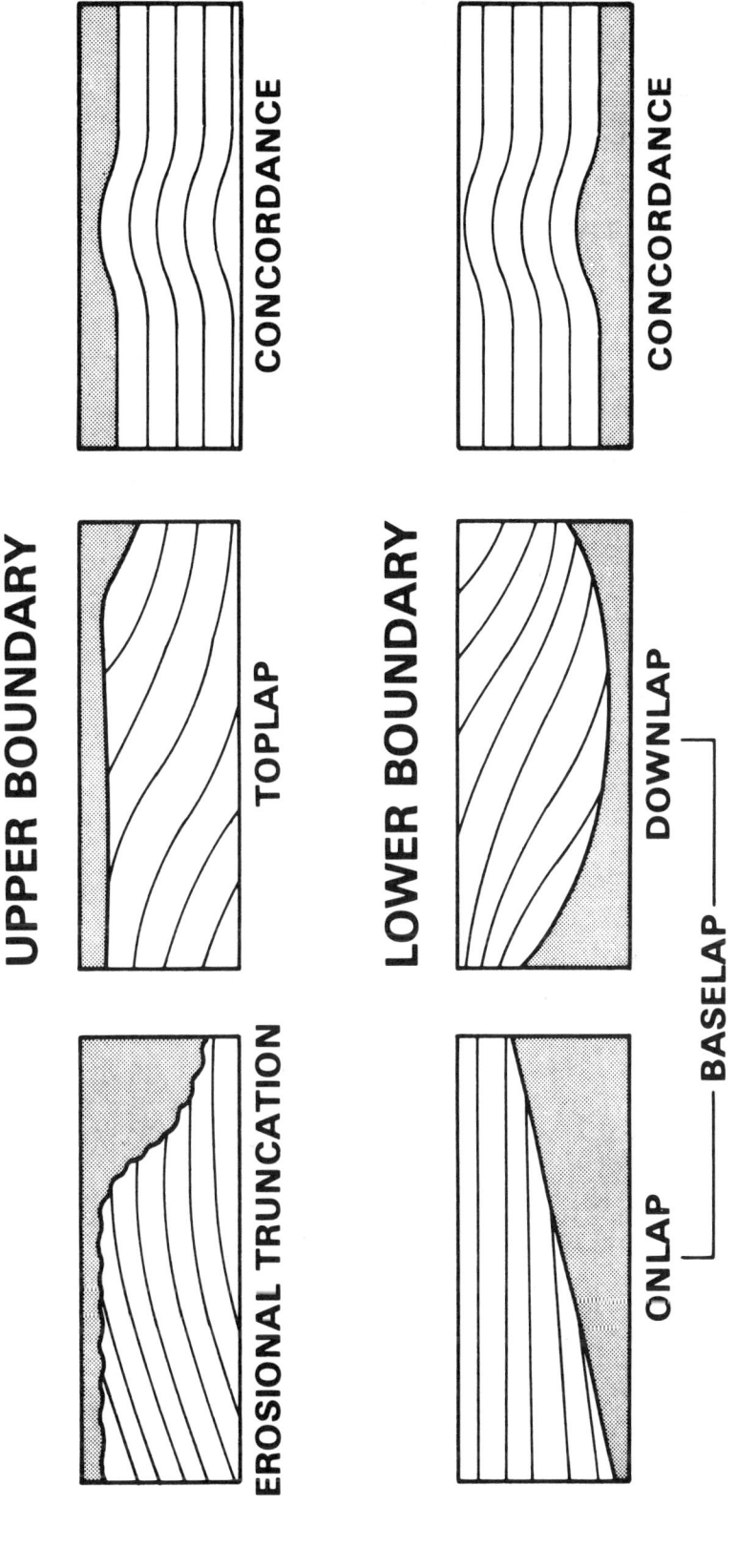

Fig. 11. Nature of cycle terminations at boundaries of seismic facies units.

dle, and lower zones correspond to Rich's (1951) undaform, clinoform, and fondoform topographic environments, respectively. Typically, beds of the middle zone are thicker and more steeply dipping than those of the upper or lower zones.

THREE-DIMENSIONAL ASPECTS OF SEISMIC FACIES UNITS

Specific reflection patterns can be areally delimited and described as seismic facies units. The upper and lower boundaries of these units may be identified by the termination of several seismic cycles against a common reflection, or they may be identified by conformable reflections that bound a particular reflection pattern.

The variety of cycle terminations at the top and base of seismic facies units is limited (Figure 11). Truncated cycles may represent toplap termination, in the case of oblique configurations, or erosional truncation; when data are not satisfactory to separate these two types, the cycles may be termed simply truncation. At the base of seismic facies units, discordant relationships are defined as onlap, downlap, or simply baselap if structural complications make distinction of onlap and downlap impossible.

When seismic facies units are bounded by onlap, downlap, or truncation cycle termination patterns, or the conformable extension of those patterns, they possess the sequence criteria described by Vail et al (part 6, 1977). As such, these boundaries may represent local changes in source or depositional environment, or they may be related to cyclic changes in sea level. From analysis of reflection cycle termination patterns, it is often possible to determine the presence and extent of erosion and to gain considerable information on the paleotopography that existed at the time of deposition.

Typical external geometries of seismic facies units for sand-shale sediments are listed in Table 2 and illustrated diagrammatically in Figure 12. This list, which undoubtedly will be expanded in the future,

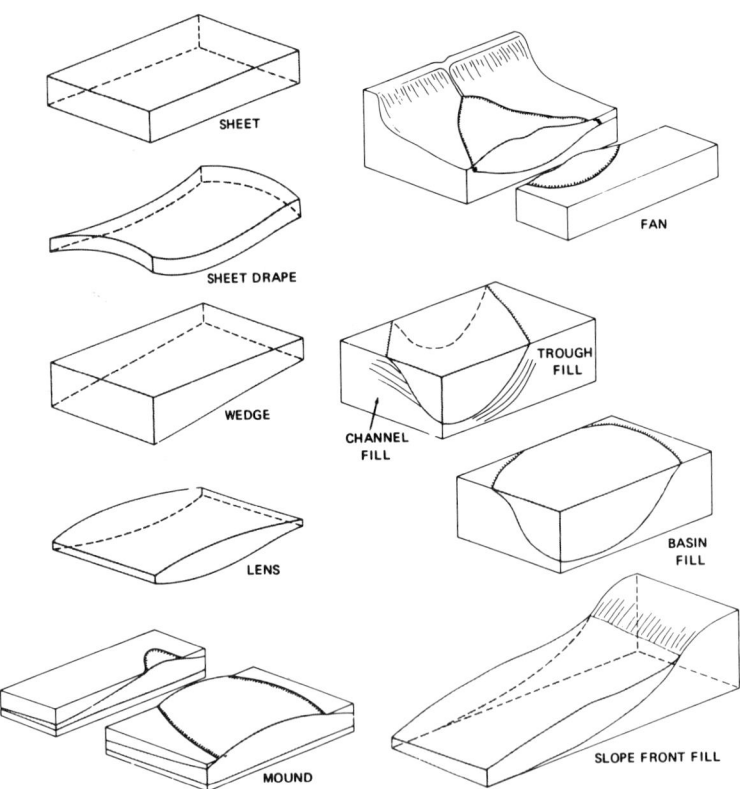

FIG. 12. Diagrammatic examples of external form types.

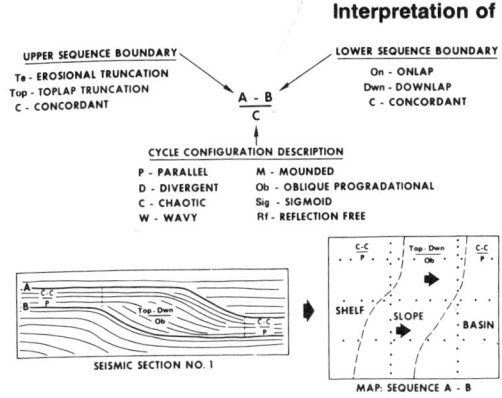

Fig. 13. Coding system for mapping reflection patterns.

has been intentionally kept brief to include only those forms that current experience shows are significant.

In classifying terrigenous clastic seismic facies types, one should be careful to separate objective descriptions of the individual seismic elements from the environmental and sand-potential interpretation of the combined set of elements. Table 2 shows the summary chart and terminology used for classification. Note that the objective and interpretive fields are separated by a vertical double line. Two different interpreters working the same data probably would produce nearly identical charts for the objective portion, even though they could differ significantly on the interpretation of sedimentary facies and depositional environment.

Within the facies units, a few basic reflection configurations such as parallel, divergent, sigmoid, oblique, complex sigmoid-oblique, chaotic, and reflection-free adequately describe most observed configurations. Frequently, however, a particular configuration in an area is best described by adding a modifier, such as wavy, irregular, etc., to the basic term. A few suggested configuration modifiers are listed in Table 2.

Interpretation of sections from a grid of seismic lines allows seismic facies units to be mapped and related to time-equivalent unconformities and other seismic facies units on a regional scale. The form in which the units are mapped varies greatly, depending on the intent and ingenuity of the interpreter. One mapping technique that is particularly helpful uses a coding system described by J. N. Bubb and others in an unpublished report. This system combines the relationships at the upper and lower boundaries and the configuration of reflections within the unit. The coding system has the form

$$\frac{A-B}{C},$$

where

A = relationship of reflections to upper boundary,
B = relationship of reflections to lower boundary, and
C = internal reflection configuration.

Use of this code to make facies maps (Figure 13) is helpful in determining the external form and lateral associations of seismic facies units. An example of the application of this procedure is shown later.

DEPOSITIONAL ENERGY CONCEPT

Once the clastic seismic facies types are described and mapped, an interpretation of sedimentary processes and environmental facies allows the interpreter to predict the sand potential of the facies. If the environmental facies has sufficiently high energy to transport and deposit significant quantities of sand, then the facies can be considered sand prone. Conversely, if depositional energies are low and insufficient to develop significant sand accumulations, then the facies is shale prone. Unfortunately, many specific examples of a sand-prone facies will prove to consist entirely of shale and silts because no source of sand was available for transport to the area of deposition. Thus, the criteria of high and low energy apply only to the potential for sand deposition. Other techniques, for example, calibration of interval velocity for sand prediction, must be used to verify the actual presence of bedded sand.

Having looked at some basic relationships between stratigraphy and seismic reflections, let us consider some documented examples of these relationships as applied in various depositional settings.

EXAMPLES OF SEISMIC FACIES TYPES

Depositional settings

A regional environmental framework of shelf, shelf margin and prograded slope, and basin slope and basin floor provides a useful gross subdivision for classification of seismic facies units. The shelf environment consists of maritime sequences composed of marine and non-marine sediments. Nonmarine sediments in these sequences (Mitchum et al, 1976) are interbedded with, or pass laterally into, marine sediments. In maritime sequences, sea level has been a controlling factor in establishing fluvial base level as well as the nature of the neritic sediments. Shelf environments are characterized by general parallelism of reflections.

Table 3. Description of clastic sediments facies type.

Depositional framework interpretation	Seismic facies unit	Environmental facies interpretation	External form of facies unit	Reflection geometry at boundaries	Reflection configuration	Lateral relations	Other seismic facies parameters		
							Amplitude	Continuity	Frequency (cycle breadth)
Shelf	High-continuity and high-amplitude	Typically shallow marine clastics deposited primarily by wave transport processes; could also be fluvial clastics interbedded with widespread marsh deposits	Sheet or wedge	Concordant at top with concordant to gentle onlap or downlap at the base	Parallel to divergent	May grade laterally to all other shelf facies or to undaform portion of slope-progradational facies	Relatively high, but variable	Relatively continuous	Variable, some broad cycles
Shelf	Low-amplitude	Marine clastics deposited by low-energy turbidity currents and by wave transport	Sheet or wedge	Concordant at top with concordant to gentle onlap or downlap at the base	Parallel to divergent	May grade laterally to slope progradational facies or to high-continuity and high-amplitude shelf facies	Very low to low	Discontinuous to continuous	Variable but lacks extremes of high-continuity and high-amplitude facies
		Fluvial and nearshore clastics deposited by fluvial and wave transport process				May grade laterally to high-continuity and high-amplitude facies or to low-continuity and variable amplitude shelf facies	Low	Discontinuous to moderately continuous	
Shelf	Low-continuity and variable amplitude	Dominantly non-marine clastics. Deposited by river currents and associated marginal marine transport processes	Sheet or wedge	Concordant at top with concordant to gentle onlap or downlap at the base	Parallel to divergent	Commonly grades laterally to high-continuity and high-amplitude facies or to sand-prone low-amplitude facies	Low to high; quite variable with frequent bursts of high-amplitude	Generally discontinuous to moderately continuous	Quite variable

Interpretation of Depositional Facies

Table 3. Description of clastic sediments facies type. (Cont.)

Depositional framework interpretation	Seismic facies unit	Environmental facies interpretation	External form of facies unit	Reflection geometry at boundaries	Reflection configuration	Lateral relations	Other seismic facies parameters		
							Amplitude	Continuity	Frequency (cycle breadth)
Shelf	Mounded	Shelf delta complex	Low mound or elongate mound	Concordant at top with gentle downlap at base	Mounded to sigmoid and divergent	Commonly grades laterally to high-continuity and high-amplitude facies or to undaform portion of slope-progradational facies	Variable but relatively low. Bursts of discontinuous high-amplitude are common	Discontinuous to moderately continuous	Quite variable. Some narrow cycles
Shelf-margin and prograded-slope	Sigmoid-progradational	Clay muds deposited by low-energy turbidity currents and by hemipelagic deposition from low-velocity water currents. Shelfal undaform portions may involve wave and even fluvial transport processes.	Elongate lens to subtle fan	Concordant at the top with downlap at the base	Sigmoid along depositional dip and parallel to sub-parallel along depositional strike	May grade laterally or vertically to oblique-progradational facies. Commonly onlapped by onlap-fill facies. Undaform part merges with shelf facies and fondoform part may grade to sheet-drape facies.	Generally moderate to high, relatively uniform	Normally continuous	Varies parallel to dip with broadest cycles associated with thicker bends of the middle clinoforming zone. Cycle breadth is uniform on sections parallel to depositional strike.
Shelf-margin and prograded slope	Oblique-progradational	Sediment complex usually deposited in shelf margin deltaic environment;	Fan	Concordant at top if undaform cycles present	Parallel	Commonly merges downdip with deep basin turbidites, mass transport	Moderate to high	Generally continuous	Relatively uniform
	Undaform								
	Upper cliroform				Oblique along depositional dip and parallel or		Moderate to high	Generally moderately continuous	Fairly uniform

Table 3. Description of clastic sediments facies type. (Cont.)

Depositional framework interpretation	Seismic facies unit	Environmental facies interpretation	External form of facies unit	Reflection geometry at boundaries	Reflection configuration	Lateral relations	Other seismic facies parameters		
							Amplitude	Continuity	Frequency (cycle breadth)
Shelf-margin and prograded slope (cont.)	Middle and lower clinoform	includes delta plain, delta front and prodelta processes. May also be formed in deep water associated with strong bottom currents.		Toplap truncation at the top	gently oblique to sigmoid parallel to depositional strike	and hemipelagic facies. Frequently onlapped by onlapping fill-facies. May grade laterally or vertically to sigmoid-progradational facies. Undaform portion merges with parallel layered shelf facies	Variable. Generally lower than other subzones	Discontinuous to moderate; increases toward lower clinoform zone	Cycle breadth decreases rapidly downdip as beds thin
	Fondoform			Downlap at the base			Generally moderate to low	Continuous	Relatively narrow cycle breadth decreases basinward
Basin-slope and basin-floor	Sheet-drape	Deep marine hemipelagic clays and oozes	Sheet-drape	Concordant at top and concordant or very slight onlap at base	Parallel	Commonly interbedded with turbidite sands and silts, and grades to gently divergent fondoform sediments of prograding complexes	Commonly relatively low to moderate	Continuous	Normally uniformly narrow
Basin slope and basin floor	Onlapping fill	Relatively low velocity turbidity current deposits	Basin trough, channel, and slope front fill	Onlap at the base and usually concordant at the top		Commonly grades to mounded onlap or chaotic fill facies. Alternation with other fill facies is common	Variable	Commonly continuous	Cycle breadth increases into fill center tends to be relatively narrow

Table 3. Description of clastic sediments facies type. (Cont.)

Basin slope and basin floor	Mounded or lapping fill	Relatively high-velocity turbidity current deposits	Mounded: basin trough, channel, and slope front fill	Onlap at the base and concordant or erosional truncation at top	Irregularly mounded to parallel	Commonly grades to onlap fill or chaotic fill facies. Alternation with other fill facies is common	Variable; decreases as continuity decreases	Discontinuous to moderately continuous. Generally less than nonmounded onlap fill facies	Cycle breadth increases in fill center
Basin slope and basin floor	Chaotic fill	Gravity mass transport and high energy turbidity current sediments. Lithology is function of upslope sediment source	Basin trough, channel, and slope front fill. Degree of mounding variable. Wavy subparallel chaotic pattern tends to be associated with smoother and lower mounds than contorted and discordant chaotic patterns	Unit onlaps at base but individual onlap terminations are rare because of reflection pattern. Where preserved, reflection segments at the top may show concordance or erosional truncation. Mass transport gouge is common at base of contorted discordant pattern	Chaotic and contorted	May grade slopeward to lower and middle clinoform subzone of oblique progradational facies. Also may lie downdip of prominent detachment scars. Alternation with other fill facies is common	Ranges from low to high in contorted discordant patterns and is generally low in wavy subparallel chaotic fill patterns	Very discontinuous short segments may occur in the contorted discordant patterns	Variable in contorted discordant facies reflecting internal heterogeneity. More uniform in wavy sub parallel chaotic fill pattern

Table 3. Description of clastic sediments facies type. (Cont.)

Basin slope and basin floor	Mounded (fan complex)	Deep-water sediment complex commonly located at mouth of submarine canyon. Composed of turbidites, mass-movement and hemipelagic deposits associated with major subaerial drainage systems	Fan	Onlap of overlying units. Downlap at base	Extremely varied Complex mounded	Located near submarine canyon. Commonly grades basinward to sheet drape facies	Variable but tends to be low. Frequently decreases rapidly with increasing depth, suggesting high energy absorption	Tends to be discontinuous	Highly variable
Basin slope and basin floor	Mounded (contourite)	Deep-water sediment complex formed by deposition from deep-marine currents. Possibly composed primarily of fine-grained clastics	Elongate mound	Truncated and contorted at top. Downlap at base	Asymmetric mounds	Thins and grades into basin floor facies	Variable	Variable	Variable

It should be recognized that this parallel configuration is in part the result of limited reflection resolution; higher resolution may reveal more complicated reflection configurations within these patterns.

The shelf-margin and prograded-slope environment typically contains thick marine sediments and has water depths sufficient for development of complex arrangements of sigmoid and oblique prograding reflection patterns. The basin-slope and basin-floor environment includes a variety of deep basin facies as well as nonprograding-slope facies and facies that extend from the slope into the basin deep. Deep basin is used here for any basin, whether inland or along a continental margin, which had water depths sufficient for seismic definition of a slope environment; consequently, water depths at the time of deposition may range from bathyal to abyssal in the typical example.

We cannot present seismic section examples for all seismic facies types that are illustrated diagrammatically for each environment. Instead, we will present seismic examples of one or two facies types for each depositional setting.

Our current state of knowledge of those seismic facies units that are high energy and those that are low energy is summarized in Table 3. These facies units are discussed in more detail under headings corresponding to the regional depositional settings of shelf, shelf margin and prograded slope, and basin slope and basin floor.

Shelf seismic facies

Shelf sedimentary environments, as broadly defined here, typically range from neritic to totally nonmarine. Sediments deposited in these environments tend to generate parallel to gently divergent reflection configurations having a widespread sheet, or wedge-shaped, external form (Figure 14, Table 3). One facies, which is characterized by a broad, low-relief mound composed of gently mounded to sigmoid and divergent reflections, forms an exception to these generalizations.

Reflections are normally concordant at the top and vary from concordant to gently onlapping and occasionally downlapping at the base. Prediction of depositional energy and sand content in shelf seismic facies units must rely heavily on analysis of variations in reflection amplitude, continuity, and cycle breadth and on areal relations with other units. The active transport-deposition modes in this environment include fluvial processes, as well as wave (surf-zone reworking and longshore drift) and other marine current and flow processes.

High-continuity and high-amplitude facies (interbedded high- and low-energy deposits).—High continuity of reflections in this shelf facies suggests continuous beds deposited in a relatively widespread and uniform environment, and high-amplitude reflections are interpreted to indicate interbedding of shales with relatively thick sands, silts, or carbonates. Normally, deposits in this facies consist of neritic marine sediments; however, there are instances in which fluvial sediments with interbedded, widespread marsh clays and coals also generate high-continuity and high-amplitude reflections. Cycle breadth varies, and it is common to find broad, high-amplitude cycles alternating with cycles having narrow cycle breadth,

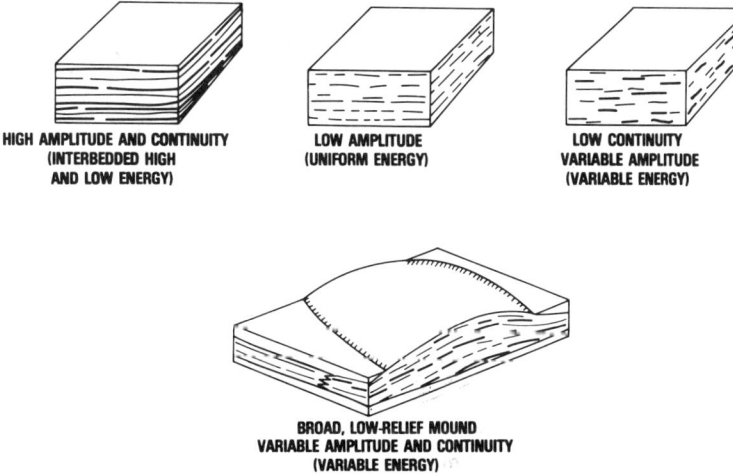

Fig. 14. Shelf seismic facies types.

suggesting vertical variations in bed thickness. Laterally, high-amplitude and high-continuity reflections may grade into any of the other shelf facies as well as into the undaform portion of the prograded-slope facies.

An example in which this facies contains significant sand is typified by Cretaceous sediments in the San Juan basin. Thirty wells provide a great deal of control for analysis of the Cretaceous portion of the San Juan section (Figure 5). Abundant, thick (50-200 ft) nearshore and shoreline sands generate continuous, high-amplitude reflections (Figure 6) characteristic of this facies. The reflections lie in a pattern of parallel reflections extending across the entire section. To the southwest, the sands grade into a nonmarine facies characterized by numerous discontinuous sands and marsh shales (low-continuity, variable-amplitude seismic facies); to the northeast, they grade into an open-marine calcareous shale facies (low-amplitude seismic facies).

In the San Juan basin, the sediments characterized by high-continuity and high-amplitude seismic facies are beach and shoreface deposits. Consequently, wave transport and deposition processes, including surf-zone reworking and longshore drift, appear dominant in formation of this facies.

Low-amplitude facies (uniform-energy deposits, either high or low energy).—Low-amplitude reflection zones on a seismic section indicate either beds too thin to be resolved by seismic methods or a zone of one predominant lithologic type. Consequently, the low-amplitude facies may be either sand prone (high energy) or shale prone (low energy). Knowledge of the regional geology, lateral seismic facies relationships, or other lithologic data is required to identify the correct lithology.

A shale-prone facies, characterized by low-amplitude reflections, typically grades shoreward to a silt or sand-prone facies, characterized by high-continuity and high-amplitude reflections and basinward to a prograded-slope facies. In contrast, a sand-prone facies of low-amplitude reflections would tend to grade landward to nonmarine facies of low-continuity and variable-amplitude reflections and seaward to a marine facies of high-continuity and high-amplitude reflections. There are, of course, notable exceptions to this principle, but it is useful in distinguishing whether low-amplitude seismic facies are indicative of massive sand or massive shale.

The San Juan basin provides a good example of the low-amplitude seismic facies. On the San Juan section (Figure 6), a vertical and lateral gradation from the low-amplitude shale-prone facies, through the high-continuity and high-amplitude interbedded sand facies, to a low-continuity and variable-amplitude nonmarine sand-prone facies is well illustrated. At the northeast end of the section, the vertical upward succession is (1) low-amplitude reflections from the Mancos formation marine shale; (2) high-amplitude and continuous reflections from the shoreline sands; (3) low-continuity, relatively low-amplitude reflections from the nonmarine sands, silts, and shales; (4) two high-amplitude and continuous reflections from shoreline sands; and (5) low-amplitude reflections from the Lewis shale. Laterally, the Lewis shale grades into nonmarine equivalents with a repetition of the succession of seismic facies types (1), (2), and (3) listed above.

Low-amplitude sand-prone seismic facies and low-amplitude shale-prone seismic facies are produced by contrasting depositional processes. Massive sands associated with low-amplitude reflections tend to be nearshore to fluvial sands that are transported and deposited by high-energy fluvial and wave transport processes. In contrast, massive shales associated with low-amplitude reflections tend to be deposited farther offshore or along sand-poor shorelines, and are associated with wave and low-velocity turbidity current or turbid layer transport processes.

Low-continuity and variable-amplitude facies (variable energy, usually has high-energy beds).—Sediments deposited by fluvial currents are generally less continuous than sediments deposited under marine conditions. In addition, numerous discontinuous sands in many nonmarine environments provide good velocity contrasts with encasing shales and coals. Consequently, the shelfal seismic facies, characterized by reflections having poor lateral continuity and bursts of high amplitude, are rather typically composed of nonmarine sediment types. Although nonmarine sediments may occasionally generate high-amplitude and relatively high-continuity reflections (for example, where thick coals are interbedded with sands or silts), we rarely see marine sediments generate the discontinuous high-amplitude reflections typical of this facies without concurrent evidence for large-scale mounding or marine cut-and-fill deposition.

This seismic facies type may occur in widespread sheets or may show prominent wedging with basal onlap where sediment supply is sufficient to maintain fill of a rapidly subsiding basin. Reflections are typically parallel in the sheet forms and divergent in the wedge forms. Cycle breadth may be quite variable, probably reflecting the complex variations in bed thickness typical of nonmarine deposition. This facies

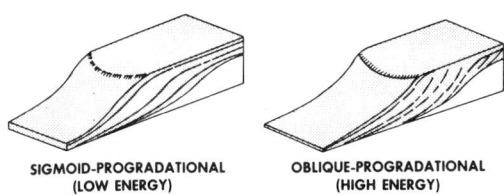

FIG. 15. Shelf-margin and prograded-slope seismic facies types.

commonly grades seaward into interbedded sand-prone facies, characterized by high-continuity and high-amplitude reflections, and sometimes grades into facies composed of thick and relatively massive sandstones, characterized by low-amplitude reflections. Nonmarine sediments in the San Juan basin (Figure 6) have been discussed above.

Broad, low-relief mound facies (variable energy, usually has high-energy beds).—This shelf facies, which has been noted at only a few locations, is interpreted as reflecting a complex of delta lobes formed on a subsiding shelf. Its key distinguishing feature is a broad, gently mounded external form. Reflections form gently mounded to sigmoid and divergent patterns. Reflections are concordant at the top and downlap gently in an arcuate pattern at the base. Laterally, this unit may grade into any of the parallel-shelf facies or into the undaform portion of the prograded-slope facies. No example is shown for this seismic facies type.

Shelf-margin and prograded-slope seismic facies

The next major class of clastic seismic facies units is associated with shelf-margin and prograded-slope types of deposition. Two principal facies, defined exclusively by reflection configuration, are recognized in this environment (Table 3, Figure 15). These are the oblique-progradational facies and sigmoid-progradational facies. Both of these facies are characterized by downlapping reflections at their base. Downlap represents the outbuilding of sediments from relatively shallow into deeper water with accompanying thinning of the outer toes of individual beds, commonly below seismic resolution. Ordinarily, the upper portions of these patterns consist of sediments deposited in fluvial to neritic environments, but examples have been documented where these patterns originate from sediments deposited entirely in bathyal and deeper water depths, presumably through the action of deep-water currents. The oblique geometry suggests outbuilding of the middle or foreset beds from a common upper depositional surface during a period of stability of sea level relative to the land surface, whereas the sigmoid pattern suggests simultaneous outbuilding and upbuilding of the upper or topset beds during a period of rising sea level and/or subsidence of the underlying deposits.

Oblique-progradational facies (typically high energy in updip portions).—The distinguishing feature of this facies is a prominent oblique reflection configuration when viewed parallel to depositional dip. Reflections terminate by toplap truncation at or near the upper surface (Figure 15) and by downlap at the base. Depositional dips may approach 10 degrees and are significantly steeper than dips of the sigmoid-progradational facies. Parallel to depositional strike, reflections may be parallel or may show low-angle oblique or sigmoid patterns. Small channels are often related to this configuration and show best on depositional strike sections. Studies of this facies in delta and delta-front Pleistocene sediments of the Gulf Coast indicate that it is commonly deposited during low-stands of sea level. However, studies of geologically older examples suggest that it can also form during still-stands of sea level or where sea level is slowly rising (if the supply of sediment is sufficient to overwhelm the effects of the relative rise).

Where this seismic pattern occurs on the shelf margin, it is characteristic of fluvial deltas and associated coastal-plain sediments and, consequently, contains high-energy deposits. The undaform zone (Figure 10), corresponding with a delta-plain environment, and the upper part of the clinoform zone, corresponding with a delta-front environment, have good sand potential. In contrast, the lower clinoform and fondoform zones, corresponding to a prodelta environment, are typically shale prone. Previous studies that have documented these relationships include Ewing et al (1963) and Lehner (1969). In some instances, the fondoform portion of oblique seismic facies units may contain turbidite deposits, and thus may contain sands interbedded with marine shales.

Reflection amplitude, continuity, and cycle breadth vary depending on position in the oblique configuration. Fondoform and undaform reflections have good continuity; fondoform reflections have moderate amplitude and undaform reflections have moderate or high amplitude. Cycle breadth decreases basinward and may become quite narrow. Clinoform zone reflections are quite variable but show a general decrease in continuity and amplitude from upper clinoform to lower clinoform. Some of the highest amplitudes observed in oblique facies occur in the upper clinoform zone. This probably represents maximum interfingering of shallow-water sands and silts with delta-front and prodelta clays. Cycle breadth in

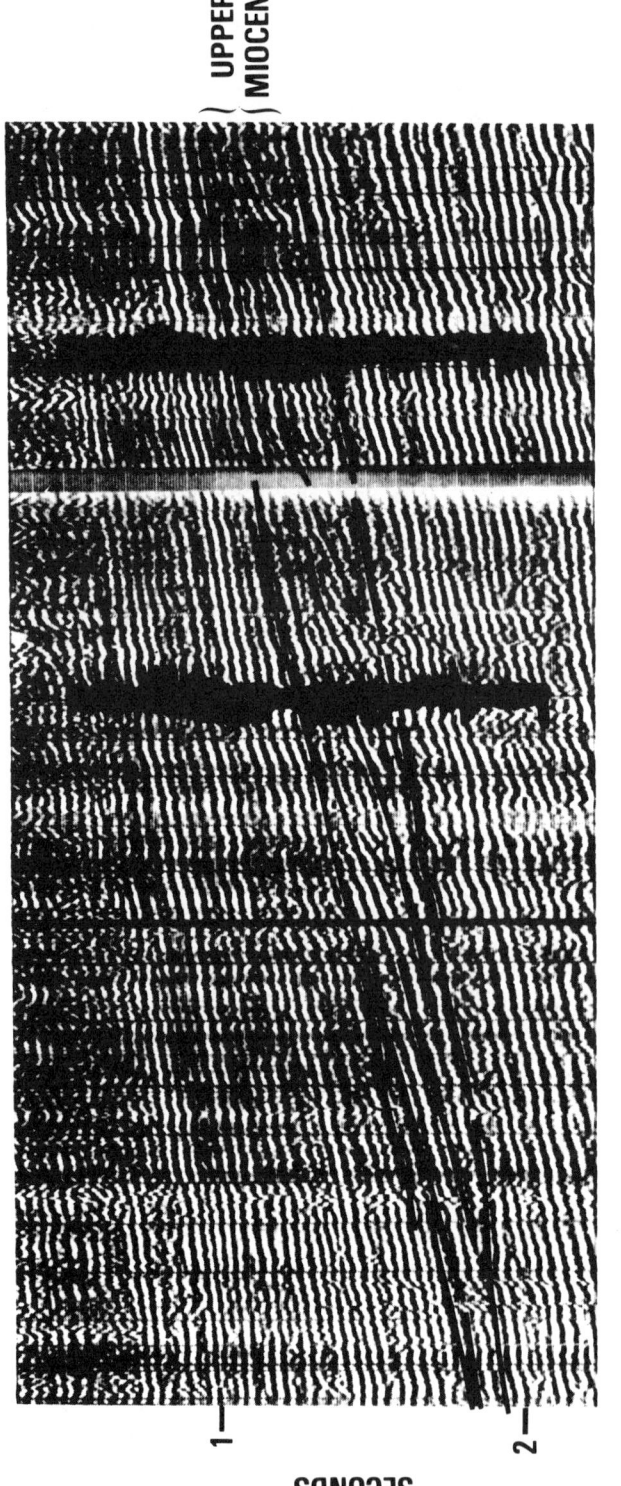

FIG. 16. Oblique-prograding seismic facies, Miocene example. Upper Miocene deposits form a prominent, basinward-thinning oblique-prograding facies unit. Two wells located along the line contained abundant reservoir-quality sand in this facies.

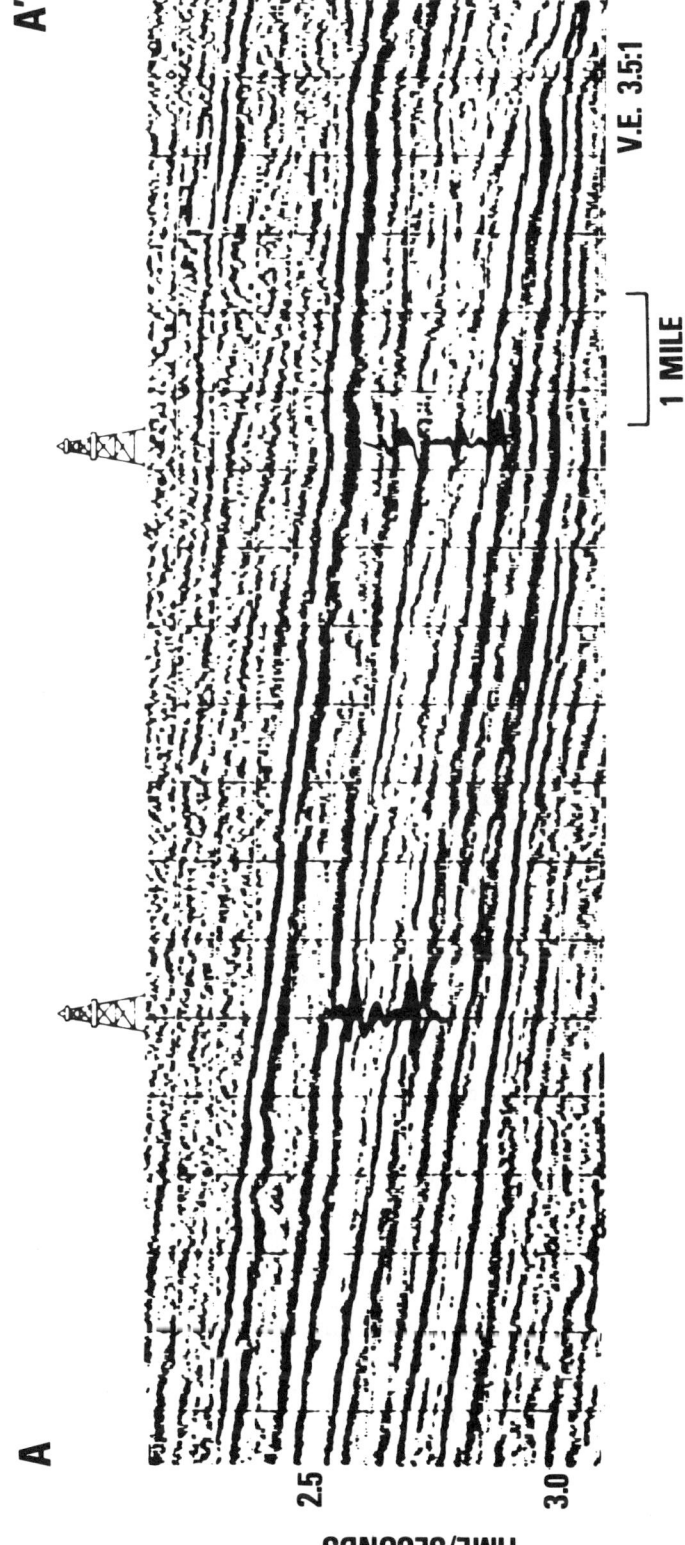

FIG. 17. Oblique-prograding seismic facies. The two wells located along this line contain only minor amounts of sand in the upper portion of the oblique-prograding reflection pattern.

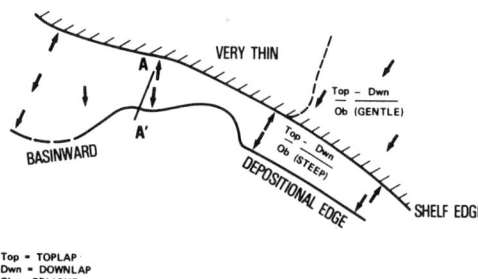

FIG. 18. Seismic facies map made for the oblique-prograding facies illustrated in Figure 17, showing the location of seismic section A-A' (Figure 17).

the clinoform zone decreases significantly downdip as beds thin.

Oblique units are frequently fan-shaped, and it is not unusual to find multiple fans forming major sedimentary complexes. Downdip, the facies may be associated with mass transport facies that slumped from the front of the growing complex, or may be onlapped by various fill facies. The oblique facies may also grade laterally and vertically into sigmoid-progradational facies.

The upper Miocene in the San Joaquin Valley in California provides a well documented example of a sand-rich oblique facies as shown by seismic data. Examination of a seismic section on Figure 16 shows zones of essentially parallel reflections sandwiching a zone of oblique reflections. Tracing individual reflections, we see that cycles terminate against a common upper surface by toplap truncation and against a common lower surface by downlap. This upper Miocene sequence thickens from 0.2 sec on the left to 0.4 sec on the right, changing from a dominantly fondoform position to a position of thickest development. The undaform zone appears to be two to three cycles thick in this example.

The upper Miocene sequence contains an unusual proportion of sand. The sand extends from the undaform zone to outer portions of the clinoform zone, as is indicated by logs of two wells lying along the line (Figure 16). Units above and below are shaley, with the upper unit showing prominent onlap onto the clinoform front of the upper Miocene deposits.

Figure 17 presents another example of an oblique-progradational pattern. The oblique pattern is recognized from cycles that are inclined more steeply in a downdip direction than the reflections above or below. The downlap pattern of cycle terminations is visible at the base, and a series of cycles terminate at the top by toplap truncation. Although we interpret this as a high-energy environment, the two wells shown on Figure 17 contain only thin sands at the top of the unit. The thinness of these sands suggests that a large source of coarse clastics was not available for transportation into this depositional setting.

By studying sections from a grid of seismic lines, we can map the areal distribution of seismic facies units and their interpreted environments. Figure 18 shows a map made for the seismic sequence containing the oblique-prograding unit of Figure 17. Seismic section A-A' (Figure 16) is located on the map. The seismic reflection pattern has been abbreviated as follows: "Top" designates toplap or foreset cycle terminations at the top, "Dwn" designates downlapping cycles at the base, and "Ob" refers to the oblique-prograding reflection configuration. One facies unit lies basinward of an underlying shelf edge. Its oblique pattern is significantly steeper than the oblique pattern in the facies unit lying landward of the shelf edge. Depositional environments in both oblique-prograding areas (basinward and landward) are interpreted as being moderately high-energy. The difference in inclination of the prograding cycles is a rather direct indication of the shallower depositional water depths in the shelf area. The area marked "very thin" is a region where the sequence thins to one or two cycles; this is too thin for detailed seismic facies analysis.

Sigmoid-progradational facies (low energy).— Sigmoid-progradational units are characterized by gentle sigmoid (S-shaped) reflections along depositional dip. The reflections downlap at the base and are concordant with the top of the unit (Figures 10 and 15). On seismic lines parallel to depositional strike, these reflections are usually parallel and concordant with unit boundaries. Sediments of this facies are deposited on the slope along continental margins with undaform portions extending onto shelf-margin areas. Considerable upbuilding produces thick undaform deposits, which require some combination of a rising eustatic sea level or a subsiding land level. Fine-grained clastics dominate in this facies and are most likely deposited from low-energy turbidity currents and from low-velocity currents as hemipelagic deposits. Undaform sedimentation may involve wave or even fluvial transport processes and provides some possibility of coarser reservoir clastics in the undaform environment.

Typically, reflections characterizing this facies have moderate to high amplitude and high continuity. Cycle breadth, while uniform on sections parallel to depositional strike, varies parallel to dip, with the broadest cycles correlative with the thicker beds of the middle clinoform zone.

Most commonly, a sigmoid-progradational facies forms a lens, elongated parallel to depositional strike. In addition to frequent association with the oblique facies, this unit is commonly onlapped by various chaotic- and onlapping-fill facies. Undaform reflections merge with parallel-shelf facies types, and fondoform reflections may grade into sheet-drape facies of the basin floor. The sigmoid-progradational facies is not illustrated with an example.

Basin-slope and basin-floor seismic facies.—Three groups of seismic facies units (Figure 19) that are common on the basin slope and basin floor are (1) sheet-drape facies units, (2) onlapping- and chaotic-fill facies units, and (3) mounded facies units. Current documentation is sketchy for some of these facies types, and this classification will be refined and probably extended in the next several years.

These facies, especially those in groups (2) and (3) above, overlap from the basin floor onto the slope environment. For example, mounded submarine fan complexes characteristically extend from the slope at their apex well out into the basinal environment. Also, the slope frequently contains numerous local basins that are filled by the various onlapping- and chaotic-fill patterns. To further complicate the classification problem, onlapping- and chaotic-fill units may be incorporated within and form a portion of submarine fan complex.

Sheet-drape facies (low energy).—The parallel reflections characteristic of this facies drape over contemporaneous topography with only very gradual changes in thickness or reflection character (Figure 19) and suggest uniform deposition independent of the bottom relief. This pattern is strongly indicative of deep-marine hemipelagic clays and oozes with almost no potential for sand development.

This facies forms moderately thin, widespread sheets of parallel reflections that are normally concordant at top and base; however, upon occasion reflections do onlap contemporaneous topography at the base. Reflection continuity is very high, and cycle breadth is normally uniformly narrow. Amplitude is variable but is commonly low because of the uniform lithology of the unit. Hemipelagic clays and oozes of the parallel sheet-drape facies are frequently found interbedded with turbidite and mass-movement sands and silts, and may pinch out very gradually laterally, or may grade to gently divergent fondoform sediments of prograding units.

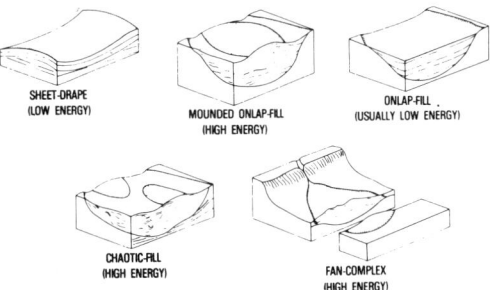

FIG. 19. Basin-slope and basin-floor seismic facies types.

Chaotic- and onlapping-fill facies.—Chaotic- and onlapping-fill seismic facies units are deposited in topographic lows on the slope and basin floor. The lows may be a basin-floor plain, local basins, channels or troughs, or areas of prominent topographic flattening in the configuration of the slope. These facies are most likely deposited by high-density turbidity current and mass transport processes; however, this may be an oversimplification of a more complex series of processes that include such things as grain flow of sands and coarse clastics. Presence of displaced, relatively shallow-water fauna is common in cores of all types of chaotic facies units.

The two characteristics most indicative of high-energy and consequently of sand-prone units appear to be increasing irregularity of reflection pattern and character, and mounding of external form. Hence, chaotic and mounded onlapping units are interpreted as higher energy deposits. The final control of the lithology of this facies rests in the source area for the sediments. No process of any energy level can transport sand from shallower into deeper water if sand is not present in the source area. Our documentation is still fragmentary for this suite of facies types, and conclusions presented in the next three sections must be considered as preliminary.

(a) Chaotic-fill facies (high energy).—Chaotic-fill seismic facies units are characterized by a mounded external form, by location in topographic lows, and by an internal pattern of contorted and discordant to wavy-subparallel reflections (Figure 19). Mass transport slump and creep and high-energy turbidity current processes are thought responsible for transportation and deposition of this facies. The composition of slump deposits is entirely dependent on the type of material at the source area, and even if the slump deposits contain some sand, they may not be sufficiently winnowed to produce clean sands.

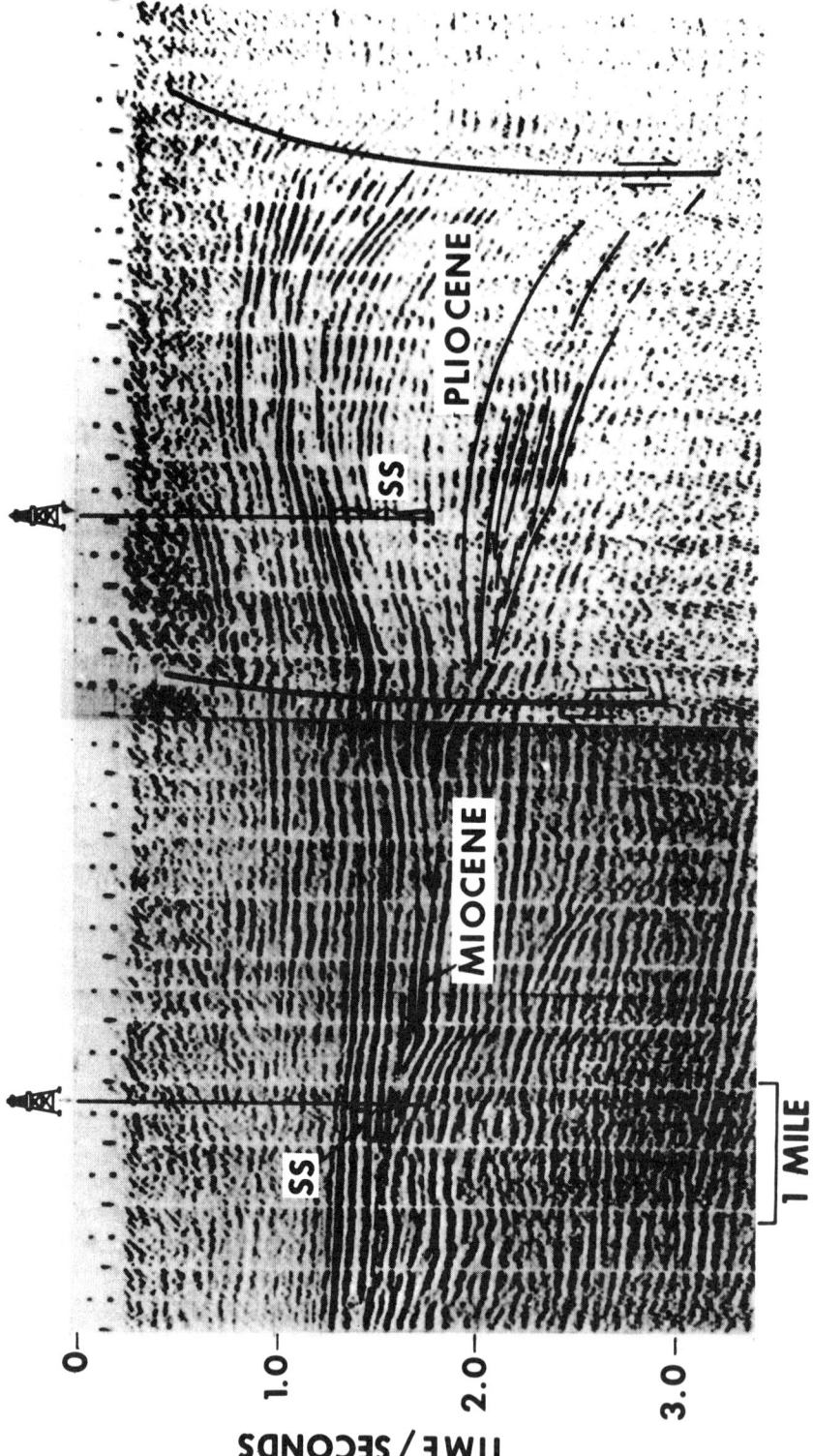

Fig. 20. Mounded onlap-fill facies, Pliocene example. Although located in a structurally complex area, this sand-rich mound is believed to be largely depositional in origin.

Chaotic units are thick in topographic lows and onlap out of these lows, although the lack of coherent reflections obviously obscures any onlap of individual cycles. The top of the unit is usually hummocky. Thickness of these units is quite variable. Chaotic zones 1000-ft thick and more are common and some exceed 6000 ft in thickness in the Pleistocene of the Gulf Coast.

Contorted internal reflections of chaotic-fill units originate from bent and folded beds that have retained some coherence during down-slope en masse transportation. (Chaotic patterns also occur in complex and stacked cut-and-fill submarine fan sediments.) Diffractions originating from these segmented masses are common.

The interpretation of a mass transport origin for a chaotic-fill unit is based on evidence for broken and contorted beds, evidence for updip detachment scars and erosional gouge below the unit, and maps that indicate the unit has lost section upslope and gained section downdip.

Reflection amplitude, continuity, and cycle breadth seem to reflect the extent of homogenization in mass transport types of chaotic units. Where homogenization is incomplete and reflections are contorted, amplitude varies from low to high and reflects the original sediment velocity contrasts. In contrast, amplitude is very low in the wavy-subparallel pattern.

(b) Mounded onlapping-fill facies (high energy).—Mounded onlapping-fill facies units are believed to be deposited by high-energy turbidity currents. Principal characteristics are a mounded form and prominent onlap away from topographic lows. Internal reflections are irregularly parallel to divergent and tend to be discontinuous. Reflection amplitude is quite variable; amplitude tends to decrease as continuity decreases. Normally, cycle breadth increases toward the center of the topographic low.

This facies is commonly intermediate to and laterally gradational with both the chaotic-fill facies and the onlapping-fill facies. It may also show frequent vertical interbedding with these facies. Individual units of this facies range in thickness from a few tens of feet to over 1000 ft and, when interbedded with chaotic- and onlapping-fill facies, attain thicknesses of several thousands of feet.

The onlapping nature of this facies, together with its tendency to fill lows selectively in the depositional topography, suggests that it was deposited by some sort of gravity-controlled flow along the bottom (Hersey, 1965). Where the seismic pattern is formed by discontinuous and divergent reflections that indicate an increased rate of bed thickening into topographic lows, we believe that the deposits contain discontinuous beds and probably reflect a less uniform and possibly high-velocity mode of transportation and deposition. All things considered, turbidity current flow seems the most likely origin for this facies type. Discontinuous and irregular patterns with minor mounding probably represent deposits formed by higher velocity turbidity currents, quite possibly capable of transporting sands, if sands are available at the source of the flow.

A large Pliocene turbidite fan that was deposited in offshore California is primarily formed of onlapping mounds (Figure 20). These mounds are interbedded with parallel-layered units of relatively uniform thickness which are interpreted as pelagic shales and turbidites deposited during intermittent times of low turbidite influx. Two wells drilled on this line contain about 40-50 percent sand in the mounded interval. Subsequent structural uplift has altered the original depositional form; however, careful inspection will reveal the internal mounding of the unit.

(c) Onlapping-fill facies (predominantly low energy).—This basically low-energy facies does not show the mounding characteristic of mounded onlap- and chaotic-fill facies, and associated reflections tend to have a much more uniform parallel to gently divergent pattern, with high continuity and moderately high amplitude. Cycle breadth also tends to be more uniform and narrow, with a slight broadening into the topographic low. Vertical and lateral gradation and interbedding with other fill facies are common for this unit.

As in the mounded-fill facies, the onlapping nature and tendency to fill lows suggest that the onlapping-fill facies is deposited by some type of gravity-controlled flow along the bottom. Parallel patterns of continuous reflections created by widespread parallel strata are probably the expression of deposits formed by low-velocity turbidity currents. These patterns can be expected to indicate mainly clay and silt. Thin sandy units below the resolution of seismic reflections may occur, interbedded with the shales and silts more typical of this seismic facies; presumably, such thin sands result from intermittent high-energy turbidity current transport from a sand-rich source area.

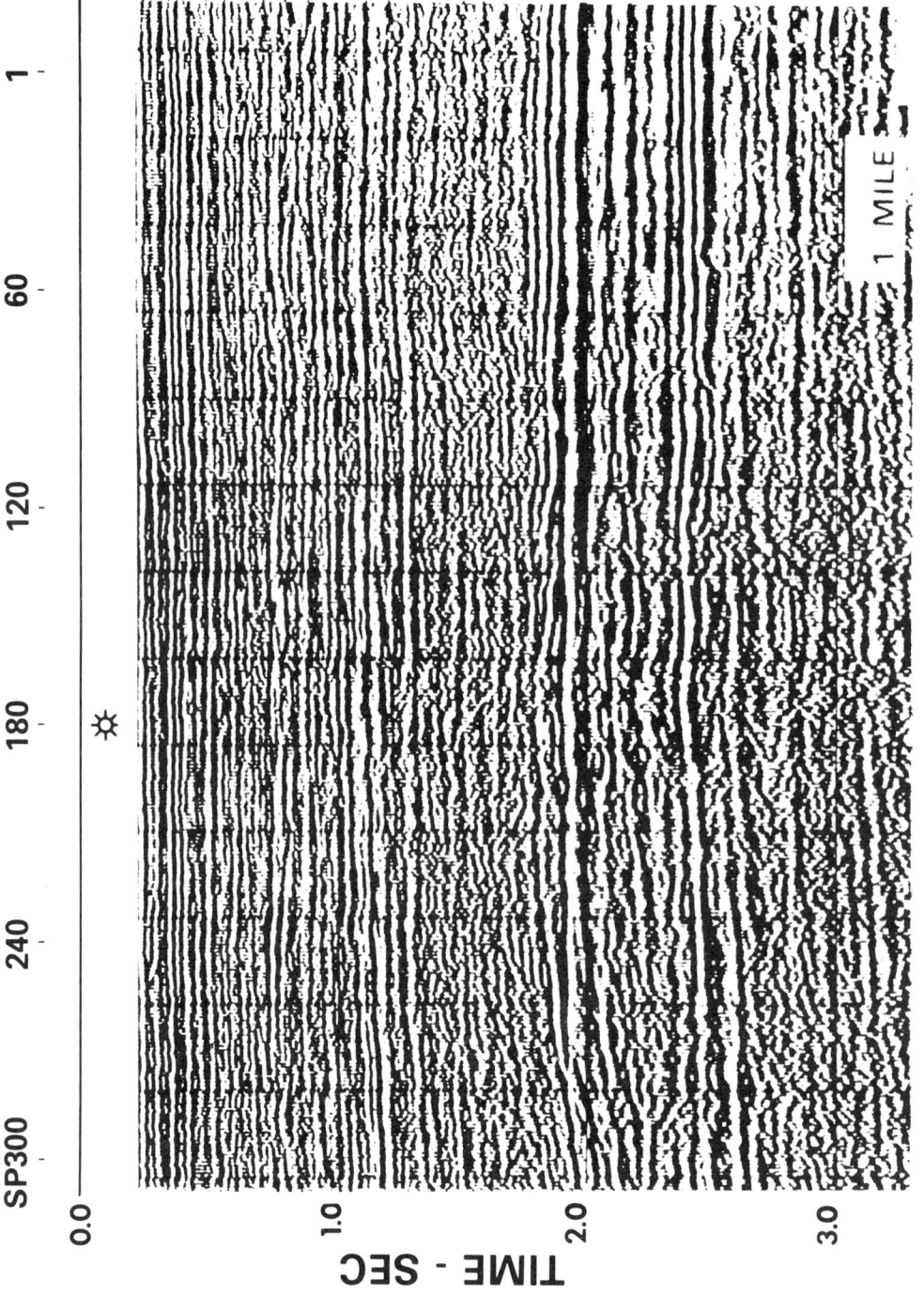

FIG. 21. Fan-complex seismic facies, Tertiary example. Mounded external form of fan is well developed on left at about 1.8 sec. Internal patterns are obscured by fluid contact just above 2.0 sec.

Mounded facies.—Large, complex piles of sediments are created in the basin-slope and basin-floor environment by gravity transport of sediments through submarine canyons and by sediment transport in bottom-flowing ocean currents. In this paper, those sediment accumulations correlated with submarine canyons are called fan complexes, whereas those formed by ocean currents are called contourites (depositional anticlines of Malloy, 1968).

(a) Fan complex facies (predominantly high energy).—Submarine fan complexes are distinguished by their fan-shaped external form, complex internal patterns, and common association with major subaerial and submarine drainage systems. These fans may contain good reservoir sands, depending on the nature of clastic sediments available to their drainage systems and on the distance and processes by which the sediments are transported to the head of and through submarine canyons.

Low continuity of reflections seems the most consistent internal seismic characteristic of fan complexes. This feature results from the original depositional discontinuity of beds and from post-depositional cut-and-fill as the fan channels shift and younger fan units pile onto older units. This makes seismic recognition of individual leveed channels or suprafans, except for the youngest, quite difficult.

Reflection characteristics recognized within the submarine fan complex include parallel, divergent, and chaotic reflections, and some tendency for reflection-free zones where the fan is very thick. Further, large fan complexes may incorporate and include individual chaotic- and onlapping-fill units. Any additional generalizations seem premature at present. A seismic example of a submarine fan is shown in Figure 21.

(b) Mounded contourite facies (predominately low energy).—Contourite mounds have been recognized on continental slopes and ocean basin floors with increasing frequency in the last few years. It has been suggested that this facies unit type is formed by deposition from major bottom-flowing oceanic currents, and the sediments have been termed contourites (Heezen et al, 1966; Schneider et al, 1967). The bottom-flowing currents probably rarely encounter good sand sources, and consequently, these depositional mounds have a low probability of containing good sands. An example is not shown for this facies type.

THE FUTURE OF SEISMIC FACIES INTERPRETATION

There are four areas in which we expect major improvement in seismic facies analysis. First, we need to improve our knowledge of the types and genesis of major nonmarine and marine sedimentary facies to obtain better documentation of reflection characteristics associated with these facies. Second, routine computer procedures to model, store, and map a variety of seismic facies data promise to be very useful. Third, quantification of such reflection parameters as amplitude, frequency, polarity, and continuity holds advantages for objective prediction of facies. Finally, close-spaced interval velocity has great capability for the prediction of lithology, especially if it is used in conjunction with reflection pattern methods of seismic facies analysis.

ACKNOWLEDGMENTS

A large number of explorationists have contributed to our understanding of seismic facies analysis. We first need to thank P. R. Vail for his inspirational help and leadership. Major contributions have been made by R. M. Mitchum, J. N. Bubb, H. J. Yorston, and F. Branisa. We are also grateful for suggestions from D. G. Blair, Dirk Gralka, W. G. Hatlelid, M. O. Nordberg, R. G. Todd, and J. M. Watson. We thank J. A. Jones for his help in preparing computer models for Figures 2, 3, and 7.

Although the original report from which this paper is taken contained examples of each facies type, space and confidentiality limitations confined us to one or two documented examples for each depositional setting.

REFERENCES

Ewing, J., Le Pichon, X., and Ewing, M., 1963, Upper stratification of Hudson apron region: J. Geophys. Res., v. 68, p. 6303–6316.

Heezen, B. C., Hollister, C. D., and Ruddiman, W. F., 1966, Shaping of the continental rise by deep geostrophic contour currents: Science, v. 152, p. 502–508.

Hersey, J. B., 1965, Sediment ponding in the deep sea: GSA Bull., v. 76, p. 1251–1260.

Lehner, P., 1969, Salt tectonics and Pleistocene sediments on continental slope of northern Gulf of Mexico: AAPG Bull., v. 53, p. 2431–2479.

Malloy, R. J., 1968, Depositional anticlines versus tectonic "reverse drag": Gulf Coast Assoc. Geol. Socs. Trans., v. 18, p. 114–123.

McDonal, F. J., Angona, F. A., Mills, R. L., Sengbush, R. L., Van Nostrand, R. G., and White, J. E., 1958, Attenuation of shear and compressional waves in Pierre shale: Geophysics, v. 23, p. 421–439.

Rich, J. L., 1951, Three critical environments of deposition and criteria for recognition of rocks deposited in each of them: GSA Bull., v. 62, p. 1–20.

Schneider, E. D., Fox, P. J., Hollister, C. O., Needham, H. D., and Heezen, B. C., 1967, Further evidence of contour currents in the western North Atlantic: Earth and Planet. Sci. Let., v. 2, p. 351–359.

Schoenberger, M., and Levin, F. K., 1974, Apparent attenuation due to intrabed multiples: Geophysics, v. 39, p. 278–291.

Tucker, P. M., and Yorston, H. J., 1973, Pitfalls in seismic interpretation: Tulsa, SEG Monograph Series, no. 2, 50 p.

Vail, P. R., Mitchum, R. M., Jr., Todd, R. G., Widmier, J. M., Thompson, S., III, Sangree, J. B., Bubb, J. N., and Hatlelid, W. G., 1977, Seismic stratigraphy and global changes of sea level, in Seismic stratigraphy—applications to hydrocarbon exploration AAPG Mem. 26.

Seismic Stratigraphic Model of Depositional Platform Margin, Eastern Anadarko Basin, Oklahoma[1]

WILLIAM E. GALLOWAY,[2] MARSHALL S. YANCEY,[3] and ARTHUR P. WHIPPLE[3]

Abstract Three-dimensional stratigraphic analysis of cratonic-basin margins has demonstrated complex genetic interrelations between shelf, shelf-edge, and basinal facies. Application of seismic stratigraphic modeling has proved useful in analyzing the geometry of platform-margin deposits of the Pennsylvanian Hoxbar Group (Missourian) in the eastern Anadarko basin in Oklahoma.

Seismic modeling requires four principal steps: (1) tabulation of petrophysical parameters of the lithologies included in the model; (2) construction of a series of model stratigraphic sequences along a line of section; (3) generation of synthetic seismograms for each model sequence; and (4) comparison of the synthetic traces with corresponding field traces. Results of such a model study, combined with subsurface geologic data, suggest an interpretation of Hoxbar platform evolution incorporating two outbuilding or progradational depositional episodes separated by an upbuilding depositional episode.

INTRODUCTION

The combination of seismic data and physical stratigraphic interpretation is becoming an increasingly important aspect of subsurface exploration. Significant examples of the application of seismic data to regional facies analysis are presented by Sangree and Widmier (1977), Mitchum et al (1977), and Stuart and Caughey (1977). The objective of this paper is to demonstrate the potential for detailed facies delineation through the use of iterative seismic models.

Utility of the seismic tool for stratigraphic interpretation in a mature exploration province depends directly on the explorationist's ability to relate specific waveforms to specific lithologic units. Seismic modeling is one workable method of matching waveforms and their lateral changes to corresponding stratigraphic facies and facies boundaries. Because simple seismic models can be constructed by use of a conventional synthetic seismogram program, modeling capability is now available to a large segment of the exploration industry.

Both one- and two-dimensional modeling techniques can be applied, depending on the geologic problem. One-dimensional model programs compute a synthetic seismogram of a vertically stacked series of velocity (or velocity and density) slabs. Such programs are considered one-dimensional because they assume a straight, vertical ray path at all reflective interfaces (Fig. 1A). All interfaces are assumed to be horizontal. Two-dimensional programs, which can be based on several different theoretical approaches, can reproduce the effects of bed dip and curvature (Fig. 1B). Use of a two-dimensional program is necessary when geometric characteristics of the geologic section are in question. Both one- and two-dimensional modeling programs produce only a simplified approximation of the earth response, but model results commonly compare favorably with field data.

Applications of modeling are illustrated by reviewing the use of seismic data in reconstruction and interpretation of a Pennsylvanian shelf edge in the Anadarko basin, Oklahoma. The specific problems involved in the reconstruction include: (1) correlation of shelf and basinal facies across a shelf edge; and (2) spatial delineation of the framework sandstone and limestone facies, leading to an interpreted depositional history.

GEOLOGIC FRAMEWORK AND DATA BASE

The study area lies in Canadian County, Oklahoma, on the eastern shelf of the Anadarko basin, which was a major locus of deposition during Late Pennsylvanian time (Fig. 2). Several clastic sequences prograded westward across central Oklahoma at that time, filling the basin with repetitive sequences of mudstone, sandstone, and limestone.

A major facies change occurs within the upper part of the Hoxbar Group (Missourian) along a north-south line that can be traced at least 25 mi (40 km; Fig. 2). The western platform sequence of interbedded shelf limestone, shale, and lenticular, discontinuous sandstone changes abruptly west-

---

© Copyright 1977. The American Association of Petroleum Geologists. All rights reserved.

[1] Read before the Association, April 8, 1975. Manuscript received, August 23, 1976; accepted, November 10, 1976. This paper also will appear in AAPG Memoir 26. Published with permission of the Acting Director, Bureau of Economic Geology, The University of Texas at Austin.

[2] Texas Bureau of Economic Geology, Austin, Texas 78712.

[3] Continental Oil Company, Ponca City, Oklahoma 74601.

This paper was prepared while the senior writer was a member of the Exploration Research Division, Continental Oil Company. The writers thank Continental for support and for permission to publish this paper. Figures were drafted by the cartographic section, Bureau of Economic Geology, under the supervision of James W. Macon.

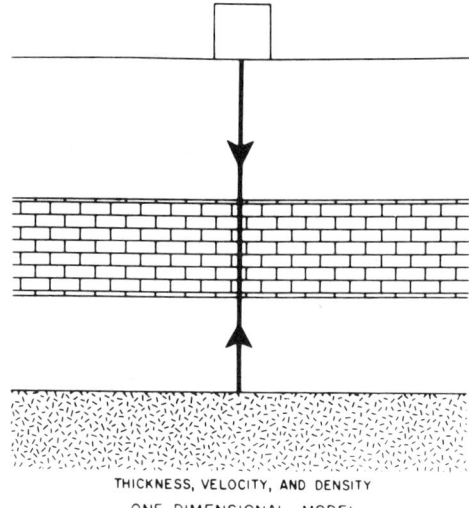

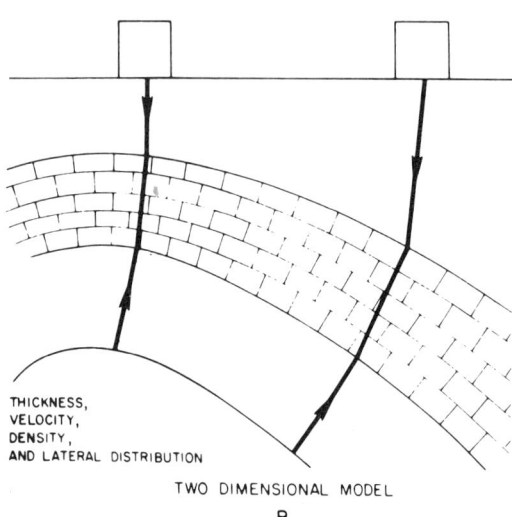

FIG. 1—Characteristics of one-dimensional, **A**, and two-dimensional seismic models, **B**. In **A**, seismic wave penetrates geologic section and returns to source/receiver along same straight-line ray path. Model uses thickness, internal velocity, and density of various layers to compute reflection character. In **B**, ray paths are refracted across dipping interfaces. Thus, effects of bed configuration are reproduced by modeling, and cross-sectional views may be obtained.

ward into an expanded sequence dominated by sandstone and mudstone (Fig. 3). Although well control is moderately dense, the abruptness of the facies change precludes definite correlation of stratigraphic units across the facies change. Consequently, subsurface nomenclature differs in shelf and basinal areas. Named units commonly used in subsurface correlation include the Belle City, Haskell, and Tonkawa limestones on the shelf, and the Medrano, lower Wade, and upper Wade sandstones in the basin (Fig. 3).

In addition to the conventional electric-log net, a grid of 12 intersecting common-depth-point (CDP) seismic lines overlies the line of facies change (Fig. 2). Sonic logs are available for a few of the wells; two sonic logs typical of shelf and basinal stratigraphic sequences were selected for synthetic seismogram computation. Though the sonic logs selected are not of wells located directly on the log cross section or any of the seismic lines, the resultant synthetic seismograms were easily correlated with the field records. The Tonkawa limestone is a particularly good marker that extends across the map area. Thus, specific seismic events can be related to specific stratigraphic units (Fig. 3), and the geometry of the seismic reflectors can be used immediately to aid in interpretation of the facies change. As expected, the facies change produced a distinctive seismic configuration, as shown on field-record B (Fig. 3). The wedge-shaped pattern shown on this and all other lines crossing the facies change suggests a progradational shelf edge, or depositional platform margin, and supports the preliminary interpretation. Other east-west lines were used for refining the map of the shelf edge.

SEISMIC MODELING

Once stratigraphic units are correlated with specific events on the seismic field traces, the grid of seismic data becomes a significant additional source of stratigraphic data that, within the limits of data quality and band width, can be used to solve problems of correlation and facies delineation. Seismic modeling, which in its simplest form is the intentional manipulation of available sonic logs and derived synthetic seismograms, provides the tool needed to utilize this expanded data base.

One-Dimensional Modeling

The initial step in seismic stratigraphic analysis of the shelf edge consisted of the construction of several one-dimensional models of selected field traces on line B to determine relations between lithologic units and their seismic response. The interval modeled extends from the Tonkawa limestone at the top through a continuous calcareous shale unit at the base of the west-east cross section (Fig. 3). Thus all units potentially affected by

---

[4]Addition of density data produced little significant change in the synthetic. Velocity variation dominated the seismic response; thus density was not used as a variable in subsequent work.

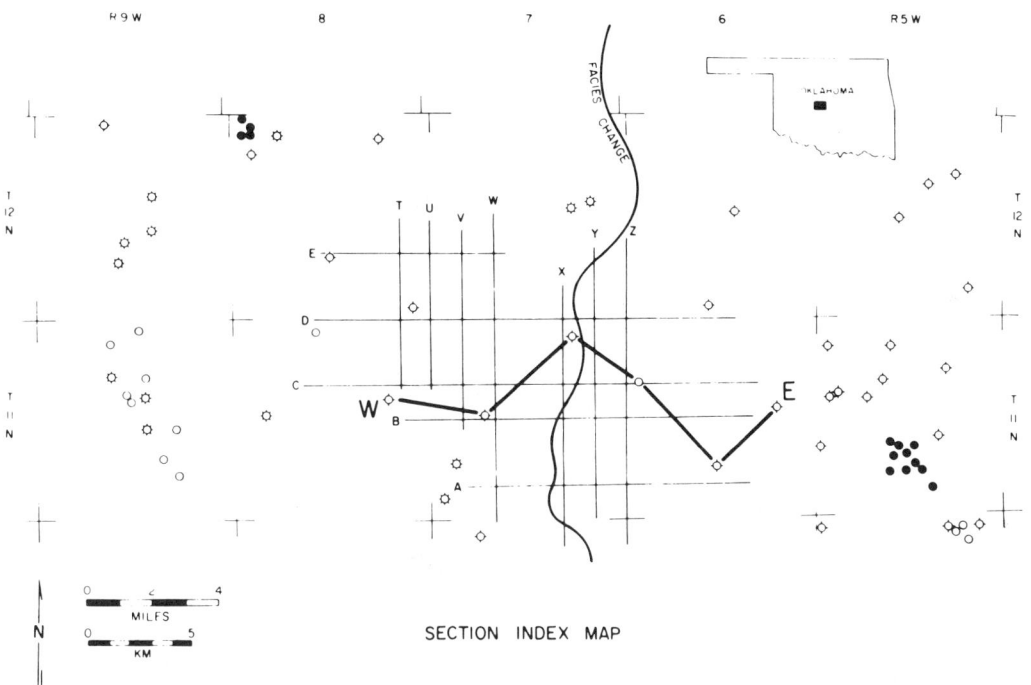

FIG. 2—Index map showing location and trend of upper Hoxbar facies change, and location of well and seismic data used in study.

the facies change are included between upper and lower reflectors.

The modeling process included three basic steps.

1. Four seismic traces in line B representative of different vertical stratigraphic sequences were selected: (MO-1) basin, (MO-2) shelf-basin transition, (MO-3) shelf edge, and (MO-4) shelf. Locations of these traces are shown in Figure 3. Wells C and E were projected into the seismic line to provide rock-unit thickness data for models 3 and 4 (MO-3, MO-4), respectively. Unit thicknesses were projected updip slightly from well A for model 1. No well data were directly applicable to the transition-zone model 2 (MO-2); bed thickness and sequences were based on the initial geologic interpretation of the geometry of the shelf-to-basin transition. In summary, thickness and vertical sequence of beds at the location of the four selected traces were tabulated.

2. Velocity-versus-lithologic data were tabulated using all available sonic logs in the study area. Velocity (V_1) is dependent on rock type (matrix) and porosity; however, in the section studied, velocity data are quite consistent for each lithologic unit. Shale velocity ranged from 12,000 to 13,000 ft/sec (3,600 to 3,960 m/sec), sandstone velocity averaged 14,000 to 15,000 ft/sec (4,270 to 4,570 m/sec), and limestone velocity ranged from 17,000 to 19,000 ft/sec (5,180 to 5,790 m/sec). The wide range of velocities provides optimum conditions for bed definition and makes the seismic response relatively insensitive to errors in velocity selection. (Tests indicated that variations in velocity of several hundred feet per second had no discernible effect on the modeled seismic response.) The velocity data were combined with bed-thickness data to produce four velocity-slab profiles, which provided the input to the synthetic seismogram program. In effect, four simple sonic logs were constructed. Data on the lithologic-sequences and velocity slabs used to produce the models are shown in Figure 4 (the vertical scale has been converted to two-way travel time by the modeling program so that the schematic sections and velocity slabs can be compared directly to the resultant seismic traces).

3. After processing and filtering to a band width comparable to field data (8 to 45 Hz), the preselected field trace was compared with the model trace, and successive iterations, with appropriate changes in slab thickness or velocity, were run where necessary to improve the match. Events on the model trace are displaced vertically from the units they represent because reflections are generated at interfaces between beds.

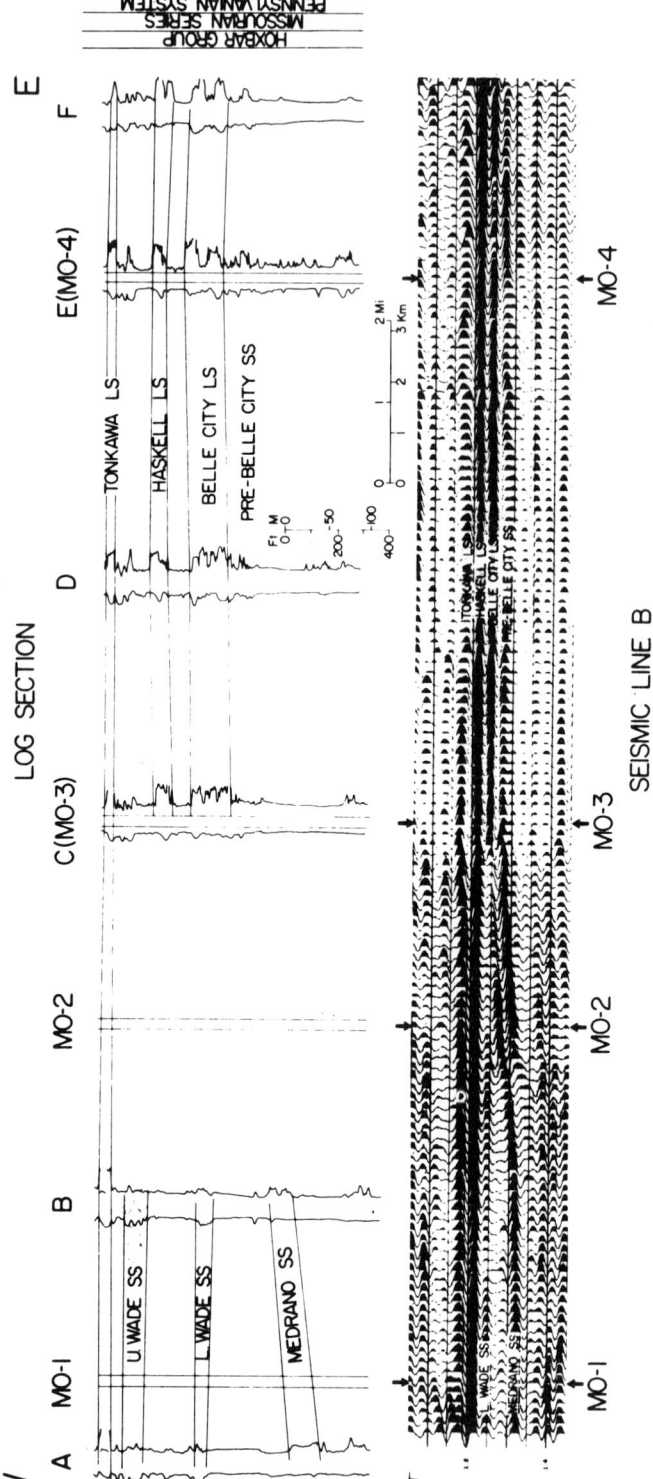

FIG. 3—West-east electric-log cross section and equivalent seismic-section B across upper Hoxbar Group facies change. Location of sections is shown on Figure 2. Log section shows abrupt change from mixed-carbonate and clastic section of shelf into expanded clastic section. At same position, equivalent seismic events display equally abrupt lateral changes. Locations of modeled traces (MO-1, etc.) are indicated by arrows on seismic section and by columnar section on geologic cross section.

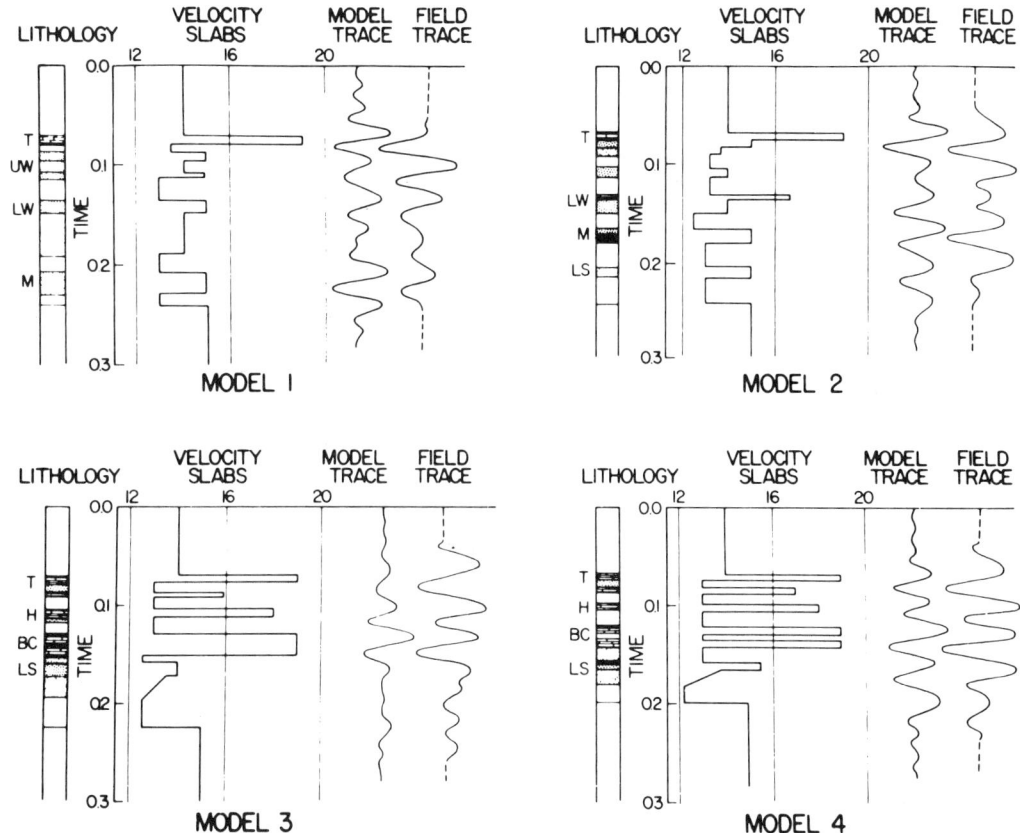

FIG. 4—One-dimensional models of selected seismic traces. All data including lithologic section are displayed in time domain for visual comparison. Velocity-slab plot graphs interval velocity (in ft/sec) used for each lithologic unit. Models are considered adequate when all major events are reproduced with correct peak-to-peak spacing. T, Tonkawa; UW, upper Wade; LW, lower Wade; M, Medrano; LS, lower sandstone; H, Haskell; BC, Belle City.

Review of the final models (Fig. 4) reveals several pertinent features.

Model 1—Peaks are produced by the lower Wade and Medrano sandstones, which are high-velocity slabs in the middle or lower velocity mudstone. The upper Wade also is underlain by a thick transition zone of sandy, calcareous mudstone of intermediate velocity. A strong "pseudo-peak" is produced by the upper Wade sandstone in combination with side lobes of the Tonkawa and lower Wade events. This peak is even more pronounced on the field data, suggesting further augmentation by interbed multiple energy not reproduced by the synthetic modeling program.

Model 4—At the other extreme, the shelf model also shows good agreement between model and field traces. Strong peaks are generated by the Belle City and Haskell limestones (Fig. 4). The thin limestone and lenticular sandstones just below the Tonkawa are lost between the two more prominent reflectors. The thin, calcareous cap on the lower sandstone amplifies its response and partially negates the smoothing effect of the basal transition zone into underlying, low-velocity shale.

Model 3—Model 3 is similar to model 4 except for the absence of a shale break within the Belle City and the absence of a calcareous cap above the lower sandstone. The effect of the lower transition zone is pronounced, and a sharp trough is absent below the lower sandstone event.

Model 2—The rapid change from trace to trace makes selection of a single trace for modeling somewhat arbitrary. However, several iterations showed that the observed seismic events can be generated by four intermediate-velocity layers (about 14,000 to 15,000 ft/sec or 4,270 to 4,570 m/sec) at the stratigraphic levels shown. Tracing these events laterally indicated that the first peak directly below the Tonkawa peak grades basinward into the "pseudopeak" at about upper Wade level, the second peak becomes the lower Wade event basinward, the third peak drops in the section and becomes the Medrano event of model 1, and the lowest peak rises slightly shelfward to become the lower sandstone event of models 3 and 4. Thus, velocity data and lateral correlation suggest that the layers of the transitional model are composed of mixed sandstone-limestone beds and mudstone.

Integration of models, logs, and geometry of the seismic events at this point could reasonably answer many questions about the shelf-to-slope transition. Construction of a two-dimensional model, however, will add additional detail and test the limits of seismic stratigraphic interpretation.

Two-Dimensional Modeling

Two-dimensional seismic modeling necessitates construction of a geologic cross section in the plane of the seismic section with adequate detail to reproduce the lateral and vertical distribution of seismic events. Although this sounds like a formidable task, experience with one-dimensional models provides the needed "feel" for the seismic response of various types of changes expected within the framework of the study area. For example, thickness of limestone units has little effect on waveform because of the extreme velocity contrasts with surrounding facies. However, some sensitivity to thickness of sandstone units might be expected because of their intermediate velocities and because their associated waveforms show significant variability on the field data (compare lower Wade and Belle City events in Fig. 3).

To test seismic response to changing sandstone-bed thickness, another seismic-stratigraphic technique was employed. A sonic log from the basin sequence was digitized and a series of models constructed by substituting different thicknesses of intermediate-velocity material at the levels of the lower Wade and Medrano sandstones. Results are shown in Figure 5. The synthetic seismograms show that, in the range from 0 to 100 ft (0 to 30 m), greater thickness of lower Wade sandstone results in increasing peak amplitude of the appropriate event. Within this geologically reasonable range of thicknesses, the seismic data should reflect details of the sandstone-unit geometry. In contrast, as little as 50 ft (15 m) of Medrano sandstone produces a large peak that changes little as successive increments of high-velocity material are added. This result is supported by the continuity and uniform amplitude of the Medrano event on the field section.

Using the seismic section as a guide, and combining data from one-dimensional models and logs, it was possible to construct the geologic cross section shown at the top of Figure 6. As with all interpretive cross sections, this one is somewhat schematic; but, because the section must show the bed geometry that is digitized and put into the computer, certain geologic conventions, such as sawtooth interfingering of rock units, are not used. Most fine detail (such as interbedding of units that are inches or a few feet thick) cannot be resolved by seismic waves of the frequencies retrieved in the field (8 to 45 Hz in this example); where thick transition zones occur, separate velocity slabs were drawn. The thick wedge of 14,000-ft/sec (4,570 m/sec) material below the lower Wade sandstone is a good example of a transition zone. Interestingly, a heterogeneous sequence of sand, silt, and mud also may have a higher velocity than the overlying porous sandstone; this is seen in the lower sandstone interval, which has a velocity of 13,800 ft/sec (4,206 m/sec) and is underlain by a 15,000 ft/sec (4,572 m/sec) transitional unit.

Output from computations for the model section is shown in the lower two panels of Figure 6. The impulse-response plot displays the distribution, polarity, and strength of reflecting interfaces in the vertical time domain. Filtering at the appropriate band width produces a model section that should closely resemble the field data if the input cross section is accurate.

Comparison of this final model with the field section illustrates the detail that can be reproduced by modern modeling technology combined with appropriate geologic input (Fig. 6).

1. A strong, continuous Tonkawa peak extends across the section and provides a reference datum for other underlying events.

2. A strong, continuous peak is generated on the shelf by the Haskell limestone, but continues uninterrupted across the shelf edge and into the basin. West of the shelf edge, the peak is generated by a combination of sidelobe and multiple energy and is not related to a particular velocity interface. The model section accurately reproduces this "pseudoevent."

3. A strong peak is generated on the shelf by the Belle City Limestone.

4. A peak of variable amplitude is produced by the lower Wade sandstone, which laps up slightly against the Belle City and dies out. The peak is underlain by a broad quiet zone.

5. The Medrano sandstone produces a strong, continuous peak, which laps up into the slope and grades across a low-amplitude transition zone into the Belle City event.

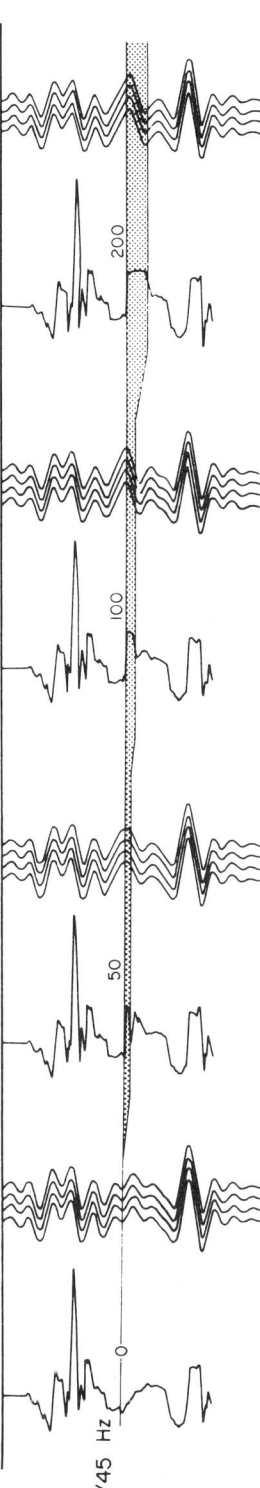
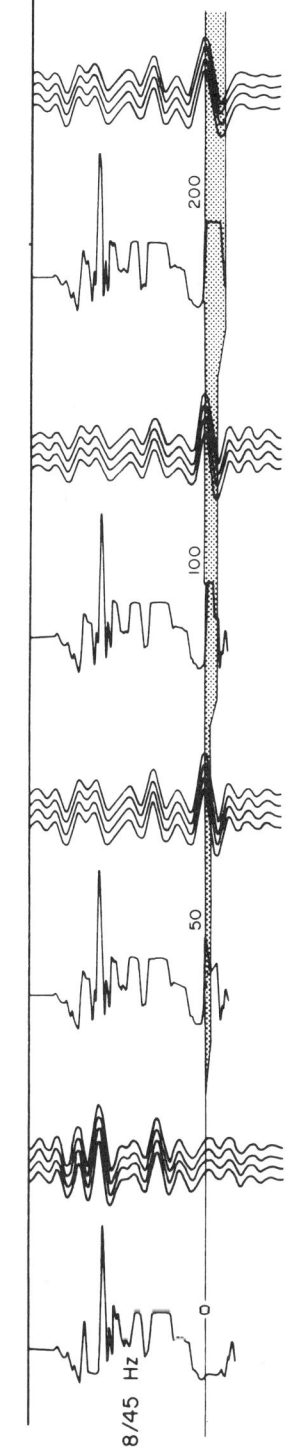

FIG. 5—Effect of increasing sandstone thickness on seismic response, lower Wade and Medrano sandstones. Model seismograms are produced by inserting increments of material with appropriate velocity at stratigraphic position of sandstone unit on digitized sonic log.

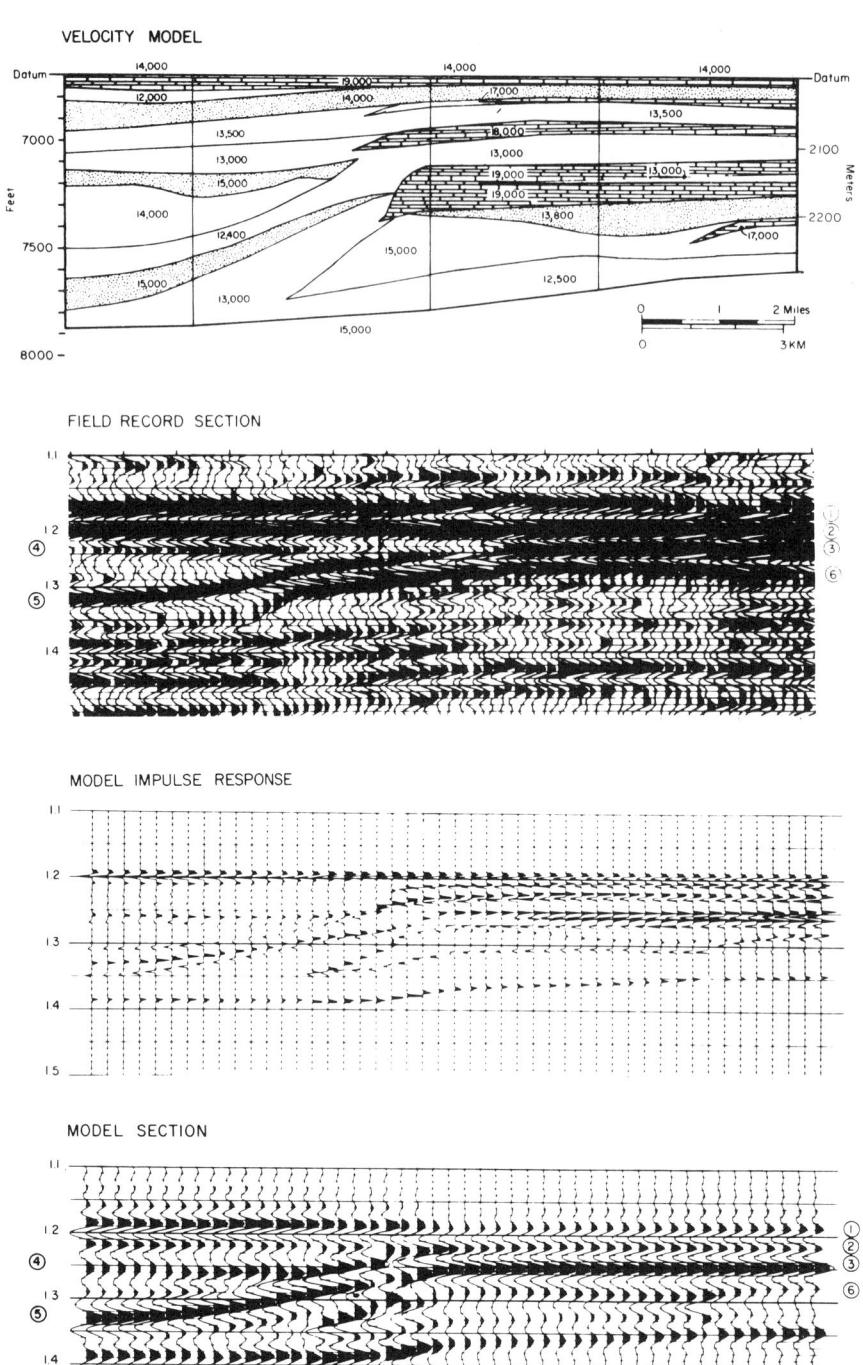

FIG. 6—Geologic-model cross section, equivalent part of field-record section, impulse-response plot of geologic section in vertical time domain, and computer-generated synthetic seismogram of geologic model. All principal events on field section are reproduced by synthetic-model section. Numbers beside seismic events correspond to usage in text.

6. A peak which dips down and dies out at the shelf edge, and which is underlain by a quiet zone, is produced by the lower, progradational sandstone.

Results of the model support the validity of the interpretive section; correlation problems are solved, and the spatial distribution of lithofacies across the shelf edge is depicted.

GEOLOGIC INTERPRETATION

The ultimate objective of modeling is the integration of seismic data with geologic information to test, refine, and document the interpreter's concepts about the three-dimensional distribution of lithofacies. Knowledge of lithofacies geometry, composition, and position in the basin of deposition is, in turn, the foundation of genetic stratigraphic analysis.

Correlation shows that the lower, progradational sandstone-shale sequence, Belle City Limestone, Medrano sandstone, and lower Wade sandstone form the shelf margin. Three-dimensional relations of these units suggest that three principal depositional episodes (terminology of Frazier, 1974) were responsible for development of the margin (Fig. 7). The lower sandstone and lower Wade episodes (I and III) are progradational or outbuilding episodes; the Medrano–Belle City episode (II) is primarily an upbuilding episode on both the shelf and basin. Isolith maps of facies within each of these episodes (Fig. 8) isolate genetically equivalent units and are the basis for the following interpretations.

Episode I

A major, fluvial-dominated delta system prograded westward into the Anadarko basin producing a progradational wedge as much as 500 ft (152 m) thick and consisting of a prodelta-mud platform capped by dip-oriented, anastomosing sandstone units (Fig. 8A). Framework sands were deposited in distributary-channel, mouth-bar, and delta-margin environments. Progradation of the delta system produced considerable differential relief between the shelf and the basin, thus forming the foundation of the shelf edge.

Episode II

Deltaic deposition ceased in the study area, perhaps as a result of upstream diversion of the fluvial system, and compactional subsidence permitted transgressive reworking of the deltaic deposits. As conditions stabilized, the shallow-water platform became the site of carbonate generation and accumulation, particularly at the shelf edge. Strike-oriented isopach trends (Fig. 8B) suggest biohermal growth. During the same episode, additional terrigenous sediment was introduced into the basin, primarily south of the study area. However, the rate of influx was slow and, instead of building a progradational mud platform, finer sediments were dispersed into the basin; sands, which also contain significant amounts of carbonate detritus, were transported downslope to form an extensive submarine apron at the base of the slope. Geometry (Fig. 8C), stratigraphic position, and compositional differences all support interpretation of the Medrano as a submarine-fan complex. Thin sandstones that trend across the shelf within the Belle City (as shown on Fig. 8C) may have provided some sand to the shelf edge, but the Medrano depocenter was south of the study area.

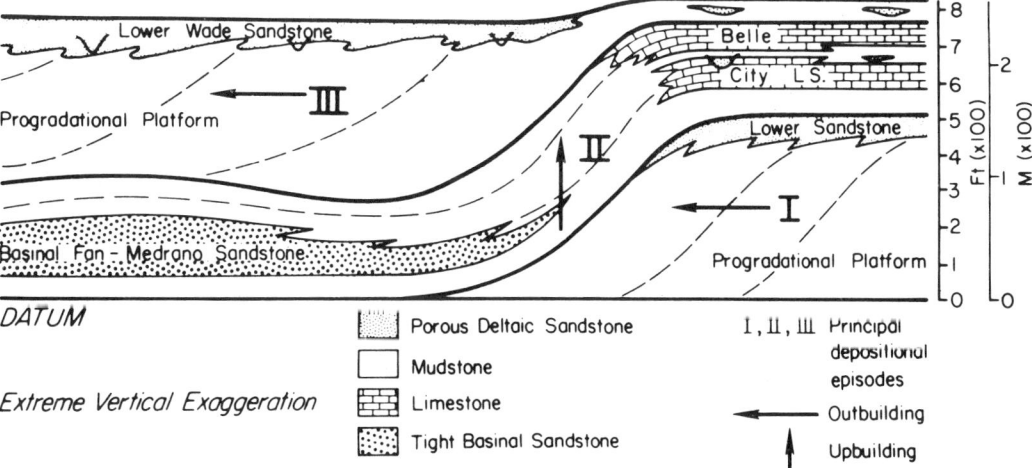

FIG. 7—Three depositional episodes producing upper Hoxbar shelf margin. No horizontal scale.

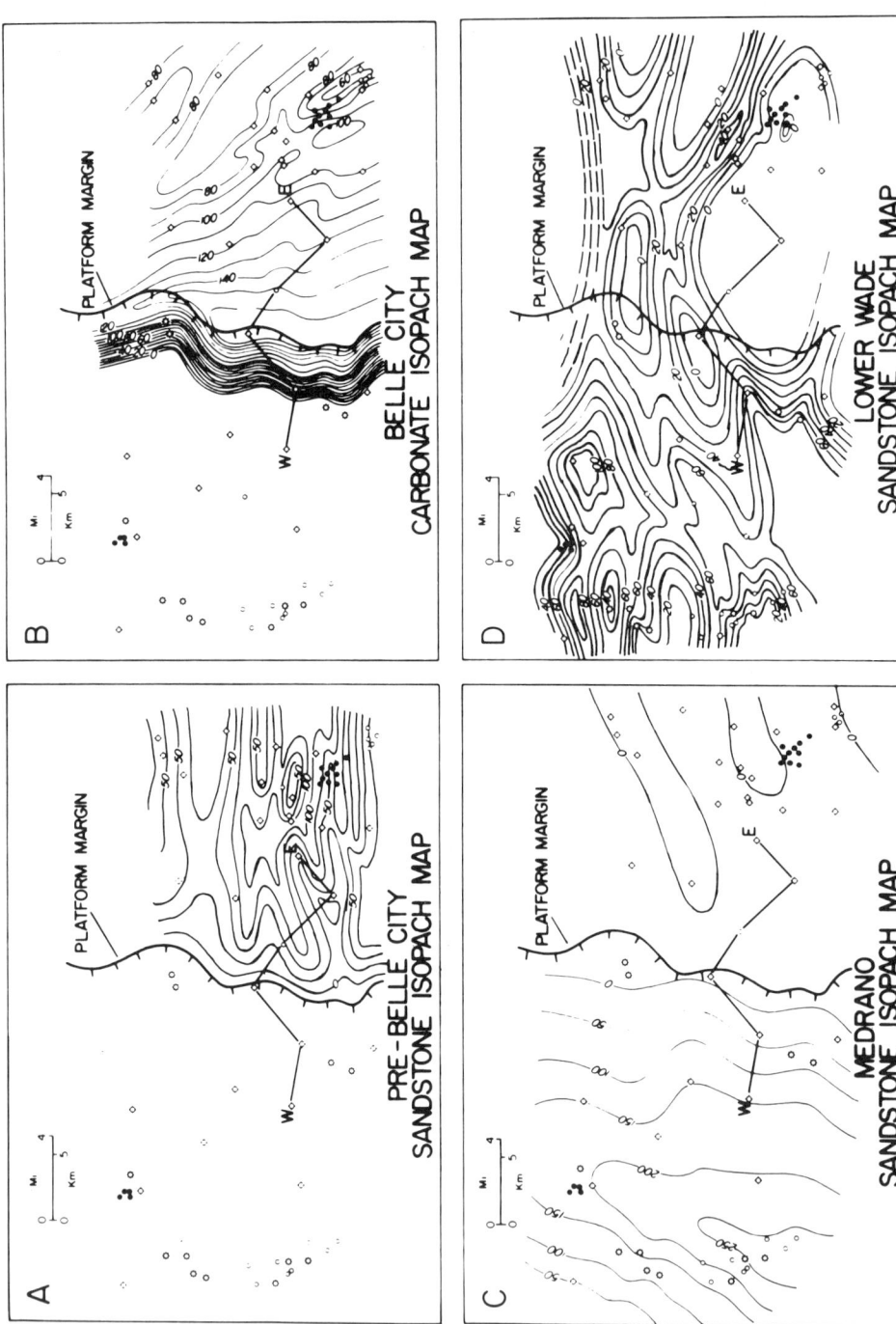

FIG. 8—Isopach maps of framework facies composing each depositional episode. Location of west-east log cross section is shown for reference. A. Dip-oriented delta-platform sandstones of progradational episode I. B and C. Shelf carbonate unit and uplapping submarine-fan sandstone of Belle City and Medrano; both are part of upbuilding episode II. D. Dip-oriented distributary-channel and delta-margin sandstones of progradational episode III, which buried Belle City shelf-edge carbonate complex and formed new topographic break west of map area.

Episode III

Influx of terrigenous sediment resulted in a renewed period of delta growth and progradation beyond the old platform edge (Fig. 8D). Water remained shallow across the platform, resulting in thin delta-margin sequences and consequent incision by major distributary channels into underlying limestone and shale formed during episode II. Beyond the shelf, the delta system was more intensely reworked by waves and progradation was slower—a common evolutionary trend in cratonic delta systems (Galloway, 1975). The abundance of carbonate debris in the deltaic sandstones and underlying prodelta muds further indicates that progradation was slow and interrupted locally at least twice by temporary destructional phases. Progradation during episode III extended several miles beyond the mapped area.

CONCLUSIONS

The examples given herein illustrate some of the possible techniques and applications of seismic modeling in genetic stratigraphic analysis. In the Wade-Medrano shelf edge, seismic response is complex. Reflection events may accurately portray lithologic continuity or mislead with their continuity; they may be sensitive or insensitive to changes in bed thickness; and they may be the product of a single lithologic interface or a cumulative result of side lobes, multiples, and complex interface distribution. Only when these various possibilities are sorted out does it become possible to use seismic data to test, document, and amplify detailed stratigraphic interpretation. Accurate three-dimensional description of lithofacies thus becomes a powerful tool for reconstruction of the depositional history of a basin.

REFERENCES CITED

Frazier, D. E., 1974, Depositional episodes: their relationship to the Quaternary stratigraphic framework in the northwestern portion of the Gulf basin: Texas Bur. Econ. Geology Circ. 74-1, 28 p.

Galloway, W. E., 1975, Process framework for describing the morphologic and stratigraphic evolution of deltaic depositional systems, in M. L. Broussard, ed., Deltas, models for exploration: Houston Geol. Soc., p. 87-98.

Mitchum, R. M., Jr., et al, 1977, Stratigraphic interpretation of seismic reflection patterns in depositional sequences: AAPG Mem. 26.

Sangree, J. B., and J. M. Widmier, 1977, Interpretation of clastic depositional facies from seismic data: AAPG Mem. 26.

Stuart, C. J., and C. A. Caughey, 1977, Seismic facies and sedimentology of terrigenous Pleistocene deposits in northwest and central Gulf of Mexico: AAPG Mem. 26.

Exploration for Oil Accumulations in Entrada Sandstone, San Juan Basin, New Mexico[1]

RICHARD R. VINCELETTE[2] and WILLIAM E. CHITTUM[3]

ABSTRACT

Recent exploration activity in the San Juan basin of northwestern New Mexico has resulted in the discovery of new oil fields in the Entrada Sandstone of Jurassic age. The major trapping element is provided by topographic relief in excess of 100 ft (30 m) on top of the Entrada. Preliminary analyses indicate that the topographic relief was created by preserved eolian sand dunes which were formed in a topographic basin which then became the sight of a large lake in which limestones and anhydrites of the Todilto Formation were deposited over the Entrada sands. The organic-rich limestones of the Todilto provide the most likely source for the oil found in the underlying Entrada.

Analysis of stratigraphy, oil shows, source-rock potential, and porosity distribution led to the selection of an initial exploration area located along the southeastern flank of the San Juan basin. Seismic model studies, confirmed by an experimental seismic program, indicated that the topographic relief on top of the Entrada could be mapped seismically. An extensive seismic and drilling program has resulted in the discovery of six new oil pools.

In addition to the topographic relief on top of the Entrada, other factors which control the oil accumulations include local structural conditions, hydrodynamics, source-rock and oil-migration history, and porosity-permeability relations in the Entrada. The knowledge gained from this exploration program should aid in future exploration for Entrada oil fields in the San Juan basin, and encourage exploration for similar stratigraphic traps in other basins.

INTRODUCTION

An intensive exploration program initiated in 1974 along the south flank of the San Juan basin in northwestern New Mexico has resulted in the discovery of six new oil fields. Production is from the Entrada Sandstone of Jurassic age. The following report outlines the basic exploration concepts and geologic-geophysical techniques used in exploring for these fields. In addition, the various geologic factors which control the oil accumulations in the Entrada are analyzed.

The exploration program was initiated on the premise that stratigraphic traps caused by topographic relief on top of the Entrada Sandstone were potentially present over a large area of the San Juan basin, and that these features might be mappable with sophisticated geophysical techniques.

INITIAL INVESTIGATION

The San Juan basin has long been known as a prolific oil and gas province. Major gas production has been obtained from shallow Upper Cretaceous sandstones, and significant oil production has been obtained from the Cretaceous Gallup and Dakota Sandstones from both structural and stratigraphic traps. In addition, major production has been obtained from Pennsylvanian carbonates on the west flank of the basin.

However, very little exploration effort had been devoted to the Jurassic Entrada Sandstone in the basin even though one Entrada field, Media, had been discovered in 1953 (Fig. 1). Although this field had produced nearly 700,000 bbl of oil by the end of 1974, drilling density was less than one Entrada penetration per two townships along the southern flank of the basin where the field was located.

Production from only one field for a given reservoir in a basin is naturally an interesting and anomalous occurrence. Therefore, the initial step in the investigation was to analyze the Media field.

Media Field

The Media oil field, located in T19N, R3W, Sandoval County, New Mexico, was discovered in 1953 as a result of a seismic exploration program along the surface axis of the Media structural nose. The discovery well, the Magnolia Hutchinson-Federal 1, in the NW¼ SW¼,

---

© Copyright 1981. The American Association of Petroleum Geologists. All rights reserved.

[1]Manuscript received, March 19, 1981; accepted, August 31, 1981.
[2]Independent Exploration Geologist, Bozeman, Montana.
[3]Lewis Energy Corp., Denver, Colorado.

The principal companies involved in the exploration program described in this paper were Filon Exploration, Trend Exploration, Jordan Oil and Gas, and Dome Petroleum. Tom Jordan suggested the initial study which ultimately led to the initiation of the exploration program. The following individuals contributed immeasurably to the geologic and geophysical understanding of the project: B. H. Berrong, A. C. Bryant, S. S. Dark, Jr., E. D. Dolly, K. H. Hayes, F. F. Meissner, and P. Van Altena. Illustrations were drafted by Ira Watson, Cecilia Vaniman, and Jacquie Jardine. The writers take full responsibility for the conclusions and interpretations shown in this report.

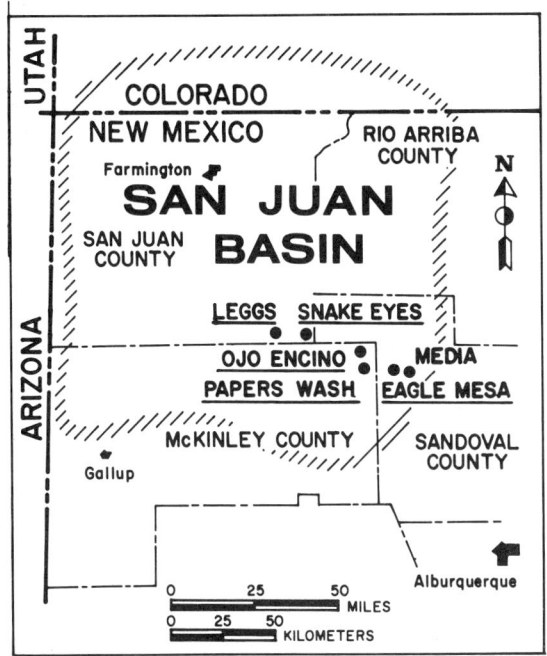

FIG. 1—Index map of San Juan basin showing location of Entrada oil fields.

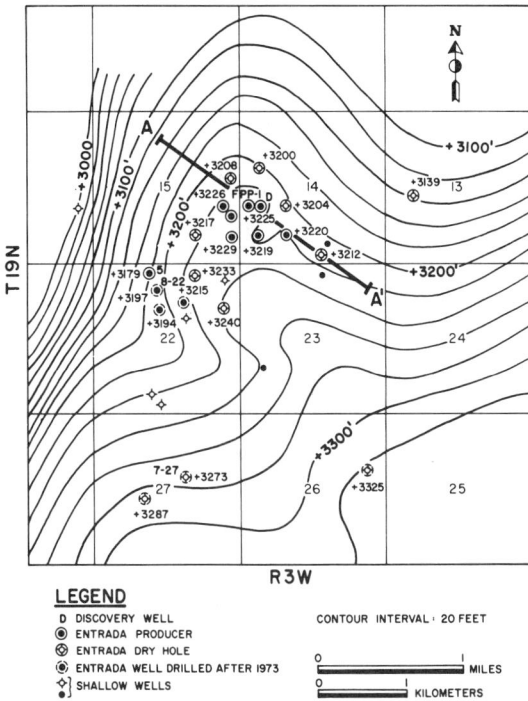

FIG. 2—Structure map on top of Cretaceous Sanostee marker, Media field.

Sec. 14, T19N, R3W, was drilled as a basement test on a seismic closure, and was ultimately completed in the Entrada for a flowing potential of 78 bbl of oil per day at a depth of 5,220 ft (1,591 m). Several other offset wells met with failure, and the discovery well was abandoned after water production increased to the point where further production was uneconomic. Cumulative production for this well was only 14,196 bbl of oil. Published articles indicate that the field was presumed to be a structural, as opposed to stratigraphic, accumulation (Ostrander, 1957, p. 138).

The field languished until 1969 when a program of development drilling was successfully initiated. One of these development wells, the Fluid Power Pump FPP 1, a 330-ft (100 m) west offset to the discovery well, was completed on pump for 500 bbl of oil plus 1,500 bbl of water per day. Cumulative oil production from this well exceeds 450,000 bbl of oil, adequately demonstrating the Entrada potential.

Sporadic development of the field over the following years was plagued by problems of high water cut and high pour point (90°F, 32°C) oil. However, the introduction of high-volume, down-hole pumps capable of lifting large quantities of fluid to the surface, and an improvement in the price of oil, made Media appear to be an economic venture and justified the exploration for similar fields in the basin.

An analysis of the subsurface stratigraphic and structural relations at Media revealed that, although located on a structural nose, critical closure for the field appeared to be controlled by topographic relief on top of the Entrada Sandstone. The Entrada is regionally 100 to 120 ft (30 to 37 m) thick, but in the discovery well at Media it has a thickness of 215 ft (66 m).

A series of maps and cross sections was prepared to illustrate the anomalous structural and stratigraphic relations present at Media. Figure 2 illustrates the structure on top of a marker bed (Sanostee) in the Upper Cretaceous, approximately 1,600 ft (488 m) above the Entrada. At this level the structure is that of a nose, plunging gently to the north, with a fairly steep west flank. Wells drilled after 1973, which postdate the initial investigations, did not significantly modify the earlier interpretations but served to confirm several of the conclusions reached.

The Entrada structure map in Figure 3 indicates at least 85 ft (26 m) of structural closure on top of the Entrada in the Media area; two oil pools are present at the Entrada level, Media and Southwest Media, separated by an intervening low. The isopach map (Fig. 4) from the Sanostee marker to the top of the Entrada illustrates the nature of the thinning between these units and shows 100 ft (30 m) of thinning over the crest of the Entrada high. To the extent that the base of the Entrada parallels the Sanostee marker, this isopach would also reflect the possible topographic relief on top of the Entrada, as illustrated in Figure 5.

The anhydrite section of the Jurassic Todilto Formation, which directly overlies the Entrada, also shows significant thickness variations in the Media field. This

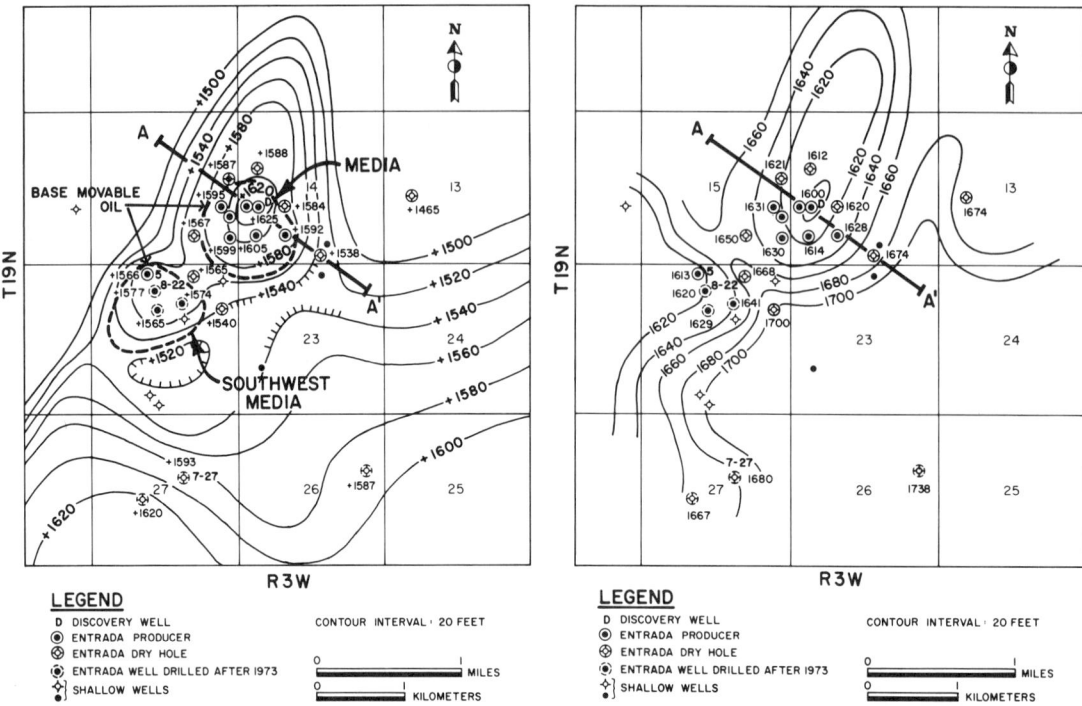

FIG. 3—Structure map on top of Entrada Sandstone, Media field.

FIG. 4—Isopach map from Sanostee marker to top of Entrada Sandstone, Media field.

unit ranges in thickness from 9 to 32 ft (3 to 10 m) over the crest of the Entrada high and thickens to 97 to 104 ft (30 to 32 m) in the structurally low wells which bound the field along its southern margin (Fig. 6).

At the time of the initial study, the only well to penetrate the entire Entrada section at Media was the discovery well. Without additional well penetrations in the field proper, it was not possible to prove conclusively that these isopach variations were due primarily to Entrada topography. Other possibilities such as paleostructure, which could have folded the base of the Entrada parallel with the top of the Entrada, could not be eliminated from these data alone. However, regional analysis of the Entrada-Todilto relations substantiated the concept that thick Entrada sands were typically overlain by thin Todilto and, conversely, thin Entrada sands were overlain by thick Todilto anhydrite.

Subsequent drilling near Media has confirmed this observation. The Petro-Lewis Federal 7-27, drilled in 1977 in the SW¼ NE¼, Sec. 27, T19N, R3W, 2 mi south of the discovery well at Media, penetrated 98 ft (30 m) of Todilto followed by only 124 ft (38 m) of Entrada Sandstone (Fig. 6). Although this well is regionally updip from the Media field, being nearly 50 ft (15 m) higher structurally at the Sanostee level, it is 32 ft (10 m) low structurally to the Media discovery well at the top of the Entrada but is again structurally high at the base of the Entrada. Thus, the difference in structural elevation at the top of the Entrada appears to be controlled primarily by differences in the thickness of the Entrada Sandstone modified by later structural tilt.

Additional studies have also revealed that postdepositional solution of the anhydrite in the Todilto Formation has considerably modified the original Entrada-Todilto relations. The general association of thin Todilto with thick Entrada is consistently valid in the area, but variations in the Todilto thickness of up to 50 ft (15 m) appear to be due to postdepositional anhydrite solution. The effect of this solution in the Media field can be seen in the cross section of Figure 5, where post-lower Morrison solution has caused a localized structural sag in units overlying the Todilto.

Another important observation concerning the oil accumulation at Media was that evidence of a tilted oil/water contact existed in the field. A pronounced southwest tilt was established by log analyses and production data, indicating that hydrodynamic factors could have played an important roll in the oil accumulations at Media. The effect of this tilt is apparent in the Entrada structure map (Fig. 3). The tilt has caused oil accumulations to be concentrated only in the southern half of the structural closure, although the fields appear to be full to the hydrodynamic spill point located on the southwest plunge of the structure.

Regardless of the complications caused by structural deformation and hydrodynamic modifications, the fact that topographic relief on top of the Entrada in excess of 100 ft (30 m) was a critical element in the oil trap at

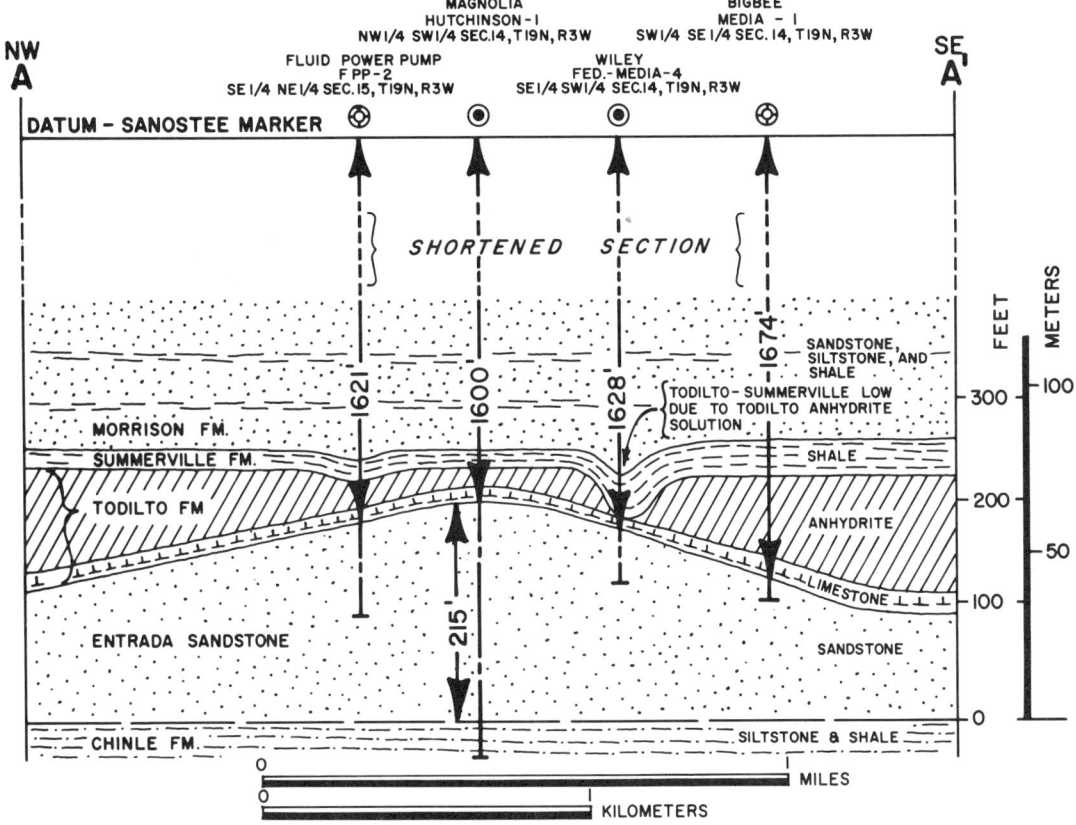

FIG. 5—Stratigraphic cross section AA', Media field. Location of section is shown in Figure 3. Cross section was constructed using Cretaceous Sanostee marker as horizontal datum, and drawing base of Entrada Sandstone parallel with Sanostee. Cross section is shortened vertically approximately 1,600 ft (488 m) from Sanostee to base of Morrison Formation for ease of viewing.

Media was an intriguing discovery, particularly when the possibility existed that this magnitude of relief might be mappable seismically.

Once an understanding of the nature of the oil accumulation at Media was obtained and the decision reached that such an accumulation could be produced economically, the next phase of the initial investigation centered on the two-fold problem of whether or not similar (or analog) accumulations could be present elsewhere in the basin and, if so, could they be located in the subsurface by geophysical means.

To determine the potential for similar fields in the San Juan basin and to outline the most favorable exploration area, several factors were analyzed. These included: (1) evidence of additional topographic highs in the Entrada, (2) distribution of oil shows in the Entrada, (3) porosity trends in the Entrada sandstones, and (4) source rock potential of the Todilto Formation.

Evidence of additional topographic highs—To determine whether other anomalous topographic highs were present in the Entrada, a regional analysis of Todilto and Entrada thickness data was made from well and outcrop control. Previous sparse drilling showed a fairly high frequency of anomalously thick Entrada intervals and/or anomalously thin Todilto intervals along the southeastern flank of the San Juan basin. In the area ultimately chosen for exploration (Fig. 7), eight Entrada tests, excluding Media, out of a total of 28 Entrada penetrations had anomalously thin Todilto intervals of less than 50 ft (15 m). Regional Todilto thickness in this same area averages approximately 100 ft (33 m). Of these eight tests, only two penetrated the entire Entrada interval, but in these two wells the Entrada Sandstone ranged in thickness from 193 to 224 ft (59 to 68 m). This contrasts with the average thickness of 110 ft (34 m) for the regional Entrada in this area.

Wells with anomalously thin Todilto intervals and/or anomalously thick Entrada intervals are identified by solid triangles in Figure 7. This same map also shows thickness data for most of the Todilto-Entrada tests in the San Juan basin prior to the exploration program initiated in 1974.

Regional cross section BB' (Fig. 8) illustrates the nature of the thickness changes in the Todilto-Entrada intervals in this area and demonstrates the type of stratigraphic traps thought to be present.

To the north, south, and west of the selected exploration area, the frequency of Todilto-Entrada thickness anomalies declines drastically. The regional Entrada Sandstone appears to thicken to the northeast (Fig. 7), where values in excess of 200 ft (61 m) and up to 330 ft (101 m) along the extreme northeast corner of the basin are the rule. In contrast, along the west flank of the basin near the New Mexico–Arizona state line, the Entrada thins to values as low as 36 ft (11 m). The Todilto Formation also shows regional thickness variations, ranging from 126 ft (38 m) in wells along the east flank of the basin to a zero edge along the west flank of the basin. The major thickness changes in the Todilto occur along the zero edge of the Todilto anhydrite, west of which only the basal Todilto Limestone Member and, perhaps, thin limestone facies equivalents of the thicker anhydrite are present.

In addition to the subsurface analysis, a reconnaissance study was made of the Todilto-Entrada outcrops which are present along the east flank of the San Juan basin. The Todilto-Entrada outcrops have been extensively studied in this and other areas surrounding the basin (Harshbarger et al, 1957; Anderson and Kirkland, 1960; Tanner, 1965, 1970, 1974; Stapor, 1972).

In general, the Entrada sandstones have been interpreted from outcrop studies to be largely windblown deposits characterized by thick-bedded units containing large-scale eolian cross-beds. Evidence of local water-deposited sandstone is also present, especially in the upper parts of the Entrada (Tanner, 1970, p. 286).

The overlying Todilto limestone and anhydrite is thought to have been deposited in a large saline lake centered in what is now the San Juan basin (Tanner, 1974, p. 219; Green and Pierson, 1977, p. 150).

Outcrops were examined along the southeastern margin of the basin, not so much to confirm or deny the interpretations made by others as to the environment of deposition of the Todilto-Entrada sequence, but to determine whether or not examples of topographic relief in the Entrada could be located and documented in the outcrop belt. In general, the outcropping Todilto-Entrada contact in this area is characterized by a lack of local relief. Although low-relief changes on the order of 10 ft (3 m) were observed, no topographic relief of the magnitude observed in the subsurface was found. However, the possibility that topographic relief of significant magnitude could still be present along Entrada outcrops should not be completely discounted, for vegetation and debris cover, tectonic disruptions, and local erosion of the Entrada-Todilto contact could have masked possible topographic relief in some of the areas examined.

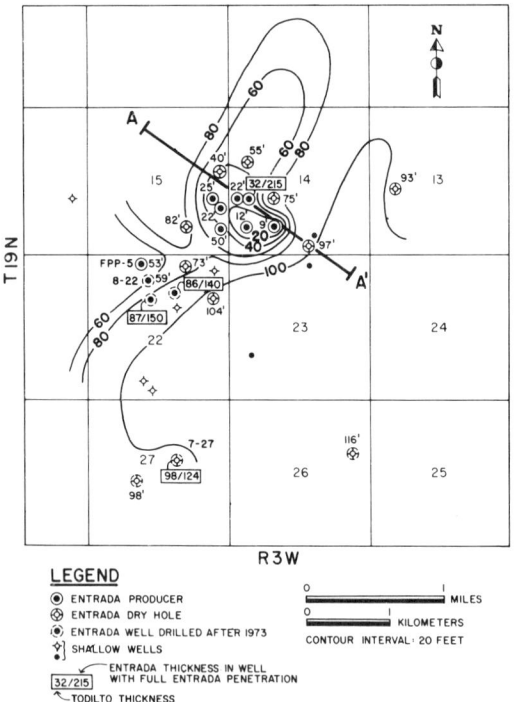

FIG. 6—Isopach map of Todilto Formation, Media field. Isopach values for underlying Entrada Sandstone are also shown where wells fully penetrated Entrada.

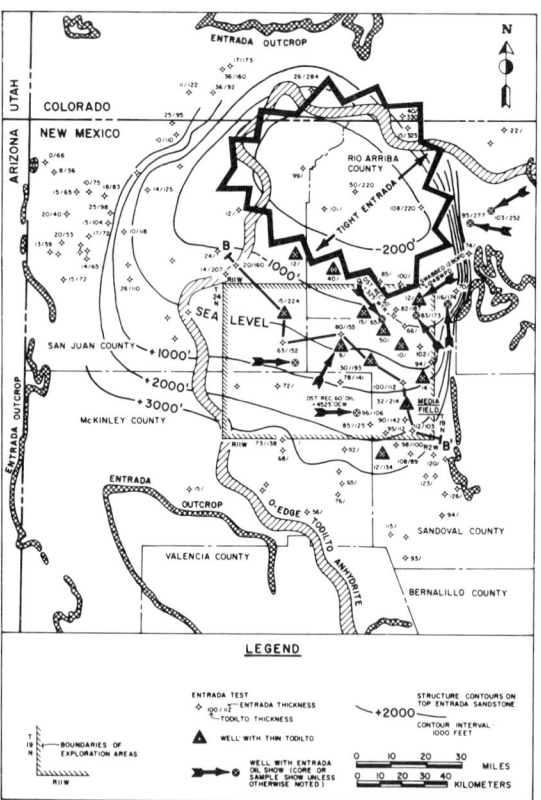

FIG. 7—Map showing wells which penetrated Entrada Sandstone in San Juan basin. All data are based on information available prior to initiation of exploration program in 1974.

Topographic relief on the order of 50 ft (15 m) on top of the Entrada has been described by Tanner (1970, p. 286) in the Ghost Ranch area (T24N, R4E) along the eastern limit of the map in Figure 7. In this area the Entrada has a maximum thickness of 295 ft (90 m) and thins both east and west over a distance of several miles (3 km) in each direction to 246 ft (75 m). Here again, the magnitude of the relief is less than that measured in the subsurface at Media, and the change in relief apparently occurs over a much longer lateral distance than at Media.

Distribution of oil shows in Entrada Sandstone—In addition to the oil field at Media, six other wells in the exploration area had oil shows in the Entrada. These shows (Fig. 7) ranged from reports of oil staining in samples or cores to the recovery of oil on drill-stem or swab tests. Two of the wells with Entrada oil shows also had anomalously thin Todilto intervals, suggesting that the shows were associated with potential stratigraphic traps in the underlying Entrada. The fact that 24% of the Entrada wildcats in the area chosen for exploration had oil shows in the Entrada indicated a reasonable chance of finding an oil accumulation in the Entrada if a trap could be identified and mapped. In contrast, very few oil shows in the Entrada had been reported outside of the boundaries of the exploration area.

Porosity trends in Entrada Sandstone—Another consideration in defining the northern, downdip limits of the prospective area was the observation that in the deeper, northeastern part of the San Juan basin, the porosity in the Entrada diminishes rapidly. Below a depth of 9,000 ft (2,700 m), approximately −2,000 ft (−600 m) subsea, the Entrada Sandstone becomes extremely tight owing to compaction and silica cement (Fig. 7). In addition to the effects of depth of burial on the porosity in the Entrada, the presence of Tertiary intrusives along the northeastern flank of the basin may also have caused high temperatures and added silica to destroy the Entrada porosity in that area.

In contrast to the tight nature of the Entrada north of the exploration area, the porosity at Media from core and log data averaged 23% in the Entrada pay zone, with an average permeability of nearly 300 md. Porosity logs from other wildcats drilled in the exploration area confirmed that similar porosities could also be expected in other Entrada fields which might be discovered.

Source-rock potential of Todilto Formation—The Entrada lacks shows of oil on the west flank of the San Juan basin. A fairly large number of wells penetrate the Entrada in this area, and some are on closed structures. There is also a similar lack of shows as the Entrada is traced farther north and south beyond the boundaries of the exploration area.

Lack of shows on the north could be attributed to a lack of Entrada porosity in the deeper parts of the basin but, along the shallow west flank, the Entrada remains porous. Therefore, some other explanation must be sought to explain the lack of shows in this area.

The basal Todilto Limestone Member, which directly overlies the Entrada, has been described (Anderson and Kirkland, 1960, p. 38) as a varved limestone, 7 to 8 ft (2 m) thick, containing alternating thin layers of calcium carbonate and organic material (sapropel). The presence of organic material in the Todilta suggests that it might be the source for the oil found in the directly underlying Entrada Sandstone. To test this hypothesis, selected cuttings of the Todilto were obtained from 28 widely scattered wells in the San Juan basin and from several outcrop locations along the east flank of the basin. These samples were analyzed for the presence of source-rock

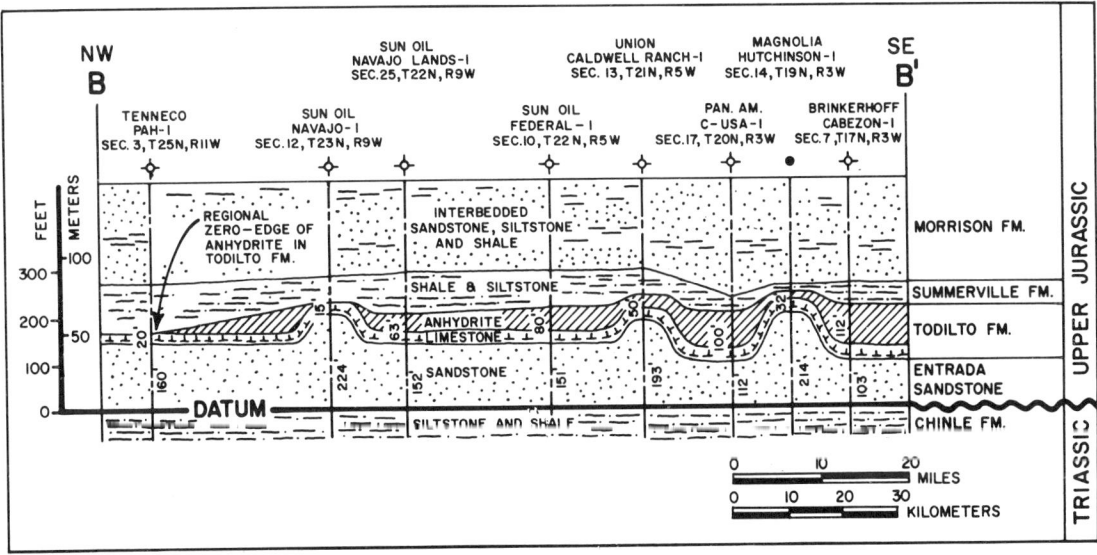

FIG. 8—Regional stratigraphic cross section BB' illustrating thickness changes in Entrada Sandstone and overlying Todilto Formation. Base of Entrada Sandstone is used as horizontal stratigraphic datum. Location of cross section is shown in Figure 7.

material. Where such organic material was present, slides were prepared and examined under the microscope to determine the thermal maturity, or level of carbonization of the organic material, on the basis of the color of the kerogen under transmitted light. Some of the samples were also subjected to pyrolysis to determine the potential hydrocarbon yield of the source rock. These analyses were made by D. M. Sparks of Geo-Logic, Inc.

These data were then collated and interpreted with respect to the distribution, maturity, and most likely migration paths of the Todilto oil (Fig. 9). There is a close correlation between the presence of organic material in the Todilto limestone and the presence of the overlying Todilto anhydrite. Outcrops of the Todilto on the west and southwest flanks of the San Juan basin are characterized by the presence of light-colored limestone containing local concentrations of algal material, probably deposited in a shallow-water, well-oxygenated environment (Tanner, 1970, p. 282). This is in contrast to the varved, organic-rich, dark-brown limestone present in the central and eastern parts of the basin beneath the anhydrite facies of the Todilto. This association was confirmed by the lack of organic material in subsurface Todilto samples from the west flank of the basin, where the anhydite is no longer present.

The lack of Entrada shows on the west flank of the San Juan basin, where the Todilto is not a source rock, lends support to the idea that the Todilto is indeed the source of the oil in the Entrada. This factor was utilized in defining the western limit of the exploration area, which corresponds, in general, to the western limit of the Todilto anhydrite facies and, presumably, to the western limit of the Todilto source rock. The eastern limit of the exploration area is controlled by the Entrada outcrops along the west flank of the Nacimiento uplift.

Analysis of the color of the kerogen in the Todilto Formation revealed a general increase in thermal maturity or organic metamorphism with increasing depth, ranging from immature source rock on the southeastern outcrop (sample T-2) to organic material which had been overcooked or carbonized to the point where only gas and condensate would be the expected product (sample T-29) at a depth of 9,300 ft (2,800 m). A zone of maximum oil generation was selected in which the kerogen was characterized by a brown to dark-brown color. Although some scatter in the data is apparent, the zone of maximum oil generation in the Todilto corresponds to a depth range of from 6,000 to 8,500 ft (1,800 to 2,600 m) or +1,000 to −2,000 ft (+300 to −600 m) structural elevation. Most of the area chosen for exploration falls within this zone (Fig. 9), with the southern, updip boundary extended to include room for some updip migration of the oil, and the northern limit being controlled more by lack of porosity in the Entrada than by source-rock considerations.

The relatively high level of organic metamorphism in Todilto samples along the northeastern margin of the basin may have been affected not only by depth of burial, but also by possibly higher temperatures created by the Tertiary intrusives which are present in the area.

Pyrolysis yields of the Todilto cuttings ranged from 0.6 to 2.4 gal/ton. These yields are rather lean but do indicate that the Todilto is capable of generating hydrocarbons. Most of the pyrolysis yields were obtained from Todilto samples which were thermally mature and which, therefore, may have already generated and lost part of their original oil by migration into the Entrada reservoirs (F. F. Meissner, personal commun.).

To determine whether or not other sources could have generated the Entrada oil, a sample of oil from the Media field was sent for analysis to the Bartlesville Energy Research Center of the U.S. Energy Research and Development Administration. This analysis was then compared to similar analyses made of other crude oils from the San Juan basin. A plot of the correlation indices of various boiling fractions of the Entrada oil from Media, compared to similar plots for crude oil obtained from Cretaceous and Pennsylvanian reservoirs in the San Juan basin (Fig. 10) shows that the Entrada oil is distinctly different from the other crude oils. Of particular interest is the much higher initial boiling point of the Entrada oil, 205°F (96°C), as compared to the initial boiling point of the Cretaceous oil, 82°F (28°C), and to that of the Pennsylvanian oil 89°F (32°C).

In addition, the values of the correlation index are

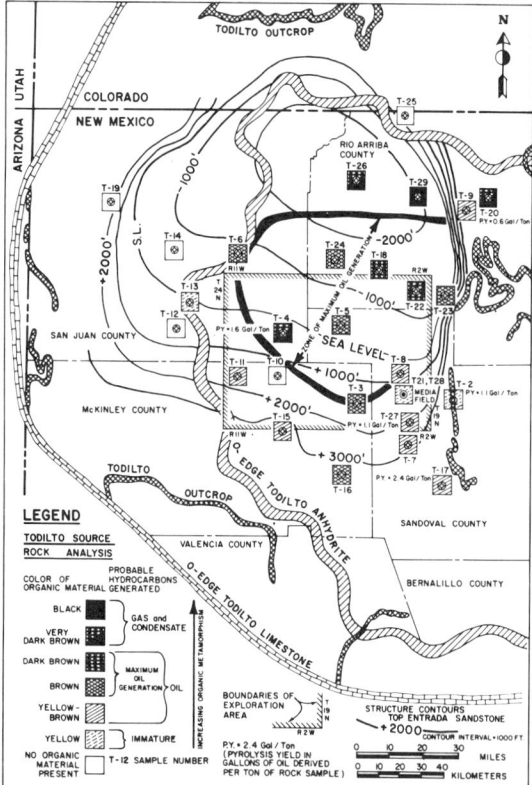

FIG. 9—Map showing results of Todilto source-rock analysis and pyrolysis yields in San Juan basin.

consistently lower for a given boiling fraction of the Entrada oil than for the Cretaceous and Pennsylvanian oils. These lower values indicate a higher paraffin content for the Entrada oil (Smith, 1940, p. 2). This is also supported by the higher pour point reported for the Entrada oil, 50 to 90°F (10 to 32°C), as compared to below 5° to 25°F (below −15 to −4°C) for the Cretaceous oil, and below 5°F (below −15°C) for the Pennsylvanian oil.

Therefore, the probability of a separate source for the Entrada oil seems indicated, and the limestones of the directly overlying Todilto Formation seem to be the most likely candidate for that source.

Selection of exploration area—As a result of the regional analysis, the conclusion was reached that stratigraphic traps in the Entrada similar to the analog field at Media could be present over a large area along the southeastern flank of the San Juan basin. Consideration of a variety of factors, including the presence of anomalous Todilto-Entrada thicknesses, oil shows, porosity distribution in the Entrada, and source-rock potential of the overlying Todilto Formation, led to the selection of an exploration area encompassing 60 townships, or nearly 1,400,000 acres (5,600 sq km) in Ts19-24N, Rs2-11W (Figs. 7, 9).

Once the exploration area had been selected, the next step was to determine what type of exploration program would be needed to locate and test the potential oil traps.

Experimental Seismic Program

Although the preliminary geologic investigations were encouraging with respect to the likelihood of other Entrada oil accumulations being present, it was obvious that owing to the limited amount of subsurface control and drilling depths to the Entrada which ranged from 5,000 to 8,000 ft (1,500 to 2,500 m), a conventional stratigraphic-drilling program would be doomed to failure. Therefore, the success of the program would have to rely on the development of geophysical techniques capable of outlining the potential stratigraphic traps in the subsurface.

Preliminary analysis of data from Media field and elsewhere suggested that high-resolution seismic might be able to map the topographic relief on top of the Entrada. The velocity and density contrast between the Todilto anhydrite-carbonate sequence (V = 18,800 ft/sec, ϱ = 2.95) and the Entrada sandstones (V = 12,200 ft/sec, ϱ = 2.3) indicated that a strong seismic signal (reflection coefficient = −0.328) should be generated from this interface. In addition, the 100-ft (30 m) relief on top of the Entrada was sufficiently large to indicate that it might be mappable.

Also, it was recognized that, as the Todilto thinned on top of the crest of the Entrada highs, the bandwidth of the seismic instrumentation was insufficient to measure such thickness in time. Therefore, it was imperative to record and process the data in a true-amplitude mode to preserve the relative strength of the Todilto reflection, which was hypothesized to yield a low-amplitude anomaly over Entrada highs.

An experimental program was designed to shoot several seismic lines over the Media field, as well as over several other subsurface Entrada anomalies, using 12-fold CDP, true-amplitude data. The results of this program, as illustrated in Figure 11, show that the Entrada anomaly at Media has a definite seismic response. In addition to the 21-msec seismic reversal present at the top of the Entrada, several other seismic characteristics of the anomaly are visible. As theorized, the amplitude of the Todilto seismic signal diminishes over the crest of the field, producing a "dull spot" where the Todilto is thinnest. Furthermore, 14 msec of isochron thinning occurs between the top of the Entrada and the shallower, overlying reflectors. This corresponds to the isopach thinning between the Entrada and the overlying Sanostee marker shown in Figure 4.

Perhaps the most interesting characteristic of the Entrada seismic anomaly is the change of the signal from within the Entrada interval from a "singlet" where the Entrada is thin to an enlarged "doublet" where the Entrada thickens in the Media field.

Thus, this preliminary seismic program proved that the Entrada topography at Media could be mapped seismically, and that other structural-stratigraphic closures in the Entrada, if they existed, could be defined seismically.

The 1,200% data acquired over Media were further processed at 900, 600, 300, and 100% to determine the minimum multiplicity required to adequately map the Entrada Formation. Although the Entrada character change was noted on all of the foregoing sections, resolution was lacking at less than 600% stack. Thus, technical and economic considerations dictated that all future data be acquired utilizing a 600% stack.

Concurrent with the experimental seismic program, preexisting seismic data in the area of interest were also purchased and, where necessary, reprocessed. Owing to original field parameters of the purchased data, the signal-to-noise was insufficient to reliably map the En-

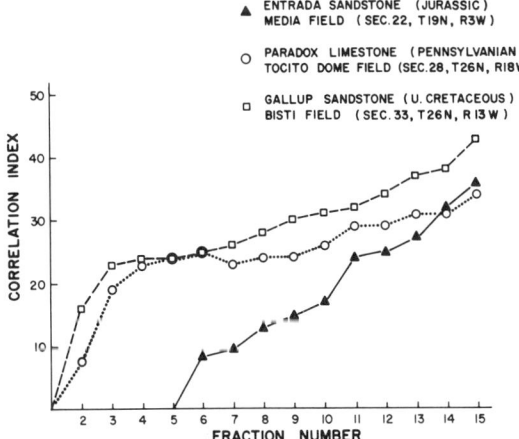

FIG. 10—Correlation-index curves comparing Entrada oil with Cretaceous and Pennsylvanian oils in San Juan basin.

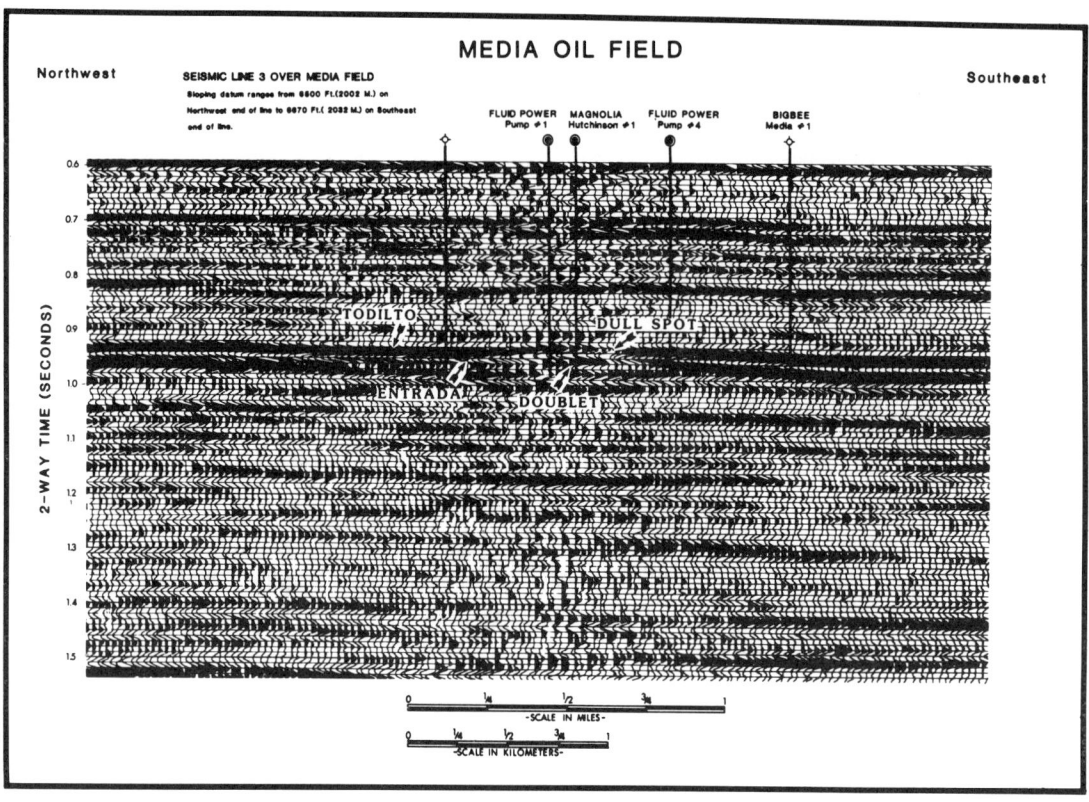

FIG. 11—Seismic line 3 over Media field. Seismic line parallels location of cross section AA' (Fig. 5) shown on Figure 3.

trada Formation, and these data were used only to locate several weak leads.

EXPLORATION PROGRAM

The area selected for the initial exploration program includes nearly 1,400,000 acres (567,000 ha.). Most of this acreage is Federal, with moderate amounts of Indian, State, and Fee leases. At the time that the exploration program was initiated, a substantial part of this acreage was already under lease to other companies, making it impossible to control the play solely by acreage acquisition. Furthermore, the checkerboard nature of available leases, combined with the uncertainty as to which acreage was most likely to contain exploration targets, dictated that seismic be utilized to delineate favorable areas prior to extensive leasing.

Initial Seismic Program

The basic premise of the initial seismic program was to shoot several long regional lines to gain a better understanding of the frequency, distribution, and configuration of the topographic anomalies in the Entrada Sandstone. The second, and ultimately the most important objective, was to delineate a few potential closures in the Entrada and to test them as quickly as possible with the drill. The basic exploration philosophy was to determine as quickly as possible by means of the drill whether or not the seismic was correctly mapping the Entrada surface and, more importantly, whether or not additional commercial oil accumulations could be found in the Entrada prior to expending large sums of time, effort, and money on a more extensive leasing and drilling program.

The reconnaissance program commenced with two contract seismic crews shooting 800 mi (1,290 km) of 600% CDP data. Owing to surface elevation variations of 2,500 ft (760 m) across the area of interest, the data were processed to a warped datum rather than to a flat datum plane. The warped datum was designed to minimize the distance from elevation of shot to datum, thus minimizing errors caused by using an improper replacement velocity. The final processed sections were displayed in VAR-wiggle format and in wiggle only. In the VAR-wiggle display (Fig. 11), the peaks and troughs were often truncated, necessitating that the interpretation be made on the wiggle display for greater accuracy. Figure 12 clearly demonstrates the doublet observed as the Entrada Sandstone thickens and the loss of amplitude of the Todilto event as the Todilto thins over the Entrada highs. For consistency the first trough under the Todilto event was picked as the top of the Entrada, and the following peak or second peak when a

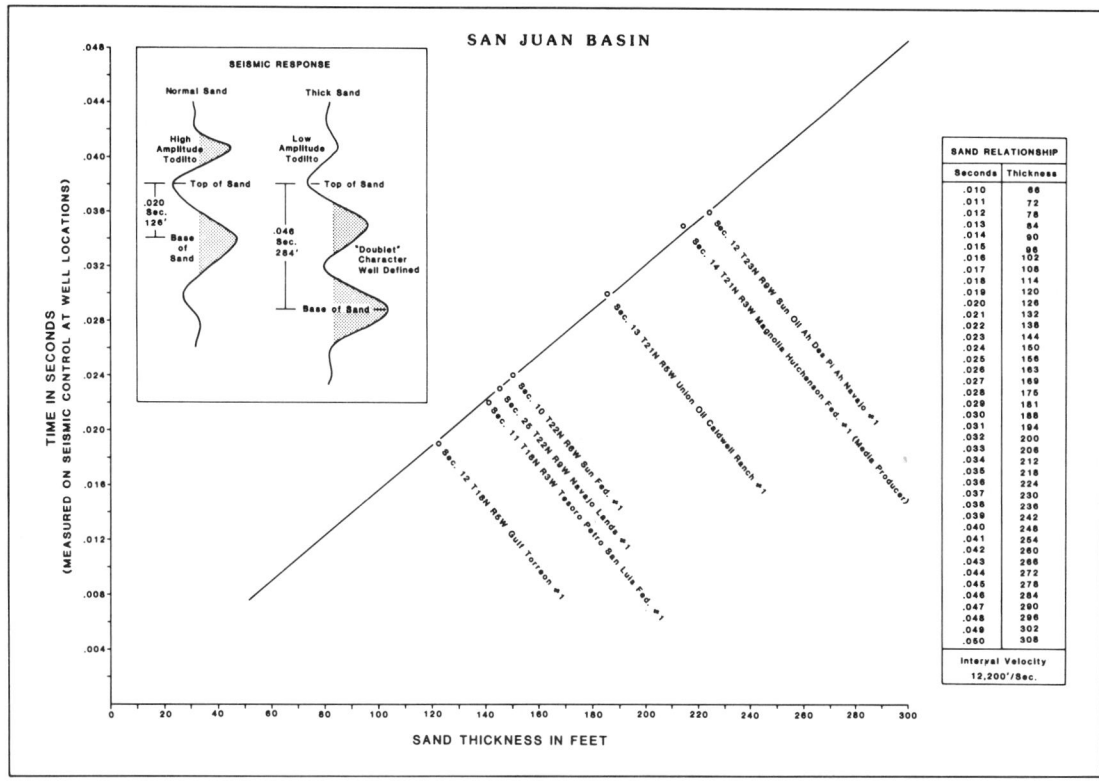

FIG. 12—Plot of Entrada sand thickness versus seismic isochron time on seismic lines recorded over control wells. Data plot along straight line indicating average seismic velocity of 12,200 ft/sec for Entrada Sandstone, which fits interval-velocity data obtained from sonic logs. Also shown is observed change in seismic signal from "normal" Entrada to "thick" Entrada.

"doublet" formed was picked as the base of the sand. The thickness of the sand was computed as: $(Je_2 - Je_1)/2 \times 12,200$ ft/sec = T_S, where Je_1 = top of Entrada sand; Je_2 = base of Entrada sand; 12,200 ft/sec = interval velocity of Entrada sand; and T_S = sand thickness.

The table in Figure 12 converts Entrada isochrons into sand thickness on the basis of the foregoing analysis. Although this empirical observation seemed to be valid, a seismic model (Fig. 13) was constructed which proved that the original observations were correct. The geologic model from which the seismic model was derived utilized known formation density and velocity data, combined with thickness changes observed in the Entrada and Todilto intervals across known Entrada sand highs. The digitized model was convolved with various minimum and zero-phase wavelets in an attempt to match the field data. The 30-hz, zero-phase model shown is an almost perfect fit to actual field data (compare with Fig. 11).

Once the signatures for sand thicks were known, they were readily identifiable, and their thickness computed to within 5% when the thickness exceeded 160 ft (49 m). Regional sands increasing in thickness to 160 ft (49 m) were subject to slightly larger errors because of limitations of the seismic data. Although the sand thickness was easily computed, constructing a structure map on top of the Entrada was more involved. Figure 14 shows a simplified block model and the resultant error in mapping in time or converting seismic times to depth without considering the effect of thickness variations in the overlying high-velocity Todilto Formation. This necessitated point plotting the Entrada depths on cross-section paper and hanging the sand isopach value from the top to determine the base of the sand. The base was then smoothed, using a least-squares fit, and the top of the Entrada adjusted accordingly. This method enhanced the ability to tie existing and proposed wells with a greater degree of accuracy than would otherwise be possible.

Numerous seismic anomalies or "leads" were encountered in the Entrada as a result of the initial reconnaissance seismic program. Some of these anomalies were then subjected to more detailed seismic work to define drillable prospects. These prospects were "graded" on the basis of four criteria (Fig. 15) as to the probable success of discovering commercial quantities of hydrocarbons. Three of the seismic anomalies thus analyzed were mapped and selected as drill sites for the initial drilling program.

Initial Drilling Program

The first test well, the Filon Federal 19-1, located in Sec. 19, T20N, R2W, Sandoval County, New Mexico, encountered a 13-ft (4 m) thick section of Todilto limestone overlying a 246-ft (75 m) thick section of Entrada sandstone. Log analysis indicated oil saturation in the top 24 ft (7 m) of the Entrada, but a straddle test over the top 13 ft (4 m) of this zone recovered only 30 ft (9 m) of free oil plus 1,900 ft (579 m) of slightly oil-cut water, and the well was abandoned.

The second well in the program, the Filon Santa Fe 27-1, tested a seismic anomaly in Sec. 27, T19N, R6W, McKinley County, New Mexico. This well encountered 67 ft (20 m) of Todilto anhydrite and limestone, followed by 164 ft (50 m) of Entrada sandstone at a depth of 5,214 ft (1,589 m). A trace of live oil was found in the drill cuttings at the top of the Entrada, but a drill-stem test recovered only 4,310 ft (1,314 m) of water with no oil shows. Therefore, this well was also abandoned.

Thus, the first two wells in the program did establish that the seismic work was accurately predicting the presence of thick Entrada sandstone, but no commerical oil accumulations were encountered. However, the results of the third well drilled were more satisfactory in that it was the discovery well for the Eagle Mesa field. This well, the Filon Federal 12-1, located in Sec. 12, T19N, R4W, Sandoval County, New Mexico, penetrated 11 ft (3 m) of Todilto limestone, followed by 214 ft (65 m) of Entrada sandstone. A drill-stem test, conducted after only 5 ft (1.5 m) of Entrada had been penetrated, recovered 825 ft (251 m) of oil and 165 ft (50 m) of mud. A 50-ft (15 m) core, which was then cut, contained 25 ft (8 m) of oil-saturated sandstone at the top followed by 8 ft (2 m) of spotty staining. The remainder of the core was water bearing. Logs indicated a 38-ft (12 m) oil column, with the highest oil saturations in the top 30 ft (9 m) of the Entrada, followed by an 8-ft (2 m) transition zone above 100% water. Average porosity in the oil zone from core analysis was 25%, and average permeability was 432 md. The oil is very similar to that produced at Media, with a 33° API gravity, and a high pour point of up to 80°F (27°C). This well was com-

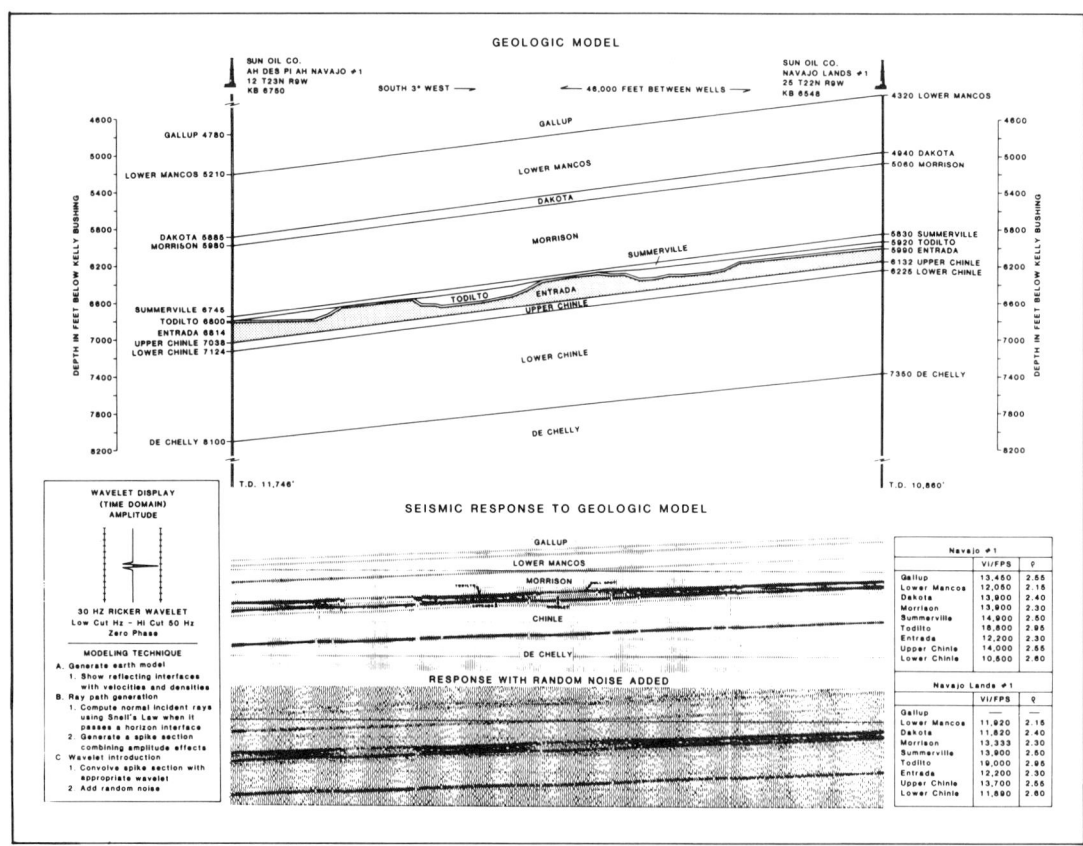

FIG. 13—Seismic model illustrating predicted seismic response over topographic highs in Entrada Sandstone. Compare predicted seismic response shown in seismic model with actual response observed on seismic line 3 (Fig. 11), which was shot over Media field.

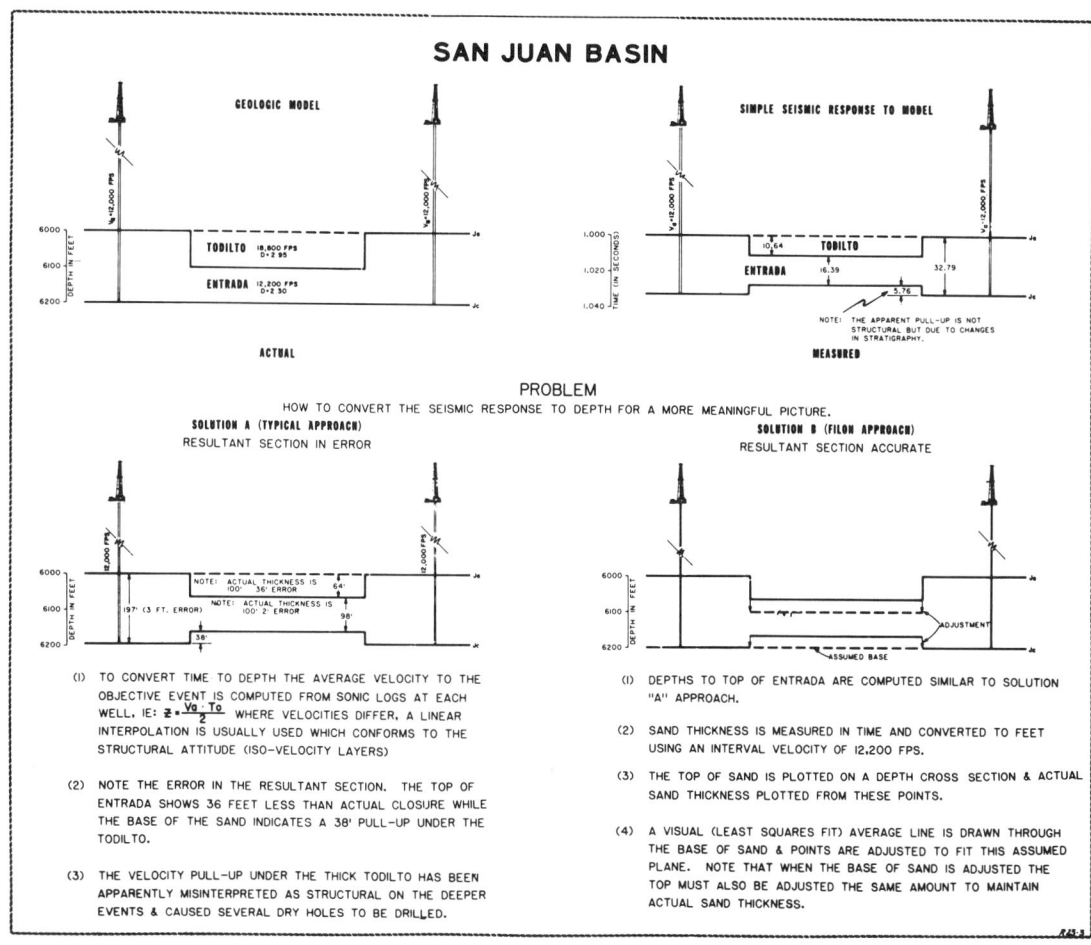

FIG. 14—Diagrams illustrating adjustments which must be made in seismic data to correct for velocity effect of Todilto Formation to obtain more accurate Entrada structure maps.

pleted in August 1975 for an initial rate of 97 bbl of oil plus 4,300 bbl of water per day by means of a large-volume, down-hole pump.

Although the high initial water cut in this discovery well was disappointing, subsequent development wells in the field were completed with substantially lower water cuts and, with high-volume pumps, production rates as high as 480 bbl of oil per day were achieved in several of the wells.

Thus, the initial three-well wildcat program proved that the seismic program was able to map the presence of anomalously thick Entrada sand zones and, in the case of Eagle Mesa field, to map a potential stratigraphic trap which did indeed contain a commercial oil accumulation.

Expanded Exploration Program

As a result of this initial success, a much more aggressive exploration program was initiated, involving

```
SAN JUAN BASIN
ANOMALY CRITERIA

1. LOW AMPLITUDE TODILTO
2. ENTRADA ISOPACH THICK
3. SANASTEE-ENTRADA THINNING
4. STRUCTURAL CLOSURE
```

CLASS "A" ANOMALY SATISFIES ALL OF THE CRITERIA
CLASS "B" ANOMALY SATISFIES 3 OF THE CRITERIA
CLASS "C" ANOMALY SATISFIES 2 OF THE CRITERIA
CLASS "D" ANOMALY SATISFIES 1 OF THE CRITERIA

FIG. 15—Chart illustrating seismic criteria used to grade Entrada seismic anomalies. Class A anomalies are considered better drilling risks than class D anomalies.

extensive leasing, seismic work, and exploration drilling.

Land acquisition—Several farmout and seismic-option agreements were made to tie up favorable lands. In addition, available fee and federal leases were obtained in prospective areas, and successful bids were entered on nominated Indian allotted lands and state tracts in favorable areas. At the height of the exploration program, in excess of 400,000 acres (161,880 ha.) were controlled in one manner or another by the exploration group.

Seismic program—A regional seismic grid was shot over the prospective fairway, and extensive detail seismic work was conducted in anomalous areas to delineate drillable prospects. The initial seismic program had established that the Entrada topographic highs tended to have their long axes oriented in a south-southwest to north-northeast direction. Therefore, the regional seismic grid was shot perpendicular to this orientation, that is, west-northwest to east-southeast, to increase the likelihood of encountering the sand highs. Cross lines were then shot along the crestal axes of the sand thicks to determine the updip (southwest) termination of the individual sand highs, and then detail seismic shooting was done in these areas to map the structural closures created by the updip terminations of the Entrada topographic ridges.

Ultimately, this program led to the shooting of more than 2,200 mi (3,500 km) of 600% CDP true-amplitude seismic data along the southeastern flank of the San Juan basin.

Exploration and development drilling—As of January 1980, 27 separate seismic closures in the Entrada have been tested by the drill in the current exploration program. Of these closures, six contained commercial oil accumulations, six others tested oil and water on drill-stem tests, and four others contained some oil staining in the top of the Entrada.

Some of the Entrada highs had more than one wildcat drilled on them in attempts to get structurally higher than wells with oil shows, or to gain additional oil column by drilling downdip in the direction of supposed

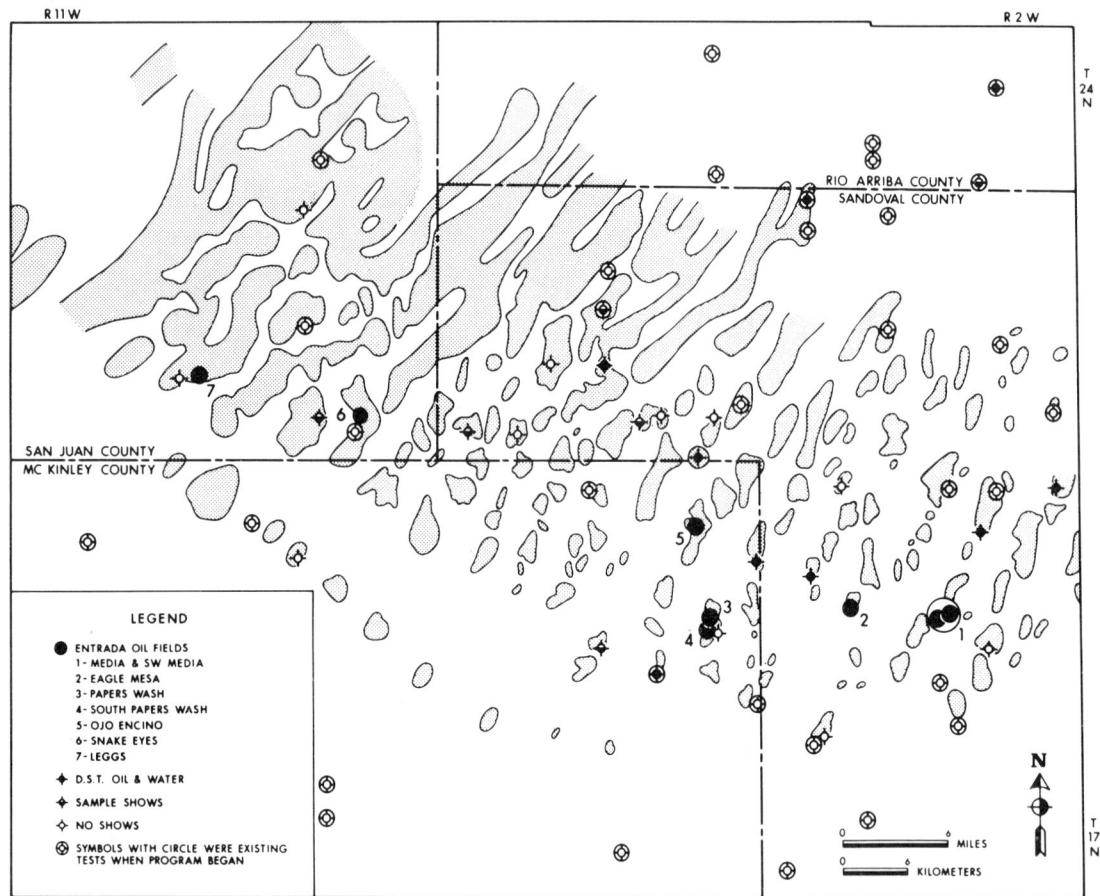

FIG. 16—Map showing general location and configuration of Entrada seismic anomalies along southwest flank of San Juan basin; areas shown by dotted pattern outline Entrada sand thicks or topographic highs. Several anomalies had more than one test, but only the initial wildcat location is shown. Drilling activity shown is as of January 1980.

hydrodynamic tilt. Unfortunately, in most cases these attempts were unsuccessful.

In addition to the six discovery wells, 10 successful development wells and six dry holes have been drilled in developing the new fields.

The map in Figure 16 schematically shows the orientation and general distribution of the Entrada topographic ridges as defined by seismic work and well control. In addition, this map also shows the general location of wells drilled in the current exploration program, as well as the location of the producing Entrada oil fields.

ENTRADA OIL FIELDS

The six new Entrada oil fields (Eagle Mesa, Papers Wash, South Papers Wash, Ojo Encino, Snake Eyes, and Leggs) are all located in a general west-northwest to east-southeast band paralleling structural strike along the southeast flank of the San Juan basin. The northwesternmost field, Leggs, is approximately 43 mi (69 km) northwest of the original Media field, which is still the easternmost of the Entrada oil pools.

Detailed illustrations and discussions of three of the new fields (Eagle Mesa, Papers Wash, and South Papers Wash) are presented in this paper. Summary data for the other fields are presented in Table 1. More comprehensive information on most of these fields may be found in the Four Corners Geological Society Publication entitled, "Oil and Gas Fields of the Four Corners Area" (Bryant, 1978a, b, c; Campbell, 1978; Reese, 1978a, b).

Eagle Mesa Field

The Eagle Mesa field, discovered in August 1975, was the first field discovered in the current exploration program. Detailed seismic work and well control in the Eagle Mesa area have defined an isolated topographic high in the Entrada which extends approximately 1.5 mi (2.4 km) north-south and 1 mi (1.6 km) east-west. Maximum vertical relief of the Entrada high, based on the Entrada sand thicknesses encountered in the producing wells, is 105 ft (32 m; Fig. 17).

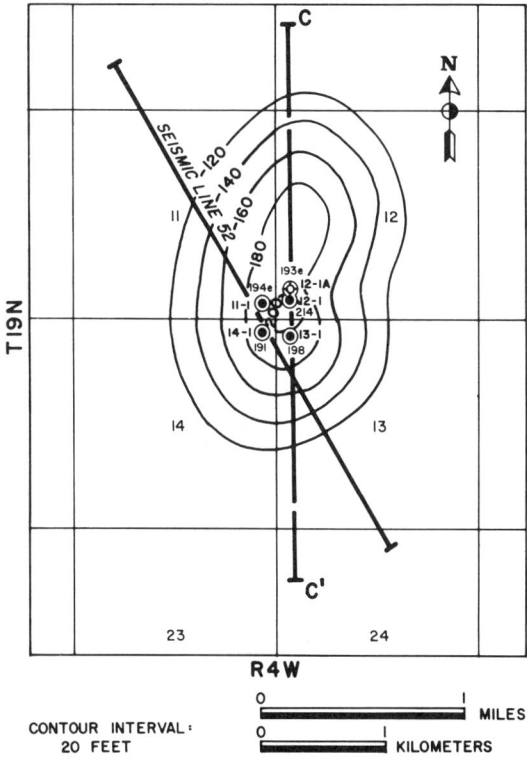

FIG. 17—Isopach map of Entrada Sandstone at Eagle Mesa field. Isopach values are based on seismic and well control.

Figure 18 shows the seismic response of the Entrada across Eagle Mesa field. On this seismic line the Entrada anomaly is characterized by 12 msec of isochron thinning between the top of the Entrada and overlying shallow seismic reflectors. In addition, 17 msec of vertical rollover at the top of the Entrada is observed, after correcting for the effects of the sloping datum used to process the seismic data. This line also displays the other seismic criteria characteristic of the Entrada sand

Table 1. Reservoir Parameters of Entrada Oil Fields

Field	Location	Year of Discovery	Number of Productive Wells	Average Depth (ft)	Average Net Pay (ft)	Average Porosity (%)	Average Permeability (md)	Cumulative Oil Production (bbl) November 1, 1980
Media	T19N, R3W	1953	7	5,250	25	23	290	987,469
Southwest Media	T19N, R3W	1972	4	5,310	18	24	360	694,524
Eagle Mesa	T19N, R4W	1975	4	5,460	23	25	430	745,290
Papers Wash	T19N, R5W	1976	5	5,170	29	25		881,255
South Papers Wash	T19N, R5W	1977	1	5,140	23	23		127,114
Ojo Encino	T20N, R5W	1976	2	5,890	20	22	180	76,440
Snake Eyes	T21N, R8W	1977	2	5,600	23	24		116,241
Leggs	T21N, R10W	1977	2	5,400	16	23		65,306

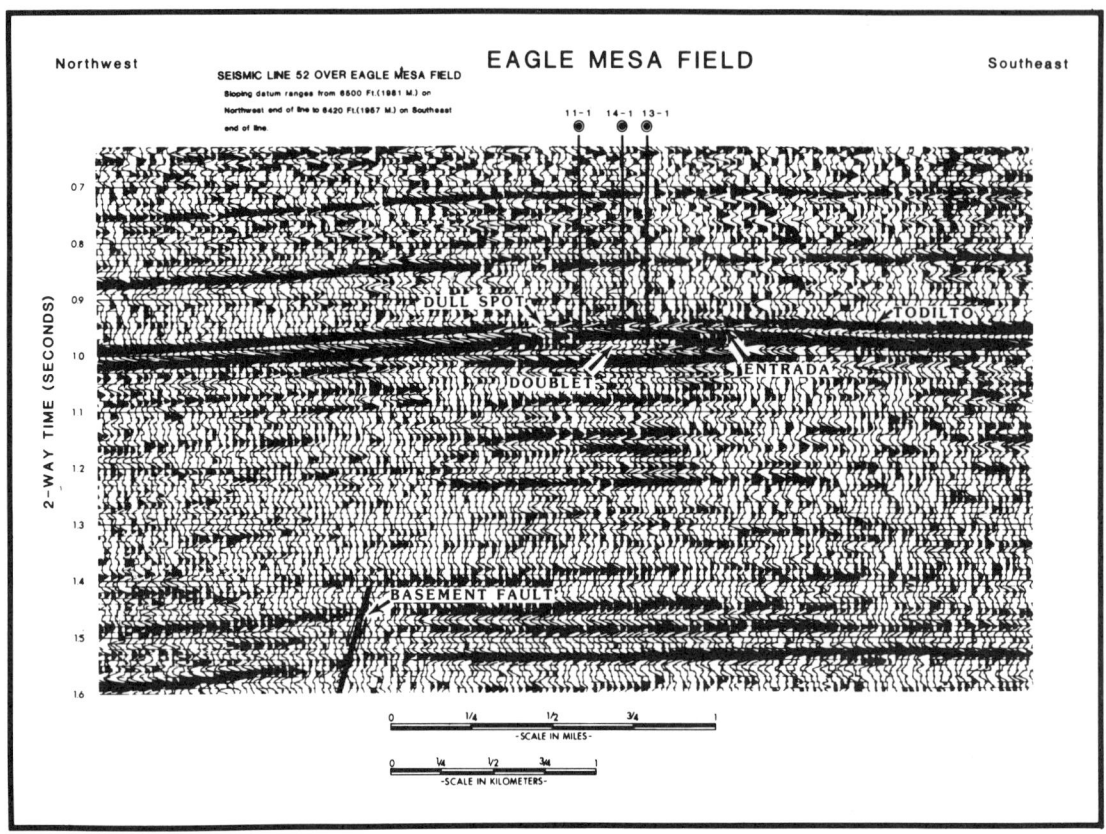

FIG. 18—Seismic line 52 over Eagle Mesa field. Location of line is shown in Figure 19.

thicks, including the development of a well-defined seismic doublet where the Entrada thickens, and a dull spot or low-amplitude anomaly where the Todilto thins over the Entrada high.

Because of the down-to-the-north, postdepositional structural tilt in this area, the resultant structural closure has been shifted to the south end of the topographic ridge. This has resulted in a structural closure encompassing approximately 460 acres (186 ha.) centered at the intersections of Secs. 11, 12, 13, and 14 of T19N, R4W, Sandoval County, New Mexico (Fig. 19).

A tilted oil/water contact, dipping approximately 60 ft/mi (11 m/km) to the south, has shifted the oil accumulation even farther down the south plunge of the structural closure. The southerly tilt of the oil/water contact (Fig. 20) has caused the 13-1 well, which is 11 ft (3 m) lower structurally than the the 12-1 well, to have the same amount of net pay, 30 ft (9 m), as the 12-1 discovery well.

A distinction is made in all of the Entrada fields between total oil column and net pay because some Entrada wells in the San Juan basin have anomalously thick transition or "flushed" zones present between the zone of 100% water saturation and the overlying zone of movable (producible) oil. All field maps and cross sections in this report show the "base of movable oil" which, on the basis of log and core analyses as well as production data, is placed at that point where log calculations indicate oil saturations greater than 40% in the movable oil zone and less than 40% in the transition or "flushed" zone. The equations utilized to calculate oil saturations from logs of Entrada wells in this area were derived from detailed analyses of core, oil, and water samples from the Eagle Mesa field and differ somewhat from the standard "Archie" or "Humble" formulas utilized in most routine log calculations where detailed rock parameters are unknown. A more detailed analysis of the presence of thick residual or "flushed" oil zones in some Entrada oil wells will be presented under the discussion of hydrodynamics in a later section of this paper.

The amount of vertical structural closure preserved at the updip termination of the Entrada topographic high at Eagle Mesa is enhanced by a local flattening of the regional structural dip. The regional dip along most of the southeast flank of the San Juan basin is about 1° at the Entrada level. However, a down-to-the-basin normal fault, developed at basement level, has locally modified structural dips at Eagle Mesa. This fault,

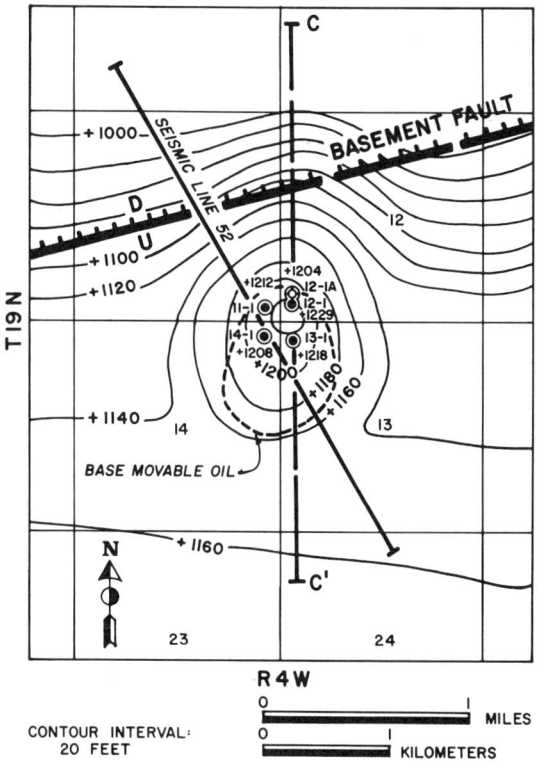

FIG. 19—Structure map on top of Entrada Sandstone at Eagle Mesa field. Structural values are based on seismic and well control.

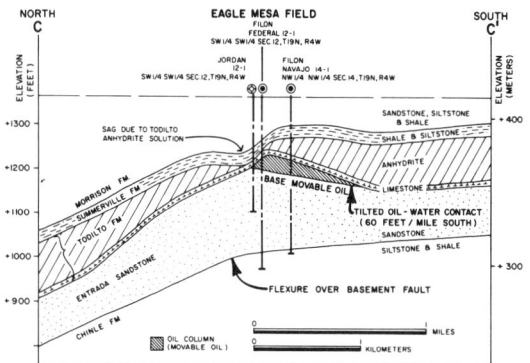

FIG. 20—Structural cross section CC' through Eagle Mesa field. Location of cross section is shown on Figure 19.

which trends west-southwest to east-northeast beneath the Entrada topographic high, has caused the regional dip at the Entrada level to steepen to about 2° where the Entrada drapes over the downthrown side of the basement fault. Conversely, the regional dip at the Entrada level flattens to less than 1/3° on the upthrown block at the point where the Entrada topographic ridge terminates on the south. This local flattening of dip has preserved 70 ft (21 m) of structural closure out of the 105 ft (32 m) of original topographic relief.

That rapid variations in the topographic relief on top of the Entrada highs can occur was established at Eagle Mesa by the drilling of the 12-1A well, located only 225 ft (69 m) north of the discovery well for the field. The original discovery well, the Filon Federal 12-1, although containing a 30-ft (9 m) column of movable oil, had a disappointing production history, in that the initial water cut averaged 96% and never improved. Believing that the high water cut might be due to uncorrectable mechanical problems, a redrill was proposed at the 12-1A location. Much to the chagrin of the supporters of this test, primarily the writers of this paper, the well came in 25 ft (8 m) low structurally to the 12-1 well at the top of the Entrada and the well was subsequently plugged and abandoned. Although the well did not penetrate the entire Entrada section, internal correlations within the Entrada and isopach and structural data from overlying units indicate that the major loss in structural elevation was due to topographic relief on top of the Entrada. The Entrada is 214 ft (65 m) thick in the 12-1 well but only approximately 193 ft (59 m) thick in the 12-1A well. This loss of 21 ft (6 m) of sand over a lateral distance of only 225 ft (69 m) indicates a dip of 5°. Had this well, rather than the original 12-1 well, been the first test drilled at Eagle Mesa, the field might never have been developed.

As with all Entrada oil wells drilled, after a period of initial flush production, all of the wells at Eagle Mesa experienced a rapid increase in water production which averaged 95% for the wells in the field after 1 year of production. This high rate of water production is largely due to the excellent vertical permeability in the Entrada, which causes rapid coning of water into the well bore from beneath the oil/water contact. Several things can improve the economics of this situation. The first is to drill an Entrada well which has low-permeability streaks separating the oil zone from the underlying water. Several wells in various fields have encountered this situation, most notably the Dome Petroleum Navajo 15-4 in the Papers Wash field. This well, which produced 120,000 bbl of oil during its first year of production, had a water cut of only 60% at the end of that time. Low porosity (14 to 16%), and low-permeability sand streaks near the base of the oil column in this well are apparently very effective in restricting water production.

Another solution to the problem of high water cuts would be to drill a well with an oil column of 50 to 60 ft (15 to 18 m). Unfortunately, no geologic or geophysical techniques have been developed to accurately predict the presence of tight streaks or thick oil columns in the Entrada. Therefore, the other solution is to utilize high-volume pumps capable of lifting 2,000 to 5,000 bbl of fluid per day. Such pumps at Eagle Mesa have enabled the field to produce a cumulative total of 745,000 bbl of oil as of November 1, 1980, and three wells in the field are currently producing at a combined rate of 335 bbl of oil plus 13,000 bbl of water per day.

Papers Wash Field

The Papers Wash field, located in T19N, R5W, McKinley County, New Mexico, contains two separate oil pools, Papers Wash in Secs. 15 and 16 and South Papers Wash in Sec. 22. Both oil accumulations are located on the same north-south-trending Entrada sand ridge but are separated by a structural low created by drape over a basement fault. This up-to-the-basin normal fault, which trends east-northeast to west-southwest through Secs. 15 and 16, has created 60 ft (18 m) of counter-regional south dip at the Entrada level. This structural drape, combined with regional structural tilt, has created an east-northeast-plunging structural nose with possibly 10 ft (3 m) of local closure at the base of the Entrada at Papers Wash (Fig. 21). The south structural dip, combined with 115 ft (35 m) of east-west reversal provided by the Entrada sand ridge, results in a structural-stratigraphic trap with approximately 45 ft (14 m) of structural closure at the top of the Entrada at Papers Wash (Fig. 22).

As of November 1, 1980, the five wells draining this pool, which covers an area of slightly more than 200 acres (81 ha.), have produced a cumulative total of 881,000 bbl of oil since the field discovery in October 1976. The wells are currently pumping a combined total of 347 bbl of oil plus 12,500 bbl of water per day.

A maximum oil column for the field of 42 ft (13 m) was encountered in the 15-4 well. The upper 30 to 36 ft (9 to 11 m) of the column appears from log calculations to be in the movable oil zone. Tight streaks at the base of the interval contain lower than normal oil saturations and make definition of the base of movable oil somewhat difficult in this well, which has produced a total of 385,000 bbl of oil and is still pumping at a rate of 180 bbl of oil plus 2,800 bbl of water per day.

Excluding the 15-4 well where porosity through the pay zone averages only 20%, the other producing wells at Papers Wash have an average porosity of 26%.

The South Papers Wash oil pool, located in Sec. 22, T19N, R5W, is separated from the Papers Wash pool by the structural low created by structural drape over the bounding basement fault. Log analyses indicate that, in contrast to the south tilt of the oil accumulations at

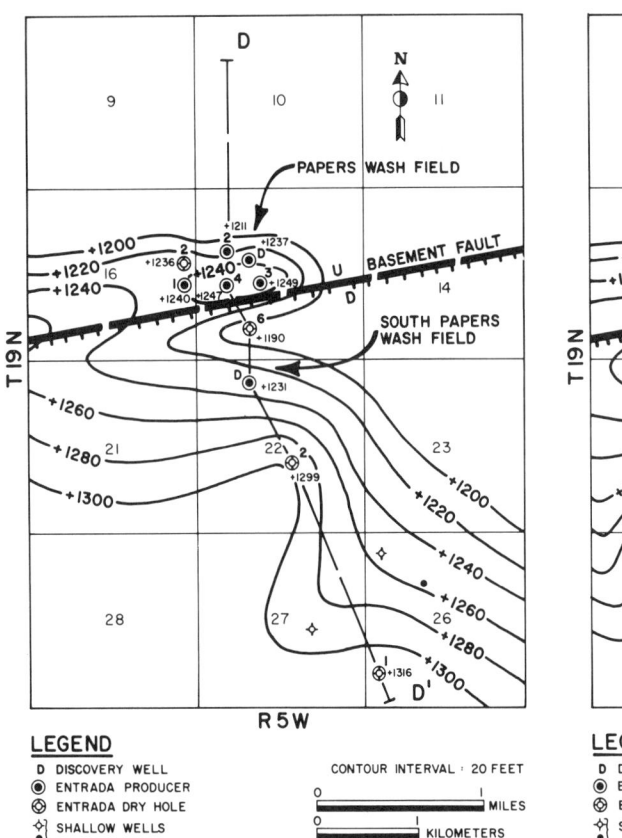

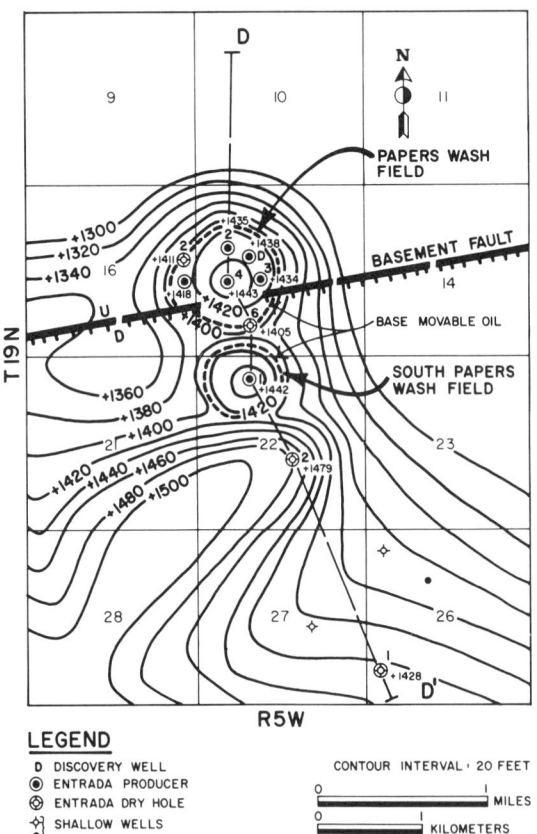

FIG. 21—Structural map on base of Entrada Sandstone, Papers Wash field.

FIG. 22—Structure map on top of Entrada Sandstone, Papers Wash field.

Eagle Mesa and Media, the oil field at Papers Wash has a nearly horizontal oil/water contact. In addition, the base of the movable oil in Papers Wash is nearly 12 ft (4 m) structurally lower than that in the South Papers Wash accumulation. This indicates that the two pools are probably separate. Both pools appear to be full to their structural spill points.

The one-well South Papers Wash field, which has produced a total of 127,000 bbl of oil since its discovery in April 1977, has an estimated vertical closure of 30 to 35 ft (9 to 11 m). Although the seismic data are not unequivocal in this area, the closure at South Papers Wash is interpreted to be due to a narrow east-west-trending channel which partly cuts through the Entrada topographic ridge just south of the South Papers Wash accumulation (see the Entrada isopach map in Fig. 23). As can be seen in comparing the Entrada isopach map in Figure 23 with the Entrada structure map in Figure 22, the Entrada sand ridge present in the Papers Wash area ultimately terminates in Sec. 28, nearly 2 mi (3.2 km) farther south than the South Papers Wash oil accumulation. Detailed seismic data indicate that regional structural dip in that area is sufficiently great to have tilted out most, if not all, of the potential structural closure at the updip termination of the ridge, leaving only a south-plunging structural nose.

The cross section in Figure 24 illustrates rather conclusively that most of the thickness changes observed in the Entrada from well control and seismic data occur at the top of the Entrada and are not due to added section at the base of the Entrada. All of the wells in the cross section penetrated the entire Entrada sand package and bottomed in the underlying Chinle Formation. The exception is the Dome Petroleum Federal 26-1 test in the SW¼ SW¼, Sec. 26, T19N, R5W, which was drilled to basement to test a deep structural anomaly in an area interpreted from seismic data to have only a thin or "regional" amount of Entrada sand. This well, which is the southernmost well in the cross section, did indeed have only a "regional" Entrada thickness of 112 ft (34 m). The overlying Todilto Formation was 90 ft (27 m) thick in this well. This contrasts to the 224 ft (68 m) of Entrada sandstone encountered in the 15-2 well at Papers Wash, which is the northernmost well on the cross section, and is approximately 2.5 mi (4 km) northwest of the 26-1 well. The Todilto is anomalously thin in the 15-2 well, being only 31 ft (9 m) thick. As can be seen from the density logs used in constructing the cross section in Figure 24, internal correlations are possible within the lower part of the Entrada. This is due to the presence of individual units of tight sand ranging in thickness from 4 to 10 ft (1 to 3 m), which alternate with more porous sand units of similar thicknesses. Similar alternation of individual sand beds, which can be traced for distances of several miles, are present in the Entrada outcrops flanking the east side of the San Juan basin.

These internal correlations make it possible to map the top of the "regional" Entrada within the thicker sand package found in the locally developed sand ridges such as the one at Papers Wash. The top of the "regional" Entrada is placed at the top of a persistent, tight sand streak approximately 4 ft (1 m) thick, which separates the generally less porous lower Entrada from the more porous upper Entrada within the Entrada sand ridges. Thus, within the topographic ridge developed at Papers Wash, the lower or "regional" Entrada varies in thickness from 104 ft (32 m) in the 15-4 well to a maximum of 116 ft (35 m) in the 22A-2 well, located in the NW¼ SE¼, Sec. 22, T19N, R5W. The regional Entrada in the 26-1 well is 112 ft (34 m) thick.

In contrast to the 12 ft (4 m) of thickness variation observed in the lower or "regional" Entrada in this area, the upper Entrada shows large thickness variations ranging from no upper Entrada in the 26-1 well to 115 ft (35 m) of upper Entrada along the crest of the topographic ridge in the 15-2 well. Although all of the development wells drilled in Papers Wash were designed to stay as close as possible to the crest of the seismically defined topographic ridge, considerable thickness variations were encountered. The 16-2 dry hole, in the SE¼ NE¼, Sec. 16, T19N, R5W, encountered 175 ft (53 m) of total Entrada, of which the upper 71 ft (22 m) appears to be "added" or upper Entrada in a position partly down the flank of the sand ridge (Fig. 23). Thus, it appears rather conclusively that most of the seismic

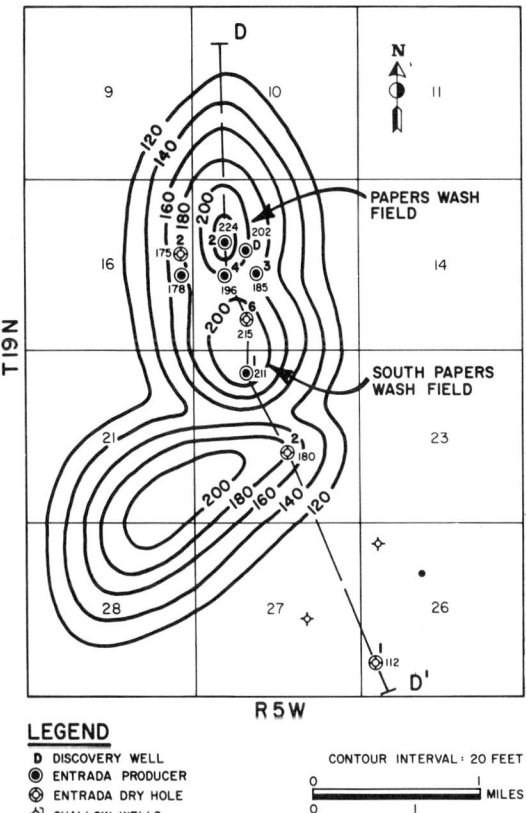

FIG. 23—Isopach map of Entrada Sandstone, Papers Wash field. Isopach values are based on seismic and well control.

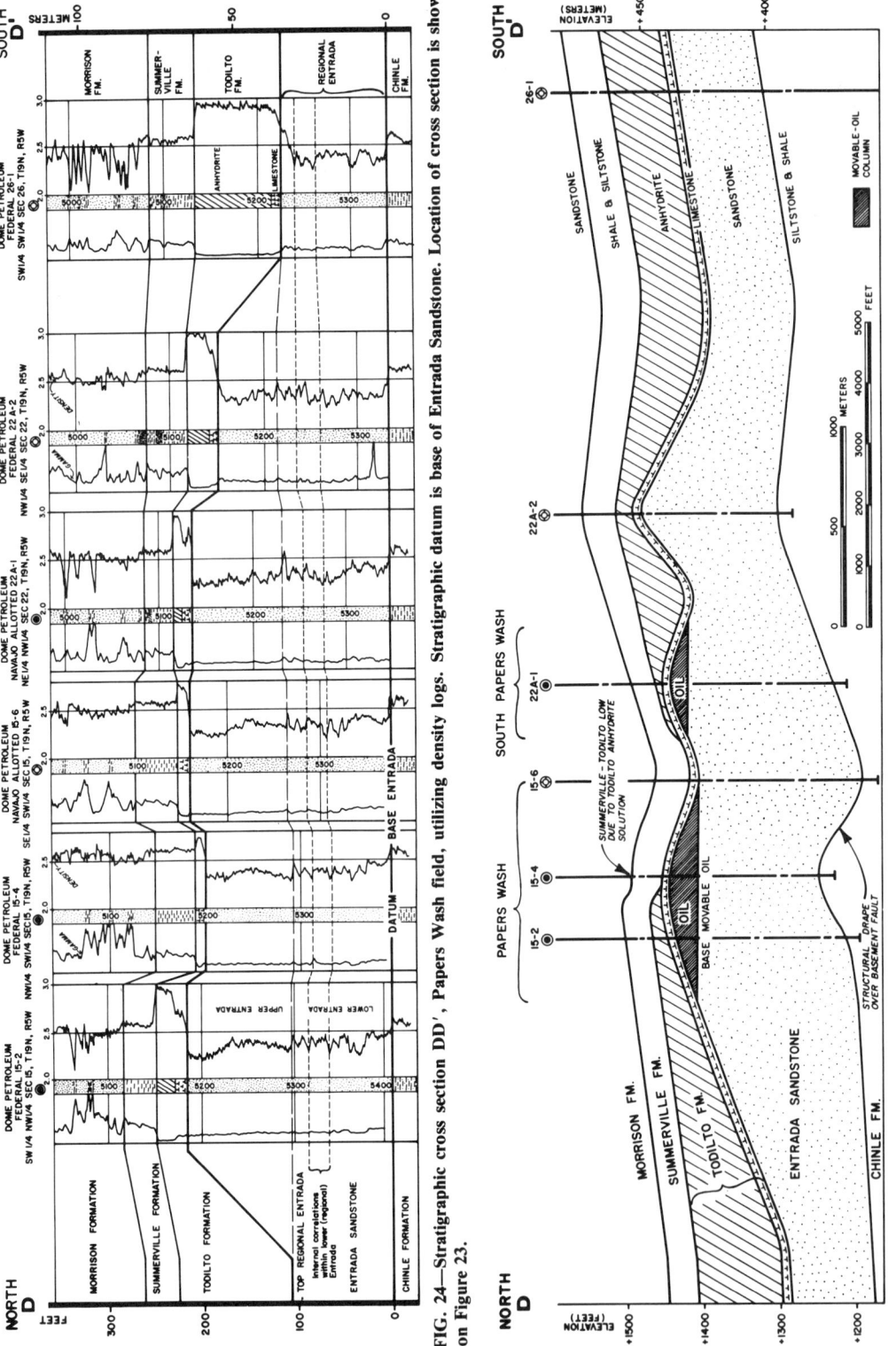

FIG. 24—Stratigraphic cross section DD', Papers Wash field, utilizing density logs. Stratigraphic datum is base of Entrada Sandstone. Location of cross section is shown on Figure 23.

FIG. 25—Structural cross section DD', Papers Wash field. Location of cross section is shown in Figure 22.

isochron thickening mapped in the Entrada is caused by topographic relief developed at the top of the Entrada sandstone.

The structural cross section in Figure 25 incorporates the same wells as the stratigraphic cross section in Figure 24 and illustrates how postdepositional, structural elements have combined with the original topographic relief in the Entrada to provide the oil traps developed at Papers Wash. The cross section of Papers Wash illustrates the structural and stratigraphic complications introduced by postdepositional anhydrite solution within the Todilto Formation in this and other Entrada oil fields. This solution has caused the top of the Todilto, as well as the overlying Summerville and basal Morrison Formations, to be 15 ft (5 m) structurally lower in the 15-4 well than in the 15-2 well, whereas both the base of the Entrada and the top of the Entrada are structurally highest in the 15-4 well. The effects of the anhydrite solution are even more pronounced at the Dakota level, approximately 1,000 ft (305 m) above the Todilto, where post-Dakota solution has dropped the Dakota in the 15-4 well to a point 27 ft (8 m) structurally lower than the Dakota in the regionally downdip 15-2 well.

Detailed analyses of the timing and magnitude of Todilto anhydrite solution in this field and elsewhere indicate that solution has occurred intermittently throughout geologic time. Where solution has occurred, anomalously thick stratigraphic intervals were created in overlying units during the time of solution to compensate for and infill the topographic lows caused by the solution. Several stages of more active solution can be documented, most notably in the uppermost Jurassic (upper Morrison) and during the Lower Cretaceous (Burrow Canyon and beneath the pre-Dakota unconformity). Unfortunately, in the Papers Wash area interpretation of the timing and history of the anhydrite solution is made difficult by isopach changes caused by paleostructural movement along the basement fault, which mask and modify the isopach changes due to anhydrite solution.

Evidence for extensive anhydrite solution in Todilto outcrops along the east flank of the San Juan basin has been presented and analyzed by Stapor (1972). Therefore, it is not surprising that evidence for such solution would also be present in the subsurface. The effect that such local solution of the high-velocity Todilto anhydrite may have on the seismic response of the Todilto-Entrada interface may be considerable, especially when the last 10 to 20 ft (3 to 6 m) of Entrada structure is as critical as it is in this area.

Other Entrada Oil Fields

As can be seen by an examination of the field data presented in Table 1, all of the other Entrada fields found to date—Ojo Encino, Snake Eyes, and Leggs—have produced substantially smaller volumes of oil than Media, Eagle Mesa, or Papers Wash. In general, these smaller fields are due to a combination of thinner oil columns and/or smaller areal extent of the oil accumulations; all are located at the updip terminations of discrete Entrada sand ridges. The size of the individual oil accumulations is controlled by a variety of factors, including the rate of regional dip at the updip termination of the sand ridge and the magnitude and direction of hydrodynamic tilt.

Snake Eyes is complicated by the presence of shallow, normal faults which provide structural traps for oil and gas accumulations in the overlying Dakota Formation. However, seismic data indicate that these faults die out with depth and apparently have little, if any, displacement at the Entrada level.

FACTORS CONTROLLING OIL ACCUMULATIONS IN ENTRADA

Many variables control the oil accumulations in the Entrada Sandstone in the San Juan basin. Of major importance are the Entrada-Todilto stratigraphic relations, postdepositional structure of the Entrada, hydrodynamics, source rock and oil-migration paths, and porosity-permeability relations within the Entrada.

Entrada-Todilto Stratigraphy

The combined seismic and drilling program for oil accumulations in the Entrada has provided a significant amount of new data on the nature, orientation, and morphology of the Entrada topographic highs which provide a critical element for the Entrada oil fields in the San Juan basin. Along the southeastern flank of the basin, these topographic features have the form of elongate ridges extending in a north-northeast to south-southwest direction. These ridges, some of which extend unbroken for as long as 15 mi (24 km), especially in the northwestern part of the exploration area, range in width from 0.5 mi to 2 mi (0.8 to 3.2 km). As shown on the Entrada anomaly map (Fig. 16), which is based on extensive seismic coverage, these topographic ridges gradually diminish southward, both in frequency and in size, attaining a move ovoid shape, with individual sand highs covering an area of from 200 to 640 acres (80 to 260 ha.). In this area of discrete, isolated sand highs, the producing Entrada oil fields have been found. The frequency, areal extent, and vertical relief of the Entrada sand highs continue to diminish southward until, ultimately, south of T18N, few sand highs remain. Conversely, as the elongate sand ridges are traced to the north, deeper into the San Juan basin, they gradually merge and coalesce into a rather uniform topography, where the Entrada sand is regionally at least 200 ft (61 m) thick, and the overlying Todilto is approximately 100 ft (30 m) in thickness.

In the area of discrete Entrada sand highs, the regional or "thin" Entrada is in general about 100 ft (30 m) thick, whereas along the crests of the highs, sand thicknesses in excess of 200 ft (61 m) are common.

The limestone and anhydrite facies of the Todilto Formation fills in the topographic lows between the Entrada sand thicks. Although postdepositional anhydrite solution has modified the original stratigraphic relations between the Entrada and the overlying Todilto, the general relation of thin Todilto, ranging from 10 to 40 ft

(3 to 12 m) over the crests of the Entrada highs, and thick Todilto, ranging from 90 to 120 ft (27 to 37 m) in areas of regional or "thin" Entrada, has proven to be consistent in the exploration area.

Data are somewhat limited as to the steepness of the flanks of the topographic ridges. However, a sufficient number of wells have been drilled partly down the flanks of these features within the developed oil fields to provide at least a minimum rate of stratigraphic dip. Detailed calculations of rate-of-thickness changes in the Entrada Sandstone between wells indicate dips ranging from 1.8 to 1.9°, that is, approximately 2° along the flanks of the sand ridges at Media, Eagle Mesa, and Papers Wash. Data are more limited for the rate of plunge along the southerly, updip terminations of the Entrada highs, but range from a calculated dip of 1.3° near the crest of the Papers Wash ridge to 1.7° for the updip termination of the ridge at Media. Local relief with dips as high as 5° can also be documented, as between the 12-1 and 12-1A wells at Eagle Mesa. Obviously, the rate of stratigraphic dip, particularly in the updip direction of these sand ridges, is of critical importance in determining how much structural closure is preserved in these features to entrap hydrocarbons.

Seismic data used to construct Entrada isopach maps prior to drilling indicated dips averaging 3° along the flanks of the Entrada ridges and 2° along the updip plunge of the ridges.

Origin of Entrada Topography

Although the shape, size, and distribution of the topographic highs in the Entrada are fairly well known from extensive seismic and drilling along the southeast flank of the San Juan basin, more information is needed before arriving at an unequivocal interpretation of their genesis. However, the preponderance of available data supports the conclusion that these features were originally formed as eolian sand dunes which were later modified by wave and/or current action in a large body of standing water, presumably a lake, which subsequently occupied the site of the original dune field.

As mentioned previously, outcrop studies of the Entrada in areas surrounding the San Juan basin have led previous researchers to conclude that the Entrada is largely a wind-blown deposit, with evidence of local water-deposited sandstones being present, especially in the upper part of the Entrada.

Tanner (1970, p. 286) presented evidence for a preserved eolian dune with approximately 50 ft (15 m) of topographic relief in outcrops along the eastern margin of the San Juan basin. In this area he observed that, where the Entrada was thickest, it was essentially all eolian, whereas, in the thinner, off-dune area the upper part of the Entrada appeared to be a lacustrine or water-laid sandstone. From this evidence he concluded that during the early part of Entrada history the area was covered by a dune field, whereas during the later phases of Entrada depositional history the area was inundated by a large lake in which a few larger dunes rose as sand hills above the lake surface (Tanner, 1970).

Limited core data support a similar conclusion for the depositional history of the Entrada along the southeastern flank of the San Juan basin. Several cores were cut in the newly discovered Entrada fields, and these cores shed considerable light on the depositional history of the Entrada. Especially instructive is the core taken from the top of the Entrada in the Filon Federal 21-2, located in the SE¼ NW¼, Sec. 21, T20N, R5W, in the Ojo Encino field.

The Entrada in this well is 219 ft (67 m) thick, indicating that it is on the crest of an Entrada topographic high. The core cut 7 ft (2 m) of Todilto limestone and then penetrated 37 ft (11 m) of Entrada sandstone, the top 25 ft (8 m) of which was oil stained. But more interesting from a stratigraphic standpoint is the fact that the top 27 ft (8 m) of the sandstone consisted of massive, horizontally bedded, uniformly sorted sandstone with good porosity and permeability (average porosity 23°, average horizontal permeability 183 md, average vertical permeability 180 md). In contrast, the bottom 10 ft (3 m) of the core contained thinly laminated, cross-bedded sandstone with cross-bedding dips as high as 21°. Although the porosity and horizontal permeability remained good in this cross-bedded, laminated zone, the vertical permeability dropped drastically (average porosity 24°, horizontal permeability 179 md, and vertical permeability 79 md). The decrease in vertical permeability is undoubtedly due to the highly laminated nature of the sand, with the individual laminae interfering with the vertical flow of fluids. The sand grains in both the upper and lower parts of the core are largely fine-grained, subrounded, highly frosted quartz grains. The thinly laminated, well-sorted, steeply dipping, cross-bedded sandstone in the lower part of the core has all of the characteristics of those deposited in an eolian environment; whereas the massive, horizontally bedded sand in the upper part of the core is more consistent with sand deposited in an aqueous environment. However, the frosted nature of the quartz grains in the upper massive zone suggests that they were derived and reworked from the underlying dune facies. A similar association of reworked sandstone in the upper part of the Entrada, underlain by cross-bedded sandstone has also been observed in cores from the Media field (Reese, 1978a, p. 410).

Whether or not the water-laid deposits in the Entrada were formed in a lacustrine or marine environment hinges largely on the interpretation of the environment of deposition of the overlying limestones and anhydrites of the Todilto Formation. Although not unchallenged (Peterson, 1972, p. 186), detailed analysis of Todilto outcrops has convinced Tanner (1974, p. 222) that the Todilto was deposited in a lacustrine environment, and that the water-laid deposits of the Entrada were also deposited in a similar environment.

The high pour point, paraffinic nature of the Entrada oil, thought to have been derived from the overlying Todilto, is very characteristic of oils derived from lacustrine source rocks in other areas of the world (Tissot and Welte, 1978, p. 394). As a consequence, the assumption is made that after the Entrada eolian dunes were formed, probably in a topographically low basin, the area became the site of a large inland lake. As the water in the Entrada-Todilto lake gradually submerged the topographic highs created by the preexisting Entrada dunes, wave and current action were able to rework the outermost parts of the sand exposed along the flanks and, ultimately, along the crests of the dunes. The effect of this reworking was to smooth and flatten the original, steep dune topography and probably accounts for the relatively low rate of dip—2°—now found along the flanks of these features.

As water depths in the lake increased, the dunes were eventually completely submerged. After submergence of the Entrada topographic highs, the area then became the depositional site for the varved, organic-rich Todilto limestone, followed by anhydrite which filled in the topographic lows between the Entrada sand ridges. One interesting observation is that the varved Todilto limestone, representative of a quiet-water depositional environment, was deposited over both the crests of the Entrada highs as well as in the intervening lows. Assuming that this thin limestone unit represents synchronous deposition, the water depth at the time must have exceeded the preserved vertical relief of the Entrada sand ridges. Therefore, the water depth in the Todilto lake along the southeastern flank of the San Juan basin must have exceeded 100 ft (30 m) during the initial stages of Todilto deposition.

Deposition of the Entrada in this type of environment can explain the absence of the Entrada topographic highs south of the exploration area. Analysis of the Todilto Formation indicates that the lake shallowed on the south and west, where the Todilto anhydrite thins and disappears, and the Todilto limestone changes from a varved, organic-rich unit to a lighter colored, shallow-water limestone. In this shallow area, wave and current action would have been more effective in destroying the Entrada dune topography until the dunes were eventually completely reworked and leveled off.

To the north, where the regional Entrada sandstone is over 200 ft (61 m) thick, perhaps even higher and thicker dunes may have originally been present. But in this area of thicker sands which infilled the depositional basin, the Entrada-Todilto lake may have again shallowed sufficiently so that the dunes were not submerged until wave and current action effectively destroyed most of the original dune topography. As a consequence, the only area where the Entrada topography survived with a large amount of vertical relief was in the deepest part of the Entrada-Todilto lake, where the water submerged the dunes rapidly enough that they were not completely destroyed by wave and current action.

The writers realize that additional work should be done to prove that the "preserved dune" hypothesis fully explains the origin of the Entrada topographic highs, and that other explanations for their origin cannot be completely discounted. However, the dune hypothesis fits all of the observed criteria obtained from the subsurface exploration program and is compatible with previous hypotheses advanced to explain the origin and environment of deposition of the Entrada Sandstone and the overlying Todilto Formation.

Postdepositional Structure

Although the original exploration concept emphasized the importance of Entrada topography in providing potential oil traps, subsequent exploration has indicated that structural considerations are equally important. Along the southeastern flank of the San Juan basin, the regional dip at the Entrada level is northward with an average of approximately 1°. This regional dip, in general, limits the area of closure to the southernmost, updip termination of the Entrada sand ridges. Both the amount of vertical closure and the areal extent of the closure are, therefore, controlled by the amount of postdepositional tilt to which the updip end of the sand ridge has been subjected. As a consequence, local changes in the regional dip can be extremely important in controlling the amount of structural closure in the Entrada sand ridges. Areas of local flattening of the regional dip are much more prospective than are areas of steep dip.

Faults and folds—Extensive seismic data in the San Juan basin show the presence of deep-seated basement faults which have had recurrent movement throughout geologic time. Local draping of the Entrada sand ridges over these basement faults has locally enhanced the amount of closure at the Entrada level. In other areas, northward steepening of the Entrada over basement faults has destroyed the potential closures.

The down-to-the-basin basement fault, which extends beneath the Entrada sand ridge at Eagle Mesa, has caused local flattening of the Entrada on the upthrown block and significantly improved the amount of closure on this feature.

Up-to-the-basin faults are less common in the basin but, as at Papers Wash, can provide critical south dip, resulting in closure within the middle of a sand ridge as opposed to closure at the ultimate updip termination of the ridge.

Thus, the controls which local structures have on enhancing, modifying, or destroying the potential closures provided by the Entrada sand ridges cannot be ignored, and must be considered as integral parts of the trapping mechanism for the Entrada oil fields in the San Juan basin.

Hydrodynamics

The initial field study of Media indicated that the oil/water contact has a pronounced southwest tilt. The extensive recharge area provided by the Entrada outcrop belt surrounding the San Juan basin, and the excellent porosity and permeability of the Entrada sands, indicated that hydrodynamic factors might be responsible for this tilt. Therefore, the hydrodynamic regime within the Entrada was analyzed and mapped on the basis of pressure data provided by drill-stem test information. This analysis revealed that hydrodynamic flow conditions were present within the Entrada, and that the gradients mapped were sufficient to cause the observed tilted oil/water contact at Media.

During the initial stages of the drilling program, new pressure information provided by drill-stem tests of the Entrada was incorporated into the hydrodynamic maps in an attempt to predict the most likely direction of the tilt of any oil accumulations encountered by the drill. All wells were drill-stem tested, whether or not oil shows were present, to obtain the required pressure data. As new fields were found, the pressure data, as well as the actual tilt observed from development wells, were incorporated into the hydrodynamic maps. In addition, several wells were drilled solely on the basis of predicted hydrodynamic tilt.

As the program progressed, it became apparent that a rather bewildering variety of tilted oil/water contacts and/or tilted interfaces at the "base of movable oil" are present within the Entrada fields. Media has a southwest tilt, Eagle Mesa has a more southerly tilt, Papers Wash and Snake Eyes have an apparently flat "base of movable oil," whereas the tilt at Ojo Encino is due west and that at Leggs is east. Rates of calculated tilt for the various fields, in general, range from 60 to 80 ft/mi (11 to 15 m/km).

In most of the fields found to date, the hydrodynamic tilt has had a destructive influence on the oil accumulations, in that the direction of tilt has had an updip component. This has caused the oil to spill out of the updip end of the trap. Consequently, in fields with tilted oil/water contacts, such as Media and Eagle Mesa, the oil occupies less space than that provided by the simple structural closure present in these fields.

Further complications in analyzing the oil/water contacts in the Entrada oil fields are introduced by the presence of apparent "flushed" zones in some wells where oil saturations are indicated on logs and cores, but only water is obtained on drill-stem or production tests. The existence of such "flushed" zones was brought home by the drilling of the "33" anomaly in Sec. 33, T21N, R5W. The first well drilled on this well-defined seismic anomaly, the 33-1 well, encountered a thick Entrada sandstone, the upper 22 ft (7 m) of which was oil bearing. Log calculations indicated that the upper 16 ft (5 m) of this zone contained movable oil. A drill-stem test across the top of the Entrada recovered 4,400 ft (1,341 m) of oil-cut water and 580 ft (177 m) of water-cut oil. The well was consequently abandoned as noncommercial. Additional seismic data recorded after the drilling of the 33-1 well revealed that the Entrada could be encountered in a structurally higher position in a location 1,400 ft (427 m) south of the 33-1 well. This location became the site of the 33-3 well, which encountered the Entrada 13 ft (4 m) structurally higher than the 33-1 well. The enthusiasm generated by this occurrence was considerably dampened, however, by a drill-stem test across the top 15 ft (5 m) of the Entrada, which recovered 4,559 ft (1,390 m) of water with no sign of oil.

Subsequent logs revealed that this well contained a 32-ft (10 m) oil column, but detailed log calculations revealed that the oil saturations in this zone averaged only 30%. This compares to calculated oil saturations exceeding 60% in the 33-1 well and in other producing field wells elsewhere in the basin. The high porosity, which averaged 22% in 33-3 well, combined with the large water recovery on drill-stem test, indicated that the low oil saturations encountered in this well could not be attributed to the presence of low-permeability, tight sands within the Entrada. Therefore, the only logical explanation appears to be that the oil zone in the 33-3 well represents a residual or "flushed" oil zone, which at one time contained movable oil which remigrated to a different location, leaving behind a zone of residual or immovable oil. The drilling of five dry holes on the "33" anomaly, in an attempt to find the "missing" oil, has resulted in considerable frustration and premature aging of the explorationists involved in the project.

As a result of the experience gained at the "33" anomaly, a review of other Entrada tests has led to the identification of other wells which contain thick columns of residual or "flushed" oil, and which produced significantly less oil than might be surmised from a casual glance at the well logs. Of particular note is the Fluid Power Pump FPP-5 well, located in the NE¼ NW¼, Sec. 22, T19N, R3W, in Southwest Media. This well has a 39-ft (12 m) oil column, defined by log and core data, but produced only 12,000 bbl of oil prior to conversion to a water-injection well. Examination of logs in this well indicate that only the top 2 to 6 ft (1 to 2 m) of the Entrada sandstone can be considered pay, whereas the remaining 31 to 37 ft (9 to 11 m) of oil column contains calculated oil saturations which average only 31%, in a rock with core porosities which average 24% and are as high as 27%. These data again suggest the presence of a long column of flushed, residual oil in this well.

Similarly, the Petro-Lewis 8-22 well, located 600 ft (183 m) south of the Fluid Power Pump-5 well, contains a total oil column of 48 ft (15 m), but only the top 14 ft (4 m) calculates as movable oil, leaving a 34-ft (10 m) zone of residual or "flushed" oil. This well was completed in 1978 for only 8 bbl of oil plus 28 bbl of water per day from the top 7 ft (2 m) of the Entrada.

Thus, it is fairly obvious that in some Entrada oil fields, remigration of the original accumulations has occurred subsequent to original emplacement. Whether such remigration is due to a change in the hydrodynamic gradient, postaccumulation structural movement, leakage out of the reservoir, or a combination of these factors has not been determined and is an obvious area of future investigation.

Source Rock, Migration, and Charge

The distribution of oil fields and oil shows in the Entrada demonstrates an excellent correlation with the presence of source rock in the overlying Todilto Formation. The fields are either overlain by mature Todilto source rocks or are located a short distance updip in a favorable location to receive an oil charge from any oil migration out of deeper parts of the basin. Where sufficient subsurface data are available, it appears that most of the fields are full to either their structural or hydrodynamic spillpoint. This raises an interesting question whenever a dry hole is drilled on a well-defined seismic prospect. In some cases, such tests have encountered no oil or thin, noncommercial accumulations. The question arises as to whether or not hydrodynamics, lack of charge, or absence of local source rock could have prevented an accumulation; or whether the feature has less structural closure than mapped. This too is an area where additional research is required.

Porosity/Permeability Relations

As mentioned previously, the decrease in porosity in the Entrada, as it is encountered in the deeper parts of the San Juan basin, acted as a deterrent to deeper exploration activities. In addition to this basic exploration parameter, local porosity/permeability variations within the Entrada have a profound influence on its productive behavior. The excellent porosity and permeability typically encountered in the Entrada results in rapid water encroachment and high water cuts in producing wells. This necessitates the use of high-volume pumps to lift sufficiently large volumes of total fluid to provide economic quantities of oil.

However, several wells have encountered low-permeability zones within the oil column, and these zones serve as a sufficient barrier to vertical water migration that a significant improvement in the water cut has been observed. Thus, local variations in porosity/permeability relations can have a significant bearing on the economics of oil production from the Entrada.

RECOMMENDATIONS

Although the current exploration program has added a substantial volume of new knowledge concerning the Entrada Sandstone, several areas of additional research could enhance the ability to find oil in this reservoir. Updated hydrodynamic studies, combined with a better understanding of the migration history of the Entrada oil, should result in the selection of areas more favorable for oil, as well as provide a better understanding of the origin and significance of the zones of residual or "flushed" oil found in some wells.

The routine use of dipmeter surveys in Entrada wells could provide important information on the magnitude and orientation of cross-bedding dips within the Entrada, which should aid in a better understanding of the genesis of the Entrada topography. In addition, dipmeter analyses could provide critical information on the direction and magnitude of dip at the Todilto-Entrada interface, which could be utilized to project updip areas of potentially thicker oil columns from wells with shows or thin oil columns.

From a seismic standpoint, a greater emphasis should be placed on obtaining crestal lines parallel with the long axes of the elongate topographic ridges present with some areas of the Entrada. Potential traps could exist in the downdip parts of these elongate topographic highs owing to local culminations superimposed on the larger ridges, such as at South Papers Wash, and owing to fault- or fold-modified closures, such as at Papers Wash.

Continued improvement in geophysical instrumentation should aid in future exploration efforts. Improved seismic resolution due to the generation and recording of higher frequencies could be enhanced by the use of lighter explosive charges and closer station spacing. Specific prospects, covering a limited area, should be processed to a flat datum using true-amplitude techniques. Prospects whose structures are complicated by faulting and drape folding could be delineated more accurately by migrating the seismic data. Three-dimensional recording and processing techniques could be used in place of multiple seismic lines to construct a more nearly accurate structural picture at the updip termination of the sand ridges.

Application of these various exploration techniques and studies could result in the discovery of additional Entrada oil fields in the San Juan basin, as well as aid in the exploration for similar features elsewhere.

CONCLUSIONS

Preserved eolian sand dunes in the Entrada Sandstone have provided unusual stratigraphic traps for oil accumulations in the San Juan basin. Similar features may exist in other eolian deposits elsewhere and should provide worthwhile exploration targets.

An experimental seismic program, combined with seismic-model studies, indicated that the topographic relief on top of the Entrada should be mappable seismically. An extensive seismic and drilling program has demonstrated the successful use of seismic as an exploration tool for these stratigraphic traps and has resulted in the discovery of six new oil fields.

In addition to the topographic relief on top of the Entrada, which locally exceeds 100 ft (30 m), other factors which control the oil accumulations include local structural conditions, hydrodynamics, source-rock and oil-migration history, and local porosity/permeability variations in the Entrada.

Understanding of the various geologic parameters which control the oil accumulations in the Entrada, combined with a sophisticated exploration program, should result in the finding of additional oil fields in the San Juan basin as well as in similar depositional settings worldwide.

REFERENCES CITED

Anderson, R. Y., and D. W. Kirkland, 1960, Origin, varves, and cycles of Jurassic Todilto Formation, New Mexico: AAPG Bull., v. 44, p. 37-52.

Bryant, A. C., 1978a, Ojo Encino field, in Oil and gas fields of the Four Corners area: Four Corners Geol. Soc., v. 2, p. 435-436.

―― 1978b, Papers Wash field, in Oil and gas fields of the Four Corners area: Four Corners Geol. Soc., v. 2, p. 451-452.

―― 1978c, Snake Eyes field, in Oil and gas fields of the Four Corners area: Four Corners Geol. Soc., v. 2, p. 500-502.

Campbell, J. P., 1978, Eagle Mesa field, in Oil and gas fields of the Four Corners area: Four Corners Geol. Soc., v. 1, p. 285-286.

Green, M. W., and C. T. Pierson, 1977, A summary of the stratigraphy and depositional environments of Jurassic and related rocks in the San Juan basin, Arizona, Colorado, and New Mexico, in San Juan basin III, northwestern New Mexico: New Mexico Geol. Soc. 28th Field Conf. Guidebook, p. 147-152.

Harshbarger, J. W., C. A. Repenning, and J. H. Irwin, 1957, Stratigraphy of the uppermost Triassic and Jurassic rocks of the Navajo country (Colorado Plateau): U.S. Geol. Survey Prof. Paper 291, 74 p.

Ostrander, R. E., 1957, Media field, Sandoval County, New Mexico, *in* Geology of southwestern San Juan basin: Four Corners Geol. Soc. 2d Field Conf. Guidebook, p. 138-140.

Peterson, J. A., 1972, Jurassic system, *in* Geologic atlas of the Rocky Mountain region: Rocky Mountain Assoc. of Geologists, p. 177-189.

Reese, V. R., 1978a, Media field, *in* Oil and gas fields of the Four Corners area, v. 2: Four Corners Geol. Soc., p. 410-412.

———— 1978b, Southwest Media field, *in* Oil and gas fields of the Four Corners area, v. 2: Four Corners Geol. Soc., p. 413-415.

Smith, H. M., 1940, Correlation index to aid in interpreting crude-oil analyses: U.S. Bur. Mines Tech. Paper 610, 34 p.

Stapor, F. W., Jr., 1972, Origin of the Todilto gypsum mounds in the Ghost Ranch area, north central New Mexico: Mountain Geologist, v. 9, p. 59-64.

Tanner, W. F., 1965, Upper Jurassic paleogeography of the Four Corners area: Jour. Sed. Petrology, v. 35, p. 564-574.

———— 1970, Triassic-Jurassic lakes in New Mexico: Mountain Geologist, v. 7, p. 281-289.

———— 1974, History of Mesozoic lakes of northern New Mexico, *in* Ghost Ranch, central-northern New Mexico: New Mexico Geol. Soc. 25th Field Conf. Guidebook, p. 219-223.

Tissot, B. P., and D. H. Welte, 1978, Petroleum formation and occurrence: Berlin, Springer-Verlag, 538 p.